GLUTEN-FREE CEREAL PRODUCTS AND BEVERAGES

Food Science and Technology
International Series

A complete list of books in this series appears at the end of this volume.

Gluten-free cereal products and beverages

Edited by
Elke K. Arendt and Fabio Dal Bello
Department of Food and Nutritional Sciences
University College Cork
Ireland

AMSTERDAM • BOSTON • HEIDELBERG • LONDON
NEW YORK • OXFORD • PARIS • SAN DIEGO
SAN FRANCISCO • SINGAPORE • SYDNEY • TOKYO

Academic Press is an imprint of Elsevier

ELSEVIER

Academic Press is an imprint of Elsevier
30 Corporate Drive, Suite 400, Burlington, MA 01803, USA
525 B Street, Suite 1900, San Diego, California 92101-4495, USA
84 Theobald's Road, London WC1X 8RR, UK

First edition 2008

Notice
No responsibility is assumed by the publisher for any injury and/or damage to persons
or property as a matter of products liability, negligence or otherwise, or from any use
or operation of any methods, products, instructions or ideas contained in the material
herein. Because of rapid advances in the medical sciences, in particular, independent
verification of diagnoses and drug dosages should be made

Library of Congress Cataloging-in-Publication Data
A catalog record for this book is available from the Library of Congress

British Library Cataloguing in Publication Data
A catalogue record for this book is available from the British Library

ISBN 978-0-12-373739-7

For information on all Academic Press publications
visit our web site at www.books.elsevier.com

Printed and bound by CPI Group (UK) Ltd, Croydon, CR0 4YY

Transferred to Digital Print 2011

Contents

Contributors

Elke K. Arendt (Chs 13, 15, 16, 18) Department of Food and Nutritional Sciences, University College Cork, Ireland

S. R. Bean (Ch. 5) USDA-ARS Grain Marketing and Production Research Center, Manhattan, KS 66502, USA

James N. BeMiller (Ch. 9) Whistler Center for Carbohydrate Research, Purdue University, West Lafayette, IN, USA

Emmerich Berghofer (Ch. 7) Department of Food Sciences and Technology, University of Natural Resources and Applied Life Sciences, Vienna, Austria

Joe Bogue (Ch. 17) Department of Food Business and Development, University College Cork, Cork, Ireland

Carlo Catassi (Ch. 1) Mucosal Biology Research Center and Center for Celiac Research, University of Maryland, School of Medicine, Baltimore, USA and Department of Pediatrics, Università Politecnica delle Marche, Ancona, Italy

Christophe M. Courtin (Ch. 11) Katholieke Universiteit Leuven, Laboratory of Food Chemistry and Biochemistry, Kasteelpark Arenberg 20 bus 2463, B-3001 Leuven, Belgium

Fabio Dal Bello (Ch. 13) Department of Food and Nutritional Sciences, University College Cork, Ireland

Maria De Angelis (Ch. 12) Dipartimento di Protezione delle Piante e Microbiologia Applicata, Università degli Studi di Bari, Bari, Italy

Jan A. Delcour (Ch. 11) Katholieke Universiteit Leuven, Laboratory of Food Chemistry and Biochemistry, Kasteelpark Arenberg 20 bus 2463, B-3001 Leuven, Belgium

Hertha Deutsch (Ch. 2), Association Of European Coeliac Societies, Vienna, Austria

Raffaella Di Cagno (Ch. 12) Dipartimento di Protezione delle Piante e Microbiologia Applicata, Università degli Studi di Bari, Bari, Italy

M. Naushad Emmambux (Ch. 6) Department of Food Science, University of Pretoria, Pretoria 0002, South Africa

Alessio Fasano (Ch. 1) Mucosal Biology Research Center and Center for Celiac Research, University of Maryland, School of Medicine, Baltimore, USA

Eimear Gallagher (Ch. 14) Ashtown Food Research Centre, Teagasc, Ashtown, Dublin 15, Ireland

Marco Gobbetti (Ch. 12) Dipartimento di Protezione delle Piante e Microbiologia Applicata, Università degli Studi di Bari, Bari, Italy

Hans Goesaert (Ch. 11) Katholieke Universiteit Leuven, Laboratory of Food Chemistry and Biochemistry, Kasteelpark Arenberg 20 bus 2463, B-3001 Leuven, Belgium

Heereluurt Heeres (Ch. 2) TNO Quality of Life, Zeist, The Netherlands

Florian Hübner (Ch. 16) Department of Food and Nutritional Sciences, University College Cork, Ireland

Anu Kaukovirta-Norja (Ch. 8) VTT Technical Research Centre of Finland, Tietotie 2, P.O. Box 1000, FI-02044 VTT, Finland

Alan L. Kelly (Ch. 18) Department of Food and Nutritional Sciences, University College Cork, Ireland

Stefan Kreisz (Ch. 16) TU Munich Weihenstephan, Munich, Germany

Pekka Lehtinen (Ch. 8) VTT Technical Research Centre of Finland, Tietotie 2, P.O. Box 1000, FI-02044 VTT, Finland

Cristina Marco (Ch. 4) Cereal Group, Institute of Agrochemistry and Food Technology (IATA-CSIC), PO Box 73, 46100-Burjasot, Valencia, Spain

Michelle M. Moore (Chs 13 and 18) Department of Food and Nutritional Sciences, University College Cork, Ireland

Andrew Morrissey (Ch. 13) Department of Food and Nutritional Sciences, University College Cork, Ireland

Blaise P. Nic Phiarais (Ch. 15) National Food Biotechnology Centre, National University of Ireland, Cork, Ireland

Roland Poms (Ch. 2) International Association for Cereal Science and Technology, Vienna, Austria

Carlo Giuseppe Rizzello (Ch. 12) Dipartimento di Protezione delle Piante e Microbiologia Applicata, Università degli Studi di Bari, Bari, Italy

Cristina M. Rosell (Ch. 4) Cereal Group, Institute of Agrochemistry and Food Technology (IATA-CSIC). PO Box 73, 46100-Burjasot, Valencia, Spain

T. J. Schober (Ch. 5) USDA-ARS Grain Marketing and Production Research Center, Manhattan, KS 66502, USA

Regine Schoenlechner (Ch. 7) Department of Food Sciences and Technology, University of Natural Resources and Applied Life Sciences, Vienna, Austria

Susanne Siebenhandl (Ch. 7) Department of Food Sciences and Technology, University of Natural Resources and Applied Life Sciences, Vienna, Austria

Tuula Sontag-Strohm (Ch. 8) Department of Food Technology, University of Helsinki, POB 27, 00014 University of Helsinki, Finland

Douglas Sorenson (Ch. 17) Department of Food Business and Development, University College Cork, Cork, Ireland

Constantinos E. Stathopoulos (Ch. 10) Teagasc, Moorepark Food Research Centre, Fermoy, Co. Cork, Ireland

John R. N. Taylor (Ch. 6) Department of Food Science, University of Pretoria, Pretoria 0002, South Africa

Jan-Willem van der Kamp (Ch. 2) TNO Quality of Life, Zeist, The Netherlands

Herbert Wieser (Ch. 3) German Research Centre for Food Chemistry, Lichtensbergstrasse 4, D085748 Garching, Germany

Martin Zarnkov (Ch. 16) TU Munich Weihenstephan, Munich, Germany

Preface

In genetically susceptible individuals, the ingestion of gluten and related proteins triggers an immune-mediated enteropathy known as celiac disease. Recent epidemiological studies have shown that 1 in 100 people worldwide suffer from this condition. Such a rate establishes celiac disease as one of the most common food intolerances. If patients with celiac disease eat wheat or related cereals, such as barley or rye, they have an immunological response, localized in the small intestine, which destroys mature absorptive epithelial cells on the surface of the small intestine. Currently, the only treatment for celiac disease is the total lifelong avoidance of gluten ingestion. Patients have to follow a very strict diet and avoid any products that contain wheat, rye, or barley (some authors also include oats). Avoidance of these cereals leads to a recovery from the disease and significant improvement of the intestinal mucosa and its absorptive functions. Patients with celiac disease cannot eat some common foods such as bread, pizzas, and biscuits or drink beer. However, due to the unique properties of gluten, it is a big challenge for food scientists to produce good-quality gluten-free products.

We developed this book to give the reader a chance to take a journey through all aspects related to celiac disease. As such, this book is unique in its form and we hope that it will represent an example for future works in this and related areas. We aimed to summarize and critically review the works and knowledge gained so far in the area of medicine, nutrition, and gluten-free food and beverage technology.

The book is divided into 18 chapters, covering:

- Celiac disease
- Detection and labeling of gluten-free products
- Raw materials and ingredients used for the production of gluten-free products
- Advances in the production of gluten-free products (e.g. bread, biscuits, pizza, and pasta)
- Production of gluten-free malt and beer as well as functional drinks
- Marketing and product development.

This book is meant to be a reference for food scientists developing gluten-free foods and beverages, people working with celiac patients (nutritionists), cereal scientists carrying out research in the area as well as support for undergraduate teaching in the area of cereal science, nutrition, or medicine.

This work is the result of the combined effort of nearly 40 professionals from academia. We would like to thank all the contributors for sharing their experience in their fields of expertise. The individual contributors are the people who make this book possible. We would also like to thank the editorial and production team at Elsevier for their time, effort, advice, and expertise.

We hope that readers will find this book a useful recourse for their work or studies, and that it will help in the development of high-quality gluten-free food products in the future which will improve the quality of life of people with celiac disease or wheat allergies.

Elke K. Arendt
Fabio Dal Bello

Celiac disease

Carlo Catassi and Alessio Fasano

Introduction

Celiac disease is an immune-mediated enteropathy triggered by the ingestion of gluten in genetically susceptible individuals. The major predisposing genes are located on the HLA system on chromosome 6, namely the *HLA-DQ2* and *DQ8* genes found in at least 95% of patients. Gluten is a complex mixture of storage proteins of wheat, a staple food for most populations in the world, and other cereals (rye and barley). Gluten proteins have several unique features that contribute to their immunogenic properties. They are extremely rich in the amino acids proline and glutamine. Due to the high proline content, gluten is highly resistant to proteolytic degradation within the gastrointestinal tract because gastric and pancreatic enzymes lack post-proline cleaving activity. Moreover, the high glutamine content makes gluten a good substrate for the enzyme tissue transglutaminase (tTG). Gluten proteins are now known to encode many peptides that are capable of stimulating both a T cell-mediated and an innate response. The 33-mer is a gliadin peptide of 33 residues (α2-gliadin 56–88) produced by normal gastrointestinal proteolysis, containing six partly overlapping copies of three T cell epitopes. The 33-mer is an immunodominant peptide that is a remarkably potent T cell stimulator after deamidation by tTG (Shan *et al.*, 2002).

Celiac disease is one of the most common lifelong disorders on a worldwide basis. The condition can manifest with a previously unsuspected range of clinical presentations, including the typical malabsorption syndrome (chronic diarrhea, weight loss, abdominal distention) and a spectrum of symptoms potentially affecting any organ or body system. Since celiac disease is often atypical or even silent on clinical grounds,

Gluten-Free Cereal Products and Beverages
ISBN: 9780123737397

many cases remain undiagnosed, leading to the risk of long-term complications, such as osteoporosis, infertility or cancer (Fasano and Catassi, 2001). There is a growing interest in the social dimension of celiac disease, since the burden of illness related to this condition is doubtless higher than previously thought (American Gastroentero-logical Association, 2001). Although celiac disease can present at any age, including the elderly, typical cases often manifest in early childhood. In 1888, Samuel Gee, having drawn attention to the disorder in a lecture delivered on October 5, 1887 at the Hospital for Sick Children, Great Ormond Street, London, produced his classical paper, *On the Coeliac Affection* (Gee, 1890). Dr. Gee described celiac disease as follows:

> There is a kind of chronic indigestion which is met with in persons of all ages, yet is especially apt to affect children between one and five years old. . . . Signs of the disease are yielded by the faeces; being loose, not formed, but not watery; more bulky than the food taken would seem to account for. . . .

Remarkably, he already hypothesized that foodstuff could be the trigger of the disease:

> The causes of the disease are obscure. Children who suffer from it are not all weak in constitution. Errors in diet may perhaps be a cause, but what error? Why, out of a family of children all brought up in much the same way, should one alone suffer? To regulate the food is the main part of treatment. . . . The allowance of farinaceous food must be small; highly starchy food, rice, sago, corn-flour are unfit.

Despite his great clinical acumen, Dr. Gee was not able to make the final link between gluten ingestion and celiac disease, since he concluded:

> "Malted food is better, also rusks or bread cut thin and well toasted on both sides. . . . "

Epidemiology

In the general population

In the past, celiac disease was considered a rare disorder, mostly affecting children of European origin. Indeed, this idea is still widespread, so much that in many European countries celiac disease continues to be included in the list of rare disorders protected by specific regulations of the healthcare system. On the other hand, a huge number of studies have recently shown that celiac disease is one of the commonest lifelong disorders affecting humans in many areas of the world. Currently most cases remain undiagnosed, due to the lack of typical symptoms, and can be recognized only through serological screening by sensitive tools (e.g. serum IgA class anti-transglutaminase and anti-endomysial antibodies determination) (Catassi *et al.*, 1996; Catassi, 2005). Serological screenings performed on general population samples have confirmed that the prevalence of celiac disease in Europe is very high (Catassi *et al.*, 1994; Csizmadia *et al.*, 1999; Catassi, 2005), ranging between 0.75 and 0.4% of the general population, with a trend toward higher figures (1% or more) in younger subjects and among groups that have been more isolated genetically (e.g. in Northern Ireland, Finland, and Sardinia) (Johnston *et al.*, 1998; Meloni *et al.*, 1999; Mäki *et al.*,

2003). Until recently, celiac disease was generally perceived to be less common in North America than in Europe (Green *et al.*, 2001). Should the frequency of celiac disease be lower in the USA, the existence of a protective environmental factor in that country should be postulated, since Americans and Europeans largely share a common genetic background. This epidemiological "dilemma" has recently been answered by our large US prevalence study including 4126 subjects sampled from the general population (Fasano *et al.*, 2003). The overall prevalence of celiac disease in this US population sample was 1:133, actually overlapping the European figures. Similar disease frequencies have been reported from countries mostly populated by individuals of European origin (e.g. Australia, New Zealand, and Argentina) (Cook *et al.*, 2000; Hovell *et al.*, 2001; Gomez *et al.*, 2001).

Celiac disease is not only frequent in developed countries, but it is increasingly found in areas of the developing world, such as North Africa (Bdioui *et al.*, 2006), Middle East (Shahbazkhani *et al.*, 2003), and India (Sood *et al.*, 2006). This disorder can contribute substantially to childhood morbidity and mortality in many developing countries. The highest celiac disease prevalence in the world has been described in the Saharawi, an African population of Arab-Berber origin. In a sample of 990 Saharawi children screened by endomysial antibody (EMA) testing and intestinal biopsy, we found a celiac disease prevalence of 5.6%, which is 5- to 10-fold higher than in most European countries (Catassi *et al.*, 1999). The reasons for this striking celiac disease frequency are unclear but may be primarily related to genetic factors, given the high level of consanguinity of this population. The main susceptibility genotypes, *HLA-DQ2* and *-DQ8*, exhibit one of the highest frequencies in the world in the general background Saharawi population (Catassi *et al.*, 2001). Gluten consumption is very high as well, since wheat flour is the staple food of this population. Celiac disease in the Saharawi children can be a severe disease, characterized by chronic diarrhea, stunting, anemia, and increased mortality (Rätsch and Catassi, 2001) (Figure 1.1). Treatment, like in other poor countries, is hampered by the lack of diagnostic facilities and the scarcity of commercially available gluten-free food.

The Middle East holds a special place in the history of celiac disease. The domestication of ancient grains began in Neolithic settlements from the wild progenitors *Triticum monococcum bocoticcum* and *T. monococcum uratru* in the north-eastern region (Turkey, Iran, and Iraq) and *Triticum turgidum dicoccoides* in the south-western region (Israel/Palestine, Syria, and Lebanon) of the so-called "Fertile Crescent" area. This extends from the Mediterranean Coast on its western extreme to the great Tigris–Euphrates plain eastward (Lewin, 1988). The cultivation of wheat and barley was first exploited and intensively developed in the Levant and western Zagros (Iran) some 10 000–12 000 years ago. From the Fertile Crescent, farming spread and reached the edge of Western Europe some 6000 years ago. During the 1980s Simoons theorized that this pattern of the spread of agriculture might explain the higher celiac disease incidence in some Western countries, particularly Ireland. Mapping the prevalence of the HLA-B8 antigen (the first HLA antigen known to be associated with celiac disease) across Europe he noted an east–west gradient, with a consistent increase in antigen frequency with the decreasing length of time since farming was adopted. Simoons then hypothesized that the HLA-B8 antigen

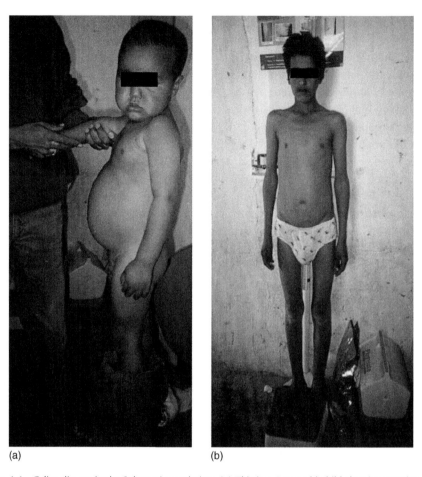

(a)   (b)

Figure 1.1 Celiac disease in the Saharawi population. (a) This is a 5-year-old child showing stunting, abdominal distention and malnutrition. (b) This is a 17-year-old boy with a severe pubertal delay and epilepsy.

may once have been prevalent throughout pre-agricultural Europe. According to this theory, the spread of wheat consumption exerted a negative selective pressure on genes associated with celiac disease, such as *HLA-B8*. Higher *HLA-B8* frequency in north-eastern Europe, and consequently higher celiac disease frequency, may therefore be attributable to a lack of exposure to cereals until relatively recently (Simoons, 1981).

This theory apparently did not survive the recent developments of both celiac disease genetics and epidemiology. On the one hand, it is now well established that the main genetic predisposition to celiac disease is not linked to *HLA-B8* but to some DQ genotypes (DQ2 and DQ8) which are in linkage disequilibrium with B8. Neither DQ2 nor DQ8 show any clear-cut east–west prevalence gradient. On the other hand, overall celiac disease prevalence is not lower in Middle East countries than in Europe, as should be the case if the longer history of agriculture tended to eliminate the genetic backbone predisposing to celiac disease.

In at-risk groups

Studies all over the world have shown that the prevalence of celiac disease is definitely increased in specific population subgroups (Figure 1.2). The risk of celiac disease in first-degree relatives has been reported to be 6–7% on average, mostly ranging from 3 to 10% (Mäki *et al.*, 1991; Corazza *et al.*, 1992; Vitoria *et al.*, 1994). In a Finnish study on 380 patients with celiac disease and 281 patients with dermatitis herpetiformis, the mean disease prevalence was 5.5%, distributed as follows: 7% among siblings, 4.5% among parents and 3.5% among children (Hervonen *et al.*, 2002). The prevalence of celiac disease is also increased in second-degree relatives (Fasano *et al.*, 2003), highlighting the importance of genetic predisposition as a risk factor. Celiac disease prevalence is increased in autoimmune diseases, especially type 1 diabetes and thyroiditis, but also in less common disorders (e.g. Addison's disease or autoimmune myocarditis). The average prevalence of celiac disease among children with type 1 diabetes is 4.5% (0.97–16.4%) (Holmes, 2002). Usually diabetes is diagnosed first, while celiac disease is often subclinical and only detectable by serological screening. The increased frequency of celiac disease in several thyroid diseases (Hashimoto's thyroiditis, Graves' disease, and primary hypothyroidism) is well established (Sategna-Guidetti *et al.*, 2001). A 3- to 5-fold increase in celiac disease prevalence has been reported in subjects with autoimmune thyroid disease (Valentino *et al.*, 1999; Hakanen *et al.*, 2001). On the other hand, celiac disease-associated hypothyroidism may sometimes lack features of an autoimmune process. Interestingly, treatment of celiac disease by gluten withdrawal may lead to normalization of subclinical hypothyroidism (Sategna-Guidetti *et al.*, 2001). The causal relationship between celiac disease and other autoimmune disorders is still a controversial issue. The two most accredited theories propose: (1) this association is secondary to a common genetic background predisposing to both celiac disease and the associated autoimmune disease or (2) untreated celiac disease leads to the onset of other autoimmune disorders in genetically susceptible individuals. This second hypothesis is supported by the evidence that tTG seems to be only one of the autoantigens involved

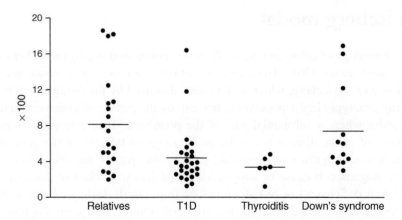

Figure 1.2 Prevalence of celiac disease in defined at-risk conditions. Each dot is the result of a study and each line is the group mean. (T1D = type 1 diabetes)

in gluten-dependent autoimmune reactions. Other autoantigens which are normally "cryptic" can be unmasked and cause a self-aggressive immunological response following the gliadin-initiated inflammatory process (Fasano and Catassi, 2001).

The phenomenon of antigen spreading has been described in well-defined natural models such as type 1 diabetes, whose clinical manifestations appear after the patient has produced an autoimmune response to various autoantigens (i.e. anti-insulin, anti-beta cell, etc.), and might also be present in celiac disease. This would explain the high incidence of autoimmune diseases and the presence of a large number of organ-specific autoantibodies in a certain number of celiac subjects on a gluten-containing diet. However, it has not been proven that an early treatment of celiac disease may prevent the development of other autoimmune disorders.

An increased frequency of celiac disease is found in some genetic diseases, especially Down's, Turner's and William's syndromes. In a multicenter Italian study on 1202 subjects with Down's syndrome, 55 celiac disease cases were found, with a prevalence of this disease association of 4.6% (Bonamico et al., 2001). In children with Down's syndrome celiac disease is not detectable on the basis of clinical findings alone and is therefore under-detected. Even when there are symptoms, they may be considered clinically insignificant or possibly attributed to Down's syndrome itself. Nevertheless, the reported amelioration of gastrointestinal complaints on a gluten-free diet (GFD) for all symptomatic patients suggest that identification and treatment can improve the quality of life for these children (Book et al., 2001). Selective IgA deficiency (total serum IgA lower than 5 mg %) predisposes to celiac disease development and this primary immunodeficiency is 10- to 16-fold more common in patients with celiac disease than the general population (Cataldo et al., 1998). Patients with selective IgA deficiency and celiac disease are missed by using the class A anti-tTG test (or any other IgA-based test, e.g. EMA) for screening purposes. For this reason it is appropriate to (1) check the total level of serum IgA in patients screened for celiac disease and (2) perform an IgG-based test (e.g. IgG-anti-tTG and/or IgG-anti-gliadin) if total IgA is lower than normal.

The iceberg model

The epidemiology of celiac disease is efficiently conceptualized by the iceberg model (Fasano and Catassi, 2001). The prevalence of celiac disease can be conceived as the overall size of the iceberg, which is not only influenced by the frequency of the predisposing genotypes in the population, but also by the pattern of gluten consumption. In countries where a substantial part of the population is of European origin, the prevalence of celiac disease is usually in the range of 0.5–1% of the general population. A sizable portion of these cases, the visible part of the celiac iceberg, are properly diagnosed because of suggestive complaints (e.g. chronic diarrhea, unexplained iron deficiency) or at-risk situations (e.g. family history of celiac disease or associated autoimmune disease). In developed countries, for each diagnosed case of celiac disease, an average of 5–10 cases remain undiagnosed (the submerged part of the iceberg), usually because of atypical, minimal or even absent complaints

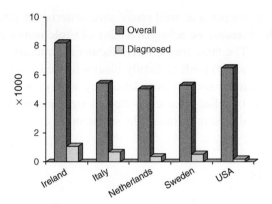

Figure 1.3 Prevalence of celiac disease in different countries, based either on clinically detected cases (white bars) or mass screening data (gray bars). In each country the gap between the gray and white bars represents the submerged part of the celiac iceberg.

(Figure 1.3). These undiagnosed cases remain untreated and are therefore exposed to the risk of long-term complications. The "water line," namely the ratio of diagnosed to undiagnosed cases, mostly depends on the physician's tendency to request serological celiac disease markers in situations of low clinical suspicion (i.e. awareness of celiac disease clinical polymorphism). It is important to realize that a number of the cases that are undiagnosed at a given time (e.g. because of lack of symptoms) can emerge later on, due to clinical deterioration (Mäki *et al.*, 2003).

How to deal with the celiac iceberg is currently a matter of debate in the scientific community. At first sight there could be good arguments in favor of mass screening: (1) celiac disease is a common disorder causing significant morbidity in the general population; (2) early detection is often difficult on a clinical basis; (3) if not recognized, the disease can manifest itself with severe complications that are difficult to manage (e.g. infertility, osteoporosis, lymphoma); (4) there is an effective treatment, the GFD; (5) sensitive and simple screening tests are available (e.g. the anti-tTG test). However, the cost/effectiveness ratio of celiac disease screening needs further clarification. Although it is well established that patients with untreated celiac disease may develop complications, the natural history of undiagnosed/untreated celiac disease, particularly the so-called "silent" form, remains unclear. This is a strong limitation, as treatment with GFD is likely to interfere heavily with the quality of life, especially in adults. Despite the high sensitivity of the serological celiac disease markers, the positive predictive value of these investigations decreases when applied to the general population. Furthermore, the appropriate age for celiac disease screening remains to be elucidated. For all these reasons, the best approach to the iceberg of undiagnosed celiac disease seems to be the serological testing of at-risk groups, a procedure defined as "case-finding" that minimizes costs and is ethically appropriate.

A primary care practice provides the best opportunity to first identify individuals who are at risk for celiac disease and need referral for definitive diagnosis. We recently undertook a multicenter, prospective, case-finding study using serological testing (IgA class anti-tTG antibody determination) of adults who were seeking medical attention from their primary care physician in the USA and Canada (Catassi *et al.*,

2007a). By applying simple and well-established criteria for celiac disease case-finding on a sample of adults, we achieved a 32- to 43-fold increase in the diagnostic rate of this condition. The most frequent risk factors for undiagnosed celiac disease were: (a) thyroid disease, (b) positive family history for celiac disease, (c) persistent gastrointestinal complaints, and (d) iron deficiency with or without anemia. Many newly diagnosed cases of celiac disease reported a long-standing history of symptoms (usually of years) that should have raised the suspicion of celiac disease well before.

Pathogenesis

Abnormalities of the jejunal mucosa are the hallmark of celiac disease (Plate 1.1). When fully expressed, the celiac enteropathy is characterized by an increase in the number of intraepithelial lymphocytes (IELs), marked crypt hypertrophy, and complete loss of villi (subtotal villous atrophy, "flat" lesion). Although a number of other conditions may cause a flat biopsy, such findings in a subject living in the Western world are almost certain to indicate celiac disease. Pathological changes are sometimes less severe and can be characterized by less extensive villous atrophy and crypt hypertrophy (partial villous atrophy) or isolated infiltration of IELs (infiltrative lesion). The celiac enteropathy is an end-stage lesion that depends on both genetic and environmental factors for expression (Fasano and Catassi, 2001) (Figure 1.4).

The concordance rate in monozygotic twins is 86%, whereas in dizygotic twins it reaches only 20%, indicating a strong influence of genetic factors (Greco *et al.*, 2002). Of these, HLA is estimated to be responsible for 40–50% of the genetic contribution in celiac disease (Sollid and Lie, 2005). Roughly 90% of patients carry the *HLA-DQ2* heterodimer (*DQA1*0501/DQB1*0201*) in *cis* (on one parental chromosome), or in *trans* (the two *DQ2* alleles being encoded on one chromosome from each parent). Most individuals that are not *HLA-DQ2*-positive express *HLA-DQ8* (*DQA1*0301/ DQB1*0302*). In a small number of patients only one of the two *DQ2* alleles is present (that is, *DQB1*0201* or, rarely, *DQA1*0501*). Different combinations of *HLA-DQ*-predisposing alleles influence the risk of disease, this being much higher in subjects

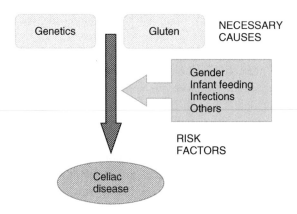

Figure 1.4 The causes of celiac disease.

showing a double copy of the *DQB*02* genes than other predisposing genotypes (e.g. *DQ2*/X or *DQ8*/X). Homozygosis for the *DQB1*0201* allele is also associated with a higher severity of the histological score and with higher risk of refractory sprue (see below) (Stepniak and Koning, 2006a). Almost all celiac disease patients carry *DQ2* or *DQ8*, yet as many as 20–30% of healthy subjects are also carriers. The presence of *HLA-DQ2* and/or *DQ8* is a necessary but not sufficient cause. At least 60% of the genetic predisposition to celiac disease is related to tens of other genes, each of them adding a small contribution to the development of the disease (Louka and Sollid, 2003).

A possible role for the myosin IXB (*MYO9B*) gene has been recently suggested by a Dutch study speculating that this unconventional myosin molecule could account for an affected integrity of the intestinal barrier (Monsuur *et al.*, 2005). Clearly, a "leaky gut" would allow an increase in the penetration of gluten peptides through the intestinal epithelium and contribute to an increased risk of breaking oral tolerance to gluten proteins. However this finding has not been confirmed in other countries, such as the UK (Hunt *et al.*, 2006) and Norway/Sweden (Amundsen *et al.*, 2006), suggesting genetic heterogeneity of different populations.

There is evidence that other genes await to be discovered on chromosomes 5 and 6. Some of those genes may actually predispose to autoimmunity in general. This would explain the increased prevalence of autoimmune diseases such as type 1 diabetes in patients with celiac disease (Stepniak and Koning, 2006a).

The cascade of pathophysiological events leading to the adaptive immune response may start with an alteration of the barrier function of the small intestinal mucosa. The upregulation of zonulin, a recently described intestinal peptide involved in tight junction regulation, seems to be responsible, at least in part, for increased gut permeability to gliadin peptides in celiac patients (Drago *et al.*, 2006). In the lamina propria, the tTG converts glutamine into the negatively charged glutamic acid, a process called deamidation. After deamidation, the affinity of gliadin peptides for the HLA molecules located on the membrane of the antigen-presenting cells (APCs) is greatly increased. The interaction between gliadin peptides and HLA molecules activates intestinal T cells. The release of proinflammatory cytokines (e.g. interferon-γ) by activated T cells may determine damage to the enterocytes, increase proliferation in the intestinal crypts and, finally, cause severe damage to the intestinal mucosa architecture (Plate 1.2). It has been recently shown that the induction of an adaptive immune response to gluten peptides is tightly related to an innate immunological mechanism (Londei *et al.*, 2005). The gliadin-derived fragment p31–43 can induce interleukin 15 (IL-15) secretion by activated dendritic cells in the lamina propria. This cytokine stimulates IELs to express NKG2D receptors and epithelial cells to express MICA molecules. Upon engagement of NKG2D receptor with MICA ligand, the IELs kill the epithelial cells, contributing to tissue destruction. The activation of intestinal dendritic cells could be triggered by intestinal infection.

Interestingly, an antibody directed against a *Rotavirus* protein has been found in subjects with active celiac disease (Zanone *et al.*, 2006). This antibody recognizes self-antigens (tTG) and is able to increase intestinal permeability and induce monocyte activation.

Infant diet and risk of celiac disease and related autoimmune disorders

Recent studies suggest that the pattern of infant nutrition may have a critical role on the development of celiac disease and other autoimmune disorders. Breastfeeding is thought to delay or reduce the risk of developing celiac disease (Akobeng *et al.*, 2006). The positive effects of breast milk can be attributed, at least in part, to its influence on the microbial colonization process of the newborn intestine.

The relationship between age at gluten introduction and the risk of celiac disease is still controversial. According to the recommendations of the European Society for Pediatric Gastroenterology and Nutrition (ESPGHAN), gluten-containing cereals should be introduced in the diet of European infants after the age of six months. However huge differences in the quantity and the quality of cereals introduced at weaning exist, even among neighboring countries (Ascher *et al.*, 1993). There is also a tendency to delay gluten introduction in infants who are at family risk of developing celiac disease. In Sweden an "epidemic" of early-onset celiac disease took place during the late 1980s and early 1990s. A retrospective analysis of this Swedish epidemic showed that the risk of celiac disease was reduced in infants introduced to gluten when still breastfed or, even better, who continued to be breastfed after gluten introduction (Ivarsson *et al.*, 2002). Based on these findings, Swedish pediatricians now recommend the introduction of gluten-containing complementary food prior to the age of six months if the mother intends to stop breastfeeding before that age.

On the other hand, prospective studies on infants at genetic risk of type 1 diabetes suggested that the risk of type 1 diabetes and celiac disease is increased either in infants started on gluten before 3–4 months or after 7 months (Norris *et al.*, 2003, 2005). The possible risk related to a late introduction of gluten (after 7 months of age) is a puzzling and counterintuitive finding that deserves further confirmations.

Clinical spectrum

The clinical spectrum of celiac disease is wide (Tables 1.1 and 1.2 and Figure 1.5). In children, the *typical* form of celiac disease is characterized by gastrointestinal manifestations starting between 6 and 24 months of age, after the introduction of gluten in the diet. Infants and young children present with impaired growth, chronic diarrhea, abdominal distention, muscle wasting and hypotonia, poor appetite and unhappy behavior. Within weeks to months of starting to ingest gluten, weight gain velocity decreases and finally weight loss can be observed. A celiac crisis, characterized by explosive watery diarrhea, marked abdominal distension, dehydration, electrolyte imbalance, hypotension, and lethargy, was more commonly described in the past, while it is now rarely observed (Fasano and Catassi, 2001).

Atypical celiac disease is usually seen in older children and features of overt malabsorption are absent. The symptoms may be intestinal or extraintestinal. Intestinal features may include recurrent abdominal pain, dental enamel defects, recurrent aphthous stomatitis, and constipation. Between 6 and 12% of patients with iron-deficiency

Table 1.1 Clinical manifestations of celiac disease

Manifestations secondary to untreated celiac disease

Celiac disease with classic symptoms	Abdominal distension
	Anorexia, irritability
	Chronic or recurrent diarrhea
	Failure to thrive or weight loss
	Vomiting
	Muscle wasting
	Celiac crisis (rare)
	Fatigue
Celiac disease with non-classic symptoms	Arthritis
	Aphthous stomatitis
	Constipation
	Dental enamel defects
	Dermatitis herpetiformis
	Hepatitis
	Iron-deficient anemia
	Pubertal delay
	Recurrent abdominal pain
	Short stature

Associated diseases (or secondary to untreated celiac disease?)

Autoimmune diseases	Type I diabetes
	Thyroiditis
	Sjogren's syndrome
	Others
Neurological and psychological disturbances	Ataxia
	Autism
	Depression
	Epilepsy with intracranial calcifications

IgA nephropathy
Infertility
Osteopenia/osteoporosis
Cancer

Genetic associated diseases
Down's syndrome
Turner's syndrome
William's syndrome
IgA deficiency

Table 1.2 Histological and clinical spectrum of celiac disease

Clinical form	Histological and clinical manifestations
Celiac disease with classic symptoms	*Fully expressed enteropathy* Intestinal symptoms
Celiac disease with non-classic symptoms	*Fully expressed enteropathy* Extra-intestinal manifestations
Silent	*Fully expressed enteropathy* Minimal complaints or symptom-free (occasionally discovered by serological screening)
Potential	*Minimal changes enteropathy or normal small intestinal mucosa* Sometimes symptomatic

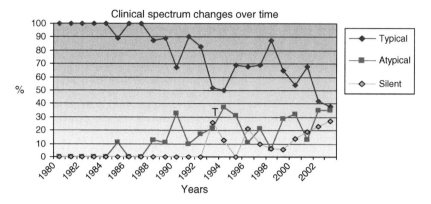

Figure 1.5 The changes of the celiac disease clinical presentation over time. This study group included all the new cases of celiac disease seen by the authors at the Department of Pediatrics of Ancona, Italy.

anemia attending a hematology clinic are found to have celiac disease. The anemia is typically resistant to oral iron therapy. Short stature and delayed puberty can be the primary manifestation in an otherwise healthy child. Celiac disease is the most common organic cause of slow growth rate and is much more common than growth hormone deficiency. Other common manifestations include chronic fatigue and isolated increase of aminotransferase serum level.

In adults with celiac disease, diarrhea is the most common symptom, but only affects just over 50% of patients, is of variable duration, and can present acutely in a previously well person (Holmes and Catassi, 2000). Lethargy and tiredness, with or without anemia, and weight loss are also common symptoms. Abdominal distention affects about one-third of patients. Peripheral neuropathy, ataxia indicating cerebellar degeneration, arthropathy, infertility and bleeding disorders are less common presentations. Dermatitis herpetiformis, a blistering skin disease, is at present regarded as a variant of celiac disease mostly affecting adult patients. In about 4% of cases, celiac disease presents during pregnancy or within weeks or months of giving birth. Celiac disease is being increasingly diagnosed in later life and, today, about 25% of cases are diagnosed in patients over 60 years of age. Contrary to common belief, 95% of these patients manage a GFD well and enjoy a much improved quality of life.

Celiac disease is defined as *silent* whenever a typical gluten-sensitive enteropathy is occasionally found in a subject who is apparently healthy. Large numbers of silent cases of celiac disease have been reported in at-risk groups (such as subjects with type 1 diabetes and first-degree relatives) and in general population samples enrolled in screening programs. An in-depth clinical examination shows that many of these "silent" cases are indeed affected with a low-grade intensity illness often associated with decreased psychophysical well-being.

A *potential* form of celiac disease is diagnosed in subjects showing positivity of EMA and/or anti-tTG antibodies, the typical HLA predisposing genotype (*DQ2* or *DQ8*), but a normal or minimally abnormal mucosal architecture (increased IEL count) at the intestinal biopsy. These cases are at risk of developing a typical celiac disease enteropathy later in life.

Complications

Osteoporosis is one of the well-known complications of untreated celiac disease (Hernandez and Green, 2006). Persistent villous atrophy is associated with low bone mineral density. Bone alterations were once thought to derive from calcium and vitamin D deficiency secondary to simple intestinal malabsorption. Recently, other causes of bone metabolism impairment have been claimed, including the interaction between cytokines and local/systemic factors influencing bone formation and reabsorption. In the pediatric population, a prompt enforcement of a GFD can lead to a satisfactory recovery of the bone mass (Barera *et al.*, 2000). Conversely, adults affected by osteoporosis secondary to celiac disease do not experience spontaneous recovery, and there are no conclusive data on the efficacy of standard therapies for osteoporosis in reducing the fracture risk. This finding stresses the need for an early diagnosis as a preventive intervention to avoid celiac disease complications.

The prevalence of *neurological and psychiatric disorders* is increased in patients with celiac disease. Gluten-sensitive neurological syndromes (ataxia, peripheral neuropathy, and other conditions) have been hypothesized in patients with various idiopathic neuropathologies, detectable anti-gliadin antibodies (AGA) and *HLA-DQ2* or *DQ7*. Further investigation of these cases has suggested a high incidence of anti-neuronal antibodies (anti-Purkinje, anti-neuronal nuclear, anti-GAD) (Hadjivassiliou *et al.*, 2003). Epilepsy is more common in celiac patients, while the existence of a syndrome characterized by epilepsy, occipital calcifications, and celiac disease is widely accepted (Gobbi *et al.*, 1992). Depression affects about 10% of celiac patients on a normal diet. The association between autism and celiac disease is still controversial, and it still remains to be established through systematic, well-designed studies whether gluten *per se* has a role in causing autistic behavior outside the context of celiac disease. Menarche is late and the menopause early in untreated celiac patients compared with those who are treated or controls (Holmes and Catassi, 2000).

Celiac disease is a cause of *infertility* in both women and men. Recurrent abortion is a feature of untreated celiac disease and successful pregnancy may ensue after gluten withdrawal. Men with celiac disease may have reversible infertility. Impotence, hypogonadism, and abnormal sperm motility and forms occur. *Ulcerative jejunoileitis* is characterized by malabsorption, almost always a flat small intestinal biopsy and chronic ulcers found mainly in the jejunum and ileum. The development of jejunoileitis may bring a patient with celiac disease to diagnosis or cause deterioration in those previously well controlled on a GFD. It may be premalignant or even a low-grade malignant condition from the onset. *Mesenteric lymph node cavitation* is a rare, serious complication that affects those with long-standing untreated celiac disease, and should also be suspected in patients who are not responding to a GFD.

In a minority of adult patients, celiac disease does not respond to treatment with a GFD. The most likely cause of non-responsiveness is continued gluten ingestion, which can be voluntary or inadvertent. Patients with celiac disease in whom the lack of compliance to a GFD has been ruled out belong to the *refractory sprue* category. An aberrant clonal intraepithelial T cell population can be found in up to 75% of patients with refractory sprue, a condition that is currently classified as cryptic enteropathy-associated T-cell lymphoma. These patients typically undergo pharmacologic therapies, including treatment with steroids, or immunosuppressants, such as azathioprine and cyclosporin (Daum *et al.*, 2005). If patients do not respond to these treatments, the ultimate treatment is total parenteral nutrition. However, none of these therapies have been subjected to rigorous controlled studies.

Celiac disease is associated with *intestinal lymphoma* and other forms of cancer, especially adenocarcinoma of the small intestine, of the pharynx, and of the esophagus. Enteropathy-associated T-cell lymphoma (EATL) is a rare form of high-grade, T-cell non-Hodgkin lymphoma (NHL) of the upper small intestine that is specifically associated with celiac disease. This NHL subtype arises in patients with either previously or concomitantly diagnosed celiac disease. In a subgroup of patients, there is progressive deterioration of a refractory form of celiac disease. EATL derives from a clonal proliferation of IELs and is often disseminated at diagnosis. Extraintestinal presentations are not uncommon in the liver/spleen, thyroid, skin, nasal sinus, and brain. The outlook for patients with EATL is poor. Recent studies indicated that: (1) celiac disease is associated with a significantly increased risk for NHL, especially of the T-cell type and primarily localized in the gut (EATL); (2) the celiac disease–lymphoma association is less common than previously thought, with a relative risk close to 3; (3) celiac disease screening is not required in patients with NHL of any primary site at the onset, unless suggested by specific findings (T-cell origin and/or primary gut localization); (4) the risk of NHL associated with clinically milder (or silent) forms could be lower than in typical cases of celiac disease. Several follow-up studies suggest that the GFD protects from cancer development, especially if started during the first years of life. Strict adherence to the GFD seems to be the only possibility of preventing a subset of rare but very aggressive forms of cancer (Catassi *et al.*, 2005).

Diagnosis

Serological testing

Although an intestinal biopsy is still considered necessary to confirm the diagnosis of celiac disease, serological tests are frequently used to identify individuals for whom the procedure is indicated (Hill *et al.*, 2005). Commercially available tests include IgA- and IgG-AGA, EMA, anti-tTG, and anti-actin antibodies. These tests are particularly helpful in individuals without gastrointestinal symptoms and those with conditions associated with celiac disease, as well as for screening asymptomatic first-degree relatives of known cases. They have also been widely used in epidemiologic studies to determine the prevalence of celiac disease.

AGA antibodies were the first serological markers of celiac disease to be widely used in clinical practice. The sensitivity of IgA-AGA among reported studies ranges between 0.52 and 1.00 in children and between 0.65 and 1.00 in adults. The specificity of IgA-AGA in children ranges between 0.92 and 0.97 and in adults between 0.71 and 0.97. The IgG-AGA is similar in sensitivity to the IgA-AGA, but the specificity is much lower, approximately 0.5. This indicates that many individuals without celiac disease express IgG-AGA antibody. False positive tests have been recorded in individuals with a variety of other gastrointestinal disorders, including esophagitis, gastritis, gastroenteritis, inflammatory bowel disease, cystic fibrosis and cow's milk protein intolerance.

EMAs are IgA class autoantibodies directed against antigens in the collagenous matrix of human and monkey tissues. The EMA test is based on an immunofluorescent technique using either monkey esophagus or human umbilical cord as substrate; the accuracy of the test is similar for either substrate. The nature of this test renders it more time consuming to perform, generally more expensive and, because the interpretation is operator-dependent, potentially more prone to errors. The sensitivity of the EMA in children ranges from 0.88 to 1.00 and in adults is reported to be 0.87 to 0.89. The specificity of the EMA in children ranges from 0.91 to 1.00 and in adults is reported to be 0.99. The EMA test may be less accurate in children under 2 years of age.

tTG was found to be the major autoantigen responsible for EMA positivity. When first introduced, the anti-tTG assays used guinea-pig protein as antigen. Subsequent cloning of the human tTG gene led to the development of ELISA assays based on the human tTG protein. The sensitivity of IgA class anti-tTG in both children and adults ranges from 0.92 to 1.00. The specificity of anti-tTG in both children and adults ranges from 0.91 to 1.00. There is evidence that anti-tTG assays using human recombinant protein and human-derived red cell tissue tTG have a higher sensitivity (0.96–1.00 versus 0.89–0.94) and specificity (0.84–1.00 versus 0.74–0.98) when compared with assays using guinea-pig protein.

Actin is a key structural protein of the cytoskeleton network that is particularly abundant in intestinal microvilli. IgA class anti-actin antibodies, detectable by either an immunofluorescence or ELISA technique, seem to contribute to villus cytoskeleton damage and to the pathogenesis of intestinal damage in celiac disease. The presence in the sera of celiac disease patients of anti-actin autoantibodies has recently been suggested as a marker of severe intestinal villous atrophy (Clemente *et al.*, 2004). Based on the current evidence and practical considerations, including accuracy, reliability and cost, measurement of IgA antibody to tTG is recommended for initial testing for celiac disease. Individuals with celiac disease who are also IgA deficient will not have abnormally elevated levels of IgA-anti-tTG or IgA-EMA. Therefore, when testing for celiac disease in subjects with symptoms suspicious for celiac disease, measurement of quantitative serum IgA can facilitate interpretation when the IgA-anti-tTG IgA is low. In individuals with known selective IgA deficiency and symptoms suggestive of celiac disease, testing with IgG-anti-tTG is recommended. Even when serological tests for celiac disease are negative, in children with chronic diarrhea or failure to thrive and in those belonging to a group at risk (e.g. selective IgA deficiency or a positive family history of celiac disease) who have symptoms compatible with celiac disease, an intestinal biopsy can be helpful to identify the unusual case of seronegative celiac disease or to detect other mucosal disorders accounting for the symptoms (Hill *et al.*, 2005).

Small intestinal biopsy

Small intestinal biopsy is the cornerstone of diagnosis and should be undertaken in all patients with suspected celiac disease. Biopsies can be obtained using a capsule with a suction-guillotine mechanism (e.g. Watson capsule). Nowadays, most biopsies in both children and adults are taken at the time of upper gastrointestinal endoscopy using standard fiber-optic instruments. Endoscopy allows multiple biopsies to be taken, which minimizes sampling error (Holmes and Catassi, 2000). The characteristic histological changes described in celiac disease include an increased number of IELs (>30 lymphocytes per 100 enterocytes), elongation of the crypts (increased crypt length), partial to total villous atrophy and a decreased villous:crypt ratio (Plate 1.1). Lamina propria changes include an increased crypt mitotic index and infiltration of plasma cells, lymphocytes, mast cells, and eosinophils.

An increase in the IELs is perhaps a more sensitive index of gluten sensitivity than the changes in villous structure, as they are found early in the course of the disease and disappear before other features of structural recovery can be detected. The histological grading system introduced by Marsh and modified by Oberhuber classifies the histological changes of celiac disease as type 0 or preinfiltrative stage (normal), type 1 or infiltrative lesion (increased IELs), type 2 or hyperplastic lesion, type 3 or destructive lesion including type 3a (partial villous atrophy), type 3b (subtotal villous atrophy), and type 3c (total villous atrophy) (Oberhuber *et al.*, 1999).

It is recommended that confirmation of the diagnosis of celiac disease requires an intestinal biopsy in all cases. Because the histological changes in celiac disease may be patchy, it is recommended that multiple biopsy specimens be obtained from the second or more distal part of the duodenum. There is good evidence that villous atrophy (Marsh type 3) is a characteristic histopathologic feature of celiac disease. The presence of infiltrative changes with crypt hyperplasia (Marsh type 2) on intestinal biopsy is compatible with celiac disease but with less clear evidence. Diagnosis in these cases is strengthened by the presence of positive serological tests (anti-tTG or EMA) for celiac disease. In the event the serological tests are negative, other conditions for the intestinal changes are to be considered and, if excluded, the diagnosis of celiac disease is reconsidered. The presence of infiltrative changes alone (Marsh type 1) on intestinal biopsy is not specific for celiac disease. Concomitant positive serological tests for celiac disease (anti-tTG or EMA) increase the likelihood that the subject has celiac disease. In circumstances where the diagnosis is uncertain additional strategies can be considered, including determination of the HLA type, repeat biopsy or a trial of treatment with a GFD and repeat serology and biopsy (Hill *et al.*, 2005). The diagnosis of celiac disease is considered definitive when there is complete symptom resolution after treatment with a strict GFD in a previously symptomatic individual with characteristic histological changes on small intestinal biopsy. A positive serological test that reverts to negative after treatment with a strict GFD in such cases is further supportive evidence for the diagnosis of celiac disease.

HLA testing

Polymerase chain reaction sequence-specific oligonucleotide typing methods are now available for the determination of alleles encoding *HLA-DQ2* and *DQ8*. The entity of the HLA-related risk (high or low) can be quantified using second-generation commercial kits allowing the complete characterization of the *HLA-DQ2* and *DQ8* genotype. Currently two major clinical applications of this test can be considered: (1) to rule out the possibility of celiac disease in at-risk subjects (e.g. first-degree relatives and patients with type 1 diabetes). Since the HLA predisposing genotype is a necessary (but not sufficient) factor for disease development, the negative predictive value of HLA typing is very high (i.e. the vast majority of subjects who are *DQ2*- and *DQ8*-negative will never develop celiac disease); (2) to rule out celiac disease in doubtful cases (celiac disease can be excluded with a 99% confidence in *DQ2*- and *DQ8*-negative subjects).

Management

The treatment of celiac disease is based on the lifelong exclusion of gluten-containing cereals from the diet. In many areas of the world, including Europe, North America, Australasia, and North Africa, gluten-rich products, such as bread and pasta, are part of the staple diet. Gluten-containing food therefore makes a substantial contribution to daily energy intake and is enjoyable to eat. The changes needed to begin and maintain a GFD are substantial and have a major impact on daily life. Thus, starting the diet is a critical step that should be handled sympathetically by experienced doctors and dietitians (Holmes and Catassi, 2000). Wheat, rye, and barley derivatives are excluded in the GFD. The exclusion of oats in the GFD is still a subject of debate. With few exceptions, clinical studies have shown that the prolonged ingestion of oats does not cause clinical or histological deterioration in children and adults with either celiac disease or dermatitis herpetiformis (Haboubi *et al.*, 2006). However many commercially available oats products are cross-contaminated by gluten-containing cereals and need to be excluded from the celiac diet.

Cereals that do not contain gluten and can be eaten include rice and maize. Other natural food, such as vegetables, salads, pulses, buckwheat, fruits, nuts, meat, fish, poultry, cheese, egg, and milk can also be eaten without limitations. A wide range of attractive and palatable gluten-free products that guarantee the absence of gluten are specifically manufactured for patients with celiac disease and may be labeled by an internationally recognized mark, the crossed ear of wheat. There are difficulties, however, in maintaining a strict GFD because of "hidden gluten" and food contamination (see below).

After starting a GFD, symptomatic patients show progressive clinical improvement that parallels the healing of the celiac enteropathy. In children the first signs of amelioration are often seen within a few days, with increased appetite and mood change, but it may take several months before symptoms disappear completely. Within 1–2 years of GFD the celiac disease-associated serum antibodies disappear and the architecture of the small intestinal mucosa normalizes (Fasano and Catassi, 2001). Patients should be followed up for life, preferably in a specialist clinic, otherwise they are most likely to stray from the GFD. Those taking gluten, either accidentally or on purpose, may well suffer ill-health and be exposed to health risks, including malignancy and osteoporosis.

Hidden gluten

Many commercial products, ready meals, and convenience foods are made with wheat flour, gluten-containing wheat proteins or gluten-containing starches added as filler, stabilizing agent, or processing aid. These include sausages, fish fingers, cheese spreads, soups, sauces, mixed seasonings, mincemeat for mince pies, and some medications and vitamin preparations. All real ales, beers, lagers, and stout should be avoided, but spirits, wines, liquors, and ciders are allowed. Whisky and malt whisky are allowed.

National celiac societies in many countries publish handbooks listing the gluten-free products that are available. These handbooks are regularly updated and are essential for celiac patients to have in their possession. It is important to remember that food lists are only applicable for use in the country in which they were compiled. Similar foods with well-known brand names may be made under franchise in different countries to slightly different recipes. They may look and taste the same, but be gluten-free in one country and not in others. It is almost impossible to maintain a "zero gluten level" diet, as gluten contamination is very common in food. Even products specifically targeted to dietary treatment of celiac disease may contain trace amounts of gluten proteins, either because of the cross-contamination of originally gluten-free cereals during their milling, storage, and manipulation, or due to the presence of wheat starch as a major ingredient.

Consequences of a low gluten intake

The effects of a low gluten intake in patients with celiac disease have been investigated in a limited number of studies. Ciclitira et al. (1984) analyzed the toxicity and time response of a gliadin dose (the major toxic fraction of gluten) in a single patient. They concluded that 10 mg produced no change, 100 mg a very slight measurable change, 500 mg moderate change and 1 g extensive damage to small intestinal morphology. The same group also reported that the ingestion of 2.4–4.8 mg per day of gluten caused no change in the jejunal biopsy morphometry of treated celiacs after either 1 or 6 weeks (Ciclitira et al., 1985). Ejderhamn et al. (1988) showed that a daily intake of 4–14 mg gliadin did not affect the morphology of the small bowel mucosa in celiacs on long-term treatment with the GFD. Recent Finnish studies indicate that an intake of 20–36 mg of daily gluten has no detectable effect on the mucosal histology (Kaukinen et al., 1999; Peräaho et al., 2003). We previously showed that a four-week challenge with 100 mg of gliadin per day caused deterioration of the small intestinal architecture and that the histological changes were more pronounced in patients challenged with 500 mg of gliadin per day (Catassi et al., 1993). Finally, a higher gluten intake (1–5 g of daily gluten), still lower than the normal gluten intake for the non-celiac population in Western countries (10–20 g per day), caused relapse of disease at a clinical, laboratory, and histological level, both in children and in adults (Jansson et al., 2001).

We recently concluded a prospective, double-blind, placebo-controlled multicenter trial to investigate the toxicity of gluten traces (10–50 mg daily) in the celiac diet. Patients were 39 adults with biopsy-proven celiac disease and on treatment with the

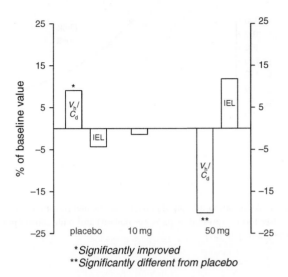

Figure 1.6 Mean changes of the intestinal morphometry indexes (V_h/C_d, villous height/crypt depth ratio; IEL, intraepithelial lymphocyte count) in celiac disease patients challenged with 0, 10 or 50 mg of daily gluten for three months.

GFD for at least 2 years. The background daily gluten intake was maintained below 5 mg. After baseline evaluation (T_0), patients were assigned to ingest daily and for 90 days a capsule containing 0 mg, 10 mg or 50 mg of gluten. Clinical, serological, and small intestine histology evaluations were performed at T_0 and after the gluten "micro-challenge" (T_1). This study disclosed a large inter-patient variability of the sensitivity to gluten traces. Some patients with celiac disease showed a clear-cut worsening of the small intestinal architecture after ingesting only 10 mg of daily gluten while others had an apparent improvement of the mucosal histology after the three-month challenge with 50 mg of daily gluten. Despite this wide individual variability we showed that 50 mg of daily gluten, if introduced for at least three months, were sufficient to cause a significant deterioration of the intestinal morphometry (decreased villous height/crypt depth ratio) in treated patients with celiac disease (Catassi *et al.*, 2007b) (Figure 1.6).

The gluten threshold issue

Establishing a safe threshold of gluten consumption for patients with celiac disease is a matter of major public health importance, particularly in light of the recent reports concerning the high prevalence of the disease worldwide (Fasano and Catassi, 2001). The recent NIH Consensus Conference position on celiac disease projected as many as 3 million people being affected by celiac disease in the USA. These findings, together with the recently approved Food Allergen Labeling and Consumer Protection Act, created a vacuum in terms of healthcare policy, food safety, legislative guidelines, and industry-related legal liability. The "gluten threshold" topic is currently under evaluation by the Codex Alimentarius, the WHO/FAO commission that is in charge of setting food standards at the international level. Currently different national positions hamper the implementation of uniform guidelines on the maximum level of gluten

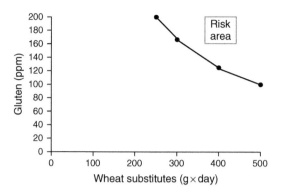

Figure 1.7 The "risky" intake of 50 mg of daily gluten can be reached by different combinations of gluten concentration in gluten-free food (expressed as parts per million) and daily consumption of wheat substitutes (expressed as grams per day).

contamination (expressed as part per million, ppm) that can be tolerated in products that are marketed for the treatment of celiac disease. This is a "hot" topic that has recently been reviewed extensively (Hischenhuber *et al.*, 2006). In Northern European countries up to 200 ppm of gluten is permitted in food for celiacs, in order to use wheat starch as ingredient. Conversely, a more prudent value of 20 ppm has been adopted in North America and Southern European countries. Based on their clinical and analytical data, Finnish experts recently advocated an "intermediate" limit of 100 ppm (Collin *et al.*, 2004).

The decision on the threshold depends not only on the minimum toxic dose, however, but also on the amount of gluten-free products consumed. The results of the micro-challenge study indicate that 200 ppm is not a safe threshold, as the harmful gluten intake of 50 mg could be reached even with a moderate consumption (250 g per day or more) of nominally gluten-free products. A 100 ppm threshold, by allowing up to 10 mg of gluten out of 100 g of product, is also probably not suitable for generalized use, especially in countries (like Italy) where consumption of wheat substitutes is occasionally as high as 500 g per day (Gibert *et al.*, 2006). The threshold of 20 ppm keeps the intake of gluten from "special celiac food" well below the amount of 50 mg, therefore allowing a safety margin for the variable gluten sensitivity and dietary habits of patients (Figure 1.7).

Novel strategies for disease prevention and treatment

The rapid progress in the elucidation of the pathophysiology of celiac disease potentially allows the development of novel strategies for disease prevention and alternative therapies. It may become possible to identify individuals at high genetic risk, particularly in families in which one member is already affected. Obviously, disease could be prevented in such individuals by avoiding the ingestion of gluten in the

diet, but more subtle approaches may also have a big impact. A more gradual introduction of lower amounts of gluten in the infant's diet may help the immune system
to cope with the dietary proteins that are clearly strong immunogens (Stepniak and
Koning, 2006a). There is also evidence that introduction of gluten while breastfeeding
has beneficial effects, which may at least partially result from reinforced protection
against pathogenic microorganisms because of maternal IgA antibodies in the breast
milk (Ivarsson *et al.*, 2002). Such approaches could thus effectively prevent celiac
disease and should be investigated for their efficacy. Several alternatives to a lifelong
GFD are now being studied. The use of bacterial prolyl oligopeptidases for degradation of gluten peptides, particularly the 33-mer, into harmless fragments has been
proposed (Shan *et al.*, 2002). A prolyl oligopeptidase from yeast that could degrade
gluten in the stomach and could prevent the activation of gluten-specific T cells in
the duodenum has been described (Stepniak and Koning, 2006b). Blocking access of
gluten peptides to the lamina propria by using a zonulin inhibitor that reduces the
paracellular permeability is a possibility that is currently being explored in the setting
of clinical research (Branski *et al.*, 2006).

The inhibition of intestinal tTG could reduce the immuno-stimulatory properties
of gluten. The use of such inhibitors could be limited because tTG is known to
participate in tissue damage repair and the issue of safety should thus be addressed.
Interfering with the binding of gluten peptides to HLA-DQ molecules is another
option (the so-called celiac "vaccine"). Specific HLA-DQ blockers would selectively
target *HLA-DQ2* and *-DQ8* molecules and leave other HLA molecules intact. Such
an approach may therefore be safe but it will be a challenge to design an effective
blocker. In addition, various other approaches such as blocking the proinflammatory
cytokine IL-15 and treatment with IL-10 have been proposed. It is doubtful, however,
whether a patient would be prepared to undergo such treatments, with many potential
side-effects, when a perfectly safe GFD is an effective alternative (Stepniak and
Koning, 2006b).

Wheat allergy

Wheat allergy is an adverse reaction to the ingestion of wheat-containing food,
caused by an immunological mechanism. This issue has recently been reviewed
extensively (Hischenhuber *et al.*, 2006). Although it is less common than celiac
disease, wheat allergy occurs more frequently than previously thought. It affects a high
proportion (10–20%) of food allergy sufferers in some populations, especially North
Europe. Pathogenic mechanisms include IgE-mediated and cell-mediated allergy. The
spectrum of gluten antigens that elicit the allergic reaction is wide, and includes α-,
β-, γ- and ω-gliadins and low molecular weight glutenins. *In vitro* cross-reactivity
between wheat, rye, and barley proteins has been demonstrated in several studies.
The reactions can be immediate or non-immediate according to the time interval
between food ingestion and the appearance of symptoms. Immediate reactions occur
within a few hours of food ingestion, and are mainly characterized by one or more of

the following symptoms: urticaria and/or angio-edema, anaphylaxis, nausea, vomiting, diarrhea, rhinitis, and bronchial obstruction. They are IgE-mediated, and are diagnosed on the basis of positive responses to prick tests, specific IgE assays, and oral provocation tests. Non-immediate reactions occur from several hours to 1 or 2 days after food intake, and are characterized by eczematous manifestations and loose stools or diarrhea. In these patients, a T cell-mediated pathogenic mechanism has been demonstrated on the basis of positive responses to patch testing with the implicated food and to oral provocations tests.

Wheat-dependent exercise-induced anaphylaxis is more common in adults than in children. This particular presentation is difficult to predict and to diagnose as the ingested wheat quantities as well as the exercise level necessary to induce the symptoms are very variable. Diagnosis of wheat allergy, as well as other food allergies, relies on the observation of clinical signs, and their timing in response to food challenge. Minimum eliciting doses are difficult to establish unequivocally, and may vary from a few milligrams to many grams from patient to patient. Wheat-allergic patients are treated the same way as patients with celiac disease, namely with the GFD. Most of these patients tolerate naturally gluten-free cereals (i.e. maize, rice and also oats), even though cross-allergy between wheat and these cereals has been described. At variance with celiac disease, wheat allergy is not always lifelong, especially in children.

Conclusions

Celiac disease is a common disorder in children as well as in adults. The spectrum of clinical presentations is wide, and currently extraintestinal manifestations (e.g. anemia or short stature) are more common than the classical malabsorption symptoms. A high degree of awareness among healthcare professionals and a "liberal" use of serological celiac disease tests can help to identify many of the non-classic cases. The primary care doctor has therefore a central role in this process of case-finding. Many key questions about this unique autoimmune condition remain unanswered. The answer to some of these questions may provide a better understanding of the pathophysiological mechanisms involved in the pathogenesis of celiac disease and, possibly of other autoimmune diseases, so paving the way to innovative treatment strategies.

References

Akobeng, A. K., Ramanan, A. V., Buchan, I., and Heller, R. F. (2006). Effect of breast feeding on risk of coeliac disease: a systematic review and meta-analysis of observational studies. *Arch. Dis. Child.* 91, 39–43.

American Gastroenterological Association (2001). Medical Position Statement: celiac sprue. *Gastroenterology* 120, 1522–1525.

Amundsen, S. S., Monsuur, A. J., Wapenaar, M. C. *et al.* (2006). Association analysis of MY09B gene polymorphisms with celiac disease in a Swedish/Norwegian cohort. *Hum. Immunol.* 67, 341–345.

Ascher, H., Holm, K., Kristiansson, B., and Maki, M. (1993). Different features of coeliac disease in two neighbouring countries. *Arch. Dis. Child.* 69, 375–380.

Barera, G., Mora, S., and Brambilla, P. *et al.* (2000). Body composition in children with celiac disease and the effects of a gluten-free diet: a prospective case-control study. *Am. J. Clin. Nutr.* 72, 71–75.

Bdioui, F., Sakly, N., Hassine, M., and Saffar, H. (2006). Prevalence of celiac disease in Tunisian blood donors. *Gastroenterol. Clin. Biol.* 30, 33–36.

Bonamico, M., Mariani, P., and Danesi, H. M. *et al.* (2001). Prevalence and clinical picture of celiac disease in italian Down syndrome patients: a multicenter study. *J Pediatr Gastroenterol Nutr* 33, 139–143.

Book, L., Hart, A., Black, J., Feolo, M., Zone, J. J., and Neuhausen, S. L. (2001). Prevalence and clinical characteristics of celiac disease in Down's syndrome in a U.S. study. *Am. J. Med. Genet.* 98, 70–74.

Branski, D., Fasano, A., and Troncone, R. (2006). Latest developments in the pathogenesis and treatment of celiac disease. *J. Pediatr.* 149, 295–300.

Cataldo, F., Marino, V., and Ventura, A. *et al.* (1998). Prevalence and clinical features of selective immunoglobulin A deficiency in coeliac disease: an Italian multicentre study. Italian Society of Paediatric Gastroenterology and Hepatology (SIGEP) and "Club del Tenue" Working Groups on Coeliac Disease. *Gut* 42, 362–365.

Catassi, C. (2005). The world map of celiac disease. *Acta Gastroenterol. Latinoam.* 35, 37–55.

Catassi, C., Rossini, M., Rätsch, I. M. *et al.* (1993). Dose dependent effects of protracted ingestion of small amounts of gliadin in coeliac disease children: a clinical and jejunal morphometric study. *Gut* 34, 1515–1519.

Catassi, C., Rätsch, I. M., Fabiani, E. *et al.* (1994). Coeliac disease in the year 2000, exploring the iceberg. *Lancet* 343, 200–203.

Catassi, C., Fabiani, E., Ratsch, I. M. *et al.* (1996). The coeliac iceberg in Italy. A multicentre antigliadin antibodies screening for coeliac disease in school-age subjects. *Acta Paediatr. Suppl.* 412, 29–35.

Catassi, C., Rätsch, I. M., Gandolfi, L. *et al.* (1999). Why is coeliac disease endemic in the people of Sahara? *Lancet* 354, 647–648.

Catassi, C., Doloretta, Macis, M., Rätsch, I. M., De Virgilis, S., and Cucca, F. (2001). The distribution of DQ genes in the Saharawi population provides only a partial explanation for the high celiac disease prevalence. *Tissue Antigens* 58, 402–406.

Catassi, C., Bearzi, I., and Holmes, G. K. (2005). Association of celiac disease and intestinal lymphomas and other cancers. *Gastroenterology* 128(4 Suppl 1): S79–86.

Catassi, C., Kryszak, D., Jacques, O. L. *et al.* (2007a). Detection of celiac disease in primary care: a multicenter case-finding study in North America. *Am. J. Gastroenterol.* 102, 1454–1460.

Catassi, C., Fabiani, E., Iacono, G. *et al.* (2007b). A prospective, double-blind, placebo-controlled trial to establish a safe gluten threshold for patients with celiac disease. *Am. J. Clin. Nutr.* 85, 160–166.

Ciclitira, P. J., Evans, D. J., Fagg, N. L. K., Lennox, E.S., and Dowling, R. H. (1984). Clinical testing of gliadin fractions in coeliac patients. *Clin. Sci.* 66, 357–364.

Ciclitira, P. J., Cerio, R., Ellis, H. J., Maxton, D., Nelufer, J. M., and Macartney, J. M. (1985). Evaluation of a gliadin-containing gluten-free product in coeliac patients. *Hum. Nutr. Clin. Nutr.* 39C: 303–308.

Clemente, M. G., Musu, M. P., Troncone, R. *et al.* (2004). Enterocyte actin autoantibody detection: a new diagnostic tool in celiac disease diagnosis: results of a multicenter study. *Am. J. Gastroenterol.* 99, 1551–1556.

Collin, P., Thorell, L., Kaukinen, K., and Mäki, M. (2004). The safe threshold for gluten contamination in gluten-free products. Can trace amounts be accepted in the treatment of celiac disease? *Aliment. Pharmacol. Ther.* 19, 1277–1283.

Cook, H. B., Burt, M. J., Collett, J. A., Whitehead, M. R., Frampton, C. M., and Chapman, B. A. (2000). Adult celiac disease: prevalence and clinical significance. *J. Gastroenterol. Hepatol.* 15, 1032–1036.

Corazza, G., Valentini, R. A., Frisoni, M. *et al.* (1992). Gliadin immune reactivity is associated with overt and latent enteropathy in relatives of celiac patients. *Gastroenterology* 103, 1517–1522.

Csizmadia, C. G. D. S., Mearin, M. L., von Blomberg, B. M. E., Brand, R., and Verloove-Vanhorick, S. P. (1999). An iceberg of childhood coeliac disease in the Netherlands. *Lancet* 353, 813–814.

Daum, S., Cellier, C., and Mulder, C. J. (2005). Refractory coeliac disease. *Best Pract. Res. Clin. Gastroenterol.* 19, 413–424.

Drago, S., El Asmar, R., Di Pierro, M. R. *et al.* (2006). Gliadin, zonulin and gut permeability: Effects on celiac and non-celiac intestinal mucosa and intestinal cell lines. *Scand. J. Gastroenterol.* 41, 408–419.

Ejderhamn, J., Veress, B., and Strandvik, B. (1988). The long term effect of continual ingestion of wheat starch-containing gluten-free products in celiac patients. In: Kumar, P. J. ed. *Coeliac Disease: One Hundred Years.* Leeds: Leeds University Press, pp. 294–297.

Fasano, A. and Catassi, C. (2001). Current approaches to diagnosis and treatment of celiac disease: an evolving spectrum. *Gastroenterology* 120, 636–651.

Fasano, A., Berti, I., and Gerarduzzi, T. *et al.* (2003). A multicenter study on the seroprevalence of celiac disease in the United States among both at risk and not at risk groups. *Arch. Intern. Med.* 163, 286–292.

Gee, S. (1890) On the coeliac affection. *St Bart. Hosp. Rep.* 24, 17–20.

Gibert, A., Espadaler, M., Angel Canela, M., Sanchez, A., Vaque, C., and Rafecas, M. (2006). Consumption of gluten-free products: should the threshold value for trace amounts of gluten be at 20, 100 or 200 p.p.m.? *Eur. J. Gastroenterol. Hepatol.* 18, 1187–1195.

Gobbi, G., Bouquet, F., Greco, L. *et al.* (1992). Coeliac disease, epilepsy, and cerebral calcifications. The Italian Working Group on Coeliac Disease and Epilepsy. *Lancet* 340, 439–443.

Gomez, J. C., Selvaggio, G. S., Viola, M., Pizarro, B., la Motta, G., and de Barrio, S. (2001). Prevalence of celiac disease in Argentina: screening of an adult population in the La Plata area. *Am. J. Gastroenterol.* 96, 2700–2704.

Greco, L., Romino, R., Coto, I. *et al.* (2002). The first large population based twin study of coeliac disease. *Gut* 50, 624–628.

Green, P. H. R., Stavropoulos, S. N., Panagi, S. *et al.* (2001). Characteristics of adult celiac disease in the USA: results of a national survey. *Am. J. Gastroenterol.* 96, 126–131.

Haboubi, N. Y., Taylor, S., and Jones, S. (2006). Coeliac disease and oats: a systematic review. *Postgrad. Med. J.* 82, 672–678.

Hadjivassiliou, M., Grunewald, R., Sharrack, B. *et al.* (2003). Gluten ataxia in perspective: epidemiology, genetic susceptibility and clinical characteristics. *Brain* 126, 685–691.

Hakanen, M., Luotola, K., Salmi, J. *et al.* (2001). Clinical and subclinical autoimmune thyroid disease in adult celiac disease. *Dig. Dis. Sci.* 46, 2631–2635.

Hernandez, L. and Green, P. H. (2006). Extraintestinal manifestations of celiac disease. *Curr. Gastroenterol. Rep.* 8, 383–389.

Hervonen, Hakanen, M., Kaukinen, K., Collin, P., and Reunala, T. (2002). First-degree relatives are frequently affected in coeliac disease and dermatitis herpetiformis. *Scand. J. Gastroenterol.* 37, 51–55.

Hill, I., Dirks, M. H., Liptak, G. S. *et al.* (2005). Guideline for the diagnosis and treatment of celiac disease in children: recommendations of the North American Society for Pediatric Gastroenterology, Hepatology and Nutrition. *J. Pediatr. Gastroenterol. Nutr.* 40, 1–19.

Hischenhuber, C., Crevel, R., Jarry, B. *et al.* (2006). Review article: safe amounts of gluten for patients with wheat allergy or celiac disease. *Aliment. Pharmacol. Ther.* 23, 559–575.

Holmes, G. K. T. (2002). Screening for coeliac disease in type 1 diabetes. *Arch. Dis. Child.* 87, 495–499.

Holmes, G. K. T. and Catassi, C. (2000). *Coeliac Disease*. Oxford: Health Press.

Hovell, C. J., Collett, J. A., Vautier, G. *et al.* (2001). High prevalence of coeliac disease in a population-based study from Western Australia: a case for screening? *Med. J. Aust.* 175, 247–250.

Hunt, K. A., Monsuur, A. J., McArdle, W. L. *et al.* (2006). Lack of association of MY09B genetic variants with coeliac disease in a British cohort. *Gut* 55, 969–972.

Ivarsson, A., Hornell, O., Stenlund, H., and Persson, A. (2002). Breast-feeding protects against celiac disease. *Am. J. Clin. Nutr.* 75, 914–921.

Jansson, U. H. G., Gudjonsdottir, A. H., Ryd, W., and Kristiansson, B. (2001). Two different doses of gluten show a dose-dependent response of enteropathy, but not of serological markers during gluten challenge in children with celiac disease. *Acta Paediatr.* 90, 255–259.

Johnston, S. D., Watson, R. G. P., McMillan, S. A., Sloan, J., and Love, A. H. G. (1998). Coeliac disease detected by screening is not silent—simply unrecognized. *Q. J. Med.* 91, 853–860.

Kaukinen, K., Collin, P., Holm, K. *et al.* (1999). Wheat starch-containing gluten-free flour products in the treatment of celiac disease and dermatitis herpetiformis. A long-term follow-up study. *Scand. J. Gastroenterol.* 34, 909–914.

Lewin, R. (1988). A revolution of ideas in agriculture origins. *Science* 240, 984–986.

Londei, M., Ciacci, C., Ricciardelli, I., Vacca, L., Quaratino, S., and Maiuri, L. (2005). Gliadin as a stimulator of innate responses in celiac disease. *Mol. Immunol.* 42, 913–918.

Louka, A. S. and Sollid, L. M. (2003). HLA in coeliac disease: unravelling the complex genetics of a complex disorder. *Tissue Antigens* 61, 105–117.

Mäki, M., Holm, K., and Lipsanen, V. *et al.* (1991). Serological markers and HLA genes among healthy first-degree relatives of patients with celiac disease. *Lancet* 338, 1350–1353.

Mäki, M., Mustalahti, K., Kokkonen, J. *et al.* (2003). Prevalence of celiac disease among children in Finland. *N. Engl. J. Med.* 348, 2517–2524.

Meloni, G., Dore, A., Fanciulli, G., Tanda, F., and Bottazzo, G. F. (1999). Subclinical coeliac disease in schoolchildren from northern Sardinia. *Lancet* 353, 37.

Monsuur, A. J., Bakker, P. I., Alizadeh, B. Z. *et al.* (2005). Myosin IXB variant increases the risk of celiac disease and points toward a primary intestinal barrier defect. *Nat. Genet.* 37, 1341–1344.

Norris, J. M., Barriga, K., Klingensmith, G. *et al.* (2003). Timing of initial cereal exposure in infancy and risk of islet autoimmunity. *JAMA* 290, 1713–1720.

Norris, J. M., Barriga, K., Hoffenberg, E. J. *et al.* (2005). Risk of celiac disease autoimmunity and timing of gluten introduction in the diet of infants at increased risk of disease. *JAMA* 293, 2343–2351.

Oberhuber, G., Granditsch, G., and Vogelsang, H. (1999). The histopathology of coeliac disease: time for a standardized report scheme for pathologists. *Eur. J. Gastroenterol. Hepatol.* 11, 1185–1194.

Peräaho, M., Kaukinen, K., Paasikivi, K. *et al.* (2003). Wheat-starch based gluten-free products in the treatment of newly detected coeliac disease. Prospective and randomised study. *Aliment. Pharmacol. Ther.* 17, 587–594.

Rätsch, I. M. and Catassi, C. (2001). Coeliac disease: a potentially treatable health problem of Saharawi refugee children. *Bull. World Health Organ.* 79, 541–545.

Sategna-Guidetti, C., Volta, U., Ciacci, C. *et al.* (2001). Prevalence of thyroid disorders in untreated adult celiac disease patients and effect of gluten withdrawal: an Italian multicenter study. *Am. J. Gastroenterol.* 96, 751–757.

Shahbazkhani, B., Malekzadeh, R., and Sotoudeh, M. *et al.* (2003). High prevalence of celiac disease in apparently healthy Iranian blood donors. *Eur. J. Gastroenterol. Hepatol.* 15, 475–478.

Shan, L., Molberg, O., Parrot, I., Hausch, F., Filiz, F., and Gray, G. M. (2002). Structural basis for gluten intolerance in celiac sprue. *Science* 297, 2275–2279.

Simoons, F. J. (1981). Celiac disease as a geographic problem. In: Walcher, D. N. and Kretchmer, N. eds. *Food, Nutrition and Evolution*. Masson: New York, pp. 179–199.

Sollid, L. M. and Lie, B. A. (2005). Celiac disease genetics current concepts and practical applications. *Clin. Gastroenterol. Hepatol.* 3, 843–851.

Sood, A., Midha, V., Sood, N., Avasthi, G., and Sehgal, A. (2006). Prevalence of celiac disease among school children in Punjab, North India. *J. Gastroenterol. Hepatol.* 21, 1622–1625.

Stepniak, D. and Koning, F. (2006a). Celiac disease—sandwiched between innate and adaptive immunity. *Hum. Immunol.* 67, 460–468.

Stepniak, D. and Koning, F. (2006b). Enzymatic gluten detoxification: the proof of the pudding is in the eating! *Trends Biotechnol.* 24, 433–434.

Valentino, R., Savastano, S., Tommaselli, A. P. *et al.* (1999). Prevalence of coeliac disease in patients with thyroid autoimmunity. *Horm. Res.* 51, 124–127.

Vitoria, J. C., Arrieta, A., Astigarraga, I., Garcia-Masdevall, D., and Rodriguez-Soriano, J. (1994). Use of serological markers as a screening test in family members of patients with celiac disease. *J. Pediatr. Gastroenterol. Nutr.* 19, 304–309.

Zanone, G., Navone, R., Lunardi, C. *et al.* (2006). In celiac disease, a subset of autoantibodies against transglutaminase binds toll-like receptor 4 and induces activation of monocytes. *PLoS Med.* 3, e358.

Labeling and regulatory issues

2

Hertha Deutsch, Roland Poms, Heereluurt Heeres,
and Jan-Willem van der Kamp

Introduction

Recent epidemiological studies have shown that 1 in 100 people worldwide have celiac disease, making it one of the most common food intolerances (Chapter 1). The only treatment for celiac disease is the total avoidance of gluten from wheat and the related proteins from barley, rye, oats, or any *Triticum* species or their cross-bred varieties. The use of oats in the dietary management of celiac disease is clarified in the Draft Revised Standard for Gluten-free Foods (Codex Committee, 2006a). In addition to celiac disease patients, people with dermatitis herpetiformis also have to exclude gluten-containing foods from the diet. Further information on the positive effect of the gluten-free diet for other diseases has been presented at sessions of the International Symposium on Coeliac Disease in recent years. Individuals with an IgE-mediated allergic reaction caused by cereal-based foods also need safe food.

Although the total exclusion of gluten-containing foods and ingredients in foodstuffs is very important to avoid health hazards, this was extremely difficult to realize in the past because of inadequate labeling directives regarding (a) compound ingredients, (b) class names, and (c) the usage of wheat gluten for technological reasons. If a compound ingredient was less than 25% of the whole food, all these compound ingredients were exempt from labeling. It was possible that a product labeled "rice-crisps" could contain undeclared wheat flour. Class names such as "starch" or "plant protein" could be used without giving any indication from which source they were derived. A severe health hazard was the so-called "clean labeling" of wheat gluten. Although wheat gluten was not a permitted food additive in European countries, wheat gluten-containing ingredients, such as soluble wheat protein products, were used without any declaration in several kinds of foodstuffs where the consumer did not expect a gluten content. The improvement of labeling directives worldwide was urgently needed.

Codex Alimentarius

Joint FAO/WHO Food Standards Programme

Ninety-eight per cent of all governments worldwide are member of the Codex Alimentarius Commission. One of the tasks of the Commission is to adopt Codex Standards, which give guidance to governments for food legislation and are mandatory for the food industry when participating in global trade. Almost all governments around the world are incorporating the Codex Standards into national legislation.

Codex Standard on food labeling

At the request of the Austrian Coeliac Society the Austrian government raised the subject of insufficient labeling of gluten-containing ingredients at the Codex Alimentarius Commission session in July 1991. The Commission decided to investigate this issue and appointed a "Working Group of Potential Allergens" to elaborate a working paper for the next session of the Codex Committee on Food Labeling for further discussion. The Association of European Coeliac Societies (AOECS), the umbrella organization of national European celiac societies, was given Observer status in the Codex and contributed to the development of the working paper. This paper contains the proposal that gluten-containing cereals and their products should always be declared. Also that other foods or ingredients, which may cause intolerance or allergy, should be added to the list. Because it covers intolerances as well as allergies, the list is called the "list of hypersensitivity." The improvements of the Codex General Standard for the Labelling of Prepackaged Foods (Joint FAO/WHO Food Standards Programme,) were discussed in the Committee on Food Labelling from 1993 till 1998 and were adopted by the Codex Alimentarius Commission in July 1999 (Codex Alimentarius Commission, 1999).

The improvements are:

1. The 25% rule of compound ingredients was abolished for the substances which are mentioned in the list of hypersensitivity. For all other ingredients the 25% rule was reduced to 5%:

   > 4.2.1.3 Where an ingredient is itself the product of two or more ingredients, such a compound ingredient may be declared, as such, in the list of ingredients, provided that it is immediately accompanied by a list, in brackets, of its ingredients

in descending order of proportion (m/m). Where a compound ingredient ... constitutes less than 5% of the food, the ingredients, other than food additives which serve a technological function in the finished product, need not be declared.
4.2.1.4 The following foods and ingredients are known to cause hypersensitivity and shall always be declared:[1]

- Cereals containing gluten; i.e., wheat, rye, barley, oats, spelt or their hybridized strains and products of these;
- Crustacea and products of these;
- Eggs and egg products;
- Fish and fish products;
- Peanuts, soybeans and products of these;
- Milk and milk products (lactose included);
- Tree nuts and nut products; and
- Sulphite in concentrations of 10 mg/kg or more.

2. Class names like "starch" or "plant protein" cannot be used any longer for substances mentioned in the list of hypersensitivity:

4.2.3.1 Except for those ingredients listed in section 4.2.1.4, and unless a general class name would be more informative, the following class names may be used:

3. If any substance from the list of hypersensitivity is used as food additive it has to be labeled without any exemption or relevance to technological functions:

4.2.4.2 A food additive carried over into foods at a level less than that required to achieve a technological function, and processing aids, are exempted from declaration in the list of ingredients. The exemption does not apply to food additives and processing aids listed in section 4.2.1.4.

With these improvements in the Codex General Standard for the Labelling of Prepackaged Foods the hazard of undeclared gluten-containing ingredients in foods is solved.

National legislation

Switzerland was the first country in Europe to adopt the Codex list of hypersensitivity plus celery and fruits for national legislation by May 1, 2002. In the European Union the labeling of gluten-containing starches was incorporated into law first, the rest of the labeling improvements followed later. Bearing in mind that the AOECS has been informing the European Commission and the members of the European Parliament about the inadequate labeling of gluten-containing ingredients in foodstuffs since 1989, it is clear that changes in legislation take some time. In 1995 the European Parliament voted for the declaration of gluten-containing starches and in March 2000 Directive 2000/13/EC was published (European Directive, 2000).

---

[1] Future additions to and/or deletions from this list will be considered by the Codex Committee on Food Labelling taking into account the advise provided by the Joint FAO/WHO Expert Committee on Food Additives (JECFA).

Labeling of gluten-containing starches and gluten-containing modified starches in the European Union

The protein content of wheat starches that are used for food for normal consumption can vary from 0.3% up to 5%. Therefore it was essential to request the correct labeling in order to distinguish gluten-containing starches or modified starches from starches gluten-free by nature. The above-mentioned Directive solved this issue in Article 6:

> However, the designation 'starch' listed in Annex I must always be complemented by the indication of its specific vegetable origin, when that ingredient may contain gluten. . . .
>
> However, the designation 'modified starch' listed in Annex II must always be complemented by the indication of its specific vegetable origin, when that ingredient may contain gluten.

Further labeling improvements in Europe

In November 2003 the European Parliament and the Council adopted Directive 2003/89/EC, which amended Directive 2000/13/EC (European Directive, 2003). Annex IIIa of this Directive is mostly in accordance with the Codex list of hypersensitivity. "Cereals containing gluten . . . and products thereof" remained the first group in the list. The difference from the Codex list is that nuts have been specified in more detail; celery, mustard, sesame seeds, and products thereof have been added to the list; sulfites have been extended to include sulfur dioxide. On December 22, 2006 lupin and molluscs and products thereof were also added to the list of "the ingredients which must under all circumstances appear on the labelling of foodstuffs" according to Directive 2006/142/EC (European Directive, 2006).

A further difference from the Codex Standard on Food Labelling is that the rule of compound ingredients has been reduced to 2%, however this is not relevant for all foods and ingredients that are listed in Annex IIIa. Cereals containing gluten and products thereof always have to be declared without any exception if the ingredient is part of a compound ingredient or has been added for technological reasons or processing purposes. Directive 2003/89 specified this issue in Article 1 (c) (iv): "substances which are not additives but are used in the same way and with the same purpose as processing aids and are still present in the finished product, even if in altered form."

In Article 2 Member States were requested:

> to bring into force, by 25 November 2004 the laws, regulations and administrative provisions necessary to permit, as from 25 November 2004, the sale of products that comply with this Directive and prohibit, as from 25 November 2005, the sale of products that do not comply with this Directive but which have been placed on the market or labelled prior to this date may, however, be sold while stocks last.

In Article 1 paragraph 10 the following instructions are given:

> . . . any substance used in production of a foodstuff and still present in the finished product, even if in altered form, and originating from ingredients listed in Annex IIIa shall be considered as an ingredient and shall be indicated on the level with a clear reference to the name of the ingredient from which it originates.

However, as a consequence, exemption of "allergen labeling" is needed to avoid confusion: If an ingredient or product has been rendered from gluten-containing cereals and the gluten content has been removed, it is misleading to list "wheat" in the ingredients of a prepackaged food. For example, wheat contains gluten but ethanol, the alcohol derived from wheat, does not.

Article 1 paragraph 11 informs that the list in Annex IIIa shall be systematically re-examined and, where necessary, updated. Updating could also include the deletion from Annex IIIa, if it has been scientifically established that some substances do not cause adverse reaction. Submissions of request for temporary labeling exemption had to be sent to the Commission before August 25, 2004. After consultation with the European Food Safety Authority the Commission adopted a list of those ingredients which shall be temporarily excluded from Annex IIIa, pending the final results of the notified studies, or at the latest until November 25, 2007. More information is given later in this chapter.

Exemption from labeling

The Codex Standard on Food Labelling and the EU Labelling Directives are valid for food for normal consumption. However, when discussing exemption from labeling, the question of possible tolerance of gluten traces has to be taken into account. This issue has been discussed in the Codex Committee on Nutrition and Foods for Special Dietary Uses, in the scientific world, in the Prolamin Working Group, and in the AOECS since 1991:

- Can people with celiac disease tolerate traces of gluten and if yes, how many?
- How many gluten traces are detectable by a reliable analytical method?
- How far can the food industry avoid traces of gluten contamination?
- How can all these issues be combined in order to allow a large variety of safe gluten-free foods available and to achieve legal protection for gluten-intolerant consumers?

It is logical that the term "gluten-free" should cover all these issues. The term "gluten-free" is described in the Draft Revised Codex Standard for Gluten-free Foods. In their "Opinions" regarding exemption from labeling the European Food Safety Authority (EFSA) referred to the Codex Standard for Gluten-free Foods.

The European Food Safety Authority

The EFSA was established to provide independent scientific advice on all matters with a direct or indirect impact on food safety. Although the Authority's main "customer" is the European Commission, the EFSA is open to respond to scientific questions from the European Parliament and the Member States and it can also initiate risk assessments on its own behalf. Risk assessment, hazard management, and the evaluation of clinical data are discussed and presented in publications and opinions. Opinions give support to legislative bodies and industries to implement measures to ensure consumer safety.

The temporary exemption list

Directive 2005/26/EC of March 21, 2005 established a list of food ingredients or substances provisionally excluded from Annex IIIa (European Directive, 2005). With regard to cereals containing gluten and products thereof, the following ingredients are exempt from compulsory labeling:

- Wheat-based glucose syrups including dextrose
- Wheat-based maltodextrins
- Glucose syrups based on barley
- Cereals used in distillates for spirits.

On May 3, 2007 the EFSA published their second "Opinions" on these subjects. With regard to cereals used in distillates for spirit "the Panel considers that distillates made from cereals are unlikely to trigger a severe allergic reaction in susceptible individuals" (European Food Safety Authority, 2007a). With regard to wheat-based glucose syrups including dextrose (European Food Safety Authority, 2007b) and wheat-based maltodextrins (European Food Safety Authority, 2007c) the EFSA came again to the conclusion that "it is unlikely to cause an adverse reaction in individuals with coeliac disease provided that the (provisional) value of gluten considered by Codex Alimentarius for foods rendered gluten-free is not exceeded."

At the last session of the Codex Committee on Nutrition and Foods for Special Dietary Uses the Committee agreed that the Codex Standard for Gluten-free Foods is only valid for dietetic food and not for foods for normal consumption. Because of the opinion of the EFSA the Codex Committee will have to take into account EFSA's opinion when discussing the threshold for rendered gluten-free foods.

In August 2006 the AOECS requested in a letter to the European Commission that the definite permission for labeling exemptions for products derived from gluten-containing cereals should only be granted if the food industry guarantees that these products are always produced below the threshold of 20 mg/kg gluten and no higher threshold should be permitted. Because the residual gluten and peptides in wheat-based glucose syrups including dextrose and wheat-based maltodextrins are far below 20 mg/kg (European Food Safety Authority, 2007b, 2007c), it is not necessary to permit the higher threshold for rendered gluten-free foods as suggested by the EFSA.

Draft Revised Codex Standard for Gluten-free Foods

Since 1992 the Codex Standard for Gluten-free Foods (Codex Stan 118–1981, amended 1983) has been under revision by the Codex Committee on Nutrition and Foods for Special Dietary Uses. The AOECS requested the revision because the Standard Stan 118–1981 covers only a minority of rendered gluten-free foods, which are wheat starch-based products, and not the large group of dietary foods made from ingredients that are gluten-free by nature (e.g. pasta or a mixture of flours to prepare bread and other dietetic foodstuffs). Research has shown that contamination in these kinds of dietetic products can be very high, therefore a threshold was requested to exclude contamination.

Further work on the analytical methods available is also needed. The revision of the Standard takes so long because of the lack of suitable analytical methods and because there is no scientific evidence regarding the question of the threshold. Progress on both these issues has been made in the last two years and it is expected that the Standard will be advanced to Step 8 in November 2007 and forwarded for adoption to the Codex Alimentarius Commission in July 2008.

Thresholds and oats

The definition of products covered by the Revised Draft Codex Standard for Gluten-free Foods was modified in October 2006 (Codex Committee, 2006a): in products gluten-free by nature the square brackets around the threshold of 20 mg/kg was deleted, which means that the threshold was accepted by the Committee. Although the threshold of 200 mg/kg for rendered gluten-free foods was reduced to 100 mg/kg, this threshold is still in square brackets, which means that this threshold will be discussed again at the next session. The AOECS requested this reduction from 200 mg/kg to 100 mg/kg in November 2005 (Association of European Coeliac Societies, 2005) to protect the health of gluten-intolerant individuals. Several wheat starch-based flour-mixes and products thereof that comply with the lower threshold are available. Gluten-free rendered wheat starch-based products have been on the market in Europe for more than 40 years and are consumed by people with celiac disease especially in the northern countries of Europe.

Both thresholds now refer to the product ready for consumption and no longer on a dry matter basis. Oats are kept in the category of gluten-containing cereals, but a footnote was added. The text of the definition for gluten-free foods is:

> a) consisting of or made only from ingredients which do not contain any prolamins from wheat, durum wheat, rye, barley, oats[2] or any *Triticum* species such as spelt (*Triticum spelta* L.,), kamut (*Triticum polonicum* L.) or their crossbred varieties with a gluten level not exceeding 20 mg/kg in total based on the foods ready for consumption;
> or
> b) consisting of ingredients from wheat, rye, barley, oats or any *Triticum* species such as spelt (*Triticum spelta* L.,), kamut (*Triticum polonicum* L.) or their crossbred varieties, which have been rendered "gluten-free; with a gluten level not exceeding (100 mg/kg) in total based on the foods ready for consumption;
> or
> c) any mixture of the two ingredients as in a) and b) with a gluten level not exceeding (100 mg/kg) in total based on the foods ready for consumption.

---

[2] Oats can be tolerated by most but not all people with coeliac disease. Therefore, the use of oats not contaminated with gluten permitted in gluten-free foods for the dietary management of coeliac disease may be determined at national level.

It is expected that the text of the footnote will be modified: It is not logical to say "not contaminated with gluten." It should be replaced by "not contaminated with wheat, rye and barley." Further on it is confusing for people with celiac disease if oats are permitted in some countries but not in other countries. The disease is the same all over the world and any individual has to find out with the assistance of a gastroenterologist whether or not she or he can tolerate oats. Scientific papers have not confirmed whether oat tolerance is a result of the slight differences in chemical structure between gliadin and avenin or whether it is connected with the individual threshold of small amounts of gluten in foods. Scientists recommend "moderate amount of oats" of 50 g per day. The very high gliadin content of wheat flour compared with the very low avenin content in oat flour could also explain the difference in tolerance.

Analytical methods

The Codex Committee on Methods of Analysis and Sampling temporarily endorsed the enzyme-linked immunoassay sorbent R5 Mendez (ELISA) method as a type 1 method in 2005. In May 2006 the Committee finally endorsed this method as a type 1, and it is described in the report of the session (Codex Committee, 2006b).

> R5 ELISA is a method based on a monoclonal antibody raised against secalin, the rye prolamin and that it was useful for detection of gluten in natural and heat-processed samples (sandwich ELISA); that the antibody reacts with the pentapeptide QQPFP, which is present in all gliadins, secalins and hordeins and that QQPFP is also present in coeliac-active epitopes; and for the detection of hydrolyzed gluten, a modification of the R5 assay (competitive ELISA) has to be applied.

More details about the method are given in Chapter 3.

It is important to have just one method in the Codex Standard for Gluten-free Foods based on the best scientific knowledge at the time. Permission for different methods or different reference materials or different antibodies may cause different results which must be avoided. It would be a very confusing situation if a product is below the threshold for gluten-free foods in one country whereas in a neighboring country the same product is not permitted to be called gluten-free because another method gives another result. Figure 3.2 shows different results obtained from the same food sample when different standards are used. This work was part of a research project initiated by the Austrian Coeliac Society.

Further Codex Standards and Guidelines

The result of the AOECS's work in sessions of the Codex Committees, ad hoc Working Group meetings and in the Codex Commission is reflected in the modification of several Codex Standards and Guidelines, which have been improved to protect the gluten-intolerant population. Additional to the Standards as mentioned before, the relevant texts of the Codex Standards and Guidelines are the following.

Codex Standard for Wheat Protein Products including Wheat Gluten

Research projects have been done in the past to use wheat protein products and wheat gluten either for coating or for technological reasons for foods gluten-free by nature. With this development the choice of gluten-free foods would have been drastically reduced. These projects have been stopped. In July 2001 the Codex Alimentarius Commission supported the request of the AOECS and the following sentences were added to the Standard (Codex Alimentarius Commission, 2001):

> Wheat gluten or wheat protein products should not be used for technological reasons e.g. coating or processing aids for foods which are gluten-free by nature[3].

Further on, a cautionary statement is permitted:

> 8.2. Cautionary statements for gluten intolerant persons shall be on the label if requested by national legislation.

Codex Standard for Cheese

In 2003 the Codex Alimentarius Commission recognized that the presence of wheat gluten and wheat protein products in cheese coatings can adversely affect the health of celiac patients (Codex Alimentarius Commission, 2003a). The Commission agreed to add a reference to the Codex Standard for Wheat Protein Products including Wheat Gluten (Codex Stan 163–1987, Rev. 1-2001) in relation to the ingredients of cheese coatings. For this purpose the Commission added the following footnote to the first bullet point of Section "Cheese coating:"

> Wheat gluten or wheat protein products should not be used for technological reasons e.g. coating or processing aids for foods which are gluten-free by nature—Codex Standard for Wheat Protein Products including Wheat Gluten (Codex Stan 163–1987, Rev. 1–2001).

Codex Standard for Chocolate and Chocolate Products

In 2003 the Codex Alimentarius Commission also agreed that no flour and starch in chocolate, except two special Spanish specialities "chocolate a la taza" and "chocolate familiar a la taza," should be permitted (Codex Alimentarius Commission, 2003b). Important for gluten-intolerant consumers are the following sentences of the Standard:

> Chocolate is the generic name for the homogenous products complying with the descriptions below and summarized in Table 1. . . . Other edible foodstuffs, excluding added flour and starch (except for products in sections 2.1.1.1 and 2.1.2.1 of this Standard) . . . may be added.

---

[3] This does not preclude the use of these products as ingredients in composite prepackaged foods provided that they are properly labeled as ingredients.

2.1.1.1 Chocolate a la taza . . . containing a maximum of 8% flour and/or starch from wheat, maize or rice.

2.1.2.1 Chocolate familiar a la taza . . . containing a maximum of 18% flour and/or starch from wheat, maize or rice.

According to above mentioned Table 1, none of the following listed chocolate types may contain flour and/or starch:

Chocolate, Sweet Chocolate, Couverture Chocolate, Milk Chocolate, Family Milk Chocolate, Milk Chocolate Couverture, White Chocolate, Gianduja Chocolate, Gianduja Milk Chocolate, Chocolate para mesa, Semi-bitter chocolate para mesa, Bitter chocolate para mesa, Chocolate Vermicelli/Chocolate Flakes, Milk Chocolate Vermicelli/Milk Chocolate Flakes, Filled Chocolate (2.2.2) , A Chocolate or Praline (2.2.3)

2.2.2 Filled chocolate is a product . . . with exception of chocolate a la taza, chocolate familiar a la taza . . . Filled Chocolate does not include Flour Confectionery, Pastry, Biscuit or Ice Cream products . . .

2.2.3 A Chocolate or Praline designates the product in a single mouthful size . . . The product shall consist of . . . with exception of chocolate a la taza, chocolate familiar a la taza . . . .

Further on it is important to note: "Assorted Chocolates (= sold in assortments) may not contain chocolate a la taza and chocolate familiar a la taza."

If any further gluten-containing ingredient is used (e.g. malt extract), these ingredients must be labeled according to the Codex Labelling Standard.

Genetically Modified Foods—Foods Derived from Biotechnology

The celiac disease population has been alarmed by research projects attempting to insert wheat genes into a rice cultivar to make rice more suitable for baking. Such developments would further restrict the diet of people with celiac disease drastically. Since 2003 gluten-intolerant consumers have been protected by three Guidelines, adopted by the Codex Alimentarius Commission (2003c):

Guideline for the Conduct of Food Safety Assessment of Foods Derived from Recombinant-DNA Plants

Two paragraphs from this guideline are important for the gluten-intolerant population:

42. The newly expressed proteins in foods derived from recombinant-DNA plants should be evaluated for any possible role in the elicitation of gluten-sensitive enteropathy, if the introduced genetic material is obtained from wheat, rye, barley, oats, or related cereal grains.

43. The transfer of genes from commonly allergenic foods and from foods known to elicit gluten-sensitive enteropathy in sensitive individuals should be avoided unless it is documented that the transferred gene does not code for an allergen or for a protein involved in gluten-sensitive enteropathy.

Guideline for the Conduct of Food Safety Assessment of Foods Produced Using Recombinant-DNA Microorganisms

This guideline also focuses on gluten intolerance:

> 47. Genes derived from known allergenic sources should be assumed to encode an allergen and be avoided unless scientific evidence demonstrates otherwise. The transfer of genes from organisms known to elicit gluten-sensitive enteropathy in sensitive individuals should be avoided unless it is documented that the transferred gene does not code for an allergen or for a protein involved in gluten-sensitive enteropathy.

Annex on Possible Allergenicity Assessment

This guideline contains a reference to para 47 of the above guideline regarding gluten-sensitive enteropathy.

Codex Standard for Processed Cereal-Based Foods for Infants and Young Children

To avoid contamination the Standard permits the claim "gluten-free:"

> 8.6.3 When the product is composed of gluten-free ingredients and food additives, the label may show the statement "gluten-free" [4].

The Standard was adopted by the Codex Alimentarius Commission in July 2006 (Codex Alimentarius Commission, 2006).

Standard for Infant Formula and Formulas for Special Medical Purposes intended for Infants

This Standard consists of two sections: Section A: Standard for Infant Formula and Section B: Formulas for Special Medical Purposes intended for Infants. In both Standards the chapter 3.1 Essential Composition includes:

> 3.1.1 All ingredients and food additives shall be gluten-free. . . .
> 3.1.3 c) Carbohydrates
>   Only precooked and/or gelatinised starches gluten-free by nature may be added to Infant Formula.

The Standard was adopted by the Codex Alimentarius Commission in July 2007 (Codex Alimentarius Commission, 2007).

---

[4] The footnote refers to the Codex Standard for Gluten-Free Foods 118–1981, which is for the time being under revision.

Food labeling and awareness

A positive result of the improvements in the worldwide Codex Standard on Food Labelling is that several governments have incorporated the mandatory labeling of gluten-containing ingredients and allergens into national legislation. With this development gluten-intolerant consumers are able to avoid the health hazard posed by "clean labeling" as described at the beginning of this chapter.

A further positive effect is that the awareness of gluten intolerance has been raised in governments and the food industry around the globe because of all the considerations during Codex sessions and the written comments from the AOECS, which had been distributed by the Codex secretariat to 98% of governments worldwide. The food industry has realized that a huge number of gluten-intolerant consumers exist who need not only a large variety of special gluten-free dietary products but also gluten-free foods for normal consumption (e.g. soups, sauces, sausages, convenience foods and further foodstuffs). In some countries (e.g. Austria) several meat product producers are using now gluten-free spices and gluten-free ingredients for their products.

In general, the increased awareness of food intolerance and food allergies and the health hazards of consuming unlabeled food ingredients has been important not only for gluten-intolerant consumers but also for all other individuals with other intolerances or food allergies, as mentioned in the list of hypersensitivity. Several studies on the so-called "allergen issue" have already been started. A work program of the European Commission's Directorate General Joint Research Centre (DG JRC) is committed to investigating the detectability of allergens in food products by ELISA, PCR, and alternative methods. Recent investigations have shown the urgent need for method validation and the availability of reference materials. This recently led to the establishment of a new working group (WG 12) in the technical committee on food horizontal methods (TC 275) of the European Committee for Standardization (CEN). Initially, the CEN will be concentrating on allergens derived from peanuts, hazelnuts, milk proteins, egg, gluten, and soybeans. The Working Group has also specified a list of matrices, in which these allergens are to be used to evaluate any test method that it considers. In addition, international collaborative efforts for the validation of test kits in various food commodities are being coordinated by the JRC. The JRC is also collaborating with other international standardization bodies such as AOAC International and with research organizations within and outside of Europe, as food allergy is a global problem.

Contamination

Because of the frequent usage of gluten-containing cereals in the whole food sector, avoidance of contamination is an important issue. Contamination of cereals that are gluten-free by nature with gluten-containing cereals may happen in the field because of alternating cultivation, during harvest, during transport, or during shipping and storage. High levels of contamination are inevitable if the same milling equipment and packing facilities for gluten-free and gluten-containing cereals are used. No legislation exists regarding the maximum levels of foreign grains in cereals. Usually 2% of other grains is the maximum limit mentioned in contracts. However, studies have detected roughly 1000 mg/kg gliadin in various samples of millet, rice, and soybean flours (Van Eckert *et al.*, 1992), 76, 250, and 570 mg/kg gliadin in rice flour, and 125 mg/kg gliadin in millet flour (Fritschy *et al.*, 1985). In a set of 28 samples of flour (rice, buckwheat, maize, millet), two severely contaminated buckwheat flours of 2000 mg/kg and 3000 mg/kg gliadin and one mixed flour (buckwheat, rice, and millet) containing 2750 mg/kg gliadin were detected (Janssen *et al.*, 1991). In oat flour up to 8.000 ppm gluten was detected because of contamination with barley (Hernando *et al.*, 2005).

Commonly available buckwheat, millet, rice, or maize flours may pose a health hazard for people with celiac disease and should be avoided when baking gluten-free bread or preparing gluten-free meals. The flours or flour-mixes with the international "Crossed-Grain-Symbol," which is the quality assurance for controlled gluten-free foods, are safe. The use of this symbol is harmonized in the AOECS in terms of thresholds, analytical method, and monitoring. The producers avoid any contamination risk by selecting their suppliers and by regularly controlling all incoming ingredients.

High levels of contamination have also been detected in breads prepared in small bakeries as a favor for their gluten-intolerant neighbors. Even if controlled gluten-free flour-mixes are used, the contamination in bread and other products can be high when production facilities have not been cleaned properly. For bakeries, separate production rooms and processing equipment are recommended.

In composite foods for normal consumption contamination can be controlled to below pre-determined levels, e.g. below 20 mg/kg when the re-work of gluten-containing products is excluded, shared processing equipment is cleaned and, more generally, an adequate hazard analysis and critical control points (HACCP) system is in operation. For controlling the cleaning from possible gluten residues, rapid gluten test-sticks have been shown to be useful.

Product liability and food safety

Regulations in the European Union for product liability are laid down in Directive 85/374/EEC (European Directive, 1985), which specifies the requirements for liability, the producer, the damage, the defective products, causal relationship and means of defence. Food allergens and gluten, both causing hypersensitivity, are considered basically in the same way in this and other regulations related to food safety; a comprehensive review on EU regulations regarding allergens is given by Heeres (2006).

The main rule is that the producer shall be liable for damage caused by a defect in his or her product. This legislation is based on strict liability for the producer, liability without fault. Each food business in the production chain, from farm to food, is a producer. The damage is restricted to damage resulting from death or by personal injuries and damage to any item of property other than the defective product itself. Products comprise all foods, food ingredients and processing aids. A product is defective when it does not provide the safety which a person is entitled to expect, taking all circumstances into account. Circumstances regarding gluten include the presentation of the food product, labeling, and the use to which it could reasonably be expected. The causal relationship between defect and damage has to be proved. This can be very difficult in practice. For example, the evidence of the presence of gluten in one of the foods eaten during a meal.

The two most relevant means of defense for a food producer are outlined below:

- The state of scientific and technical knowledge at the time when the producer put the product into circulation was not such as to enable the existence of the defect to be discovered. For instance, the allergenic properties of a certain substance were unknown at the moment of putting into circulation, but later on this substance caused allergenic reactions.
- For a manufacturer of a component, a means of defense is to demonstrate that the defect is attributable to the design of the product in which the component has been fitted or to the instructions given by the manufacturer of the product. For instance, a food producer uses an allergenic ingredient but does not label this ingredient, while the producer of the component has clearly mentioned the presence of this allergen on the ingredient label.

In the past decade much attention has been paid to generic food safety policies and regulations by World Health Organization, Food and Agriculature Organization, national authorities and the European Union, where Regulation 178/2002, called the General Food Law, was issued in 2002 (European Union, 2002).

HACCP (Hazard Analysis and Critical Control Points) became an issue of great importance in WHO/FAO. In the European Union basic principles of hygiene in foodstuffs have been laid down in Regulation 852/2004 (European Union, 2004). It states that food business operators shall put in place, implement, and maintain a permanent procedure or procedures based on the HACCP principles. The HACCP system is an instrument to help food business operators attain a higher standard of food safety. A HACCP system for gluten-free products will contain adequate measures for achieving contamination levels below the limit of 20 mg/kg. A certified HACCP system, taking into account hazards and control points related to contamination with allergens, will help a producer both with reducing the risks of contamination and the risks of high liability payments in case such a contamination unexpectedly has taken place.

Cautionary statements and disclaimers—helpful for consumers?

The word "disclaimer" means literally the refusal of liability. A disclaimer text points out to the reader the impossibility of extracting right from particular statements. Below, some examples of disclaimers and misleading practices are given, focusing on cases in which people with celiac disease are not optimally informed and are getting confused.

Article 2 of EU Directive 2000/13/EC on labeling of foods (European Directive, 2003) requests the avoidance of misleading labels:

> The labelling and methods used must not: be such as could mislead the purchaser to a material degree, particularly: as to the characteristics of the foodstuff and, in particular, as to its nature, identity, properties, composition, quantity, durability, origin or provenance, method of manufacture or production.

According to the Codex Standard on Food Labelling and also according to the EU legislation, labeling is not required for allergens when their presence is caused by cross-contamination.

However, some producers do not take these Directives into account and choose to label the presence of gluten in any case, even if a possible cross-contamination has no relevance. As an example, it caused confusion to see on a producer's homepage that only two different types of unprocessed beans are listed to be gluten-free, whereas five types are available on the market. The reason was that the subsupplier of three types was afraid of contamination and declared the raw and unprocessed beans not to be gluten-free just for liability reasons. Nobody eats unprocessed beans. Even if traces of contamination with wheat flour, for example, may occur in a container, they will be washed away before cooking the beans. Such cases have also been reported with some types of rice, lentils, etc.

Consumers read the ingredients list very carefully when they cannot tolerate a certain substance or ingredient in foodstuffs. For a few years cautionary claims such as "may contain wheat protein" or similar statements have caused confusion. People with celiac disease want to know what to do with this "information" and may ask their celiac society for advise as to whether or not they can eat this food. In one instance when a celiac society contacted a food producer asking for clarification of this "may contain" statement, it was informed that although no trace of gluten could be detected with analytical methods, even when the worst case was calculated, and that a possible contamination will be in any case far below 20 mg/kg gluten, the claim "may contain wheat protein" was maintained, based on the opinion of the legal advisors of the company. In some other cases analysis of the food was refused because the claim "may contain traces of wheat" or similar words have to be kept in any case because of the company's policy to avoid liability. With this development food labeling tends to become senseless.

Some supermarkets and producers want to provide a special service for the celiac disease population and print a gluten-free logo on foods. However the outcome is not helpful if supermarket chains create their own gluten-free logo and consumers are confronted with several logos not knowing which quality system stands behind which logo. A further confusing development is when the claim "gluten-free" is printed on the label of products where it is self-evident that the product is gluten-free. This claim has already been seen on a bottle of mineral water, which may lead to the wrong impression that gluten-containing mineral water exists! The Codex Standard for Gluten-free Foods regulates the use of the claim "gluten-free" and has to be considered. For dietary products and foods that are especially prepared for people with celiac disease (e.g. a soup that usually contains wheat flour that has been replaced by a gluten-free flour), it would be helpful if the International Gluten-free Symbol was used; this can be immediately recognized by people with celiac disease.

Conclusion

Improvements in the worldwide Codex Standard on Food Labelling and national food labeling legislations have resolved the health hazard posed by unknown gluten intake caused by insufficient declaration of gluten-containing ingredients and food additives in foodstuffs. The labeling exemption of derivatives, in which the gluten content has been removed, contributes to a better understanding when reading the label of the food, whether or not gluten-containing ingredients and/or additives have been used. Efforts should be made to exclude contamination by adequately cleaning of shared processing equipment in foods prepared for normal consumption. Statements such as "may contain . . . " should be avoided because they are not helpful for consumers in making their choice whether or not they can eat this food. The improvement of several Codex Standards and Guidelines have contributed to the safety and large variety of gluten-free foods and, as a consequence, guarantee a better quality of life for the gluten-intolerant population.

References

Association of European Coeliac Societies (2005). Codex Committee on Nutrition and Foods for Special Dietary Uses, 27th Session, Bonn, Germany, 21–25 November 2005, page 2, para 11, CRD 13, Comments from AOECS.

Codex Alimentarius Commission (1999). Report of the Twenty-third Session, Rome, 28 June–3 July 1999: para 130–140.

Codex Alimentarius Commission (2001). Report of the 24th Session, Geneva, 2–7 July 2001: Vegetable Proteins: Codex Standard for Wheat Protein Products Including Wheat Gluten; pp. 26–27, para 191–195.

Codex Alimentarius Commission (2003a). Report of the 26th Session, Rome, 30 June–7 July 2003: Amendment to the Codex General Standard for Cheese: Appendix; p. 14, para 101–102.

Codex Alimentarius Commission (2003b). Report of the 26th Session, Rome, 30 June–7 July 2003: Standard for Chocolate and Chocolate Products: p. 6, para 42.

Codex Alimentarius Commission (2003c). Report of the 26th Session, Rome, 30 June–7 July 2003: Foods Derived from Biotechnology: pp. 7–8, para 51–53.

Codex Alimentarius Commission (2006). Report of the 29th Session, Geneva, 3–7 July 2006: Standard for Processed Cereal-Based Foods for Infants and Young Children: p. 11, para 91–93.

Codex Alimentarius Commission (2007). Draft Report of the 30th Session, Rome, 2–7 July 2007: Standard for Infant Formula and Formula for Special Medical Purposes Intended for Infants: pp. 13–14, para 60–63.

Codex Committee (2006a). Draft Revised Standard for Gluten-free Foods. *Report of the 28th Session of the Codex Committee on Nutrition and Foods for Special Dietary Uses, Chiang Mai, Thailand, 30 October–3 November 2006*, p. 11, para 91–108 and pp. 72–74.

Codex Committee (2006b). Report of the 27th Session of the Codex Committee on Methods of Analysis and Sampling, Budapest, Hungary, 15–19 May 2006, p. 8, para 68–71.

European Directive (1985). Directive 85/374/EEC on product liability, Council Directive of 25 July 1985 on the approximation of the laws, regulations and administrative provisions of the Member States concerning liability for defective products, amended by Directive 1999/34/EC of the European Parliament and of the Council of 10 May 1999.

European Directive (2000). Directive 2000/13/EC of the European Parliament and of the Council of 20 March 2000 on the approximation of the laws of the Member States relating to the labelling, presentation and advertising of foodstuffs.

European Directive (2003). Directive 2003/89/EC of the European Parliament and of the Council of 10 November 2003 amending Directive 2000/13/EC as regards indication of the ingredients present in foodstuffs.

European Directive (2005). Commission Directive 2005/26/EC of 21 March 2005 establishing a list of food ingredients or substances provisionally excluded from Annex IIIa of Directive 2000/13/EC of the European Parliament and of the Council.

European Directive (2006). Commission Directive 2006/142/EC of 22 December 2006 amending Annex IIIa of Directive 2000/13/EC of the European Parliament and of the Council listing the ingredients which must under all circumstances appear on the labelling of foodstuffs.

European Food Safety Authority (2007a). Opinion of the Scientific Panel on Dietetic Products, Nutrition and Allergies on a request from the Commission related to a notification from CEPS on cereals used in distillates for spirits, pursuant to Article 6, paragraph 11 of Directive 2000/13/EC; adopted on 3 May 2007.

European Food Safety Authority (2007b). Opinion of the Scientific Panel on Dietetic Products, Nutrition and Allergies on a request from the Commission related to a notification from AAC on wheat-based glucose syrups including dextrose pursuant to Article 6, paragraph 11 of Directive 2000/13/EC; adopted on 3 May 2007.

European Food Safety Authority (2007c). Opinion of the Scientific Panel on Dietetic Products, Nutrition and Allergies on a request from the Commission related to a notification from AAC on wheat-based maltodextrins pursuant to Article 6, paragraph 11 of Directive 2000/13/EC; adopted on 3 May 2007.

European Union (2002). Regulation (EC) No 178/2002 of the European Parliament and of the Council of 28 January 2002 laying down the general principles and requirements of food law, establishing the European Food Safety Authority and laying down procedures in matters of food safety, amended by Regulation (EC) No 1642/2003 of the European Parliament and of the Council of 22 July 2003, and Commission Regulation (EC) No 575/2006 of 7 April 2006.

European Union (2004). (Corrigendum to) Regulation (EC) No 852/2004 of the European Parliament and of the Council of 29 April 2004 on the hygiene of foodstuffs.

Fritschy, F., Windemann, H., and Baumgartner, E. (1985). Bestimmung von Weizengliadinen in Lebensmitteln mittels ELISA. *Z. Lebensm. Unters. Forsch.* 181, 379–385.

Heeres, H. (2006). EU regulation of undeclared allergens in food products. In: Koppelman, S. and Hefle, S. L. eds. *Detecting Allergens in Food*. Cambridge: Woodhead Publishing Limited, pp. 378–404.

Hernando, A., Mujico, J. R., Juanas, D. *et al.* (2005). Corroboration of a massive contamination of wheat, barley and rye in oat samples by confirmatory techniques: R5 ELISA, Western blot, PCR and mass spectrometry. In: *Proceedings of the 20th Meeting of the Working Group on Prolamin Analysis and Toxicity*, 16–18 September 2005, Maikammer, Germany, pp. 29–35.

Janssen, F. W., Hägele, G. H., and de Baaij, J. A. (1991). Gluten-free products, the Dutch experience. In: *Coeliac Disease*. Dordrecht: Kluwer Academic, pp. 95–100.

Joint FAO/WHO Food Standards Programme. *Codex Alimentarius*, Food Labelling, Complete Texts, 4th edn. General Standard for the Labelling of Prepackaged Foods: para 4.2.1.3, 4.2.1.4, 4.2.3.1, and 4.2.4.2 (latest publication: Fourth Edition, Rome 2005).

Van Eckert, R., Pfannhauser, W., and Riedl, O. (1992). Vienna Food Research Institute, Vienna, Austria. Contribution to quality assessment during production of gluten-free food. *Ernährung/Nutrition* 16, 511–512.

Detection of gluten

Herbert Wieser

Introduction

Celiac disease, one of the most frequent permanent food intolerances (see Chapter 1), is induced by ingestion of storage proteins (gluten) from wheat, rye, barley, and possibly oats. The current essential therapy of celiac disease is a strict adherence to a gluten-free diet, which means a permanent withdrawal of gluten from daily food. The aim of the treatment is the prevention of small intestinal inflammations associated with damage to the mucosa and generalized malabsorption of nutrients. The total daily intake of gluten for patients with celiac disease should not exceed 20 mg. In addition to patients with celiac disease, numerous other individuals cannot tolerate gluten proteins due to IgE-mediated allergic reactions; they also have to avoid gluten-containing foods.

Gluten-sensitive people eat gluten-free food from two different categories. First, they are allowed to consume a wide range of common products such as meat, fish, milk, fruits, and vegetables. In the case of composite foods, however, it is difficult to recognize whether they are gluten-free or not. It is of great help that gluten has been incorporated into the list of foods and ingredients that are known to cause hypersensitivity, and should always be declared on the labels of prepacked foods (Codex Standard for the Labelling of Prepacked Foods, 2001). Nevertheless, gluten-sensitive people should be aware of numerous foods that contain hidden sources of gluten such as thickened sauces and soups, puddings, or sausages. Second, patients consume dietetic food that is gluten-free according to the "Codex Standard for Gluten-Free Foods" (see Chapter 2). This standard was established in 1981 and amended in 1983, however, without having a method to determine gluten (Codex Stan 118–1981, amended 1983). The only methodological point was the nitrogen content, so it was limited to the analysis of cereal starches used in the preparation of gluten-free food; the nitrogen content should be less than 0.05% on a dry matter basis. In practice, the Kjeldahl method and, more recently, the Dumas combustion method are used for nitrogen determination.

A revision of Codex Stan 118–1981 is now at step 6 of the Codex procedure (Codex document CL 2006/5-NFSDU, 2006). The most noticeable difference between the proposed new standard and the old one is that the new standard comprises all foods labeled as gluten-free, whereas the old standard is restricted to a few ingredients such as starch. A three-point definition of gluten-free foods is presented by the "Draft Revised Codex Standard for Gluten-free Foods" (2006):

Gluten-free foods are foodstuffs

a) consisting of or made only from ingredients which do not contain any prolamins from wheat or all *Triticum* species such as spelt, kamut or durum wheat, rye, barley, [oats] [*Note*: The square brackets indicate that there was insufficient information to make a final decision.] or their crossbred varieties with a gluten level not exceeding [20] mg/kg; or

b) consisting of ingredients from wheat, rye, barley, [oats], spelt or their cross-bred varieties, which have been rendered "gluten-free," with a gluten level not exceeding [200] mg/kg; or

c) any mixture of the two ingredients as in a) and b) with a gluten level not exceeding [200] mg/kg.

For the purpose of this standard, gluten is defined as a protein fraction from wheat, rye, barley, [oats] or their crossbred varieties and derivatives thereof, to which some persons are intolerant and that is insoluble in water and 0.5 mol/L NaCl.

Prolamins are defined as the fraction from gluten that can be extracted by 40–70% of ethanol. The prolamin content of gluten is generally taken as 50%. The gluten content of solid food products has to be expressed in mg/kg on a dry matter basis and that of liquid food products in mg per kg of the original product. If a limit of 20 mg gluten/kg dry product is adopted, methods will be needed with a detection limit well below this threshold, e.g. $\approx$5 mg/kg, to ensure good reliability for measurement of gluten concentration around the limit value. The definition of gluten-free foods has recently been modified by the twenty-eighth Session of the Codex Committee (see Chapter 2). Accordingly, the gluten level of 200 mg/kg for rendered gluten-free foods was reduced to [100] mg/kg[1], and gluten levels refer to the product ready for consumption, no longer to dry matter basis.

Reliable methods for the detection and quantitative determination of gluten are essential for gluten-sensitive consumers, the food industry, and food control. So far, however, only a general outline of an analytical method has been given by the Draft Revised Codex Standard, namely prolamins should be extracted from the product with 60% ethanol, and quantitated by an immunologic method. Up to now, it has not been

---

[1] The square brackets indicate that there was insufficient information to make a final decision.

possible to design such a method in detail, as the existing methods do not correspond to minimum requirements of sensitivity, selectivity, precision in repeatability and reproducibility, availability of a gluten/prolamin reference and/or they were not ring-tested and not available as commercial test kits. Moreover, problems arose with heated products such as bread and with partially hydrolyzed products such as malt products and beer.

Many laboratories have searched for solutions for accurate gluten detection and quantitation over the last 25 years. This chapter summarizes different techniques that have been developed for the quantitative determination of prolamins or gluten in foods specially produced for the diet of patients with celiac disease. It is subdivided into the description of precipitating proteins, extraction procedures, reference proteins, immunochemical and non-immunochemical methods.

The precipitating factor

Analytical methods play an important role in the assessment and maintenance of food quality, both in industry and for enforcement authorities at the national and international levels. In particular, an optimal analysis of food constituents and additives that affect the health of consumers is essential. The most important requirements for a successful analysis are an understanding of the principle, availability of adequate facilities, and a careful, meticulous execution.

Most toxic food constituents are either single compounds such as acrylamide or small groups of compounds like mycotoxins, and the methods for their detection and quantitative determination can be aligned to their special structure. In the case of gluten intolerance, however, the precipitating factor is a complex mixture of proteins that vary according to their botanical origin (e.g. cereal species, varieties), agricultural conditions under which they were produced (e.g. climate, fertilization), and food processing (e.g. heating, enzymatic degradation), and knowledge about these toxic structures is incomplete. Therefore, a profound understanding of gluten chemistry and its relation to gluten toxicity is necessary for the development and judgment of methods to determine gluten.

Chemistry of gluten proteins

Storage proteins of cereals consist of numerous components that are deposited exclusively in the endosperm of the kernels. Their only biological function is to supply the seedling with nitrogen and amino acids during germination. According to this function, they are unique in terms of their amino acid composition (high contents of glutamine and proline) and sequences (frequent repetitive units). Traditionally, cereal storage proteins have been divided into two fractions according to their solubility in alcohol–water solvents: the soluble prolamins and the insoluble glutelins (Osborne, 1907). The prolamin fractions contain monomeric and oligomeric proteins and the glutelin fractions contain polymeric proteins.

Storage proteins of celiac-toxic cereals have been investigated extensively by various separation techniques (e.g. SDS-PAGE, acid PAGE, SE-HPLC, RP-HPLC, capillary electrophoresis) and characterized by the determination of amino acid compositions, molecular weights, and partial or total amino acid sequences (reviewed by Wrigley *et al.*, 2004). The results demonstrated that wheat, rye, barley, and oats possess, in parts, homologous storage proteins, which reflect very well the botanical relationship of these cereals. According to common structures, they have been classified into three groups: (1) a high-molecular-weight (HMW) group, (2) a medium-molecular-weight (MMW) group, and (3) a low-molecular-weight (LMW) group, the latter being the major group for all four cereals (Shewry and Tatham, 1990; Wieser, 1994). Structural data for representatives of the different groups and types are presented in Table 3.1.

The HMW group contains HMW glutenin subunits (HMW-GS) (wheat), HMW secalins (rye) and D-hordeins (barley). HMW-GS and HMW secalins can be subdivided into the x- and the y-type. The molecular weights of these proteins range from around 70 to 90 kDa. Their amino acid compositions are characterized by high contents of glutamine, glycine, and proline, which together account for about 70% of total residues. They consist of three structural domains: a non-repetitive N-terminal domain of around 100 residues, a non-repetitive C-terminal domain with about 40 residues and a repetitive central domain 400–700 residues long. The central domain contains repetitive hexapeptides such as QQPGQG[2] as a backbone with inserted hexapeptides like YYPTSP and tripeptides like QQP or QPG. The non-repetitive N- and C-terminal domains contain much less glutamine, glycine, and proline and more amino acid residues with charged side-chains and, in particular, cysteine that forms interchain linkages via disulfide bonds. In a native state, the proteins of the HMW group are aggregated and hardly extractable with aqueous alcohols.

The MMW group comprises the homologous ω1,2-gliadins (wheat), ω-secalins (rye) and C-hordeins (barley) and the unique ω5-gliadins (wheat). Their molecular weights range from 40 to 50 kDa. They have an unbalanced amino acid composition characterized by high contents of glutamine, proline, and phenylalanine, which together account for about 80% of total residues. Most regions of the amino acid sequences are composed of repetitive units like (Q)QPQQPFP or (Q)QQQFP. Because cysteine is usually absent, the proteins of the MMW group occur as monomers and are completely extractable with aqueous alcohols.

The members of the LMW group can be divided into monomeric proteins including α/β- and γ-gliadins (wheat), γ-40k-secalins (rye), γ-hordeins (barley) and avenins (oats), and aggregative proteins including LMW glutenin subunits (LMW-GS) (wheat), γ-75k-secalins (rye) and B-hordeins (barley). Their molecular weights are in the range of 30–40 kDa, with the exception of γ-75k-secalins (molecular weights of about 50 kDa) and of avenins (molecular weights of about 22 kDa). All these proteins have an N-terminal domain rich in glutamine, proline, and aromatic amino acids (phenylalanine, tyrosine) and a C-terminal domain with a more balanced amino acid composition and with most of the cysteine residues. The length of both domains varies from type to type. γ-Gliadins, γ-40k-secalins, and γ-hordeins are

---

[2] One-letter code for amino acids

Table 3.1 Characterization of storage protein types of wheat, rye, barley, and oats

Group/Type	Code[a]	Residues	State[b]	Repetitive unit[c]	Partial amino acid composition (mol %)				
					Q	P	F+Y	G	C
HMW group									
HMW-GS x	Q6R2V1	815	a	QQPGQG (72×)	36	13	5.8	20	0.5
HMW-GS y	Q52JL3	637	a	QQPGQG (50×)	32	11	5.5	18	1.1
HMW-secalin x	Q94IK6	760	a	QQPGQG (66×)	34	15	6.7	20	0.5
HMW-secalin y	Q94IL4	716	a	QQPGQG (60×)	34	12	5.0	18	1.1
D-hordein	Q40054	686	a	QQPGQG (26×)	26	11	5.5	16	1.5
MMW group									
ω5-gliadin	Q40215	420	m	(Q)QQQFP (65×)	53	20	10	0.7	0.0
ω1,2-gliadin	Q6DLC7	373	m	(QP)QQPFP(42×)	42	29	9.9	0.8	0.0
ω-secalin	O04365	338	m	(Q)QPQQPFP (32×)	40	29	8.6	0.6	0.0
C-hordein	Q40055	327	m	(Q)QPQQPFP (36×)	37	29	9.4	0.6	0.0
LMW group									
α/β-gliadin	Q9M4M5	273	m	QPQFPPQQPYP(5×)	36	15	7.4	2.6	2.2
γ-gliadin	Q94G91	308	m	(Q)QPQQPFP (15×)	36	18	5.2	2.9	2.6
LMW-GS	Q52NZ4	282	a	(Q)QQPFS (11×)	32	13	5.7	3.2	2.8
γ-40k-secalin[d]	Q41320	–	m	QPQQPFP	–	–	–	–	–
γ-75k-secalin	Q9FR41	436	a	QQPQQPFP(32×)	38	22	6.1	1.6	2.1
γ-hordein	P17990	286	m	QPQQPFP (15×)	28	17	7.7	3.1	3.5
B-hordein	P06470	274	a	QQPFPQ (13×)	30	19	7.3	2.9	2.9
Avenin	Q09072	203	m	PFVQQQQ (3×)	33	11	8.4	2.0	3.9

[a] Databank Uni Prot KB/TREMBL (http://pir.georgetown.edu).
[b] a = aggregated, m = monomeric.
[c] Basic unit frequently modified by substitution, insertion and deletion of single amino acid residues.
[d] Fragment.

homologous, having frequent repetitive units such as QPQQPFP and four intrachain disulfide bonds within the C-terminal domain. α/β-Gliadins are unique for wheat; their N-terminal domain is characterized by repetitive units such as QPQPFPPQQPYP and the C-terminal domain contains three intrachain disulfide bonds. Most of the α/β- and γ-type proteins occur as monomers and are extractable with aqueous alcohols. A small number of these proteins have an odd number of cysteine residues due to point mutations in the genes and appear either in the ethanol-soluble oligomeric prolamin fraction or in the ethanol-insoluble polymeric glutelin fraction.

Avenins are the smallest proteins within the LMW group due to a shortened N-terminal domain with only three repetitive units (PFVQQQQ). The C-terminal domain is, in parts, homologous to those of α/β- and γ-types, and, in parts, unique having glutamine-rich repetitive sequences such as QPQLQQQVF. LMW-GS, γ-75k-Secalins, and B-hordeins are aggregative proteins forming at least one interchain disulfide linkage with other proteins. The N-terminal domain of LMW-GS is characterized by repetitive units such as QQPPFS, and the C-terminal domain includes three interchain disulfide bonds. One cysteine residue in the N-terminal and one cysteine in the C-terminal domain form interchain linkages.

γ-75k-Secalins are homologous to γ-40k-Secalins with the exception that the N-terminal domain is much longer and possesses a cysteine residue forming interchain linkages. B-Hordeins are homologous to γ-hordeins, but form both intra- and interchain disulfide bonds.

Toxicity testing

In vivo testing is commonly considered to be the "gold standard" for assessing celiac disease toxicity of proteins and peptides. Early studies established toxicity by feeding tests based on the production of symptoms such as diarrhea or malabsorption of fat or xylose. However, the optimal amount of gluten equivalent used to challenge patients was uncertain. In any case, 10–100 g were necessary for each patient and such large amounts were the most crucial limiting factor for feeding tests with purified proteins or peptides. By direct instillation into the small intestine followed by biopsy after several hours, the required amounts could be reduced to 1 g-equivalents of gluten. Histological measurements of villus height and the ratio of villus height to crypt depth as well as the immunochemical determination of the intraepithelial lymphocytes were shown to be reliable parameters for toxicity assessment (Fraser *et al.*, 2003; Dewar *et al.*, 2006).

Because *in vivo* tests require relative large quantities of substances and only a limited number of test patients are available, a series of *in vitro* tests has been developed. The organ culture of human intestinal tissue, which requires only milligram-equivalents of gluten, has been proposed as providing the most reliable *in vitro* approach. By measuring the enzyme activities or morphology, the tissue of a flat mucosa shows improvement in the medium alone, but not in the presence of celiac disease toxic substances. More recently, T cell lines and clones of cells from patients with celiac disease have been used to test celiac disease stimulatory effect. For example, a T cell transformation assay can be performed by the incubation of the putative antigen (approx. 10–200 μg/mL), antigen-presenting cells, T cells, and tritiated thymidine (Ellis *et al.*, 2003). After a maximum of 2 days, the incorporation of thymidine into T cells is quantified by scintillation measurement. In addition, the production of interferon-γ or interleukin 4 can be determined as parameters for celiac disease-specific stimulatory effects. Further *in vitro* tests such as organ culture tests with fetal rat or chicken intestine, leukocyte-migration inhibition factor, macrophage proagulant activity, or agglutination of leukemia K562 cells are more or less specific screening tests. However, *in vivo* testing ultimately will be necessary to evaluate conclusions on *in vitro* testing.

Toxicity of gluten proteins and peptides

Dicke (1950) was the first to establish the celiac toxicity of wheat. Shortly afterwards, it was demonstrated that rye and barley were also harmful, whereas maize, rice, and buckwheat were not (reviewed by Kasarda, 1994). To date, the toxicity of oats has been judged controversially. Fractionation of wheat flour and testing by feeding trials led to the conclusion that gluten was toxic, whereas starch and the water-soluble albumins were not. Since that time, a "gluten-free diet" has been the conventional treatment for celiac disease. Accordingly, gluten has been defined as those proteins, commonly found in wheat, triticale, rye, barley, or oats to which some individuals are intolerant (Codex Stan 118–1981). Continuing studies on protein toxicity were performed only with wheat (reviewed by Wieser, 1995). When gluten was fractionated with aqueous ethanol into the alcohol-soluble prolamins (gliadins) and the alcohol-insoluble glutelins (glutenins), toxicity tests indicated that the gliadin fraction was the most toxic factor. Subsequent *in vivo* and *in vitro* tests demonstrated that all gliadin types (α/β-, γ-, ω-gliadins) produced toxic effects. Equivalent to the gliadin fraction of wheat, the prolamin fractions and single homologous types of rye protein (secalins) and barley (hordeins) were associated with celiac toxicity without serious testing.

The toxicity of oat prolamins (avenins) has been considered controversial until today. The toxicity of wheat glutenins was described as either non-toxic, weakly toxic or as toxic as gliadins, but on very inadequate evidence. Neither type of wheat glutenins, HMW-GS and LMW-GS, had been tested until recently. *In vivo* and *in vitro* tests revealed that HMW-GS exacerbate celiac disease just as gliadins (Molberg *et al.*, 2003; Dewar *et al.*, 2006). T cell stimulation tests on peptides from LMW-GS indicated that this protein type, too, has the potential to induce a celiac-specific immune response (Vader *et al.*, 2002). In summary, all storage proteins (prolamins + glutelins) of wheat, rye, barley, and possibly oats appear to be gluten proteins as defined in Codex Stan 118–1981 and in the Draft Revised Codex Standard.

A panel of peptides either isolated from gluten protein digests or synthesized and purified has been tested for toxicity in order to find out the epitopes responsible for celiac disease (reviewed by Stern *et al.*, 2001; Anderson and Wieser, 2006). Most studies were focused on peptides from wheat gliadin and glutenin (some selected toxic peptides are shown in Table 3.2 as an example). Summarizing, *in vivo* and *in vitro* studies showed that glutamine- and proline-rich epitopes of the storage proteins are the major precipitating factors. Substitution of proline and glutamine residues as well

Table 3.2 Origin and amino acid sequences of selected celiac disease-toxic peptides from wheat gluten

Type	Sequence[a]	Test[b]	Reference
α/β	LGQQQPFPPQQPYPQPQPF	IN	Sturgess et al. (1994)
α/β	PQPQPFPSQQPY	IN	Marsh et al. (1995)
α/β	LQLQPFPQPQLPYPQPQLPY	IN	Fraser et al. (2003)
α/β	VPVPQLQPQNPSQQQPQEQVPL	OC	Wieser et al. (1986)
γ	LQPQQPFPQQPQQPYPQQPQ	TC/TG	Arentz-Hansen et al. (2002)
γ	FSQPQQQFPQP	TC/TG	Arentz-Hansen et al. (2002)
γ	PQQPFPQPQQQFPQPQQPQQ	TC/TG	Arentz-Hansen et al. (2002)
HMW	GQQGYYPTSPQQS	TC	Van de Wal et al. (1999)
HMW	QGYYPTSPQQSG	TC	Van de Wal et al. (1999)
LMW	QQQQPPFSQQQQSPFSQQQQ	TC/TG	Vader et al. (2002)
LMW	QQPPFSQQQQQPLPQ	TC/TG	Vader et al. (2002)

[a] One-letter-code for amino acids.
[b] IN, instillation test; OC, organ culture test; TC, T cell test; TG, treated with tissue transglutaminase.

as shortening active peptides to less than nine residues mostly inhibit celiac disease activity (Sollid, 2002). Results collected so far strongly indicate that methods for gluten determination should comprehend all storage protein types present in wheat, rye, barley, and possibly oats, and that the specifity of tests should be focused on glutamine- and proline-rich epitopes.

Protein extraction

The first step of gluten analysis is the extraction of gluten proteins from raw material or food. Native gluten proteins are not soluble in water or salt solution. A fraction (prolamins) is soluble in aqueous alcohols, whereas the second fraction (glutelins) remains in the insoluble residue. Total gluten proteins are soluble in aqueous alcohols after the reduction of disulfide bonds (e.g. by dithiothreitol); disaggregating agents such as urea or sodium dodecyl sulfate (SDS) accelerate the dissolving process (Wieser et al., 2006).

A previous Draft Revised Codex Standard (CX/NFSDU 00/4, 2000) presented a detailed description of the extraction procedure, whereas the recent draft (CL 2006/5-NFSDU, 2006) only refers to the R5 ELISA method (see below). According to document CX-NFSDU (2000) the determination of gluten in foodstuffs or ingredients shall be based on the determination of prolamins defined as the fraction from gluten that can be extracted by 40–70% of ethanol. The ethanol concentration of 60% is recommended for the extraction of total prolamins, because previous studies demonstrated that the optimal extraction of gliadins from wheat flour was achieved with this concentration (Wieser et al., 1994). With respect to solid foodstuffs and ingredients, products with a fat content higher than 10% have to be defatted as follows: 5 g are homogenized with a blender in 50 mL hexane and centrifuged for 30 minutes at $1500 \times g$. The supernatant is discarded and the extraction step is repeated until the sample is fat-free. In products with a fat content lower than 10%, defatting is generally not necessary. Before the extraction procedure, 5 g of the defatted or non-defatted products are dried at 60°C and milled. An aliquot of the dried sample is then homogenized with 60% ethanol in a volume 10 times its weight for 2 minutes and after 15 minutes centrifuged for 10 minutes at $1500 \times g$. The supernatant is taken off and stored, if necessary, at 4°C before determination. When a precipitate is formed, this is spun down and discarded.

In the case of liquid foodstuffs and ingredients, an aliquot is diluted with the volume of ethanol, which yields a concentration of 60% in the resulting mixture. The mixture is homogenized and further treated like solid food extracts. Matrix effects caused by different constituents of the sample can affect extraction yield and thus, the results of gluten determination. For example, binding to polyphenols such as those from tea, hops, and cocoa products decrease the yield of prolamins. Addition of casein, urea (CX-NFSDU 00/4, 2000), or gelatine (Garcia *et al.*, 2004) to the extractant is recommended to avoid underestimation of the prolamin content. Highly viscous samples such as starch-derived syrups should be diluted with the appropriate solvent to avoid matrix effects (Iametti *et al.*, 2004).

Another major problem in gluten analysis has been the incomplete extraction of prolamins from heated products, when aqueous alcohols are used. Extraction studies on heated gluten and wheat bread spiked with a gliadin standard, respectively, demonstrated that the extractability of α/β- and γ-gliadins was strongly reduced, whereas that of ω-gliadins was scarcely affected (Schofield *et al.*, 1983; Wieser, 1998). It has been postulated that Cys-containing α/β- and γ-gliadins could be bound to alcohol-insoluble glutenins due to disulfide/sulfhydryl interchain reactions. After the reduction of disulfide bonds, however, total gliadins could be completely recovered in the alcoholic extract (Wieser, 1998). The heat stability of ω-gliadins was used by Skerritt and Hill (1990) to develop an immunoassay with monoclonal antibodies against ω-gliadins. The results for gluten-containing raw material and heated products suggested that 40% ethanol was the most suitable extractant for the quantitative determination of gluten in all types of food (raw, cooked or heat-processed). Another suggestion to improve gluten extraction from heated food was the limited hydrolysis, with pepsin leading to about 90% of protein extraction in a saline buffer from both raw gluten and gluten heated at 100°C (Denery-Papini *et al.*, 2002). Antibodies directed against repetitive epitopes of α/β- and γ-gliadins were used for the quantitation of partially hydrolyzed prolamins.

The combination of a reducing agent (2-mercaptoethanol) with a disaggregating agent (guanidine), the so-called "cocktail," allowed the complete extraction of prolamins (and glutelin subunits) from both unheated and heat-processed foods (Garcia *et al.*, 2005). Figure 3.1 compares the recovery of gliadins from wheat flour and doughs heated from 22 to 230°C and extracted with 60% ethanol or cocktail. The recovery of gliadins extracted with 60% ethanol decreased from 76 to 22%, as the samples were heated up from 60 to 230°C. In contrast, when the cocktail was used, the recovery was nearly quantitative, even in the sample heated up to 230°C. After diluting the extract (e.g. 1:100), the cocktail did not affect the ELISA system

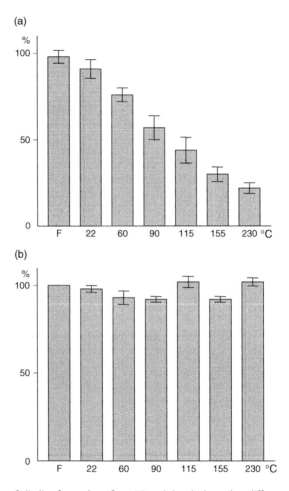

Figure 3.1 Recovery of gliadins from wheat flour (F) and doughs heated at different temperatures and extracted with (a) 60% ethanol and (b) cocktail (flour = 100%).

based on R5 monoclonal antibody. Other types of antibodies, however, might be more sensitive against the reducing agent (Ellis *et al.*, 1998; Dona *et al.*, 2004). For the extraction of both unheated and heat-treated samples, incubation for 40 minutes at 50°C was recommended. The extract was compatible with MALDI-TOF MS and Western blot besides ELISA. The cocktail always yielded either slightly similar or higher values than 60% ethanol depending on the type of foods: 1.1-fold in unheated foods, 1.4-fold in wheat starches and 3.0-fold in heated foods (Garcia *et al.*, 2005).

Comparative studies on different products (e.g. cereal and soy products, baby food, syrup, chocolate, beer) revealed that the values for prolamins extracted with 60% ethanol or cocktail might be divergent even to a much larger degree and no explanation could be found for that (Immer and Haas-Lauterbach, 2005a, 2005b; Iametti *et al.*, 2005; Laffey *et al.*, 2005; Malmheden Yman, 2006).

Reference protein

A prolamin (gluten) reference protein is essential to establish a calibration curve for the determination of the prolamin (gluten) concentration in the extracts. Moreover, the reference should be used in order to minimize interassay variations and to make possible the comparison of results from different laboratories and/or obtained with different techniques. Important criteria for a reference material are high protein content, solubility in the extraction solvent, homogeneity, stability, equivalence to celiac disease toxic proteins, and good response to measurement techniques. A number of prolamin, mostly gliadin references has been produced by different laboratories and companies for the use in their own test systems or kits. References have been isolated from different sources of cereals and chemically characterized by protein content or qualitative and quantitative protein composition. Previous studies have shown that the measured gluten content of a sample determined by ELISA methods can vary significantly depending on the origin and type of the reference used for calibration and on the test system (van Eckert, 1993; van Eckert et al., 1997; Sima et al., 1999). As an example, Figure 3.2 shows the divergent calibration curves of five different commercial reference gliadins in the same immunoassay. Therefore, Draft Revised Codex Standard CX-NFSDU 00/4 (2000) recommends that a "gold standard" should be prepared by one laboratory under strictly standardized conditions.

The European Working Group on Prolamin Analysis and Toxicity (PWG) decided to organize the preparation of a reference gliadin for collective use (van Eckert et al., 2006). Twenty-eight wheat cultivars representative of the three main European wheat-producing countries, France, UK, and Germany, were selected as starting material. The kernels were mixed and milled and the resulting flour was defatted and vacuum dried. Albumins and globulins were removed by extraction using 0.4 mol/L NaCl solution and gliadins were extracted with 60% ethanol. The gliadin extract was concentrated, desalted by ultrafiltration, freeze-dried, and homogenized.

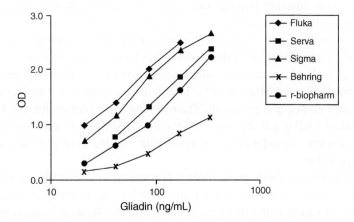

Figure 3.2 Reaction of different reference gliadins in the assay of Ridascreen® Gluten. Adapted from van Eckert et al. (1997).

The material was analyzed by different laboratories using various methods. The product was homogeneous to a high degree and completely soluble in 60% ethanol. The crude protein content (N × 5.7, Dumas) was 89.4%. RP-HPLC revealed identical protein patterns (ω-, α/β-, γ-gliadins) of the flour and the reference gliadin, demonstrating that no major gliadin components had been lost during the isolation procedure. According to the results of GP-HPLC, the reference gliadin contained 68% monomeric gliadins, 29% oligomeric HMW-gliadins and only 3% albumins and globulins. The preparation had a high homogeneity and was stable, even if it was stored at 37°C for 28 days. In summary, PWG gliadin met all criteria important for a reference material and was regarded as a suitable prolamin reference for gluten determination.

Immunochemical methods

Principles

The first decades of immunoassay development and food analysis have been extensively described by Morris and Clifford (1985) and the immunochemistry of cereal storage proteins by Skerritt (1988). Immunochemical tests are the methods of choice for gluten analysis and are recommended by the Draft Revised Codex Standard, since they provide specific and sensitive recognition of celiac disease toxic proteins as well as rapid results. Immunoassays are based on the specific reaction of antibodies (immunoglobulins) with antigens, the substance to be determined (celiac disease toxic proteins and peptides). Antibody-containing antisera are produced by immunization of animals (e.g. rabbits or mice) by the injection of the corresponding immunogen. Because only compounds with molecular weights higher than about 5000 provoke immunological activity, covalent coupling of LMW immunogens (haptens) such as peptides to a protein (e.g. bovine serum albumin) has to be done. This conjugate produces an antiserum that contains antibodies against both the hapten and the coupled protein. The antiserum obtained from the animal is tested for its specificity and, as far as possible, purified to remove undesirable specificities. These polyclonal antibodies (PAb) react with different binding sites (epitopes) of the antigen and, considering gluten analysis, the results are less influenced by cereal species or cultivar. A disadvantage is the high risk of cross-reactions with proteins from non-toxic cereals. More specific monoclonal antibodies (MAb) can be produced after immunization by the fusion of isolated splenocytes with murine myeloma cells using the technique developed by Galfre and Milstein (1981). Hybridoma, positive for antibodies against the antigen, are cloned and grown up. The resulting MAb preparation can be purified by precipitation and/or affinity chromatography. MAb have tremendous advantages due to the absolute reproducibility of specificity and the ability to produce almost unlimited quantities.

For antiserum or antibody assessment it is necessary to determine whether or not an antibody is specific for its antigen and whether the antibody cross-reacts to a greater or lesser extent with other proteins. In the main, Western immunoblotting has been used to investigate the binding of antibodies to antigens. For example, Freedman *et al.* (1988) used Western blots to characterize the binding of MAb to gliadins.

The proteins were separated by SDS-PAGE and transferred to nitrocellulose membranes using a Trans-blot cell system. The blots were incubated with the antibodies, washed, and then incubated with an enzyme-labeled second antibody directed to the first antibody and with a corresponding color developing substrate.

A crucial point of immunoassays is the quantitation of antibody–antigen binding. Older methods required the formation of a precipitate of the antibody–antigen complex. Later on, antigens were marked by different markers, such as fluorescent or luminescent dyes, stable radicals, radioisotopes (^{3}H, ^{14}C, ^{125}I), or enzymes. For radioimmunoassays (RIA), laboratories need specific equipment and a further disadvantage is that free antigens have to be separated from those bound to the antibodies. ELISA is the most frequent technique used for gluten determination. ELISA is relatively easy to perform, often cheaper than other techniques and provides rapid results. Horseradish peroxidase (substrate 2,2′-azinobis(3-ethylbenzothiazoline-6-sulfonic acid)), alkaline phosphatase (substrate 4-nitrophenylphosphate) and β-D-galactosidase (substrate 4-nitrophenyl-β-galactoside) are the most common indicator enzymes. They are available in high purity, are very stable, and their activity can be determined sensitively and precisely. The enzymes are linked to the antigen by covalent bonds, e.g. by reaction with glutaraldehyde or carbodiimide.

Two ELISA systems have been most frequently applied for gluten analysis: the sandwich ELISA and the competitive ELISA. The principle of the sandwich ELISA is shown in Figure 3.3a. Capture antibodies are immobilized onto the walls of a plastic carrier (microtiter plate). Aliquots of the sample containing the antigen to be determined are incubated in the micro cells, leading to the formation of the antibody–antigen complex (step 1). After rinses, the detection antibody labeled with an enzyme is added and a further incubation binds it to the antigen (step 2). Thus, the antigen is "sandwiched" between two antibodies. Unbound enzyme-marked antibodies are washed out. At this stage, the enzymatic substrate is added, transformed into a colored end-product and measured spectrophotometrically (step 3). The absorbance is directly proportional to the antigen concentration in the sample, which can be calculated based on a reference protein and a calibration curve. The sandwich ELISA is suitable only for large antigens, because the antigen must have at least two epitopes, which are spatially separated to bind both antigen and enzyme-labeled antigen. Therefore, for gluten analysis, the sandwich ELISA is inappropriate, when partially hydrolyzed products such as sourdough products, malt, and beer have to be analyzed.

In contrast, the competitive ELISA is suitable for the detection of small-sized antigens with only one epitope (Figure 3.3b). The assay comprises three components: (i) the antibody immobilized onto the microtiter plate, (ii) a limited and constant quantity of the enzyme-labeled antigen, and (iii) unlabeled antigen from the sample. When the components of the systems are mixed, labeled and unlabeled antigens compete for the limited number of antibody binding site (step 1). The greater the quantity of unlabeled antigen present, the smaller will be the quantity of the labeled antigen binding to the antibody. Unbound antigens are washed out and the enzymatic substrate is added and transformed to the colored end-product (step 2). The greater the quantity of the sample antigen, the fainter will be the color produced by the

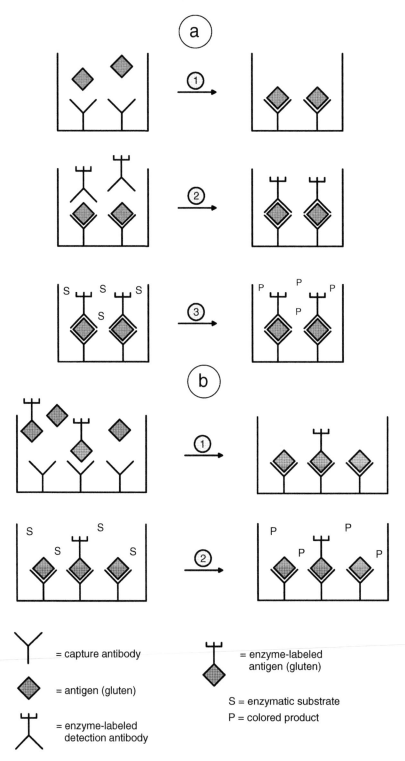

Figure 3.3 Formats for ELISA. (a) Sandwich ELISA: Step 1: Formation of antibody–antigen complex. Step 2: Binding of the enzyme-labeled antibody. Step 3: Transformation of the enzymatic substrate into a colored end-product. (b) Competitive ELISA. Step 1: Competition of unlabeled and enzyme-labeled antigens for antibody binding. Step 2: Transformation of the enzymatic substrate into a colored end-product.

enzyme-labeled antigen. Calibration curves created with reference proteins enables the quantification of the sample antigen.

Heat treatment employed in food processing can affect gluten determination not only due to the reduced extractability of prolamins (see above), but also due to changes in protein conformation, which may modify/mask epitopes recognized by the antibodies. For example, Ellis et al. (1994) described a decrease in reactivity of heated gliadin fractions in a sandwich ELISA with MAb. α/β- and γ-gliadins were more heat-labile, retaining only 33–51% of their original reactivity, whereas ω-gliadins retained 93%. In a similar approach, gliadin fractions were heated at 70–100°C for 5–20 minutes and then quantified by a competitive ELISA using four different MAb and a rabbit anti-gliadin serum (Rumbo et al., 2001). The results demonstrated that the effects of heat treatment on reactivity can vary not only in dependence on temperature and heating time, but also on the employed antibody. Therefore, antibodies used for gluten determination of heat-processed foods should be tested for their immunoreactivity towards heated gluten proteins.

Developed assays

At the beginning of the twentieth century the first immunologic studies on cereals were performed. In 1925, Lewis and Wells (1925) injected alcohol extracts from wheat in guinea-pigs, followed by second injection with an extract of either wheat flour or other cereals. They obtained an anaphylactic reaction not only using wheat, but also using rye, barley, and oats, whereas no reaction was observed with maize. The group of Berger and Freudenberg (1961) in Basel did a more systematic research on gliadin antigenicity using immunoprecipitation techniques. First attempts to identify wheat proteins in gluten-free baked products were based on the measurement of antigen–antibody precipitation by means of gel-diffusion techniques, immunoelectrophoresis and countercurrent electrophoresis (Amtliche Sammlung von Untersuchungsverfahren nach § 35 LMBG, 1984). The detection limits ranged from 1 to 50 μg protein/mL in dependence on the applied method. Much more sensitive was the RIA described by Ciclitira and Lennox (1983) using an antiserum raised in rabbits against A-gliadin, a component of α-gliadins. The antigens used in the assays were α/β-gliadins labeled with ^{125}I; antigen-antibody complexes were collected after adsorption to a *Staphylococcus aureus* cells suspension. The sensitivity of the assay as judged by competitive binding with unlabeled antigen was 1 mg of α/β-gliadins. Cross-reactivity to other wheat proteins was less than 1% and no cross-reactivity to extracts of rye, barley or oats was observed.

To date, ELISA has been the most frequent method for the quantitative determination of gluten. The efficiency and limitation of ELISA methods developed up to 1998 has been reviewed by Denery-Papini et al. (1999). The first reliable competitive and sandwich ELISA for gluten determination using PAb against whole gliadin and A-gliadin were developed by Windemann et al. (1982). This assay was very sensitive and could be carried out in the range of 1–20 ng/mL for A-gliadin and 10–300 ng/mL for whole gliadin, but did not react with ω-gliadins and proteins from other toxic cereals such as rye. Sandwich ELISA was used for the detection of gliadins in unheated and heated foods (Meier *et al.*, 1984; Fritschy *et al.*, 1985). The recovery of gliadins after extraction of unheated foods with 70% ethanol was good, except for those food containing cocoa, coffee, or tea. The amount of recovered gliadins in bread dough after heating above 80°C, however, was strongly reduced. McKillop *et al.* (1985) described a similar ELISA using a polyclonal rabbit antiserum. The detection limit was 3.3 ng gliadin and the assay detected other cereals beside wheat toxic for people with celiac disease. The PAb against whole gliadin used in a sandwich ELISA by Troncone *et al.* (1986) and Aubrecht and Toth (1995) gave cross-reactions with non-toxic rice and maize prolamins, which limited the application of these tests. Friis (1988) described a competitive ELISA using rabbit PAb against whole gliadin. The antibodies did not react with maize, millet, rice, or soy proteins, but weakly with buckwheat proteins. This assay had a detection limit of 1 ng antigen with very high degree of accuracy. More recently, Chirdo *et al.* (1995) developed a competitive ELISA with PAb against whole gliadin, which recognized all types of gliadins and HMW-GS as well as the prolamins from rye, triticale, and barley; only slight reactions were observed with oat proteins and no cross-reaction with maize, rice, and soy. The sensitivity of the test was 1 ng gliadin/mL or 1 mg gluten/1 kg flour. The test was 10–15 times less sensitive with barley and rye prolamins and 50 times less sensitive with oat prolamins.

Several ELISA were developed using PAb absorbed on the microtiter plate for the capture of antigens and MAb conjugated with horseradish peroxidase or alkaline phosphate for the measurement of antigens. A "triple" sandwich ELISA was applied by Freedman *et al.* (1987) to measure the gliadin content of foods. Rabbit polyclonal anti-gliadin IgG was used as capture antibody. The detection system consisted of the murine monoclonal supernatant, goat anti-mouse IgG and IgM conjugated with alkaline phosphate and *p*-nitrophenylphosphate as substrate. The assay detected all gliadin fractions and prolamins from rye, barley, and oats as well as wheat glutenins. The detection limit for whole gliadin was 0.75 mg. The same system was used by Ellis *et al.* (1994) except that MAb raised against a celiac-active peptide 54 amino acid residues long were taken as part of the detection system. The sensitivity of the assay for whole gliadin and rye prolamins was 15 ng/mL (0.3 mg/kg flour), 125 ng/mL (2.5 mg/kg) for barley prolamins and 250 ng/mL (5 mg/kg) for oat prolamins. Prolamins from non-toxic rice, maize, millet, and sorghum did not cross-react. Later on, the assay was modified to a more sensitive test by using MAb raised against a synthetic peptide 19 amino acids long corresponding to the sequence positions 31–49 of α-gliadins (Ellis *et al.*, 1998). The sensitivity of the assay was 4 ng/mL (0.08 mg/kg flour) for gliadins, 500 ng/mL for secalins and 1000 ng/mL for hordeins and avenins. The assay could detect gluten in cooked foods, although at reduced sensitivity. Prolamins from non-toxic cereals did not cross-react.

A series of assays were developed using only MAb. Theobald *et al.* (1983) were the first to report the production of MAb against cereal flour proteins, in particular against salt-soluble proteins that cause wheat allergy. A large collection of MAb against cereal proteins was generated by Skerrit and co-workers (Skerritt and Underwood, 1986). Most anti-gliadin antibodies bound to all gliadins, while several antibodies bound to small groups of gliadins. Anti-gliadin MAb were used in a simple enzyme-coupled assay to quantitate gliadins in a variety of food (Skerritt, 1985). The limit of detection for gliadins, however, was rather high (20 μg/mL). A more sensitive competitive ELISA using horseradish peroxidase-labeled MAb was described by Hill and Skerritt (1990). The antibodies were selected for specific reactions with ω-gliadins; these antibodies also bound proteins from rye and barley beside wheat proteins, and results were not affected by different varieties. Binding of these antibodies was not inhibited by heating of gluten during cooking or baking. For extraction, 40% or 70% ethanol was recommended. The sensitivity of the assay was in a range from 0.05 to 0.10 μg gliadin equivalent to 200–400 mg gluten/kg using a 1:5 dilution of a food extract. Two antibodies binding ω-gliadins, HMW-GS, and prolamins from rye and barley were used to develop a sandwich ELISA, which was patented and marketed in two forms (see below). According to the authors' description (Skerritt *et al.*, 1991), the "Gluten Lab Test" was the first method able to quantify gluten in all types of uncooked, cooked, and processed foods. The second, "Rapid Gluten Test Kit," provided rapid, qualitative, or semi-quantitative results and was suitable for home use or in process quality control by food and wheat starch manufacturers.

Chirdo *et al.* (1998) developed assays of high sensitivity using three MAb raised against gliadins with variable formats (competitive ELISA, sequential competitive ELISA, and sandwich ELISA). Biotinylated antibodies were used in two of the assays. Two of the antibodies reacted broadly with gliadins, secalins, and hordeins, and the third only with gliadins; reactions with proteins from soy, rice, and maize proteins were not observed. In dependence on the systems and antibodies used, the detection limits were in a range from 1 to 20 ng gliadin/mL with a 1:50 dilution. The use of the biotin–streptavidin interaction as signal amplification system was found to be very useful for gliadin quantification.

A mixture of MAb raised against ethanol extracts of wheat, rye, or oat flours were tested in a sandwich ELISA by Sorell *et al.* (1998). Two antibodies were used as capture antibodies and a third one conjugated to horseradish peroxidase as detection antibody. Due to the wide specificity, this combination of antibodies ensured a high cross-reactivity with toxic prolamins, and permitted the recognition of gliadins, secalins and hordeins to the same extent in the 3–200 ng range per mL extract (detection limit ≈1.5 ng/mL), while the sensitivity against avenins was much lower (detection limit ≈12 ng/mL). Heating of prolamins in solution (120°C, 30 minutes) did not affect the quantitative measurement, demonstrating that the epitopes recognized by the antibodies were not denatured by this treatment. Thus, the reduced extractability of prolamins appeared to be the major problem for the analysis of heated products. The same group developed a sandwich ELISA based on a single MAb (R5) raised against ω-secalins. R5 was used as both the capture and the detection antibody, the latter labeled with horseradish peroxidase (Valdes *et al.*, 2003). R5 ELISA was equally sensitive to wheat, rye, and barley prolamins, while cross-reactions with oat, maize, and rice proteins were not observed. The detection limit was 1.5 mg gliadin/mL corresponding to 3.2 mg gluten/kg; the reproducibility was ± 8.7% and the repeatability 7.7%. The assay was compatible with the cocktail extraction procedure for heat-processed foods (see above). The epitope specifity of R5 was characterized by tests for binding to synthetic peptides spanning in overlapping

Table 3.3 Analysis of beer production: comparison of sandwich and competitive R5 ELISA[a]

Beers	Origin	Sandwich (cocktail)	Competitive (60% ethanol)
1	Spain	6	30
2	Spain	16	76
3	Czech Republic	<3	24
4	Czech Republic	6	102
5	Belgium	181	833
6	Belgium	1113	4053
7	Germany	2410	4530
8	Germany	22	66
9	Ireland	26	49
10	Ireland	26	101
11	Mexico	<3	11
12	USA	<3	16
13	Germany	10	76
14	Germany	14	88
15	Germany	<3	74
16	Germany	8	98
17	Germany	52	212

[a] Values in milligrams gluten/kilogram.
Adapted from Hernando *et al.* (2005) and Immer and Hans-Lauterbach (2005b).

manner sequences of gliadins (Kahlenberg *et al.*, 2006). In a luminiscence assay, R5 bound all peptides from the N-terminal domain of α/β-type gliadins known to be toxic for celiac disease patients (see Table 3.2). Sequences such as QQPFP, QQQFP, LQPFP, and QLPFP were bound most strongly. Recently, a competitive ELISA using R5 MAb was developed (Ferre *et al.*, 2004). In contrast to the sandwich ELISA, this system detects also small, but still toxic peptides derived from prolamins and is designed especially for the analysis of partially hydrolyzed products such as malt extracts and beers. Table 3.3 shows the high efficiency of the competitive ELISA in measuring gluten in beers with factors around 2- to 17-fold as compared with the sandwich ELISA (Hernando *et al.*, 2005; Immer and Haas-Lauterbach, 2005b). Commercial test kits based on R5 MAb have been developed and ring-tested (see below).

A novel competitive ELISA based on the use of a MAb raised against a toxic peptide from α-gliadins was described by Bermudo Redondo *et al.* (2005). This assay was shown to be specific to celiac disease toxic prolamins, capable of the detection of hydrolyzates and compatible with typical extraction agents. The detection limit was 0.3 mg gluten/kg, and the reproducibility was $\pm 3.6\%$.

Until recently, the immunochemical determination was focused on the detection of prolamins and the prolamin content was multiplied by the factor 2 to obtain gluten. This calculation proposed by the Draft Revised Codex Standard is questionable, because the ratio of prolamins (defined as storage proteins soluble in 60% ethanol) to glutelins (defined as storage proteins insoluble in 60% ethanol) can be extremely different from the proposed ratio of 1. Examples are the variations within common wheat cultivars (prolamin/glutelin = 1.7–3.1) (Wieser and Kieffer, 2001), wheat species (1.8–6.6) (Wieser, 2000), rye cultivars (6.3–8.2) (Gellrich *et al.*, 2003), barley cultivars (0.5–2.5) (Wieser, unpublished), and wheat starches (0.2–4.9) (Wieser and Seilmeier, 2003). For these reasons, an accurate method for the quantitative determination of glutelins beside prolamins is necessary. Following results showing that peptides from HMW- and LMW-GS have the potential to stimulate T cells of celiac disease patients (Van de Wal *et al.*, 1999; Vader *et al.*, 2002; Molberg *et al.*, 2003) and that HMW-GS have been shown to be toxic *in vivo* (Dewar *et al.*, 2006), Ellis *et al.* (2006, 2007) produced murine MAb raised against HMW-GS 1Dx5 and 1Dy10. The results demonstrated that a single MAb is sufficient to measure both HMW-GS. Immunoblots indicated that this antibody did not react with gliadins. According to the authors' suggestion, such MAb could be used in a cocktail ELISA system in combination with anti-gliadin antibodies. Spaenij-Dekking *et al.* (2004, 2006) presented an immunoassay based on MAb that recognized celiac disease toxic gluten peptides. MAb specific for T-cell epitopes from α/β-gliadins, γ-gliadins, LMW-GS, and HMW-GS were generated. Assays with these antibodies detected T-cell stimulatory epitopes in different backgrounds. In addition, both intact proteins and small protein fragments could be analyzed, since the assays were based on competition.

In summary, numerous ELISAs based on PAb or MAb have been developed for gluten quantitation. Most of them, however, do not correspond to all requirements necessary for common acceptance with respect to specificity, sensitivity, and precision. Only few assays have been ring-tested and are commercially available.

Commercial ELISA kits

The common availability of assays is an important precondition for Codex regulations, but only few developed ELISA tests were transferred to commercial test kits. A sandwich ELISA using PAb has been offered by Riedel-de Haen AG (Seelze, Germany; article no. 45213). The antiserum is produced by the immunization of rabbits with a mixture of native gliadins from different wheat cultivars and gliadins from the same varieties denatured by heat (Weisgerber, 1998). The microtiter plates (polystyrol) are coated with PAb. Detection antibodies are labeled with horseradish peroxidase and the substrate solution contains tetramethylbenzidine/peroxide. A gliadin standard from a mixture of 13 German wheat cultivars extracted with 70% ethanol is used. To prevent matrix effects, sample extracts are diluted 1:5000 prior to analysis. The detection limit is 100 mg gluten/kg food and, consequently, too high for the control of foods gluten-free by nature. The sample is prepared in 1 hour and the run performed in 2.5 hours.

The sandwich ELISA developed by Skerritt and Hill (1990) has been marketed by several companies, e.g. Cortecs (UK), Transia (France), and R-Biopharm (Germany). Two MAb against ω-gliadins are bound to the wells and another antibody conjugated to horseradish peroxidase and the substrate percarbamide is used for detection. The chromogens are either 2,2'-azinobis(3-ethylbenzothiazoline-6-sulfonic acid) (Cortecs) or tetramethylbenzidine (Transia, R-Biopharm). The gliadin standard has been prepared by the extraction of wheat flour (cultivar Timgalen) with 40% ethanol. Samples have to be extracted with 40% ethanol and a dilution of 1:100 is recommended. The detection limit is around 10 ng gliadin/mL and the sensitivity indicated by the manufacturers ranges from 20 to 160 mg gluten/kg. The assay has been successfully ring-tested (Skerritt and Hill, 1991a) and validated by the Association of Official Analytical Chemists (AOAC). The analyses of a number of samples (buckwheat and rice flours, corn and wheat starches) spiked with gliadin, however, revealed that the results varied to a great extent, when different commercial ELISA kits based on the same MAb against ω-gliadins were used (Sima et al., 1999). Therefore, the authors concluded that it was not possible to decide which kit reliably determined the gliadin content in an unknown sample.

To improve compliance with a gluten-free diet a simple prototype test kit for use at home has been developed (Skerritt and Hill, 1991b). The food is extracted with diluted hydrochloric acid and one drop of the extract transferred to an antibody-coated tube; the enzyme-labeled antibody is added and after 3 minutes, the tube is washed and color developer is added. The reaction is stopped after 2 minutes by the addition of sulfuric acid. The home kit was compared with the quantitative laboratory kit, and the qualitative agreement was very good. The kit could distinguish foods containing traces of gluten (acceptable for a gluten-free diet) from those with a slightly higher, but unacceptable gluten content.

A serious disadvantage of MAb against ω-gliadins is the fact that the proportions of this protein type are relatively low according to total storage proteins of wheat, rye, and barley, and strongly dependent on varieties. For example, the quantitative determination of gliadin types in 16 wheat varieties revealed a range from 6.2 to 20.0% according to total gliadins, so the method bears a considerable source of systematic errors (Wieser et al., 1994). This was confirmed by studies on gliadin fractions from different wheat species (common and durum wheat, spelt, emmer, einkorn) (Seilmeier and Wieser, 2003). The calibration curves based on the same protein level of the gliadin fractions and the kit reference gliadin differed widely, so that the gliadin content of the fractions were, in parts, either strongly underestimated or overestimated.

The sandwich ELISA based on R5 MAb (Valdes et al., 2003) has been marketed by R-Biopharm (Germany) and Ingenasa (Spain). R-Biopharm offers four different kits (Immer and Lauterbach, 2003) to detect prolamins from wheat, rye, and barley. All systems are adapted to the reference gliadin described by van Eckert et al. (2006) and two extraction methods are proposed (1 g sample/10 mL): (i) the normal extraction with 60% ethanol and (ii) the extraction with the so-called cocktail (see above) especially for heat-denatured samples. Ridascreen® Test Gliadin (R 7001) recommends 3×30 minute incubation time and provides six reference concentrations starting with 5 ng/mL. Ridascreen® Fast Gliadin (R 7002) recommends 3×10 minute incubation time and provides five reference concentrations starting with 10 ng/mL. The detection limit for both tests has been found to be 5 and 10 mg gluten/kg, respectively. The third test, Rida® Quick Gliadin (R 7003) is based on a lateral flow technique and is delivered as a kit including a stick, in which the MAb has been immobilized (Garcia et al., 2002). The stick is put into the diluted sample extract, and, after 5 minutes, a red line appears, if the sample contains the corresponding prolamins. The assay has a sensitivity of around 10 mg gluten/kg. This assay is particularly suited for a swab method to check environment such as machines or tables for prolamin contamination. The stick kit (Stick Gluten) has also been marketed by Operon, S.A. (Cuarte de Huerva, Zaragoza, Spain). Recently, a fourth test was introduced by R-Biopharm, Ridascreen® Gliadin Competitive (R 7011). The system is based on a competitive ELISA using R5 MAb. This assay detects also small peptides derived from prolamins and has been developed especially for products with partially hydrolyzed prolamins such as malt extracts and beers (see above). Ingenasa has marketed two ELISA systems corresponding to those of R-Biopharm R 7001 and R 7002, Ingezim Gluten (Ranz *et al.*, 2004) and Ingezim SEMIQ (Ranz *et al.*, 2005).

Two kits (Ridascreen Gliadin and Ingenasa Ingezim Gluten) were included in a ring test (Mendez *et al.*, 2005). Twenty laboratories participated in a coded form to evaluate the gliadin content of 12 encoded samples (spiked or contaminated with gliadins) in two runs using each extract in three dilutions. The statistics of the data obtained led to a variance of repeatability of 11–25% and of reproducibility of 23–47% for both kits. They could clearly distinguish between negative and gliadin-containing samples. Both kits were valid to determine gliadin contamination and guaranteed a sensitivity of 3.0 mg gluten/kg. In 2005, the R5 ELISA was endorsed as type I method by the Codex Committee of Methods of Analysis and Sampling (CCMAS) and is recommended by the recent Draft Revised Codex Standard CL 2006/5-NFSDU (2006).

Electrochemical sensors

Recently, the Institut für Mikrotechnik Mainz (IMM) funded by the EU project CD-CHEF developed a chip system to determine the gluten content of foods (www.imm-mainz.de). For the detection, various formats of ELISA and enzyme-linked oligonucleotide assays (ELONA) have been generated that are able to recognize gluten proteins. All sensors have in common that the receptor molecules (antibodies or aptamers) are immobilized on a substrate surface. After binding of the antigens, an enzymatic reaction is triggered, resulting in a fluorescent or electrochemical signal. For the optical detection, a chip has been designed, in which many beam-guiding components are integrated. Two electrodes (gold layer as working electrode and Ag/AgCl layer as reference and counter electrode) have been implemented in the chip for the amperometric sensor. Future studies have to prove the useability of these chip systems.

Polymerase chain reaction

The polymerase chain reaction (PCR) is based on the determination of a specific DNA. In comparison with protein analysis, DNA analysis is more sensitive by several orders of magnitude. A few molecules of any DNA sequence can be multiplied by a factor of 10^6 to 10^8 in a very short time. PCR can be applied also for heated products, because DNA is considerably more heat-stable than proteins. The first step of PCR is DNA extraction and heating, which causes denaturation and separation into single strands. Then, primers (oligodeoxynucleotides having base sequences complementary to a portion of the target DNA) are added and hybridized with the complementary segments of the single strands. By the addition of the four deoxynucleoside 5′-triphosphates and a thermostable DNA polymerase, new complementary strands are synthesized. The DNA is amplified by repeating the steps 20–30 times and can be analyzed electrophoretically (qualitative PCR). For quantitative PCR, also called "real-time PCR," oligodeoxynucleotides labeled with a fluorescent or enzyme marker are used and quantitation is performed by measuring the intensity of fluorescence or color. The calibration is done with standard DNA fragments. Both qualitative and quantitative PCR can be done automatically by means of a DNA-Thermal Cycler.

The group of Lüthy in Berne, Switzerland, was the first to apply PCR for gluten analysis. Allmann *et al.* (1993) used primers specific for highly conserved eukaryote DNA sequences to prove isolated nucleic acid substrate accessibility to PCR amplification. Subsequently, a highly repetitive and specific genomic wheat DNA segment was amplified by PCR for wheat detection. This assay was tested with 35 different food samples ranging from bakery additives to heated and processed food samples. Wheat starch even having a very low gliadin content reacted strongly positive and pure gliadin or gluten used as additives could not be detected. PCR and ELISA (Ridascreen® Gluten Kit) were compared by the analysis of oat samples spiked with wheat (Köppel *et al.*, 1998). The results demonstrated that the wheat PCR system was about 10 times more sensitive than the ELISA system provided that the isolated DNA was amplifiable.

A quantitative competitive (QC-) PCR system was developed by Dahinden *et al.* (2001) to detect wheat, rye and barley contamination. This system simultaneously recognized wheat, rye, and barley DNA on the basis of a non-coding region of chloroplast *trnL* gene. An internal DNA standard was constructed by adding 20 bp to the original PCR product. The QC-PCR system was applied to 15 commercially available products labeled as gluten-free and compared with ELISA (Ridascreen® Gluten Kit). Both methods yielded identical results for most cases and were proposed to support each other in testing gluten-free products.

Real-time PCR using melting curve analysis for product identification were established by Sandberg *et al.* (2003) to specifically discriminate wheat, rye, barley, and oat contamination in food samples. The PCR method gave a good correlation with ELISA (Transia Plate Gluten). The advantages of using melting curve analysis over gel electrophoresis were that the analysis was performed in the same closed capillary used for amplification, thus the risk of contamination between samples was eliminated. Henterich *et al.* (2003) introduced a one-step real-time immuno-PCR for gliadin detection. In this technique, R5 MAb was conjugated with an oligonucleotide; the sensitivity of gliadin analysis was increased more than 30-fold above the level reached by ELISA.

Three different real-time PCR systems for measuring wheat, rye, and barley DNA were developed by Mujico *et al.* (2004, 2005). The combination of these systems allowed the discrimination of not only the type of cereal, but also the determination of the proportion of wheat, rye, and barley contamination in oat samples. The data presented in this study indicated that most oat samples analyzed were contaminated, mainly by barley. The comparison of PCR and R5 ELISA used or the analysis of maize flours and unheated food samples revealed a good linear correlation (Mujico and Mendez, 2006). In summary, the developed quantitative PCR systems were recommended as a highly sensitive tool for gluten analysis complementary to immunological methods (e.g. ELISA and Western blot). DNA from hydrolyzed products such as beers, syrups, and malt extracts, however, were not detected by the PCR system.

Mass spectrometry

In recent years, matrix-assisted laser desorption/ionization time-of-flight mass spectrometry (MALDI-TOF MS) has become an important method to determine molecular masses of proteins. This technique allows the simultaneous measurement of masses from 1000 to 100 000 without chromatographic purification in the low picomol range within few minutes. Thus, not only intact proteins, but also protein hydrolyzates can be analyzed. MALDI-TOF MS may be divided into three parts: binding of the analyte to the matrix, ionization and desorption of the analyte by a laser, separation and detection of the analyte by a mass spectrometer. The matrix (e.g. sinapinic acid) is dissolved in an appropriate volatile solvent, mixed with the analyte, spotted onto a metal plate and dried under vacuum. A laser light (mostly a pulsed nitrogen laser of wavelength 337 nm) is fired at the spot; thereupon, the analyte is ionized and carried into the vapor phase. Multiple laser shots are used to improve the signal-to-noise ratio. The type of mass spectrometer mostly used in combination with MALDI is the TOF mass spectrometer. The ions released by the laser are accelerated by a short high-voltage impulse and then separated according to their mass (m) to charge (z) ratios (m/z) by measuring the time (microseconds) it takes for the ionized analyte to traverse an evacuated field-free drift tube. The heavier ions are slower than the higher ones. Separated ions arriving at the end of the tube are detected by an appropriate recorder that produces a signal upon impact of each ion group. The digitized data generated from successive laser shots are summed up, yielding a TOF mass spectrum.

The group of Mendez in Madrid was the first to use MALDI-TOF MS for the identification of celiac disease-toxic prolamins (Mendez et al., 1995). The high resolution and sensitivity of this technique allowed protonated molecular masses of gliadins, secalins, hordeins, and avenins displaying typical mass patterns to be solved. The authors proposed that MALDI-TOF MS is a useful alternative technology for the identification of gluten in food. The sample preparation was shown to be quite easy (Camafeita et al., 1997a). The procedure consisted only of mixing the prolamin-containing alcohol extract with a detergent (octyl-β-D-glucopyranoside) and an appropriate matrix (sinapinic acid in acetonitrile (30%)/trifluoroacetic acid (0.1%)). An aliquot of this mixture was deposited on a stainless steel probe tip, dried and measured on a MALDI-TOF mass spectrometer. The equipment was externally calibrated with a mixture of bovine serum albumin and horse heart cytochrome C. The detection limit for gliadins was found to be 0.01 mg/mL extract. The use of a reducing agent for prolamin extraction was not a handicap for analysis by MALDI-TOF MS.

Thirty food samples (wheat breads and starches, gluten-free food) were simultaneously analyzed by MALDI-TOF MS and a laboratory-made sandwich ELISA (Camafeita *et al.*, 1997b). The MS results revealed a linear response in the range of 4–100 mg gliadin/kg and a good correlation with those of ELISA. The comparison of celiac disease-toxic prolamins showed that gliadins, secalins, hordeins, and avenins had characteristic mass profiles that allowed the discrimination of cereal species (Camafeita *et al.*, 1998). Different cultivars of wheat, rye, and oats were nearly identical, while barley hordeins yielded different cultivar dependent mass profiles.

Recently, MALDI-TOF MS was used to characterize gluten-derived peptides in beer (Iametti *et al.*, 2005, 2006), which were not detectable by sandwich ELISA. The most relevant differences between the MS profiles of peptides in beers were found in the low-mass region (<5000). Beer produced in different countries had widely different peptide profiles, suggesting that manufacturing practices might play a major role in determining the presence and quantity of celiac disease-toxic peptides. The authors proposed that detailed analysis using amino acid sequencing by HPLC-tandem MS are necessary to clarify the nature and origin and their possible toxicity for patients with celiac disease.

Altogether, MALDI-TOF MS is a highly valuable non-immunological approach for the detection and quantitative determination of gluten in foods. Its limitation lies in expensive equipment so that only a few specialized laboratories are able to perform analyses. The service of such laboratories should be particularly used to confirm the reliability of immunochemical methods and to analyze selected suspicious samples.

Column chromatography

Column chromatography has been used for a long time to characterize, separate, and quantitate cereal protein fractions. In particular, gel permeation (GP) chromatography, separating according to different molecular weights, and reversed-phase (RP) chromatography, separating according to different hydrophobicities, have been widely used. Applications on the basis of HPLC (Kruger and Bietz, 1994) have considerably reduced the time of analysis (often less than 30 minutes). The detection and quantitation of proteins eluted from the column is carried out by UV absorbance in the range of 200–220 nm. At these wavelengths, the absorbance units are highly correlated with protein quantity (Wieser *et al.*, 1998). The detection limit is around 1–2 μg protein. A disadvantage is that the detection technique cannot differentiate between gluten and non-gluten proteins and is, therefore, not applicable for the analysis of complex foods. Nevertheless, column chromatography is a valuable aid, for example, to determine composition and quantity of a reference protein (van Eckert *et al.*, 2006) or to judge the results of other methods (Wieser *et al.*, 1994).

In special cases, however, column chromatography can be applied for gluten determination. GP-HPLC on Superdex 200 HR was used to quantify both gliadins and total gluten in a series of wheat starches by means of the following steps (Wieser and Antes, 2002): extraction of 1 g with 10 mL of 60% ethanol (gliadins) or 50%

2-propanol plus reducing agent (total gluten), centrifugation, drying of 4 mL supernatant in a vacuum centrifuge, dissolving in 500 μL elution solvent, injection of 100–200 μL, UV absorbance at 210 or 205 nm. The analyzed 23 starch samples had gliadin contents between 15 and 574 mg/kg (Wieser and Seilmeier, 2003). The average coefficient of variation resulting from two determinations was ±2.6%. According to the ratio of gliadins to glutenins, strong differences were found (0.2–4.9), demonstrating that the calculation gliadin ×2 = gluten proposed by the Draft Revised Codex Standard is not justified. In addition to wheat starch, other raw materials used for the production of gluten-free food were tested. Gluten determination was, in principle, possible for apple fiber, buckwheat groats, spice mixture, chestnut, millet, and rice flour. Skim milk powder and maize flour, however, contained components that prevent correct analysis by means of GP-HPLC. In conclusion, column chromatography can serve as an alternative method for gluten analysis in special cases and can help to control other methods.

Conclusions and future trends

Since the recognition that wheat gluten exacerbates celiac disease (Dicke, 1950), comparatively slow progress has been made in developing methods for the quantitative determination of celiac disease-toxic proteins. Around 30 years passed until the first Codex Standard for Gluten-Free Foods was established in 1981, which was limited to the analysis of wheat starch by means of nitrogen determination. A revision of this standard is still under way and the Draft Revised Codex Standard has arrived at step 6 of the Codex procedure. The general major problem is that the analyte (gluten proteins) is incompletely defined with respect to protein composition and toxicity, and an aggravating factor is that gluten proteins are often subjected to denaturation or proteolysis during food processing. Moreover, the choice of an appropriate reference protein is pivotal in getting accurate results. Numerous analytical methods based on immunochemistry, PCR, MS, or HPLC have been developed during the last 25 years, but only a couple have met the minimum requirements of sensitivity, selectivity, precision, speed, and availability. Therefore, the Draft Revised Codex Standard still gives only a general outline of a method and recommends an immunochemical approach.

ELISA has been the most frequently used immunoassay and different test systems have been marketed and were, in parts, successfully ring-tested. The sandwich ELISA developed by Skerritt and Hill (1990) containing MAb against heat-stable ω-gliadin has been in use for many years. This method, however, bears a considerable source of systematic error due to the different proportions of ω-type proteins in wheat, rye, and barley species and varieties. More recently, a sandwich ELISA was developed based on the MAb R5 that recognizes celiac disease-toxic epitopes of prolamins from wheat, rye, and barley. The detection limit is 3 mg gluten/kg and the test does not cross-react with oats and non-toxic cereals. Problems arising from heat-treatment or matrix effects can be solved by this assay. For hydrolyzed products, an alternative competitive ELISA is offered. The test kit includes PWG gliadin as a reference protein, which has been recently produced for collective use. In 2005, the R5 ELISA was endorsed as type I method by the Codex Committee of Methods of Analysis and Sampling (CCMAS).

Though the recent Revised Draft Codex Standard and the proposed method are not perfect, they represent an important progress compared to Codex Stan 118–1981. For the near future, ELISA will remain the first choice for gluten analysis, but alternative methods will be necessary to control ELISA results. Two points remain serious problems. First, there is no final decision on the celiac disease toxicity of oats. Thus, it is an open question whether anti-avenin antibodies and a reference avenin should be included in the test system. Second, the calculation gluten $= 2\times$ prolamin is invalid, since the ratio of prolamins to glutelins is strongly dependent on cereal species and varieties and is different in products derived from cereals. Recent research work has shown that wheat glutelins exacerbate celiac disease just as prolamins, and corresponding proteins from rye and barley may also be toxic. However, there are currently no agreed antibodies and reference proteins for glutelins. A future task will, therefore, be the development of a method for the determination of both prolamins and glutelins and the production of a reference material that contains all types of storage proteins from wheat, rye, barley, and possibly oats.

Sources of further information and advice

CODEX document CL 2006/5-NFSDU (2006). Draft Revised Standard for Gluten-Free Foods. Joint FAO/WHO Food Standards Programme. Codex Alimentarius Commission. *Codex Standard*. Rome: WHO.

Denery-Papini, S., Nicolas, Y., and Popineau, Y. (1999). Efficiency and limitations of immunochemical assays for the testing of gluten-free foods. *J. Cereal Sci.* 30, 121–131.

Morris, B. A. and Clifford, M. N. (1985). *Immunoassays in Food Analysis*. London, New York: Elsevier Applied Science Publishers.

Skerritt, J. H. (1988). Immunochemistry of cereal grain storage proteins. *Adv. Cereal Sci. Technol.* 9, 263–338.

Stern, M. ed. (1998–2006). *Proceedings of the 12th–20th Meetings of the Working Group on Prolamin Analysis and Toxicity*. Zwickau: Verlag Wissenschaftliche Scripten.

References

Allmann, M., Candrian, U., Höfelein, C., and Lüthy, J. (1993). Polymerase chain reaction (PCR): a possible alternative to immunochemical methods assuring safety and quality of food. *Z. Lebensm.-Wiss. Untersuch. Forsch.* 196, 248–251.

Amtliche Sammlung von Untersuchungsverfahren nach § 35 LMBG (1984). *Untersuchung von Lebensmitteln. Immunologischer Nachweis von Proteinen in Backwaren (einschließlich Brot und glutenfreie Backwaren) und Süßwaren*. Berlin, Wien, Zürich: Beuth-Verlag GmbH.

Anderson, R. P. and Wieser, H. (2006). Medical applications of gluten-composition knowledge. In: Wrigley, C., Bekes, F., and Bushuk, W. eds. *Gliadin and Glutenin— the Unique Balance of Wheat Quality*. St. Paul, MN: American Association of Cereal Chemists, pp. 387–409.

Arentz-Hansen, H., McAdam, S. N., Molberg, Ø. *et al.* (2002). Celiac lesion T cells recognize epitopes that cluster in regions of gliadins rich in praline residues. *Gastroenterology* 123, 803–809.

Aubrecht, E. and Toth, A. (1995). Investigation of gliadin content of wheat flour by ELISA method. *Acta Aliment.* 24, 23–29.

Berger, E. and Freudenberg, E. (1961). Bemerkungen über die antigenen Eigenschaften von Abbaustufen des Gliadins. *Ann. Paediatr.* 196, 238–243.

Bermudo Redondo, M. C., Griffin, P. B., Garzon Rasanz, M., Ellis, H. J., and Ciclitira, P. J. (2005). Monoclonal antibody-based competitive assay for the sensitive detection of coeliac disease toxic prolamins. *Anal. Chim. Acta* 551, 105–114.

Camafeita, E., Alfonso, P., Acevedo, B., and Mendez, E. (1997a). Sample preparation optimization for the analysis of gliadins in food by matrix-assisted laser desorption/ionization time-of-flight mass spectrometry. *J. Mass Spectrom.* 32, 444–449.

Camafeita, E., Alfonso, P., Mothes, T., and Mendez, E. (1997b). Matrix-assisted laser desorption/ionization time-of-flight mass spectrometric micro-analysis: the first non-immunological alternative attempt to quantify gluten gliadins in food samples. *J. Mass Spectrom.* 32, 940–947.

Camafeita, E., Solis, J., Alfonso, P., Lopez, J. A., Sorell, L., and Mendez, E. (1998). Selective identification by matrix-assisted laser desorption/ionization time-of-flight mass spectrometry of different types of gluten in foods made with cereal mixtures. *J. Chromatogr. A* 823, 299–306.

Chirdo, F. G., Anon, M. C., and Fossati, C. A. (1995). Optimization of a competitive ELISA with polyclonal antibodies for quantification of prolamins in foods. *Food Agric. Immunol.* 7, 333–343.

Chirdo, F. G., Anon, M. C., and Fossati, C. A. (1998). Development of high-sensitive enzyme immunoassays for gliadins quantification using the strepavidin-biotin amplification system. *Food Agric. Immunol.* 10, 143–155.

Ciclitira, P. J. and Lennox, E. S. (1983). A radioimmunoassay for α- and β-gliadins. *Clin. Sci.* 64, 655–659.

Codex document CX/NFSDU 00/4 (2000). Draft revised standard for gluten-free foods. Joint FAO/WHO Food Standards Programme. Codex Alimentarius Commission. *Codex Standard*. Rome: WHO.

Codex document CL 2006/5-NFSDU (2006). Draft revised standard for gluten-free foods. Joint FAO/WHO Food Standards Programme. Codex Alimentarius Commission. *Codex Standard*. Rome: WHO.

Codex Stan 118–1981 (1981). Codex Standard for Gluten-Free Foods. Joint FAO/WHO Food Standards Programme. Codex Alimentarius Commission. *Codex Standard*. Rome: WHO; p. 118.

Codex Standard for the Labelling of Prepacked Foods (2001). Joint FAO/WHO Food Standards Programme. Codex Alimentarius Comission. *Codex Standard*. Rome: WHO.

Dahinden, I., von Büren, M., and Lüthy, J. (2001). A quantitative competitive PCR system to detect contamination of wheat, barley or rye in gluten-free food for coeliac patients. *Eur. Food Res. Technol.* 212, 228–233.

Denery-Papini, S., Nicolas, Y., and Popineau, Y. (1999). Efficiency and limitations of immunochemical assays for the testing of gluten-free foods. *J. Cereal Sci.* 30, 121–131.

Denery-Papini, S., Boucherie, B., Larré, C. *et al.* (2002). Measurement of raw, heated and modified gluten after limited hydrolysis. In: Stern, M. ed. *Proceedings of the 16th Meeting of the Working Group on Prolamin Analysis and Toxicity.* Zwickau: Verlag Wissenschaftliche Scripten, pp. 71–73.

Dewar, D. H., Amato, M., Ellis, H. J. *et al.* (2006). The toxicity of high molecular weight glutenin subunits of wheat to patients with coeliac disease. *Eur. J. Gastroenterol. Hepatol.* 18, 483–491.

Dicke, W. K. (1950). Coeliac disease. Investigation of the harmful effects of certain types of cereals on the patients with coeliac disease. PhD thesis, University of Utrecht.

Dona, V. V., Fossati, C. A., and Chirdo, F. G. (2004). Interference of denaturing and reducing agents on gliadin/antibody interaction. In: Stern, M. ed. *Proceedings of the 18th Meeting of the Working Group on Prolamin Analysis and Toxicity.* Zwickau: Verlag Wissenschaftliche Scripten, pp. 51–57.

Ellis, H. J., Doyle, A. P., Wieser, H., Sturgess, R. P., Day, P., and Ciclitira, P. J. (1994). Measurement of gluten using a monoclonal antibody to a sequenced peptide of α-gliadin from the coeliac-activating domain I. *J. Biochem. Biophys. Methods* 28, 77–82.

Ellis, H. J., Rosen-Bronson, S., O'Reilly, N., and Ciclitira, P. J. (1998). Measurement of gluten using a monoclonal antibody to a coeliac toxic peptide of A gliadin. *Gut* 43, 190–195.

Ellis, H. J., Pollock, E. L., Engel, W., Fraser, J. S., Rosen-Bronson, S., Wieser, H., and Ciclitira, P. J. (2003). Investigation of the putative immunodominant T cell epitopes in coeliac disease. *Gut* 52, 211–217.

Ellis, H. J., Dewar, D. H., Gonzales-Cinca, N., Wieser, H., O'Sullivan, C., and Ciclitira, P. J. (2006). Production of murine monoclonal antibodies to toxic gluten peptides and proteins, for use in ELISA. In: Stern, M., ed. *Proceedings of the 20th Meeting of the Working Group on Prolamin Analysis and Toxicity.* Zwickau: Verlag Wissenschaftliche Scripten, pp. 53–57.

Ellis, H. J., Dewar, D. H., Gonzales-Cinca, N. *et al.* (2007). Characterisation of monoclonal antibodies raised against HMW glutenin subunits. In: Stern, M. ed. *Proceedings of the 21st Meeting of the Working Group on Prolamin Analysis and Toxicity.* Zwickau: Verlag Wissenschaftliche Scripten (in press).

Ferre, S., Garcia, E., and Mendez, E. (2004). Measurement of hydrolysed gliadins by a competive ELISA based on monoclonal antibody R5, analysis of syrups and beers. In: Stern, M. ed. *Proceedings of the 18th Meeting of the Working Group on Prolamin Analysis and Toxicity.* Zwickau: Verlag Wissenschaftliche Scripten, pp. 65–69.

Fraser, J. S., Engel, W., Ellis, H. J. *et al.* (2003). Coeliac disease: *in-vivo* toxicity of the putative immunodominant epitope. *Gut* 52, 1698–1702.

Freedman, A. R., Galfre, G., Gal, E., Ellis, H. J., and Ciclitira, P. J. (1987). Monoclonal antibody ELISA to quantitate wheat gliadin contamination in gluten-free foods. *J. Immunol. Methods* 98, 123–127.

Freedman, A. R., Galfre, G., Gal, E., Ellis, H. J., and Ciclitira, P. J. (1988). Western immunoblotting of cereal proteins with monoclonal antibodies to wheat gliadin to investigate coeliac disease. *Int. Arch. Allergy Appl. Immunol.* 85, 346–350.

Friis, S. U. (1988). Enzyme-linked immunoasorbent assay for quantitation of cereal proteins toxic in coeliac disease. *Clin. Chim. Acta* 178, 261–270.

Fritschy, F., Windemann, H., and Baumgartnerm, E. (1985). Quantitative determination of wheat gliadins in foods by enzyme-linked immunosorbent assay. *Z. Lebensm. Untersuch. Forsch.* 181, 379–385.

Galfre, G. and Milstein, C. (1981). Preparation of monoclonal antibodies: strategies and procedures. *Methods Enzymol.* 73, 3–75.

Garcia, E., Hernando, A., Toribio, T., Genzor, C., and Mendez, E. (2002). Test immunochromatographic rapid assay: a rapid, highly sensitive and semi-quantitative test for the detection of gluten in foodstuffs. In: *Proceeding of the 16th Meeting of the Working Group on Polamin Analysis and Toxicity*. Zwickau: Verlag Wissenschaftliche Scripten, pp. 55–64.

Garcia, E., Hernando, A., Mujico, J. R., Lombardia, M., and Mendez, E. (2004). Matrix effects in the extraction and detection of gliadins in foods by R5 ELISA and MALDI-TOF mass spectrometry. In: Stern, M. ed. *Proceedings of the 18th Meeting of the Working Group on Prolamin Analysis and Toxicity*. Zwickau: Verlag Wissenschaftliche Scripten, pp. 59–64.

Garcia, E., Llorente, M., Hernando, A., Kieffer, R., Wieser, H., and Mendez, E. (2005). Development of a general procedure for complete extraction of gliadins from heat processed and unheated foods. *Eur. J. Gastroenterol. Hepatol.* 17, 529–539.

Gellrich, C., Schieberle, P., and Wieser, H. (2003). Biochemical characterization and quantification of the storage protein (secalin) types in rye flour. *Cereal Chem.* 80, 102–109.

Henterich, N., Osman, A. A., Mendez, E., and Mothes, T. (2003). Assay of gliadin by real-time immunopolymerase chain reaction. *Nahrung* 47, 345–348.

Hernando, A., Garcia, E., Llorente, M. *et al.* (2005). Measurements of hydrolysed gliadins in malts, breakfast cereals, heated/hydrolysed foods, whiskies and beers by means of a new competitive R5 ELISA. In: Stern, M. ed. *Proceedings of the 19th Meeting of the Working Group on Prolamin Analaysis and Toxicity*. Zwickau: Verlag Wissenschaftliche Scripten, pp. 31–37.

Hill, A. S. and Skerritt, J. H. (1990). Determination of gluten in foods using a monoclonal antibody-based competition enzyme immunoassay. *Food Agric. Immunol.* 2, 21–35.

Iametti, S., Cappelletti, C., Oldani, A., Scafuri, L., and Bonomi, F. (2004). Improved protocols for ELISA determination of gliadin in glucose syrups. *Cereal Chem.* 81, 15–18.

Iametti, S., Bonomi, F., Ferranti, P., Picariello, G., and Gabrovska, D. (2005). Characterization of gliadin content in beer by using different approaches. In: Stern, M. ed. *Proceedings of the 19th Meeting of the Working Group on Prolamin Analysis and Toxicity*. Zwickau: Verlag Wissenschaftliche Scripten, pp. 73–78.

Iametti, S., Bonomi, F., Ferranti, P., de Martino, A., and Picariello, G. (2006). Characterization of peptides and proteins in beer by different approaches. In: Stern, M. ed. *Proceedings of the 20th Meeting of the Working Group on Prolamin Analysis and Toxicity*. Zwickau: Verlag Wissenschaftliche Scripten, pp. 47–52.

Immer, U. and Haas-Lauterbach, S. (2003). Ridascreen®/Rida® gliadin test systems. In: Stern, M. ed. *Proceedings of the 17th Meeting of the Prolamin Working Group on Prolamin Analysis and Toxicity*. Zwickau: Verlag Wissenschaftliche Scripten, pp. 45–52.

Immer, U. and Haas-Lauterbach, S. (2005a). The question of extraction procedures. In: Stern, M. ed. *Proceedings of the 19th Meeting of the Working Group on Prolamin Analysis and Toxicity*. Zwickau: Verlag Wissenschaftliche Scripten, pp. 45–52.

Immer, U. and Haas-Lauterbach, S. (2005b). Sandwich ELISA versus competitive ELISA: which approach is the more appropriate? In: Stern, M. ed. *Proceedings of the 19th Meeting of the Working Group on Prolamin Analysis and Toxicity*. Zwickau: Verlag Wissenschaftliche Scripten, pp. 53–62.

Kahlenberg, F., Sanchez, D., Lachmann, I., Tuckova, L., Tlaskalova, H., Mendez, E., and Mothes, T. (2006). Monoclonal antibody R5 for detection of putatively coeliac-toxic gliadin peptides. *Eur. Food Res. Technol.* 222, 78–82.

Kasarda, D. D. (1994). Toxic cereal grains in coeliac disease. In: Feighery, C. and O'Farrelly, C. eds. *Gastrointestinal Immunology and Gluten-Sensitive Disease*. Dublin: Oak Tree Press, pp. 203–220.

Köppel, E., Stadler, M., Lüthy, J., and Hübner, P. (1998). Detection of wheat contamination in oats by polymerase chain reaction (PCR) and enzyme-linked immunosorbent assay (ELISA). *Z. Lebensm. Untersuch. Forsch.* 206, 399–403.

Kruger, E. and Bietz, J. A. (1994). *HPLC—High-Performance Liquid Chromatography of Cereal and Legume Proteins*. St. Paul, MN: American Association of Cereal Chemists.

Laffey, C., Madden, N., Fogarty, T., amd Burke, P. (2005). Gluten testing: an Irish perspective. In: Stern, M. ed. *Proceedings of the 19th Meeting of the Working Group on Prolamin Analysis and Toxicity*. Zwickau: Verlag Wissenschaftliche Scripten, pp. 63–68.

Lewis, J. H. and Wells, H. G. (1925). The immunological properties of alcohol-soluble vegetable proteins. *J. Biol. Chem.* 66, 37–48.

Malmheden Yman, I. (2006). Detection of gluten/cereals in baby food samples—collaborative study. In: Stern, M. ed. *Proceedings of the 20th Meeting of the Working Group on Prolamin Analysis and Toxicity*. Zwickau: Verlag Wissenschaftliche Scripten, pp. 65–74.

Marsh, M. N., Morgan, S., Ensari, A. *et al.* (1995). In-vivo acitivty of peptides 31–43, 44–55, 56–68 of α-gliadin in gluten sensitive enteropathy (GSE). *Gastroenterology* 108, A871.

McKillop, D. F., Goslin, J. P., Stevens, F. M., and Fottrell, P. F. (1985). Enzyme immunoassay of gliadin in food. *Biochem. Soc. Trans.* 13, 486–487.

Meier, P., Windemann, H., and Baumgartner, E. (1984). Zur Bestimmung des α-Gliadin-Gehaltes in glutenhaltigen und 'glutenfreien' erhitzten Lebensmitteln. *Z. Lebensm. Untersuch. Forsch.* 178, 361–365.

Mendez, E., Camafeita, E., Sebastian, J. S. *et al.* (1995). Direct identification of wheat gliadins and related cereal prolamins by matrix-assisted laser desorption/ionization time-of-flight mass spectrometry. *J. Mass Spectrom.* (Spec. Issue), S123–S128.

Mendez, E., Vela, C., Immer, U., and Janssen, F. W. (2005). Report of a collaborative trial to investigate the performance of the R5 enzyme linked immunoassay to determine gliadin in gluten-free food. *Eur. J. Gastroenterol. Hepatol.* 17, 1053–1063.

Molberg, Ø., Solheim Flaete, N., Jensen, T. *et al.* (2003). Intestinal T-cell responses to high-molecular-weight glutenins in celiac disease. *Gastroenterology* 125, 337–344.

Morris, B. A. and Clifford, M. N. (1985). *Immunoassays in Food Analysis.* London: Elsevier Applied Science.

Mujico, J. R., Lombardia, M., and Mendez, E. (2004). Detection of wheat DNA in foods by a quantitative real-time PCR system: can the measurement of wheat DNA be used as a non-immunological and complementary tool in gluten technology? In: Stern, M. ed. *Proceedings of the 18th Meeting of the Working Group on Prolamin Analysis and Toxicity.* Zwickau: Verlag Wissenschaftliche Scripten, pp. 91–98.

Mujico, J. R., Hernando, A., Lombardia, M. *et al.* (2005). Quantification of wheat, barley and rye contamination in oat samples by real-time PCR. In: Stern, M. ed. *Proceedings of the 19th Meeting of the Working Group on Prolamin Analaysis and Toxicity.* Zwickau: Verlag Wissenschaftliche Scripten, pp. 87–94.

Mujico, J. R. and Mendez, E. (2006). Simultaneous detection/quantification of wheat, barley and rye DNA by a new quantitative real-time PCR system. In: Stern, M. ed. *Proceedings of the 20th Meeting of the Working Group on Prolamin Analaysis and Toxicity.* Zwickau: Verlag Wissenschaftliche Scripten, pp. 39–45.

Osborne, T. B. (1907). The proteins of the wheat kernel. Publication 84. Carnegie Inst., Washington, DC.

Ranz, A. I., Venteo, A., Vela, C., and Sanz, A. (2004). Ingezim gluten. Immunoenzymatic assay for gluten detection using monoclonal antibody R5. In: Stern, M. ed. *Proceedings of the 18th Meeting of the Working Group on Prolamin Analaysis and Toxicity.* Zwickau: Verlag Wissenschaftliche Scripten, pp. 37–49.

Ranz, A. I., Venteo, A., Cano, M. J., Vela, C., and Sanz, A. (2005). Development of a new and rapid semiquantitative method for gliadin detection using R5 antibody. In: Stern, M. ed. *Proceedings of the 19th Meeting of the Working Group on Prolamin Analaysis and Toxicity.* Zwickau: Verlag Wissenschaftliche Scripten, pp. 39–44.

Rumbo, M., Chirdo, F. G., Fossati, C. A., and Anon, M. C. (2001). Analysis of the effects of heat treatment on gliadin immunochemical quantification using a panel of anti-prolamin antibodies. *J. Agric. Food Chem.* 49, 5719–5726.

Sandberg, M., Lundberg, L., Ferm, M., and Malmheden Yman, I. (2003). Real time PCR for the detection and discrimination of cereal contamination in gluten free foods. *Eur. Food Res. Technol.* 217, 344–349.

Schofield, J. D., Bottlomley, R. C., Timms, M. F., and Booth, M. R. (1983). The effect of heat on wheat gluten and the involvement of sulphydryl-disulphide interchange reactions. *J. Cereal Sci.* 1, 241–253.

Seilmeier, W. and Wieser, H. (2003). Comparative investigations of gluten proteins from different wheat species. IV. Reactivity of gliadin fractions and components from different wheat species in a commercial immunoassay. *Eur. Food Res. Technol.* 217, 360–364.

Shewry, P. R. and Tatham, A. S. (1990). The prolamin storage proteins of cereal seeds: structure and evolution. *Biochem. J.* 267, 1–12.

Sima, A., van Eckert, R., and Pfannhauser, W. (1999). Vergleich unterschiedlicher kommerzieller ELISA-Testsysteme zur Bestimmung von Gluten. *Lebensmittelchemie* 53, 40.

Skerritt, J. H. (1985). A sensitive monoclonal-antibody-based test for gluten detection: quantitative immunoassay. *J. Sci. Food Agric.* 36, 987–994.

Skerritt, J. H. (1988). Immunochemistry of cereal grain storage proteins. *Adv. Cereal Sci. Technol.* 9, 263–338.

Skerritt, J. H. and Hill, A. S. (1990). Monoclonal antibody sandwich enzyme immunoassays for determination of gluten in foods. *J. Agric. Food Chem.* 38, 1771–1778.

Skerritt, J. H. and Hill, A. S. (1991a). Enzyme immunoassay for determination of gluten in foods: collaborative study. *J. AOAC* 74, 257–264.

Skerritt, J. H. and Hill, A. S. (1991b). Self-management of dietary compliance in coeliac disease by means of ELISA "home test" to detect gluten. *The Lancet* 337, 379–382.

Skerritt, J. H. and Underwood, P. A. (1986). Specifity characteristics of monoclonal antibodies to wheat grain storage proteins. *Biochim. Biophys. Acta* 874, 245–254.

Skerritt, J. H., Devery, J. M., and Hill, A. S. (1991). Chemistry, celiac-toxicity and detection of gluten and related prolamins in foods. *Panminerva Med.* 33, 65–74.

Sollid, L. M. (2002). Coeliac disease: dissecting a complex inflammatory disorder. *Nat. Rev. Immunol.* 2, 647–655.

Sorell, L., Lopez, J. A., Valdes, I. *et al.* (1998). An innovative sandwich ELISA system based on an antibody cocktail for gluten analysis. *FEBS Lett.* 439, 46–50.

Spaenij-Dekking, E. H. A., Kooy-Winkelaar, E. M. C., Nieuwenhuizen, W. F., Drijfhout, J. W., and Koning, F. (2004). A novel and sensitive method for the detection of T cell stimulatory epitopes of α/β- and γ-gliadins. *Gut* 53, 1267–1273.

Spaenij-Dekking, L., Kooy-Winkelaar, Y., Stepniak, D., Edens, L., and Koning, F. (2006). Detection and degradation of gluten. In: Stern, M. ed. *Proceedings of the 20th Meeting of the Working Group on Prolamin Analaysis and Toxicity.* Zwickau: Verlag Wissenschaftliche Scripten, pp. 59–64.

Stern, M., Ciclitira, P. J., van Eckert, R. *et al.* (2001). Analysis and clinical effects of gluten in coeliac disease. *Eur. J. Gastroenterol. Hepatol.* 13, 741–747.

Sturgess, R., Day, P., Ellis, H. J. *et al.* (1994). Wheat peptide challenge in coeliac disease. *Lancet* 334, 758–761.

Theobald, K., Bohn, A., Thiel, M., Ulmer, W. T., and König, W. (1983). Production of monoclonal antibodies against wheat flour components. *Int. Arch. Allergy Appl. Immunol.* 72, 84–86.

Troncone, R., Vitale, M., Donatiello, A., Farris, E., Rossi, G., and Auricchio, S. (1986). A sandwich enzyme immunoassay for wheat gliadin. *J. Immunol. Methods* 92, 21–23.

Vader, W., Kooy, Y., van Veelen, P. *et al.* (2002). The gluten response in children with celiac disease is directed towards multiple gliadin and glutenin peptides. *Gastroenterology* 122, 1729–1737.

Valdes, I., Garcia, E., Llorente, M., and Mendez, E. (2003). Innovative approach to low-level gluten determination in foods using a novel sandwich enzyme-linked immunosorbent assay protocol. *Eur. J. Gastroenterol. Hepatol.* 15, 465–474.

Van de Wal, Y., Kooy, Y. M. C., van Veelen, P. *et al.* (1999). Glutenin is involved in the gluten-driven mucosal T cell response. *Eur. J. Immunol.* 29, 3133–3139.

Van Eckert, R. (1993). Methodological and practical experience in gluten analysis. *Ernährung/Nutrition* 17, 163–165.

Van Eckert, R., Scharf, M., Wald, T., and Pfannhauser, W. (1997). Determination of proteins with ELISA-methods: doubtful quantitative results? In: Amado, R. and Battaglia, R. eds. *Authenticity and Adulteration of Food—the Analytical Approach.* Proceedings of the 9th European Conference on Food Chemistry, FECS Event No. 220, Vol. 1. Zürich: Swiss Society of Food and Environmental Chemistry, pp. 263–268.

Van Eckert, R., Berghofer, E., Ciclitira, P. J. *et al.* (2006). Towards a new gliadin reference material—isolation and characterisation. *J. Cereal Sci.* 43, 331–341.

Weisgerber, C. (1998). ELISA for the detection of gliadin in food. In: Stern, M. ed. *Proceedings of the 12th Meeting of the Working Group on Prolamin Analaysis and Toxicity.* Zwickau: Eigenverlag, p. 59.

Wieser, H. (1994). Cereal protein chemistry. In: Feighery, C. and O'Farrelly, C. eds. *Gastrointestinal Immunology and Gluten-Sensitive Disease.* Dublin: Oak Free Press, pp. 191–202.

Wieser, H. (1995). The precipitating factor in celiac disease. In: Howdle, P. D., ed. *Bailliere's Clinical Gastroenterology*, Vol. 9: *Coeliac Disease.* London: Bailliere Tindall, pp. 191–207.

Wieser, H. (1998). Investigations on the extractability of gluten proteins from wheat bread in comparison with flour. *Z. Lebensm. Untersuch. Forsch.* A207, 128–132.

Wieser, H. (2000). Comparative investigations of gluten proteins from different wheat species. I. Qualitative and quantitative composition of gluten protein types. *Eur. Food Res. Technol.* 211, 262–268.

Wieser, H. and Antes, S. (2002). Development of a non-immunochemical method for the quantitative determination of gluten in wheat starch. In: Stern, M. ed. *Proceedings of the 16th Meeting of the Working Group on Prolamin Analaysis and Toxicity.* Zwickau: Verlag Wissenschaftliche Scripten, pp. 19–23.

Wieser, H. and Kieffer, R. (2001). Correlations of the amount of gluten protein types to the technological properties of wheat flours determined on a micro-scale. *J. Cereal Sci.* 34, 19–27.

Wieser, H. and Seilmeier, W. (2003). Determination of gliadin and gluten in wheat starch by means of alcohol extraction and gel permeation chromatography. In: Stern, M. ed. *Proceedings of the 17th Meeting of the Working Group on Prolamin Analaysis and Toxicity.* Zwickau: Verlag Wissenschaftliche Scripten, pp. 53–57.

Wieser, H., Belitz, H.-D., Idar, D., and Ashkenazi, A. (1986). Coeliac activity of the gliadin peptides CT-1 and CT-2. *Z. Lebensm. Untersuch. Forsch.* 182, 115–117.

Wieser, H., Seilmeier, W., and Belitz H.-D. (1994). Quantitative determination of gliadin subgroups from different wheat cultivars. *J. Cereal Sci.* 19, 149–155.

Wieser, H., Antes, S., and Seilmeier, W. (1998). Quantitative determination of gluten protein types in wheat flour by reversed-phase high-performance liquid chromatography. *Cereal Chem.* 75, 644–650.

Wieser, H., Bushuk, W., and MacRitchie, F. (2006). The polymeric glutenins. In: Wrigley, C., Bekes, F., and Bushuk, W. eds. *Gliadin and Glutenin: the Unique Balance of Wheat Quality.* St. Paul, MN: American Association of Cereal Chemists, pp. 213–240.

Windemann, H., Fritschy, F., and Baumgartner, E. (1982). Enzyme-linked immunosorbent assay for wheat α-gliadin and whole gliadin. *Biochim. Biophys. Acta* 709, 110–121.

Wrigley, C., Corke, H., and Walker, C. E. (2004). *Encyclopedia of Grain Science*, Vol. 1–3. Amsterdam: Elsevier Academic Press.

Rice

4

Cristina M. Rosell and Cristina Marco

Introduction

Rice has been, throughout history, one of the most important foods in the human diet and one of the most extended cereal crops (9% of the total cultivated soil). In fact, rice has probably fed more people in history than any other crop. Even today, rice grains sustain two-thirds of the world's population, approximately 2.5 billion people. However, around the world, the contribution that rice makes to diet differs and the types of processing involved are also quite different. Rice is mainly consumed as white grain, but in the last decade dozens of products containing rice as an ingredient have appeared on the food market. Two different species of rice are cultivated: *Oryza sativa* and *Oryza glaberrima*, and there are around 22 wild species. *Oryza sativa* originated in the wet tropic of Asia, but is now cultivated around the world, whereas *Oryza glaberrima* has been cultivated in West Africa for the last 3500 years.

Rice accounts for 29% of the world's total cereal production, and is comparable to the production of wheat and corn. Cultivation is concentrated in the developing countries, mainly around East and Middle Asia, where 91% of the total world production is located (FAOSTAT, 2007) (Figure 4.1). China is the world's largest rice producer (30%), followed by India (21%), Indonesia (9%), and Bangladesh (6%). The rest of Asia, America, and Africa produce 37%, 5%, and 3%, respectively of the total world rice production. The amount of rice and rice-based products available for human consumption in the different countries is almost parallel to the rice production. With minor exceptions, practically all the rice production is consumed within the producers' countries. The highest daily rice consumption is observed in Myanmar, with 795 g per capita. The average daily consumption of rice in the Asian countries is 285 g per capita, ahead of the 44 g per capita of rice consumed in the developed countries. Nowadays, there are three big models of rice consumption (Infocomm, 2007): the Asian model, with an average yearly consumption higher than

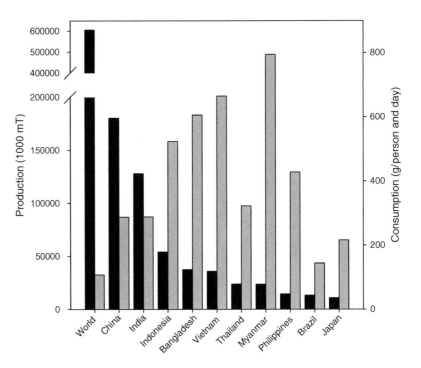

Figure 4.1 Paddy rice production (black bars) and consumption (gray bars) among the world's largest rice producers. Values are expressed in grams per person and per day. Data from FAOSTAT (2004).

80 kg per person; the subtropic developing countries model with a consumption between 30 and 60 kg per person; and the Occident model with a consumption lower than 10 kg per person. In the last decade, rice consumption experienced a steady decrease in the developed countries; this tendency promoted the development of new and innovative rice-based products. Remarkably, in 2000 over 400 new products containing rice were placed on the market (Wilkinson and Champagne, 2004), as a result of new initiatives designed to increase rice consumption.

Rice provides 27% of the total energy intake in the developing countries, and only 4% of the total energy intake in the developed countries. Like other cereals, rice is a cheap source of protein, and in developing countries, rice supplies 20% of the dietary protein intake.

The composition of the rice grain depends on the cultivars grown, environmental factors, and processing. Rice can be cultivated in diverse conditions, although it grows faster in wet and warm environments. Rice grains can be short, medium, or long. They can be sticky or glutinous and non-sticky, and can be a variety of different colors, including black to red and brown; some species are even aromatic. Rice grains without the hull are brown due to the color of the three-pericarp layers that cover the grain. The rice grain is rich in complex carbohydrates, and represents a source of proteins, minerals, and vitamins, mainly B vitamins, and does not contain cholesterol (Table 4.1). The chemical composition of the rice grain changes during milling. Removal of the outer bran layers causes a loss of proteins, fats, and a large percentage of the fiber, vitamins, and minerals. Iron, phosphorous, potassium, and

Table 4.1 Rice composition and energetic value of rice grain and rice flour (referred to 100 g)

		Rough rice	Milled rice	Whole meal flour	White flour
Carbohydrates (%)		77.2	79.9	76.5	80.1
Proteins (%)		7.9	7.1	7.2	5.9
Dietetic fiber (%)		3.5	1.3	4.6	2.4
Lipids (%)		2.9	0.7	2.8	1.4
Minerals (%)		1.5	0.6	1.5	0.6
Lipids	Saturated (g)	0.6	0.7	2.8	1.4
	Monounsaturated (g)	1.1	0.2	1.0	0.4
	Polyunsaturated (g)	1.0	0.2	1.0	0.4
Minerals	Calcium (mg)	23.0	28.0	11.0	10.0
	Iron (mg)	1.5	0.8	2.0	0.4
	Magnesium (mg)	143.0	25.0	112.0	35.0
	Phosphorus (mg)	333.0	115.0	337.0	98.0
	Potasium (mg)	223.0	115.0	289.0	76.0
	Sodium (mg)	7.0	5.0	8.0	0.0
Vitamins	Vitamin E (mg)	1.2	0.1	1.2	0.1
	Vitamin K (mg)	1.9	0.1	—	0.0
	Thiamine (mg)	0.4	0.1	0.4	0.1
	Riboflavin (mg)	0.1	0.0	0.1	0.0
	Niacin (mg)	5.1	1.6	6.3	2.6
	Pyridoxine (mg)	0.5	0.2	0.7	0.4
	Folate (μg)	20.0	8.0	16.0	4.0
	Pantothenic acid (mg)	1.5	1.0	1.6	0.8

magnesium are the most important minerals in this cereal. The hull represents 20% of the grain and is composed of silica and hemicelluloses (Champagne *et al.*, 2004).

Carbohydrates are the most abundant component in rice, with starch contents of approximately 80% (14% moisture). Rice starch is a glucose polymer composed of amylose and amylopectin in different proportions depending on the rice variety. The content of starch in the rice grain increases from the surface to the core, and thus milled rice is rich in starch. Rice starch is considered not allergenic because of the hypoallergenic proteins present. Starch determines the physical properties and functionality of the rice grains, and these properties are greatly dependent on the amylose/amylopectin ratio. Amylopectin is the branched polymer and is more abundant than the linear polymer (amylose). However, amylose has received more attention from the scientific community because it is considered an indicator of cooking quality. Rice starch that lacks amylose is called "waxy," because of its mutation at the waxy locus, or "glutinous," due to its opaque appearance. Complete information about rice starch structure and functional properties has been recently reviewed by Fitzgerald (2004).

Protein is the second most abundant constituent of milled rice, ranging from 6.3 to 7.1 g of N × 5.95. Protein concentration decreases from the surface to the center of the kernel, and they are deficient in the essential amino acid lysine. The albumin, globulin, prolamin, and glutelin content is unique among the cereals, with a high concentration of glutelins and a low concentration of prolamins (Hamaker, 1994). This characteristic determines the high content of lysine when compared with other cereals. The most abundant essential amino acids are glutamic acid, aspartic acid, leucine, and arginine, followed by alanine, valine, phenylalanine, and serine. Lipids are minor components, but they contribute to the nutritional, sensoric, and functional characteristics, since they form complexes with the amylose chains. Rice lipids are classified as starchy or non-starchy lipids. The majority of the lipids are non-starchy lipids, and they are located in the aleurone layer and germ. They comprise neutral lipids, with a small amount of glycolipids and phospholipids. Recently, some minor lipids have been related to the role of rice in the prevention of chronic diseases such as cancer and heart diseases (Watkins *et al.*, 1990).

Production of rice flours and their properties

Rice flour production

Milling

Rice is harvested and threshed to produce the so-called "paddy" or "rough" rice, where the kernel is still within the hull or husk. As for wheat, milling is the usual method used to process rice, although the term "milling" in the rice industry is used for a process that is completely different from wheat milling. Wheat is milled to obtain flour, whereas milling of rice comprises the removal of the husk, stripping the bran of the endosperm, and finally removing broken and altered kernels. Milling of rice drastically affects its composition.

In developed countries rice milling involves a very sophisticated process. Initially, the paddy rice is cleaned through coarse screens to remove straw, stones and other foreign objects that are larger than the rice kernel. This process is repeated using fine screens in order to remove small weed seeds, sand, stones and other objects smaller than the rice kernels. Stones are separated from the rice in specific gravity tables that separate the product by density. Any metallic particles are removed by magnetic separators. After this cleaning step, the husk is removed by passing the rice through two spinning rubber rollers, which rotate in opposite directions at different speeds (Bond, 2004). Brown rice is obtained after de-hulling. This product can be either eaten as it is, milled into white rice, or processed to obtain different products and by-products. The brown color is due to the presence of bran layers, which are rich in minerals and vitamins. Milled rice, also known as milled white rice, polished rice, or polished white rice, is obtained after removing the bran and germ from brown rice. There are many machines and methods designed for milling rice, but often an abrasive system, followed by frictional and polishing systems are used. During the first step, 95% of the bran is removed in an abrasive whitener, by contact of the grain over an abrasive surface. Subsequently, the bran layers that remain on the grain are removed by friction between the grains using a friction whitener. The degree of milling does not increase linearly with the milling time (Lamberts *et al.*, 2007). A change during the milling process is observed, which is attributed to the different hardness of the bran; the hardness of the bran decreases from outer to inner layers. However, the different endosperm fractions have a similar hardness. The transition from the bran to the endosperm is reached when the degree of milling is approximately 9%.

The color of rice, an important quality parameter, is related to the degree of milling, since the distribution of the pigments is not uniform in the grain. The brightness of the raw kernels and rice flours increases according to the degree of milling, until the bran and the outer endosperm are removed. The bran and outer endosperm contain more red and yellow pigments than the middle and core endosperm. However, these pigments are uniformly distributed in the middle and core endosperm.

Most rice is consumed as a grain. However, rice kernels can be cracked in the field or during drying or milling processes. Often, these cracks lead to the breaking of the kernel, generating broken rice. Rice milling can yield from 4 to 40% broken kernels depending on the incoming rice quality and the milling equipment. Broken kernels are separated from the whole kernels by indent graders, because they tend to go mushy during cooking, thus decreasing the quality of the table rice. Broken kernels can be further separated into various sizes according to their final use (brewing, screening, flour milling). In some countries broken rice is sold as it is, but at lower price than the milled rice. Broken rice is also used for the production of beer, high fructose syrup, flour and high protein flour, starch, maltodextrins, glucose syrup, feed for livestock, spirits, or distilled liquors.

In conclusion, the milling of paddy rice produces milled rice, broken rice, rice bran and hulls and husks. Numerous products with added value have been developed from rice, such as convenience processed rice forms (parboiled, germinated, etc.), rice flour, puffed and crisped rice, breakfast cereals and snacks (Barber and Benedito, 1970; Nguyen and Tran, 2000; Wilkinson and Champagne, 2004).

Grinding

Broken kernels of rice can be ground into flour using three different methods (Yeh, 2004): (1) Wet grinding consists first in soaking the broken kernels in water. After draining, the kernels are ground in the presence of water, in order to reduce the amount of damaged starch. The excess water is removed by drying and the flour is again reground, yielding the wet rice flour. This product is used in the production of different Asian specialities such as Japanese cake, Taiwanese cake, Indian fermented foods, etc. (2) Wet grinding in the presence of 0.3–0.5% NaOH is used for the production of rice starch and rice maltodextrins and syrups. (3) Semi-dry grinding also involves soaking, draining, and grinding without using any excess of water. The semi-dry flour has similar applications to the wet rice flour. Dry grinding is also possible; in this case broken kernels are directly ground to different sizes. Dry rice flour is used for baking, baby foods, extrusion-cooked products and for the production of high-protein flour.

Rice flour properties

Rice varieties can be classified according to their original cultivation area, grain size, and amylose content. Indica rice has been grown in India, Bangladesh, Vietnam, Thailand, Pakistan, etc., while Japonica rice has been cultivated in Japan, Korea as well as northern and central regions of China. Based on the grain size, rice can be classified as long (longer than 6.6 mm), medium (between 5.5 and 6.6 mm), or short (shorter than 5.5 mm). The amylose content differs between waxy (less than 1%

amylose) and non-waxy (higher than 10% amylose) rice. Rice is mainly consumed as polished rice and, thus, primary differences among different types of rice rely on their cooking characteristics, although they also differ in their physico-chemical properties (Vasudeva et al., 2000). Rice flour can be obtained from complete grains, but it is usually produced from the kernels broken during the milling process because their cost is lower than that of the whole milled kernels. Usually rice flours have the same chemical composition as parent-milled kernels. The characteristics of the rice flours are governed by inherent cultivar's variations, environmental variation, the grinding methods, and their previous treatments.

Rice flours mainly differ in the amylose content, which determines the gelatinization temperature, and in their general pasting behavior and viscoelastic properties (Fan and Marks, 1998; Singh et al., 2000; Meadows, 2002; Saif et al., 2003; Rosell and Gómez, 2006). Analysis of the pasting behavior is a useful method to characterize the properties of the rice flour. Although the amylograph was the equipment traditionally used, in recent years it has been replaced by the rapid viscoanalyzer (RVA), since the latter allows a better understanding of the pasting properties with high precision, sensitivity, and rapidity (Meadows, 2002; Gujral et al., 2003a). The pasting properties of rice flours greatly depend on the cultivars; in fact, rice breeders frequently use the RVA as an index of rice quality. Rice flours from Bomba and Thaibonet cultivars show higher pasting temperatures, which correspond to high gelatinization temperatures and lower peak viscosities, resulting from their high amylose content. These properties are typically attributed to long grains (Rosell and Gómez, 2006; Rosell and Collar, 2007). However, Bomba has very short grains that during cooking behave as long grains. Bomba rice grains show a low viscosity breakdown during high temperature holding cycles and a marked increase in viscosity during cooling that corresponds to a tendency to retrograde. Therefore, grain length alone cannot be used to represent the pasting properties of a rice. In contrast, Bahia and Senia have higher peak viscosities and lower pasting temperatures, and both show similar behavior during heating and cooling.

Near-infrared spectroscopy is a rapid technique for determining the protein and amylose content (Miryeong et al., 2004). A new piece of equipment that has just appeared in the market is the Mixolab (developed by Chopin). The Mixolab allows the mixing and pasting properties of the flours (i.e. flour behavior under mechanical and thermal constraints) to be determined (Bonet et al., 2006; Rosell et al., 2007). From the plot obtained, it is possible to extrapolate useful information. The first part of the curve, before the heating cycle starts, allows the water absorption of the flour to be determined. The target of a torque of 1.1 Nm approximately corresponds to 500 BU obtained with the Brabender Farinograph. In the second part of the curve, similar results can be obtained as those commonly originating from the RVA. However, the Mixolab works with dough systems, whereas RVA analysis is performed on suspensions. The different slopes of the curve during the assay are related to different properties of the flour: speed of the weakening of the protein network due to heating (α); gelatinization rate (β); and enzymatic degradation speed (γ). For example, Mixolab allowed the effect of water addition on rice flour properties to be determined (Figure 4.2; Tables 4.2 and 4.3). As the water addition increased, a decrease in the dough consistency was

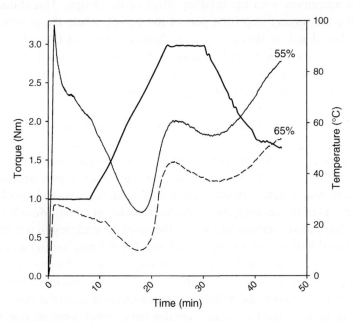

Figure 4.2 Mixolab analysis of rice dough behavior during mixing, heating, and cooling when different amounts of water are added. Temperature: thick line. Numbers are referred to the amount of water (expressed in percentage, flour basis) added for dough mixing.

Table 4.2 Effect of the addition of different amounts of water on rice flour behavior as studied by the Mixolab device

Water level (%, flour basis)	Development time (min)	Maximum consistency (Nm)	Amplitude (Nm)	Stability (min)	Minimum torque (Nm)	Peak torque (Nm)	Torque at the end of heating (Nm)	Final torque (Nm)
55.0	1.18	3.25	0.276	0.9	0.82	2.01	1.81	2.78
65.0	1.62	0.93	0.051	1.12	0.33	1.48	1.22	1.78

Table 4.3 Derived parameters obtained from the Mixolab curves of rice flour doughs with different hydrations

Water level (%, flour basis)	Derived parameters				
	Cooking stability	Setback (Nm)	α (Nm/min)	β (Nm/min)	γ (Nm/min)
55.0	0.90	0.97	−0.156	0.340	−0.018
65.0	0.82	0.56	−0.090	0.384	−0.042

detected, in agreement with the dilution effect of the dough. This difference was higher during the mixing step (first part of the curve), where the proteins play the main role when dough is affected by a mechanical constraint (Rosell *et al.*, 2007). However, during the heating and cooling cycles, the differences between samples with different water content decreased. The amount of water present in the system, although limited for starch gelatinization, was sufficient to gelatinize a large amount of starch.

In order to achieve a suitable consistency for breadmaking, rice flour doughs require very high hydration compared with wheat flour doughs. The addition of large quantities of water leads to considerable improvement of the dough behavior during mixing (i.e. higher stability). During heating–cooling cycles, higher hydrated doughs result in lower peak torque (related to the starch gelatinization) and also lower final torque at the end of the cooling, due to the starch dilution effect. More hydrated rice doughs also have lower setback (related to the amylose tendency to retrograde), due to the amylose dilution, whereas the rate of starch gelatinization (α) and enzymatic degradation speed (γ) increase because of the large amount of water available.

Since the pasting properties influence the behavior during baking, careful selection of the rice cultivars used in the grinding is recommended when the flour is destined for bakery. In general, the long-grain varieties have higher amylose content, higher gelatinization temperatures, and a greater tendency to retrograde than medium- or short-grain varieties. Rice flours can also be obtained from waxy rice varieties. These waxy varieties have an amylose content lower than 1% and low gelatinization temperatures (61–62°C). Even if their properties are not appropriated for baking, these flours can be used as minor ingredients (Bean *et al.*, 1984).

Environmental variation also plays a significant role in determining the pasting properties of rice flour. In fact, Minh-Chau-Dang and Copeland (2004) studied the effect of growth season and location on the pasting properties of three different rice cultivars (Doongara, Langi, and Kyeena). Genotype, growth season, and growth location all affected the pasting behavior of the rice flour. The amylose content of individual cultivars was significantly higher in the coolest growing season, resulting in RVA traces with lower peak viscosities and higher setbacks than samples with lower amylose contents. When the same rice cultivar was grown in different locations in the same season, no significant differences could be detected in the chemical composition of the resulting flours. However, significant differences were found in the pasting behavior, indicating that the environment influences the pasting behavior of rice flour. In conclusion, it was found that the pasting behavior of rice flour is related to genotype and influenced by environmental factors that result in minor changes in the grains that are not detected by chemical analyses.

The physical properties of rice flour are also affected by the time elapsed between harvest and milling, as well as by the temperature used in the drying process prior to storage. The influence of rice moisture content at harvesting on the rice flour properties was investigated for long- and medium-grain rice (Linfeng-Wang *et al.*, 2004). Peak viscosity of the flour, an indicator of rice functionality and performance, increased as the rice harvest moisture content decreased, although the rate of increase was influenced by the rice cultivar and growing location. Moreover, duration and temperature of storage were found to significantly affect the enthalpies and temperature of gelatinization and retrogradation of rice flour (Fan and Marks, 1999). Recently, Zhou *et al.* (2003) observed that the time and temperature of rice storage also influence the RVA pasting curves of rice flours from different cultivars. In this work, a change in the protein profile was observed, with a positive correlation between increase in the amount of high molecular weight peptides and storage time. In fact, the changes in the structure and properties of oryzenin, rather than the starch, are responsible for the modification of the rice physical properties associated with storage (Teo *et al.*, 2000; Patindol *et al.*, 2003).

The conditions of the milling process are very important, since the number of drying steps and temperature during tempering affect the rough rice quality (Correa *et al.*, 2007). Two- or three-step drying reduces the percentage of fissured kernels compared with one-step drying (Aquerreta *et al.*, 2007). The tempering at high temperature (60°C) also reduced the percentage of fissured kernels independently of the number of drying steps.

The grinding method employed will also affect the functional properties of the rice flour. The method and type of mill determine the particle size of the rice flour and also the amount of starch damage. Nishita and Bean (1982) reported a comparative study about the properties of rice flours obtained with different mills. Roller mills led to rice flours with medium granulometry that showed good performance in the bakery. In contrast, the burr mills yielded excessive coarse flour that produced low-quality breads. The use of hammer mills led to finer particles with high levels of damaged starch that are not adequate for breadmaking, but could be used in cake production. With regard to pasting properties, the flours with greater particle size showed lower peak viscosity and final viscosity at 50°C (low tendency to retrograde), while the flours with medium or small particle size did not show any significant changes in their properties.

The enthalpy values obtained by differential scanning calorimetry (DSC) are indicative of the degree of the starch damage occurring during grinding. Lower enthalpy values are related to higher damage of the starch. Thermally stimulated luminescence (TSL) can also be used as a thermal analysis technique, since the observed TSL is due to permanent phase change in the specimens (Murthy *et al.*, 2007). Irradiated rice flour can be detected by this technique since the generation of electron/hole traps results in an increase in TSL peak intensity.

Another luminescence technique is photoluminescence (Katsumata *et al.*, 2005). Peak intensity of photoluminescence varies depending on the variety and source of the rice. Two-dimensional (2D) images of photoluminescence allow blended rice from different species, contamination, and foreign objects to be detected, making this technique potentially useful for non-destructive and quick evaluation of rice products for quality control purposes (Katsumata *et al.*, 2005).

The rheological properties of the flour are also influenced by the temperature, moisture, and lipid content. Dautant *et al.* (2007) found that at constant moisture content, the viscosity decreases when the temperature increases and also when the shear rate increases, regardless of the temperature applied. This decrease in viscosity with the increase in the shear rate demonstrates the pseudoplastic nature of the material. An increase in the moisture or in the lipid content (up to 5%) also results in a decrease in viscosity. During extrusion processes rice flour is cooked. Therefore, it is important to take into account the viscous behavior of the rice flour in order to establish the best processing conditions to be applied in the food industry, since this will affect the quality of the end-products.

Rice flours are usually obtained from polished or milled kernels, although sometimes brown rice is employed for grinding. Flours obtained from brown kernels have a 13–17°C higher temperature of gelatinization and around 40% greater gelatinization enthalpy than their milled counterparts (Normand and Marshall, 1989). Flours from brown rice contain high amount of fiber and vitamins, predominant in the outer layers of the kernel. These compounds confer special organoleptic properties (color, texture and taste) on the baked products. However, brown rice flours have a very short shelf-life. This is due to the presence of active lipase and lipooxygenase, and thus to the release of free fatty acids, which start going rancid, imparting a bitter taste to the products. The stability of these flours can be increased by reducing the temperature and humidity during storage, or using inert atmospheres; however, these modifications affect the cost of the product. As an alternative, brown rice flour can be obtained by adding milled bran at appropriate levels to already ground rice. In this case, the bran can be chemically or physically treated previously to ensure its stability and to extend the shelf-life (Champagne *et al.*, 1991; Champagne and Grimm, 1995). A different approach is to remove the bran fat. Rice bran, defatted or not, can also be used as a source of fiber and vitamins in wheat-based products (Lima *et al.*, 2002).

Rice flours can also be obtained from cooked rice, a process which modifies the rheological behavior in steady and dynamic shears (Chun and Yoo, 2004). Processes such as parboiling the kernels before milling can modify the physico-chemical characteristics of rice. In the parboiling process, paddy rice is soaked and steamed under pressure in order to gelatinize the starch within the kernel. After cooling, slow drying reduces the formation of cracks. The conversion of the starch from a crystalline to an amorphous state favors the migration of nutrients from the bran layer to the starchy endosperm. Parboiled rice thus has higher levels of nutrients (vitamins and minerals) and different sensory properties. Flours obtained from parboiled rice produce soft and sticky doughs, due to the low water retention capacity and the high susceptibility to amylase attack. Therefore, these flours are not suitable for breadmaking, but they can be used in small concentrations for cake production, where the short process time reduces the activity of the amylases. Pre-gelatinized rice flour can be obtained by extrusion, puffing, or roasting. All these treatments negatively affect the rheological properties of the rice flour, leading to sticky doughs and low-volume breads (Bean and Nishita, 1985).

Production and characterization of gluten-free cereal products based on rice

Rice is mainly consumed as milled rice, although fiber and minerals are lost during the milling process. As a food ingredient, rice confers creaminess, crunchiness, and firmness to the final product. Beside the common use as table rice, rice can be used for the production of beer, baby foods, breakfast cereals, snacks, confections, desserts, as well as bakery products. The increased use of rice in food processing is the result of increasing consumer demands for healthier and more convenient products, as well as a growing interest in ethnic products. Moreover, rice-based products represent the solution for consumers with allergenic problems. In addition, husks, hulls, and bran are used as energy sources, fillers for polymeric composite, and raw materials for the production of nutraceuticals and protein concentrates.

Dry rice breakfast cereals include rice flakes, oven, gun or extruder-puffed rice, shredded-rice cereals, and multigrain cereals. These products are prepared by pressure-cooking in the presence of sugar, salt, flavorings, and sufficient water. Rice flakes are prepared in a similar way to wheat and corn flakes: the rice is cooked and coated with nutritious ingredients (skimmed milk) and then partially dried, tempered and passed through flaking rolls before toasting in an oven (Wilkinson and Champagne, 2004).

Rice snacks include granola, breakfast, and energy bars (Juliano and Hicks, 1996). Some snacks are designed as functional foods (e.g., they can help to reduce cholesterol levels). A number of these products are aimed at children, women, and other specific groups. Rice flour is used in many Asian snacks, since it is the most cultivated cereal in these countries. Rice noodles are obtained by extrusion and rice flour with a high amylose content is usually used. The process consists of partial cooking of the dough, kneading and forming, final cooking, and drying. Rice noodles are consumed as main foods, soups, or snacks. According to the production process, cakes can be divided into pastry, unleavened, dry or fermented cakes (Rosell and Gómez, 2006). Finally, crackers can be obtained using non-waxy rice (i.e. senbei) or waxy rice (i.e. arare). Rice flour is extensively used for the production of infant food formulas due to its digestibility and hypoallergenic properties. A partial acid or enzymatic (using starch-hydrolyzing enzymes) hydrolysis of the rice flour is applied in order to increase the concentration of free sugars, contributing to the sweet taste and consistency (Cantoni, 1967).

Rice flour is increasingly used in baking as a substitute for wheat for the preparation of products intended for wheat-intolerant people or those with celiac disease. It is the most suitable cereal grain flour for the production of gluten-free products due to its bland taste, white color, digestibility, and hypoallergenic properties (Neumann and Bruemmer, 1997). In addition, other attributes such as the low content of protein and sodium, the low levels of prolamins and the presence of easily digested carbohydrates make rice the best cereal for patients suffering from allergies. However, in spite of the numerous advantages of rice flour, rice proteins have relatively poor functional properties for food processing. Due to their hydrophobic nature, rice proteins are insoluble and unable to form the viscoelastic dough necessary to hold the carbon dioxide produced during proofing of yeast-leavened bread-like products. The low content of prolamins in rice flours results in the lack of formation of a protein network when rice flour is kneaded with water. As a consequence, the carbon dioxide produced during fermentation cannot be retained, leading to a product with low specific volume and a very compact crumb (Plate 4.1) which does not resemble the soft and open structure of common wheat bread (He and Hoseney, 1991).

To improve the quality of bread, structuring agents, such as xanthan gum and carboxymethylcellulose (CMC), are commonly added to gluten-free bread formulations (Kulp *et al.*, 1974). Recently, pectin, CMC, agarose, xanthan, or oat β-glucans were used in gluten-free formulations based on rice flour, corn starch, and sodium caseinate (Lazaridou *et al.*, 2007). With the exception of xanthan gum, the presence of these hydrocolloids resulted in breads with higher volume. Finally, breads supplemented with 2% CMC received the best score on sensory testing. Among the cellulose derivatives, hydroxypropylmethylcellulose (HPMC) seems to be a suitable gluten substitute in rice bread formula due to its gas retention capacity and its properties as a crumb-structuring agent (Nishita *et al.*, 1976; Ylimaki *et al.*, 1988; Gujral *et al.*, 2003a). Upon addition of HPMC, the consistency and rheological properties of rice doughs closely resemble those of wheat doughs (Plate 4.2) (Sivaramakrishnan *et al.*, 2004). The presence of 4% (flour basis) HPMC leads to a significant increase in bread volume and loaf structure (Plate 4.3). Other gums, such as locust bean gum, guar gum, carrageenan, xanthan gum, and agar, have been tested as gluten replacers in rice bread (Kang *et al.*, 1997; Cato *et al.*, 2004; Lazaridou *et al.*, 2007). In general, the volume of rice breads increases with the addition of hydrocolloids except for xanthan; however, increasing the level of hydrocolloids from 1% to 2% results in a decrease in loaf volume, except for pectin. High values of crumb porosity are obtained when 1% CMC and β-glucans or 2% pectin are added, whereas high crumb elasticity is induced by CMC or pectin addition.

Addition of hydrocolloids has allowed the production of rice breads with a loaf-specific volume comparable to that of wheat bread, however sensory appearance and crumb texture are still poor. Improvement of the crumb texture has recently been achieved with the addition of vegetable seed oil (Gujral *et al.*, 2003a). In contrast to wheat flour, rice flour is not responsive to the presence of dough conditioners or enzymes (Nishita *et al.*, 1976), probably due to the hydrophobic nature of rice proteins. However, recent studies have shown the usefulness of some enzymes in the making of rice-based products. For example, addition of cyclodextrin glycosyl transferase (CGTase) led to rice loaves with very soft crumbs (Gujral *et al.*, 2003a, 2003b). This enzyme acts as an α-amylase that can use hydrolysis products to produce cyclodextrins that can form complexes with a variety of solid, liquid, and gaseous compounds. The improving effect of CGTase in rice breadmaking was due to the formation of complexes between lipids and proteins with the cyclodextrins. The addition of CGTase also helped to extend the shelf-life of rice bread, acting as anti-staling agent through its hydrolyzing and cyclizing activity (Gujral *et al.*, 2003b). Another enzyme with anti-staling activity is α-amylase. This is an endo-enzyme that randomly hydrolyzes the α-1,4 glucosidic linkages in polysaccharides, resulting in short chains that can be fermented by yeast. α-Amylase of intermediate thermostability has been shown to improve the shelf-life of gluten-free bread, increasing the crumb softness and elasticity of bread (Novozymes, 2004). Laccase (*p*-diphenol oxygen oxidoreductase) is an oxidative enzyme that catalyzes the oxidative gelation of feruloylated arabinoxylans by dimerization of their ferulic esters (Figueroa-Espinoza *et al.*, 1998; Labat *et al.*, 2001). The addition of small levels of laccase (1.5 U/g flour) improved the rice bread specific volume, but increasing the enzyme levels showed detrimental effects, particularly regarding crumb hardness (Gujral and Rosell, unpublished results).

Other enzymes with good potential in rice bread formulation are glucose oxidase and transglutaminase (Gujral and Rosell, 2004a, 2004b). These enzymes promote the formation of a protein network by catalyzing inter- and intra-molecular cross-links between the rice proteins. However, the protein network catalyzed by enzymatic treatment of rice proteins does not completely meet the gluten functionality, and thus a reduced amount of hydrocolloids is still needed (Gujral and Rosell, 2004a, 2004b). When considering transglutaminase applications, addition of an external source of proteins has been suggested in order to increase the amount of lysine residues, which are the limiting factor of the cross-linking reaction (Moore et al., 2006). Exogenous protein sources such as soybean flour, skim milk powder, or egg powder were therefore added (12.5% composite flour basis) to a gluten-free bread formulation (containing rice flour, potato starch, corn flour, xanthan gum) in the presence of increasing levels of transglutaminase. Confocal laser scanning micrographs of the bread crumbs confirmed the cross-linking of dairy proteins, although high amounts of transglutaminase (10 U/g of protein) were needed, probably because of the thermodynamic incompatibility between polar and apolar surfaces of milk proteins (Moore et al., 2006). The compatibility between rice flour proteins and different protein isolates (pea, soybean, egg albumen, and whey proteins) in the cross-linking reaction catalyzed by transglutaminase has been evaluated studying the rice dough behavior subjected to small deformations (Marco and Rosell, 2007). The elastic modulus recorded in the oscillatory tests was significantly affected by both the protein isolates and the transglutaminase. The extent of the effect was dependent on the protein source; pea and soybean proteins increased the elastic modulus, whereas egg albumen and whey protein decreased it.

A deeper evaluation of the cross-linking in the presence of soybean proteins, by using different electrophoretic techniques, indicated that the main protein fractions involved in these interactions were both β-conglycinin and glycinin of soybean as well as the glutelins of rice flour, although albumins and globulins were also cross-linked (Marco, Pérez, León, Rosell, results unpublished). The interaction between rice and soybean proteins was intensified by the formation of new intermolecular covalent bonds catalyzed by transglutaminase, and also by the indirect formation of disulfide bonds between proteins. Concerning the pea proteins, main protein fractions involved in the interactions were the albumins and globulins from the pea protein isolate and rice flour, but also the glutelins were cross-linked (Marco et al., 2007). In conclusion, the studies carried out with different protein sources have shown that their combination with network-forming enzymes has great potential for enhancing the structure of gluten-free products. Recently, chemical modifications of rice flour allowed the production of rice bread with similar texture characteristics to those of wheat bread (Nabeshima and El-Dash, 2004).

Some bread specialties have been adapted to obtain gluten-free products addressed to people with gluten intolerance. This is the case of chapatti, an unleavened bread made from whole wheat in India. The use of different hydrocolloids (HPMC, guar gum, xanthan gum, or locust bean gum) and α-amylase in the formulation of rice flour chapattis improved the texture by keeping the extensibility during storage (Gujral et al., 2004c). In addition, hydrocolloids and α-amylase delayed the amylopectin retrogradation, keeping the freshness of the chapattis during longer period.

A different approach to the production of gluten-free bread is to use rice flour blended with other flours and different starches (Gallagher et al., 2004). Complex formulations including corn starch, brown rice, soy and buckwheat flour have been proposed (Moore et al., 2004). Using these recipes, breads were brittle after 2 days of storage, although this effect was reduced when dairy products such as skimmed milk powder were included in the formulation. In addition, a combination of rice flour (45%) with corn (35%) and cassava (20%) starches gave a good gluten-free bread with uniform and well-distributed cells over the crumb as well as a pleasant flavor and appearance (Lopez et al., 2004). Gluten-free breads of good quality were also obtained using small amounts of rice flour (about 17.2%) and using corn starch (74.2%) and cassava starch (8.6%) (Sanchez et al., 2002). Finally, blends of buckwheat and rice flours in the presence of hydrogenated vegetable fat also have the potential to give gluten-free breads with good sensory attributes (Moreira et al., 2004).

Future trends

Rice is an important source of energy, providing 26% of the total energy intake in developing countries, although it provides only 4% of the total energy intake in the developed world. In developing countries, rice supplies 20% of the dietary protein intake, but because of its incomplete amino acid profile and the limited levels of micronutrients (especially in milled rice), the use of rice as staple food may lead to malnutrition. Patients with celiac disease already tend to have malnutrition, since the immunological reaction induced by gluten ingestion produces damages the mucous membrane of the small intestine, reducing its nutrient absorption capacity. In addition, most gluten-free products are low in micronutrients, which increases the risk of deficiencies. In order to improve the nutritional quality of gluten-free products based on rice, other protein sources can be added. Dairy and soybean proteins are the most used. Legume proteins are a good supplement for cereal-based foods, since both legume and cereal proteins are complementary in essential amino acids. Nowadays, different techniques for rice fortification have been developed in order to add essential vitamins and minerals to the grain (Nunes et al., 1991; Hoffpauer and Wright, 1994; Rosell, 2004). Alternatively, specific minerals can be added to the products during the manufacturing process. For example, Kiskini et al. (2007) obtained gluten-free bread fortified with iron (incorporated as ferric pyrophosphate) that presented good sensory and nutritional characteristics. However, these compounds might affect the sensory quality of the products, and therefore particular attention has to be given to the form and amount of added compounds.

Rice is mainly consumed as milled rice, although brown rice has better nutritional value. Brown rice is obtained after de-hulling, and the brown color is due to the presence of bran layers, which are rich in minerals and vitamins. Brown rice contains more nutritional components than the ordinary milled rice grains (e.g. dietary fibers, phytic acids, E and B vitamins, and γ-aminobutyric acid (GABA)). All these compounds are present in the bran layers and germ that are removed during polishing or milling (Champagne *et al.*, 1991, 2004; Champagne and Grimm, 1995). Despite the nutritional benefits linked to its consumption, brown rice is not considered suitable for table rice because it has to be cooked in a pressure rice cooker, and also because of its dark appearance and hard texture. Moreover, when the husk is removed from rice, the bran layer starts going rancid, contributing to the bitter taste of the brown rice. This is why brown rice is mainly used for fermentation purposes, or in materials for food processing.

The use of germination in grains started some decades ago, mainly applied to wheat and soybean (Finney, 1978; Tkachuk, 1979). Germinated brown rice arose following research into the development of new value-added products from rice. In 1994, Saikusa *et al.* found that GABA levels increased significantly when brown rice was soaked in water at 40°C for 8–24 hours. An increase in dietary GABA intake has been found to lower blood pressure, improve sleep and the autonomic disorder associated with the menopausal or presenile period, and can even suppress liver damage (Okada *et al.*, 2000; Tadashi *et al.*, 2000; Jeon *et al.*, 2003). In Japan, germinated brown rice was launched on the market in 1995. Since then, it has increased in popularity within the Japanese population, and numerous industries have emerged in Japan related to its production. During the last decade, 49 items related to germinated brown rice have been patented. The basic procedure for obtaining pre-germinated brown rice consists in the selection of good brown rice, which then is soaked for around 20 hours at 30–40°C. This product is washed slightly before cooking, and is marketed either dry or wet (i.e. 15 or 30% moisture, respectively). During the germination process, saccharification softens the endosperm and dormant enzymes are activated, leading to an increase in the amount of digestible compounds (Manna *et al.*, 1995). In addition, the mineral content changes, resulting in an increase of GABA, free amino acids, dietary fiber, inositols, ferulic acid, phytic acid, tocotrienols, magnesium, potassium, zinc, γ-oryzanol, and prolylendopeptidase inhibitor (Kayahara and Tsukahara, 2000; Ohisa *et al.*, 2003; Ohtsubo *et al.*, 2005). Germinated brown rice can be cooked in an ordinary rice cooker, giving a soft product with easier chewiness. Moreover, it can be used as a raw material in the production of various foods, including germinated brown rice balls, soup, bread, doughnuts, cookies, and rice burgers (Ito and Ishikawa, 2004).

Sources of further information and advice

Champagne, E. T. ed. (2004). *Rice: Chemistry and Technology*. St. Paul, MN: American Association of Cereal Chemists Inc.

Rosell, C. M. (2007). Enzymatic manipulation of gluten-free bread. In: Gallagher, E. ed. *Gluten-free Food Science and Technology*. Oxford: Blackwell Publishing.

Rosell, C. M. and Collar, C. (2007). Rice based products. In: Hui, Y. H. ed. *Handbook of Food Products Manufacturing*. Weinheim: Wiley-VCH.

Rosell, C. M. and Gómez, M. (2006). Rice. In: Hui, Y. H. ed. *Bakery Products: Science and Technology*. Ames, Iowa: Blackwell Publishing, pp. 123–133.

Wrigley, C., Corke, H., and Walker, C. eds. (2004). *Encyclopedia of Grains Science*. Oxford: Elsevier Science.

References

Aquerreta, J., Iguaz, A., Arroqui, C., and Vírseda, P. (2007). Effect of high temperature intermittent drying and tempering on rough rice quality. *J. Food Eng.* 80, 611–618.

Barber, S. and Benedito de Barber, C. (1970). Posibilidades de industrializacion del arroz en Espana. IV. La calidad del arroz para la fabricacion de productos preparados. *Rev. Agroquim. Tecnol. Aliment.* 10, 18–26.

Bean, M. M. and Nishita, K. D. (1985). Rice flours for baking. In: Juliano, B. O. ed. *Rice: Chemistry and Technology*. St. Paul. MN: American Association of Cereal Chemists.

Bean, M. M., Esser, C. A., and Nishita, K. D. (1984). Some physicochemical and food application characteristics of California waxy rice varieties. *Cereal Chem.* 61, 475–480.

Bond, N. (2004). Rice milling. In: Champagne, E. T., ed. *Rice: Chemistry and Technology*, 3rd edn. St. Paul, MN: American Association of Cereal Chemists, pp. 283–300.

Bonet, A., Blaszczak, W., and Rosell, C. M. (2006). Formation of homopolymers and heteropolymers between wheat flour and several protein sources by transglutaminase catalyzed crosslinking. *Cereal Chem.* 83, 655–662.

Cantoni, G., inventor and assignee (1967). Nov 22. Foodstuffs derived from rice. GB patent 1,092,245.

Cato, L., Gan, J. J., Rafael, L. G. B., and Small, D. M. (2004). Gluten free breads using rice flour and hydrocolloid gums. *Food Aust.* 56, 75–78.

Champagne, E. T. and Grimm, C. C. (1995). Stabilization of brown rice products using ethanol vapors as an antioxidant delivery system. *Cereal Chem.* 72, 255–258.

Champagne, E. T., Hron, R. J., and Abraham, G. (1991). Stabilizing brown rice products by aqueous ethanol extraction. *Cereal Chem.* 68, 267–271.

Champagne, E. T., Wood, D. F., Juliano, B. O., and Bechtel, D. B. (2004). The rice grain and its gross composition. In: Champagne, E. T., ed. *Rice: Chemistry and Technology*, 3rd edn. St. Paul, MN: American Association of Cereal Chemists, pp. 77–107.

Chun, S. Y. and Yoo, B. (2004). Rheological behavior of cooked rice flour dispersions in steady and dynamic shear. *J. Food Eng.* 65, 363–370.

Correa, P. C., da Silva, F. S., Jaren, C., Afonso, P. C., and Arana, I. (2007). Physical and mechanical properties in rice processing. *J. Food Eng.* 79, 137–142.

Dautant, F. J., Simancas, K., Sandoval, A. J., and Müller, A. J. (2007). Effect of temperature, moisture and lipid content on the rheological properties of rice flour. *J. Food Eng.* 78, 1159–1166.

FAOSTAT (2007). http://faostat.fao.org/site/340/default.aspx. This website provides statistics on commodities, food supply, food balance sheets, food aid, population, and the Codex Alimentarius

Fan, J. and Marks, B. P. (1998). Retrogradation kinetics of rice flours as influenced by cultivar. *Cereal Chem.* 75, 153–155.

Fan, J. and Marks, B. P. (1999). Effects of rough rice storage conditions on gelatinization and retrogradation properties of rice flours. *Cereal Chem.* 76, 894–897.

Figueroa-Espinoza, M. C., Morel, M. H., and Rouau, X. (1998). Effect of lysine, tyrosine, cysteine, and glutathione on the oxidative cross-linking of feruloylated arabinoxylans by a fungal laccase. *J. Agric. Food Chem.* 46, 2583–2589.

Finney, P. L. (1978). Potential for the use of germinated wheat and soybeans to enhance human nutrition. *Adv. Exp. Med. Biol.* 105. *Nutr. Improv. Food Feed Proteins*, 681–701.

Fitzgerald, M. (2004). Starch. In: Champagne, E. T. ed. *Rice: Chemistry and Technology*, 3rd edn. St. Paul, MN: American Association of Cereal Chemists, pp. 109–141.

Gallagher, E., Gormley, T. R., and Arendt, E. K. (2004). Recent advances in the formulation of gluten free cereal-based products. *Trends Food Sci. Technol.* 15, 143–152.

Gujral, H. S. and Rosell, C. M. (2004a). Improvement of the breadmaking quality of rice flour by glucose oxidase. *Food Res. Int.* 37, 75–81.

Gujral, H. S. and Rosell, C. M. (2004b). Functionality of rice flour with a microbial transglutaminase. *J. Cereal Sci.* 39, 225–230.

Gujral, H. S., Guardiola, I., Carbonell, J. V., and Rosell, C. M. (2003a). Effect of cyclodextrin glycosyl transferase on dough rheology and bread quality from rice flour. *J. Agric. Food Chem.* 51, 3814–3818.

Gujral, H. S., Haros, M., and Rosell, C. M. (2003b). Starch hydrolyzing enzymes for retarding the staling of rice bread. *Cereal Chem.* 80, 750–754.

Gujral, H. S., Haros, M., and Rosell, C. M. (2004c). Improving the texture and delaying staling in rice flour chapati with hydrocolloids and α-amylase. *J. Food Eng.* 65, 89–94.

Hamaker, B. R. (1994). The influence of rice proteins in rice quality. In: Marshall, W. E. and Wadsworth, J. I. eds. *Rice Science and Technology*. New York: Marcel Dekker, pp. 177–194.

He, H. and Hoseney, R. C. (1991). Gas retention of different cereal flours. *Cereal Chem.*, 68, 334–336.

Hoffpauer, D. W. and Wright III, S. L. (1994). Iron enrichment of rice. In: Marshall, W. E. and Wadsworth, J. I. eds. *Rice Science and Technology*. New York: Marcel Dekker, pp. 195–204.

Infocomm (2007). Rice information (http://r0.unctad.org/infocomm/).

Ito, S. and Ishikawa, Y. (2004). Marketing of value-added rice products in Japan: germinated brown rice and rice bread. *FAO Rice Conference* 04/CRS.7. http://www.hatsuga.com/DOMER/english/en/GBRRB.html.

Jeon, T. I., Hwang, S. G., Lim, B. O., and Park, D. K. (2003). Extracts of *Phellinus linteus* grown on germinated brown rice suppress liver damage induced by carbon tetrachloride in rats. *Biotechnol. Lett.* 25, 2093–2096.

Juliano, B. O. and Hicks, P. A. (1996). Rice functional properties and rice food products. *Food Rev. Int.* 12, 71–103.

Kang, M. Y., Choi, Y. H., and Choi, H. C. (1997). Effects of gums, fats and glutens adding on processing and quality of milled rice bread. *Korean J. Food Sci. and Technol.* 29, 700–704.

Katsumata, T., Suzuki, T., Aizawa, H., and Matashige, E. (2005). Two dimensional imaging of photoluminescence from rice for quick and non-destructive evaluation. In: Voet, M., Willsch, R., Ecke, W., Jones, J., and Culshaw, B. eds. *17th International Conference on Optical Fiber Sensors. Proceedings of the SPIE*, Vol. 5855, pp. 423–426.

Kayahara, H. and Tsukahara, K. (2000). Flavor, health and nutritional quality of pregerminated brown rice. *2000 International Chemical Congress of Pacific Basin Societies in Hawaii*.

Kiskini, A., Argiri, K., Kalogeropoulos, M. *et al.* (2007). Sensory characteristics and iron dialyzability of gluten-free bread fortified with iron. *Food Chem.* 102, 309–316.

Kulp, K., Hepburn, F. N., and Lehmann, T. A. (1974). Preparation of bread without gluten. *Bakers Digest* 48, 34–37, 58.

Labat, E., Morel, M. H., and Rouau, X. (2001). Effect of laccase and manganese peroxidase on wheat gluten and pentosans during mixing. *Food Hydrocolloids* 15, 47–52.

Lamberts, L., De Bie, E., Vandeputte, G. E. *et al.* (2007). Effect of milling on colour and nutritional properties of rice. *Food Chem.* 100, 1496–1503.

Lazaridou, A., Duta, D., Papageorgiou, M., Belc, N., and Biliaderis, C. G. (2007). Effects of hydrocolloids on dough rheology and bread quality parameters in gluten-free formulations. *J. Food Eng.* 79, 1033–1047.

Lima, I., Guraya, H., and Champagne, E. (2002). The functional effectiveness of reprocessed rice bran as an ingredient in bakery products. *Nahrung* 46, 112–117.

Linfeng-Wang, Siebenmorgen, T. J., Matsler, A. D., and Bautista, R. C. (2004). Effects of rough rice moisture content at harvest on peak viscosity. *Cereal Chem.* 81, 389–391.

Lopez, A. C. B., Pereira, A. J. G., and Junqueira, R. G. (2004). Flour mixture of rice flour, corn and cassava starch in the production of gluten free white bread. *Brazilian Arch. Biol. Technol.* 47, 63–70.

Manna, K. M., Naing, K. M., and Pe, H. (1995). Amylase activity of some roots and sprouted cereals and beans. *Food Nutr. Bull.* 16, 1–4.

Marco, C. and Rosell, C. M. (2007). Modification of rice proteins functionality by crosslinking with different protein isolates. *J. Food Eng.* DOI:10.1016/j.jfoodeng.2007.05.003.

Marco, C., Pérez, G., Ribotta, P., and Rosell, C. M. (2007). Effect of microbial transglutaminase on the protein fractions of rice, pea and their blends. *J. Sci. Food Agric.* (in press).

Meadows, F. (2002). Pasting process in rice flour using Rapid Visco Analyser curves and first derivatives. *Cereal Chem.* 79, 559–562.

Minh-Chau-Dang, J. and Copeland, L. (2004). Genotype and environmental influences on pasting properties of rice flour. *Cereal Chem.* 81, 486–489.

Miryeong, S., Barton, F. E., McClung, A. M., and Champagne, E. T. (2004). Near-infrared spectroscopy for determination of protein and amylose in rice flour through use of derivatives. *Cereal Chem.* 81, 341–344.

Moore, M. M., Schober, T. J., Dockery, P., and Arendt, E. K. (2004). Textural comparisons of gluten-free and wheat-based doughs, batters, and breads. *Cereal Chem.* 81, 567–575.

Moore, M. M., Heinbockel, M., Dockery, P., Ulmer, H. M., and Arendt, E. K. (2006). Network formation in gluten-free bread with application of transglutaminase. *Cereal Chem.* 83, 28–36.

Moreira, R., Severo-Da-Rosa, C., and Miranda, M. Z. (2004). Elaboration of bread without gluten for the celiac disease carriers. *Alimentaria* 354, 91–94.

Murthy, K. V. R., Rey, L., and Belon, P. (2007). Photoluminescence and thermally stimulated luminescence characteristics of rice flour. *J. Luminesc.* 122–123, 279–283.

Nabeshima, E. H. and El-Dash, A. A. (2004). Chemical modification of rice flour as alternative for utilization of rice processing by-products. *Bol. Centr. Pesquisa Process. Aliment.* 22, 107–120.

Neumann, H. and Bruemmer, J. M. (1997). Investigations with the production of gluten free bread and roll specialities. *Getreide Mehl Brot* 51, 50–55.

Nguyen, V. N. and Tran, D. V. (2000). Rice and life. In: *FAO Rice Information*, Vol. 2. http://www.fao.org/WAICENT/FAOINFO/AGRICULT/AGPC/doc/riceinfo/Riceinfo.htm.

Nishita, K. and Bean, M. M. (1982). Grinding methods: Their impact on rice flour properties. *Cereal Chem.* 59, 46–49.

Nishita, K. D., Roberts, R. L., and Bean, M. M. (1976). Development of a yeast leavened rice bread formula. *Cereal Chem.* 53, 626–635.

Normand, F. L. and Marshall, W. E. (1989). Differential scanning calorimetry of whole grain milled rice and milled rice flour. *Cereal Chem.* 66, 317–320.

Novozymes (2004). Novamyl in gluten-free bread. Information sheet. Cereal Food 2004–43278–01.

Nunes, G. S., Gomes, J. C., Cruz, R., and Jordao, C. P. (1991). Mineral enrichment of rice during hidrotermal processing. *Arch. Biol. Technol.* 34, 571–582.

Ohisa, N., Ohno, T., and Mori, K. (2003). Free amino acid and gamma-aminobutyric acid contents of germinated rice. *J. Jpn. Soc. Food Sci. Technol. Nippon* 50, 316–318.

Ohtsubo, K., Suzuki, K., Yasui, Y., and Kasumi, T. (2005). Bio-functional components in the processed pre-germinated brown rice by a twin-screw extruder. *J. Food Composition Anal.* 18, 303–316.

Okada, T., Sugishita, T., Murakami, T. *et al.* (2000). Effect of the deffated rice germ enriched with GABA for sleeplessness, depression, autonomic disorder by oral administration. *Nippon Shokuhin Kagaku Kaishi* 47, 596–603.

Patindol, J., Wang, Y. J., Siebenmorgen, T., and Jane, J. L. (2003). Properties of flours and starches as affected by rough rice drying regime. *Cereal Chem.* 80, 30–34.

Rosell, C. M. (2004). Fortification of grain-based foods. In: Wrigley, C., Corke, H., and Walker, C.E. eds. *Encyclopedia of Grain Science*. Oxford: Elsevier Academic Press, pp. 399–405.

Rosell, C. M. and Collar, C. (2007). Rice based products. In: Hui, Y. H. ed. *Handbook of Food Products Manufacturing*. Weinheim: Wiley-VCH.

Rosell, C. M. and Gómez, M. (2006). Rice. In: Hui, Y. H., ed. *Bakery Products: Science and Technology*. Ames, Iowa: Blackwell Publishing, pp. 123–134.

Rosell, C. M., Collar, C., and Haros, M. (2007). Assessment of hydrocolloid effects on the thermo-mechanical properties of wheat using the Mixolab. *Food Hydrocolloids* 21, 452–462.

Saif, S. M. H., Lan, Y., and Sweat, V. E. (2003). Gelatinization properties of rice flour. *Int. J. Food Properties* 6, 531–542.

Sanchez, H. D., Osella, C. A., and De La Torre, M. A. (2002). Optimization of gluten free bread prepared from corn starch, rice flour and cassava starch. *J. Food Sci.* 67, 416–419.

Singh, V., Okadome, H., Toyoshima, H., Isobe, S., and Ohtsubo, K. (2000). Thermal and physicochemical properties of rice grain, flour and starch. *J. Agric. Food Chem.*, 48, 2639–2647.

Sivaramakrishnan, H. P., Senge, B., and Chattopadhyay, P. K. (2004). Rheological properties of rice dough for making rice bread. *J. Food Eng.* 62, 37–45.

Sodchit, C., Kongbangkerd, T., and Weeragul, K. (2003). Development of premix for rice flour bread using guar gum as a binder. *Food* 33, 222–232.

Tadashi, O., Sugishita, T., Murakami, T. *et al.* (2000). Effect of the deffated rice germ enriched with GABA for sleeplessness, depression, autonomic disorder by oral administration. *Nippon Shokukin Kagaku Kaishi* 47, 596–603.

Teo, C. H., Abd-Karim, A., Cheah, P. B., Norziah, M. H., and Seow, C. C. (2000). On the roles of protein and starch in the aging of non-waxy rice flour. *Food Chem.* 69, 229–236.

Tkachuk, R. (1979). Free amino acids in germinated wheat. *J. Sci. Food Agric.* 30, 53–58.

Vasudeva, S., Okadome, H., Toyoshima, H., Isobe, S., and Ohtsubo, K. (2000). Thermal and physicochemical properties of rice grain, flour and starch. *J. Agric. Food Chem.* 48, 2639–2647.

Watkins, T. R., Geller, M., Kooyenga, D. K., and Bierenbaum, M. L. (1990). Hypocholesterolemic and antioxidant effect of rice bran oil non-saponifiables hypercholesterolemic subjects. *Environ. Nutr. Interact.* 3, 115–122.

Wilkinson, H. C. and Champagne, E. T. (2004). Value-added rice products in today's market. *Cereal Foods World* 49, 134–138.

Yeh, A. I. (2004). Preparation and applications of rice flour. In: Champagne, E. T. ed. *Rice: Chemistry and Technology*, 3rd edn. St. Paul, MN: American Association of Cereal Chemists, pp. 495–540.

Ylimaki, G., Hawrysh, Z. J., Hardin, R. T., and Thomson, A. B. R. (1988). Application of response surface methodology to the development of rice flour yeast breads: Objective measurements. *J. Food Sci.* 53, 1800–1805.

Zhou, Z., Robards, K., Helliwell, S., and Blanchard, C. (2003). Effect of rice storage on pasting properties of rice flour. *Food Res. Int.* 36, 625–634.

Sorghum and maize

5

T. J. Schober and S. R. Bean

Introduction

Sorghum (*Sorghum bicolor* L. Moench) and maize (*Zea mays*) are closely related members of the subfamily Panicoideae in the family Gramineae. Sorghum originated in Central Africa with various hypotheses placing the domestication of sorghum sometime between 4500 and 1000 BC, after which it spread to Asia and India (Kimber, 2000). Sorghum is grown throughout the world, with the majority (~55%) produced in Asia and Africa. The United States produces approximately 30% of the world production, with the majority of the remainder being produced in South America (Smith, 2000; Rooney and Serna-Saldivar, 2000). Little sorghum is produced in Europe. Sorghum is an important food staple in many arid parts of the world due to its drought tolerance; it often grows where other cereal crops fail.

Maize is a major cereal grain that is grown worldwide and ranks second only to wheat in total production area and second to rice in total amount produced (Farnham *et al.*, 2003). The US is the world largest maize grower, and North America produces ~50% of total world production, followed by Brazil and China (Johnson, 2000). Maize is native to North America and may have been domesticated as early as 5000 BC in present-day Mexico, from where it eventually spread to Europe (Johnson, 2000; Farnham *et al.*, 2003).

Physical grain properties

Sorghum kernels are typically thought of as round, though most have at least one flat surface (Reichert *et al.*, 1988). Due to the genetic diversity of sorghum, kernels can vary widely in size and shape, with 1000 kernel weight for sorghum varying from 30 to 80 g (Rooney and Serna-Saldivar, 2000). Commercial sorghum hybrids on average have kernels weighing from 25 to 35 mg and are around 4 mm long, 2 mm wide, and 2.5 mm thick (Rooney and Serna-Saldivar, 2000). Anatomically, the sorghum grain is made up of the pericarp, endosperm, and the germ. Sorghum is unique in that it is the only cereal grain to have starch granules present in the pericarp. The outer edge of the sorghum endosperm is composed of the aleurone layer containing lipids, enzymes, and protein bodies. Under the aleurone layer is the outer corneous (hard, sometimes referred to as vitreous—see Hoseney 1994 for a discussion of these terms) endosperm fraction surrounding an inner floury (soft) core (Serna-Saldivar and Rooney, 1995; Rooney and Serna-Saldivar, 2000). The outer corneous endosperm is tightly packed with protein bodies covered with a continuous protein matrix (Seckinger and Wolf, 1973). Starch granules in this area of the sorghum kernel often show indentations where protein bodies were pressed into the sides of the granules (Rooney and Serna-Saldivar, 2000). In contrast, the floury endosperm in the center of the kernel is loosely packed with a discontinuous protein matrix and round starch granules (Seckinger and Wolf, 1973; Rooney and Serna-Saldivar, 2000). The relative proportions of corneous to floury endosperm can vary widely in sorghum and overall grain hardness in sorghum is often reported to be well correlated to the percent vitreosity of the kernel (Hallgren and Murty, 1983).

The outer appearance of sorghum can vary widely from white or yellow to red. Endosperm color in sorghum can be yellow to white and can influence the outer appearance of the grain in germplasm with a thin pericarp. Tannins, or proanthocyanidins, are polyphenolic compounds found in the sorghum lines with a pigmented testa. The presence of the pigmented testa, and thus tannins, is under genetic control and only sorghums with the B1/B2 genes have a pigmented testa (Waniska, 2000). It is a common myth that all sorghum lines contain tannin and often non-tannin phenolic compounds are presented as tannins. Another common myth is that the presence of tannins is linked to kernel color in sorghum; sorghum lines with a pigmented testa can have any pericarp color, including white (Waniska, 2000).

Maize kernels are the largest of the cereal grains, with kernel weights on the average from 250 to 300 mg and have a unique, flattened shape (Johnson, 2000; Watson, 2003). Like sorghum, the major components of the maize kernels are the outer pericarp layer, endosperm, and germ. Maize endosperm is the major fraction of the kernel and like sorghum contains areas of both corneous (hard) and floury (soft) endosperm. Maize kernels also have a high degree of variability and have been divided into five basic types, including: dent, flint, flour, sweet, and pop (Watson, 2003). Within each type of kernel, color can vary from yellow, white, red, to blue (Johnson, 2000; Watson, 2003). The major distinguishing factor between the various maize kernel types is the variation in its endosperm composition. Dent kernels have a floury endosperm center with corneous endosperm on the sides and back. The soft endosperm core collapses on drying to form a dent at the end of the kernel. Yellow dent corn is the type of maize most widely grown in the US (Johnson, 2000; Watson, 2003) and is used for a wide range of applications including such uses as fuel ethanol production, the production of isolated starch, animal feed, and human food products (Watson, 2003). White colored dent kernels, however, are often preferred for dry milling and corn tortilla industries as light colored food products can be made from such kernels (Johnson, 2000). Kernels of flint type maize are similar to sorghum, in that a corneous layer of endosperm surrounds a floury endosperm center. Popcorn kernels are similar to flint kernels in that they have a corneous outer endosperm layer but are generally smaller than flint kernels. Flour kernels have soft endosperm throughout the kernel and as such are easy to grind, but the overall softness of the kernels results in poor mould resistance and grain handling attributes. Sweetcorn, which is typically consumed as a vegetable, results from alterations in the conversion of sugars to starch in the kernel, increasing the sweetness of the grain as well as its texture (Johnson, 2000; Watson, 2003).

Grain hardness or endosperm texture (grain strength) is an important physical grain quality attribute that plays a role in the processing of cereal grains and in the end-use quality of cereal grain-based products such as breads and snack foods (Cagampang and Kirleis, 1984; Bettge and Morris, 2000). Grain hardness also plays a role in plant defense against molds and even possibly from insect attack (Chandrashekar and Mazhar, 1999). Therefore, grain hardness is an important economic and end-use quality trait in cereal grains. Accordingly, considerable research has been carried out to understand the biochemical basis for hardness of cereal grains. Although progress has been made for some grains, such as wheat, for maize and sorghum the exact biochemical mechanism for controlling grain strength is not well understood. The current understanding of the biochemical basis for grain strength in maize and sorghum was recently reviewed (Chandrashekar and Mazhar, 1999) and is discussed in greater detail later in this chapter.

As mentioned above, kernel hardness plays an important role in the processing of cereal grains. This is especially true in milling, where kernel hardness can influence milling yield and the quality of the milled product (Cagampang and Kirleis, 1984; Chandrashekar and Mazhar, 1999; Bettge and Morris, 2000). The relationship between kernel hardness and dry milling properties is well established (Paulsen and Hill, 1985; Peplinski *et al.*, 1992; Pan *et al.*, 1996; Shandera *et al.*, 1997). Dry milling of hard maize or sorghum kernels releases large hard endosperm particles (referred to as grist) early in the mill flow. The softer endosperm in the center of the kernel makes up other product mill streams.

In addition to dry milling, the physical characteristics of both maize and sorghum kernels influence their "nixtamalization" (alkaline processing) properties during the production of tortillas and snacks (Sahai *et al.*, 2000). Maize kernel hardness has been related to both the amount and the composition of solids lost during alkaline cooking (Pflugfelder *et al.*, 1988) and also the moisture content and texture of the final product remaining after the nixtamalization process (Serna-Saldivar *et al.*, 1993; Almeida-Dominguez *et al.*, 1997). In addition to hardness, other kernel properties and factors have been found to influence the alkaline processing quality of maize including: grain grade, bulk density, percent floaters, and the amount of cracked and broken kernels (Sahai *et al.*, 2000). Several different tests have been used to measure maize hardness and predict dry milling performance, including the following: the tangential abrasive de-hulling device (TADD), Stenvert micro-hammermill test, Wisconsin breakage tester, specific density measured with a gas pycnometer, percentage of kernels floating in a sodium nitrate solution, test weight, kernel size, and 1000 kernel weight. Of these various hardness tests, the TADD, Stenvert micro-hammermill hardness test, and percentage floaters provided the best prediction of maize grit yield, an important quality measurement (Shandera *et al.*, 1997).

As mentioned above, grain hardness has also been found to be an important factor in the nixtamalization of sorghum (Almeida-Dominguez *et al.*, 1997). Due to the variability in sorghum, the degree of decortication or pearling (removal of the outer bran layers) has also been found to impact cooking characteristics as well as final product (e.g. tortilla) color (Bedolla *et al.*, 1983; Choto *et al.*, 1985). Again, like maize, several different tests have been utilized to measure kernel hardness in sorghum. Pomeranz (1986) reported on the use of the Brabender hardness tester, Stenvert micro-hammermill test, particle size index, and near-infrared reflectance (NIR) to measure/predict hardness in sorghum. Perhaps the most widely used method for measuring grain hardness and relating it to milling performance in sorghum is the TADD (Rooney and Waniska, 2000) which abrasively grinds the outer layers of the kernels away. The amount of weight loss per unit time can then be used to calculate an abrasive hardness index (AHI) (Oomah *et al.*, 1981). The single kernel characterization system (SKCS) has also been used to measure grain hardness in sorghum (Pedersen *et al.*, 1996; Bean *et al.*, 2006). In the SKCS, the kernels are crushed between a crescent and rotor (Osborne and Anderssen, 2003); this provides a different type of hardness measurement compared with the TADD.

Slight correlations between SKCS hardness and TADD hardness values have been reported (Awika *et al.*, 2002; Bean *et al.*, 2006). Bean *et al.* (2006) compared SKCS hardness values, AHI, and kernel properties from a number of different sorghum lines and found complex relationships which suggested that many kernel factors play a role in the measurement of hardness by the TADD and SKCS. The relative proportions of corneous to floury endosperm can vary widely in sorghum and overall grain hardness in sorghum is often reported to be well correlated to the percent vitreosity of the kernel (Hallgren and Murty, 1983) using simple tests such as the percentage of kernels floated in a sodium nitrate solution. While simple to perform, as pointed out by Pedersen *et al.* (1996), vitreosity is not a measure of physical hardness, but still is a relatively reliable, rapid, and simple method for predicting hardness in sorghum.

Chemical composition

It is difficult to compare the chemical composition of cereal grains within a given type, let alone between two different types of cereals, due to the effect of environment and cultural practices (e.g. fertilization) on growth conditions, differences in the analytical methods used to measure the composition, differences in nomenclature used in describing the composition, etc. Given the above warnings, the "typical" composition of sorghum has been reported previously (Serna-Saldivar and Rooney, 1995; Rooney and Serna-Saldivar, 2000) as has that of maize (Johnson, 2000; Watson, 2003). It is clear from these works, that (like all cereal grains) the major components of sorghum and maize are proteins and starch. As such, these two classes of compounds will be discussed in greater detail below.

Sorghum prolamins

While it is often difficult to compare studies on protein composition of cereals due to differences in extraction methods, analytical methods utilized, and the nomenclature used to describe the proteins, it is clear that prolamins are the dominant type of protein in both sorghum and maize. Prolamins are storage proteins whose primary function is to serve as a nitrogen reserve for the next generation of plant. Prolamins, those proteins which are soluble in aqueous alcohols (with or without the use of reducing agents used in the extraction), contain high levels of the amino acids proline and glutamine (Belton *et al.*, 2006). In sorghum, recent studies based on improved extraction procedures show that the prolamins, called kafirins, account for roughly 70–90% of the total grain protein (Hamaker *et al.*, 1995). Kafirins have been subdivided into α, β, and γ subclasses based on their solubility, structure, and amino acid sequence (Shull *et al.*, 1991). The major kafirin is the α subclass which represents about 65–85% of the total kafirins, while the β and γ subclasses account for approximately 7–8% and 9–12%, respectively, of the prolamins (Watterson *et al.* 1993; Hamaker *et al.* 1995). In addition to these three major subclasses of prolamins, other minor subclasses such as the δ-kafirins have also been reported (Belton *et al.*, 2006). Kafirins are located primarily in spherical protein bodies in sorghum endosperm, with the α-kafirins mainly in the center of the protein bodies, and β and γ-kafirins forming the outer edges of the protein bodies. The protein bodies of sorghum are highly resistant to enzymatic digestion and to disruption by processing such as extrusion. It is currently thought the β and γ-kafirins form a highly cross-linked shell around the more easily digested α-kafirins (Hamaker and Bugusu, 2003). Kafirins are generally thought to be the most hydrophobic of the cereal prolamins based on improvements in their extraction using more non-polar solvents such as 50% tertiary-butanol compared with the more commonly used 70% ethanol. Recent reports on the free energies of hydration of kafirins seems to support this claim, as sorghum kafirins were found to be more hydrophobic than wheat prolamins (Belton *et al.*, 2006). Comparison of the water-binding capacities of kafirins and maize prolamins did not reveal major differences between the two however (Belton *et al.*, 2006).

One important characteristic feature of sorghum is that its protein digestibility decreases upon cooking, apparently through the formation of more protein cross-links during the cooking process (Duodu *et al.*, 2003). In agreement with this finding, Hamaker and Bugusu (2003) observed by laser scanning confocal microscopy that cooking causes sorghum proteins to form extended, web- and sheet-like structures. Both formation of oligomers and formation of web-like protein structures occurred to a lesser extent in maize (Duodu *et al.*, 2003; Hamaker and Bugusu, 2003). Interestingly, the water-binding capacity of cooked zein was found to decrease more than that of cooked kafirin (Belton *et al.*, 2006). Recently, highly digestible mutants of sorghum that have oddly shaped protein bodies have been discovered (Oria *et al.*, 2000); these might affect the functionality of sorghum flour from these mutants in food production. More studies are needed to confirm this hypothesis.

Maize prolamins

Overall, the proteins of maize are similar to those of sorghum. The dominant protein class is again the prolamins, which are called zeins in maize. Like the sorghum prolamins, zeins have been divided into subclasses (Esen, 1987). In fact, the sorghum subclasses were created to be analogous to those found in maize (Shull *et al.*, 1991). The major zein is the α-zein, which comprises ∼70% of total protein, followed by the β and γ subclasses at 5% and ∼20%, respectively. Other minor prolamin subclasses such as the δ have also been reported. Zeins, like kafirins, are also located in protein bodies, with the α-zeins located primarily in the center of the protein bodies, and β- and γ-zeins on the outer edges (Lawton and Wilson, 2003). Isolated zeins are available commercially and are mainly used for coatings on food products, although historically they have had a number of uses (Lawton, 2000). Isolated zeins have also been found to be able to form viscoelastic dough when mixed at high temperatures (Lawton, 1992). What role, if any, this may play in the development of gluten-free foods is not currently known.

Proteins and kernel hardness

Research has indicated that the endosperm proteins of maize and sorghum play a role in the hardness of these grains (Wall and Bietz, 1987; Wallace *et al.*, 1990; Mazhar and Chandrashekar, 1993; Mazhar and Chandrashekar, 1995; Pratt *et al.*, 1995; Dombrink-Kurtzman and Bietz, 1997; Chandrashekar and Mazhar, 1999). Endosperm hardness in maize and sorghum has been positively correlated with both protein content and prolamin composition (reviewed in Chandrashekar and Mazhar, 1999). Pratt *et al.* (1995) demonstrated a relationship between the levels of γ-prolamins and grain hardness in maize using reverse phase high-performance liquid chromatography. In contrast, Dombrink-Kurtzman and Bietz (1997) reported that the floury endosperm is richer in γ-zein compared with the vitreous endosperms, and that the vitreous endosperms contained up to twice the amount of α-zeins than that present in the soft endosperm. The inner portions of grains, while containing lower amounts of prolamins, contain proportionally more γ-prolamins than α-prolamins.

Mazhar and Chandrashekar (1995) postulated that both the content and distribution of α- and γ-kafirins are responsible for modifying endosperm texture, with the α-kafirin responsible for protein body size, and the γ-kafirin conferring rigidity by cross-linking the outer edges of the protein bodies. Furthermore, these authors reported that for a kernel to be hard, large protein bodies are needed (high levels of α-kafirin) with strong cross-linking (high levels of γ-kafirin). In conclusion, Chandrashekar and Mazhar (1999) described the relationship between the prolamin subclasses and grain hardness as follows: "the γ-prolamins form the cement, while the α-prolamins are the bricks."

Sorghum starch

As with all cereal grains, starch is the major component of sorghum and maize kernels. On a weight basis, 50–75% of the sorghum grain is starch (Rooney and Serna-Saldivar, 2003). Starch is located in the endosperm (both vitreous and floury) and, as noted above, in the pericarp of the kernel, which is a unique feature of sorghum (Rooney and Serna-Saldivar, 2003). Starch granules in sorghum range from 2 to 30 μm in diameter, with starch granules in corneous endosperm being polygonal and smaller than those in the floury endosperm, which are more round in shape (Serna-Saldivar and Rooney, 1995). Gelatinization temperatures of sorghum starch have been reported to vary from 71 to 80°C (Sweat *et al.*, 1984), with starch isolated from corneous endosperm having a higher gelatinization temperature than that from the floury endosperm (Cagampang and Kirleis, 1985). Corneous endosperm starch also has a higher intrinsic viscosity, and lower iodine-binding activity than that of the floury endosperm (Cagampang and Kirleis, 1985).

Starch from normal grains contains 23–30% amylose, while that from waxy sorghum has less than 5% amylose. Waxy sorghum starch differs in its properties when compared to normal starch, and has higher peak viscosity as well as water-binding capacity (Serna-Saldivar and Rooney, 1995). The digestibility of waxy sorghum starch is also reported to be higher than that of normal sorghum starch (Rooney and Pflugfelder, 1986).

Maize starch

The chemistry of maize starch has been intensely studied due to its widespread use in food and non-food applications; over 80% of the total worldwide starch production comes from maize (Johnson, 2000; Boyer and Shannon, 2003). The overall content and composition of maize starch is similar to that of sorghum starch. Waxy maize types are present as in sorghum; however, unlike sorghum, high amylose maize lines have been identified. In these types of maize, the amylose content can range from 50 to 80% (Johnson, 2000). Maize starch granules are similar in size to those of sorghum, spanning about 5 to 30 μm in diameter (Johnson, 2000). Though beyond the scope of this chapter, a number of types of modified starches can be produced from maize starch, which have unique characteristics different from those of the native starch (Johnson, 2000). The properties of such starches vary widely, and may prove useful in some instances in the production of gluten-free foods, though more research is needed in this area. As noted above, overall sorghum and maize starch are similar, but the water-binding ability of sorghum starch has been reported to be lower than that of maize starch. In addition, sorghum starch has also been reported to have a higher swelling at 90°C and lower solubility than maize starch, as well as higher peak and cold viscosities (Abd Allah *et al.*, 1987).

Milling

Dry milling

In Western countries sorghum has traditionally been used as an animal feed and, as such, the milling technology has not kept pace with that of other cereals such as wheat and maize. In Africa, much of the sorghum is processed by pounding by hand (Murty and Kumar, 1995; Munck, 1995). Hammer milling of sorghum is common practice although more sophisticated approaches have been reported (Munck, 1995). Hallgren *et al.* (1992) have reported a scheme where hard and soft endosperm fractions are separated and can thus be used for different purposes. Hard endosperm fractions can be re-milled to flour for use in food along with the soft endosperm fraction, but this causes increased starch damage which could impact the functionality of the flour as discussed later. Roller milling using equipment for wheat milling has been reported as not producing an economically viable product with desirable characteristics, though a semi-wet roller milling method has been reported that produced acceptable flours and food products (Munck, 1995).

Maize, unlike sorghum, is extensively dry milled in Western countries. In the US, three basic types of dry grind maize food flours are produced: full-fat, "bolted," and tempered-degermed (Duensing *et al.*, 2003). These products vary in the proportion of the original kernel remaining in the flour as well as in the process used to obtain the fractions. As indicated by the name, the full-fat products have most of the germ oil in the product and readily go rancid. Bolted flours are produced by sieving out parts of the mill streams and are lower in fat and fiber content than the full-fat fractions. For the tempered product, moisture is first added to the maize in order to facilitate the separation of the anatomical parts of the kernel, in particular the bran and germ. The amount of added moisture is critical and depends on several characteristics of the maize grain itself, including considerations such as cracked and damaged grains (Duensing *et al.*, 2003).

Gluten-free food production

Traditional foods

Both sorghum and maize have been used for thousands of years in human food products. As such, a diverse selection of traditional food products are available including fermented and un-fermented flat breads and porridges, rice-like products, taco shells, and tortillas (Serna-Saldivar and Rooney, 1995; Rooney and Serna-Saldivar, 2003). With few exceptions such as the tortilla, these products are not typical of Western diets. Since such products are typically made without any wheat, they are safe for people with celiac disease, and could, therefore, fill a specialty market for the celiac community.

Breads

The use of sorghum in wheat/sorghum composite breads has been studied by many scientists (Munck, 1995), not so much as a food for people with celiac disease, for which they would not be suitable, but more as research into breads that could reduce the expensive importation of wheat into parts of Africa (Satin, 1988). For this second purpose, maize has not been used since it is not a staple crop in Africa. Little research is available on the production of breads from maize. This may be due in part to the distinctive flavor of maize, but may also lie in the fact that maize is widely used in products such as tortillas and for the isolation of starch, which is itself a major food product. Thus, "value added" research and utilization of maize in wheat-free foods has not been a major area of emphasis of research for maize. As mentioned above, much of the work on sorghum bread production has stemmed from a need to reduce wheat imports into Africa, where maize is not a staple crop and therefore would not be a viable alternative to wheat.

In addition to composite breads, several researchers have reported on the production of gluten-free bread from sorghum and much of this work is reviewed by Taylor and Dewar (2001). A more recent review has summarized early works along with current reports on sorghum bread production and provided a detailed understanding of wheat-free bread production from sorghum (Taylor et al., 2006). One of the first comprehensive studies on the production of sorghum bread was reported by Hart et al. (1970). These researchers produced a basic recipe and in subsequent studies tested the effect of adding different gums, starches, enzymes, emulsifiers, and shortening to the recipe (Hart et al., 1970). They also tested the use of sourdough fermentation. In these studies a soft batter containing $\sim$100–150% water on flour weight basis, was required to obtain sufficient rise. The addition of methylcellulose was reported to improve bread quality by increasing gas retention and preventing loafs from collapsing, with 2% of 4000 cps hydroxypropyl-methylcellose (Methocel) optimum. When isolated starches were combined with the methylcellulose, improvements in oven rise and crumb structure were found. The type of added starch was not critical, as starches from sorghum, modified and waxy sorghum, maize, cassava, arrowroot, potato all produced similar results. Adding α-amylases, proteases and emulsifiers was found to weaken the crumb structure, but using shortening together with methylcellulose and methylcellulose-derivatives softened the loaves. The addition of a sourdough fermentation process did not improve sorghum bread quality.

The addition of xanthan gum was reported to produce sorghum bread with acceptable quality, but the right technique for its addition was important to get good results (Satin, 1988). Soaking the xanthan gum in water before adding it to the dough resulted in improved bread quality relative to dry addition. In addition to xanthan gum and the additives tested by Hart et al. (1970), scientists have used pre-gelatinized cassava starch (Olatunji et al., 1992b; Hugo et al., 1997) to improve sorghum bread quality. Pre-gelatinizing the cassava starch was found to be an important element and specific volumes as high as 3.3 cm^3/g were reported (Hugo et al., 1997). Cauvain (1998) suggested several complicated formulations for sorghum bread which contained either skim milk powder, sodium carboxymethyl cellulose, baking powder and soy flour or 50% maize starch, skim milk powder, sodium carboxymethyl cellulose and dried egg albumen in addition to sorghum flour, yeast, salt and water in order to produce acceptable sorghum-based breads.

To evaluate differences in the intrinsic breadmaking quality of sorghum hybrids, Schober et al. (2005) compared nine selected sorghum hybrids and a commercial sorghum flour for sorghum bread production. In this study, an extrusion cell was used to standardize batter consistency to a constant value in a similar fashion to using a farinograph to adjust water levels in wheat dough. Another important feature in this study was proofing of breads to a constant height rather than proof time. This was done due to difficulties in achieving reproducible proofing of gluten-free bread using a constant time, even when highly controlled conditions were used. A simple formulation was used, similar to that of Olatunji *et al.* (1992a), which was based on sorghum flour and maize starch (70/30) plus water, salt, sugar, and yeast. While bread volume and height were not affected by the hybrid used, considerable differences were found with regard to crumb grain and texture. The amount of mechanically damaged starch in the flour (highest in the samples with the hardest kernels) was identified as a key factor explaining these differences, with higher starch damage resulting in a coarser crumb structure. Most likely, damaged starch was more easily degraded by endogenous amylases, resulting in a higher amount of fermentable sugars and a weaker starch gel.

In addition, two hybrids with the most different crumb grain were selected and were used to produce bread containing xanthan gum, skim milk powder and various water levels added to the base formulation. Differences in crumb grain between the hybrids were maintained at various combinations of xanthan gum, skim milk powder and water. It was also noted that while the xanthan gum and skim milk powder improved the appearance of the bread crust, they had negative effects on overall bread quality.

Rye pentosans have also been suggested as an additive to improve gluten-free bread quality, including sorghum-based breads (Casier *et al.*, 1977). The addition of rye pentosans produced sorghum breads with acceptable volumes and reportedly improved staling properties. However, it should be pointed out that rye prolamins (secalins) are toxic to people with celiac disease (Murray, 1999), and as such any rye pentosans isolated for use in breads targeted to the celiac populations have to be completely free of secalins.

Little research appears to have been done on the production of maize bread. Olatunji *et al.* (1992a) produced maize bread using the same formulation they used for sorghum. Sanni *et al.* (1998) and Edema *et al.* (2005) produced sour maize breads using maize flour and maize starch (70:30) as well as different maize flours, soy flours or blends of maize and soy (Edema *et al.*, 2005). Leavening of these breads was obtained by using mixed cultures of lactic acid bacteria and yeast. Salt, fat, sugar, and high water levels were added. Specific volumes of these breads were low, however, indicating that these products would represent specialty bread for most gluten-free markets.

Cakes and cookies

In addition to breads, cakes and cookies can also be produced from sorghum and maize flour. However, as noted above for bread, the literature contains more reports on the use of sorghum in these types of products, and as such sorghum will be emphasized in the following discussions.

Oyidi (1976) reported the successful production of cake and biscuits from sorghum flour (obtained from an unusual sorghum mutant). Similarly, Olatunji *et al.* (1992a) developed a cake recipe for cake using sorghum or maize and cassava starch (70/30). Similarly, Oyidi (1976) reported the successful production of cake and biscuits from sorghum flour (obtained from an unusual sorghum mutant). Badi and Hoseney (1976) studied the production of cookies from 100% sorghum flour. The cookies produced lacked the desired spread and top cracks and were described by the authors as "tough, hard, gritty, and mealy in texture and taste." These authors went on to identify the lipid composition as partly responsible for the low quality of the cookies. By adding wheat-flour lipids to defatted sorghum flour the top grain and spread of the cookies was improved. The sorghum cookies were also improved by adding unrefined soybean lecithin or refined lecithin plus monoglycerides. Additional improvement in the quality of the cookies was found when the sorghum flour was hydrated for several hours with malt syrup or water and then air dried, and by increasing the pH of the cookie dough using sodium carbonate (in contrast to Badi and Hoseney, 1976). Morad *et al.* (1984) found that sugar cookies made from 100% sorghum flour generally had the highest spread factor (width to thickness) relative to cookies from a commercial wheat cookie flour and cookie flour/sorghum mixtures.

Snack foods

Several types of gluten-free snack foods can be produced from both sorghum and maize. Maize is a preferred flour source for the production of extruded snack foods. The production of such snack foods offers a relatively straightforward method for producing products for people with celiac disease.

Sorghum snack foods

High-quality tortilla chips can be easily produced from white food-grade sorghum by reducing the lime concentration used in their production as well as the cooking and steeping time relative to that used in maize tortilla chip production (Serna-Saldivar *et al.*, 1988). A snack food with a light crunchy texture was prepared from sorghum by deep-fat frying dried kernels (pellets) that had been cooked under alkaline conditions (Suhendro *et al.*, 1998). This product was based on a similar Indonesian food made from whole maize using the same procedure. The optimized process for production of the sorghum product consisted of first autoclaving the grains for 60 minutes at 120°C followed by rinsing and drying the grains to 9% moisture and finally deep-fat frying at 220°C. By comparing different sorghum samples, it was found that samples with intermediate to soft endosperm expanded more during the process than those with hard endosperm. The use of a waxy sorghum to produce this snack product resulted in poor quality and is not recommended.

Young *et al.* (1990) produced a rice-like product from sorghum by parboiling before decortication. Parboiling was found to increase the yield of the decorticated grain and reduced kernel breakage as well as increased firmness and reduced stickiness of the final cooked kernels.

Maize snack foods, breakfast foods, and other food products

The production of snack foods from maize is a major industry, and a wide range of snack products are commercially produced from maize. In 2000, snack products alone from maize reached $5.6 billion in value. Extruded snack products are a major use of maize. By extrusion cooking of maize grits or meal, products such as curls, puffs, and balls can be produced (Rooney and Serna-Saldivar, 2003). Many factors control the final extruded product, including the composition of the maize used, the extrusion parameters and the shape of the die. In addition to extruded snacks, fried snack products are also produced from maize. These are typically produced from alkaline processed maize and consist of items such as tortilla chips.

Another major use of maize is in the production of breakfast cereals. Breakfast cereals can be produced into "flakes, shreds, granules, puffs, or other forms" (Rooney and Serna-Saldivar, 2003). Extrusion and flaking are two of the major processes used to produce breakfast cereals. While a number of commercial corn-based snack products and breakfast cereals are available, they may contain added wheat or barley malt. People with celiac disease should therefore be careful when consuming such products. However, wheat-free maize snacks and cereals should be easy to produce, specifically for the celiac community if necessary.

Sorghum noodles and pasta

Suhendro *et al.* (2000) produced sorghum noodles from decorticated sorghum flour, water, and salt by preheating, extrusion, and drying. Heterowaxy sorghum produced noodles of inferior quality relative to normal sorghum. The noodles were sticky, soft, and had a high dry matter loss during cooking. Increased amylopectin and reduced amylose content in the heterowaxy sorghum limited retrogradation. The authors further reported that the timing of amylose dispersion (solubilization), formation of noodles, and amylose retrogradation was critical as suggested by effects of the preheating and drying methods. Flour particle size was also critical, with finer flour producing better quality noodles. Good-quality noodles resulted when processing conditions were optimized and when the noodles were cooked properly (Suhendro *et al.*, 2000).

Future trends

The studies reported on in this chapter show that significant research has been done on developing wheat-free foods from sorghum and maize. To date, much of this research has focused on development of product formulas and testing the effects of additives such as hydrocolloids on product quality. Little work has been done, however, on modifying the proteins and/or starches in sorghum and maize flour to improve their functionality in wheat-free food products. Many methods are available for modifying properties of proteins, for example, their solubility, charge, molecular weight, etc. More research is needed though to understand the science and technology behind the production of products like wheat-free breads and pastas so that targeted modifications of proteins and starch can be done to achieve the desired improvements in wheat-free food quality. Breeding of sorghum and maize with specific traits for the production of wheat-free foods such as breads would be beneficial. However, as with modifying proteins and starch, more research is needed to identify breeding targets.

Further information and advice

This chapter has provided a summary of the properties of sorghum and maize and the wheat-free food products that can be produced from them. To provide a base for anyone wishing to make sorghum and maize breads, this section will discuss how one might go about developing a sorghum pan bread. Assume that the formulation will start with 100% sorghum flour. In order to avoid associated problems with bran, one might decide to use decorticated sorghum flour, although this would reduce the health benefits. The next factor in developing a formula would be to consider factors influencing the crumb structure. Amylose is needed for quick retrogradation to facilitate setting, so waxy sorghum flour should be avoided as the base flour. The amount of water must also be determined; a good starting place would be to start with 100% water on a flour basis (which falls in the range of 80–110% used in most gluten-free bread formulae). For a simple formula, other basic ingredients would include salt, sugar, and yeast.

The next important factor to consider is the consistency of the batter, which in the basic formula would depend to a large extent on the amount of mechanically damaged starch. The amount of damaged starch will influence the amount of water needed. A flour with high levels of damaged starch may produce too thick a batter so that more water has to be added. However, the resulting bread may have a coarse crumb, large holes in the crumb and/or a collapsed bottom layer—both effects of the excessive degradation of damaged starch granules by amylases in combination with too much water. Adding pure starch (30% is a good starting place) will dilute the damaged starch in the sorghum flour and promote structure formation in the crumb. Pre-gelatinized starch would not be the best choice in this situation (high starch damage in the flour) as it would bind even more water. The amount of water would probably need to be adjusted when the pure starch is included in the formulation. If a higher loaf volume, more regular crumb structure, and slower staling rate are wanted, one might add 2% hydroxypropyl methylcellulose (HPMC), which should be carefully mixed with the flour and starch, so that lumps are not formed when water is added. The amount of water may now need to be adjusted again as HPMC will bind more water in the batter. In an alternative situation where the original sorghum flour does not contain high levels of damaged starch one needs to increase the water binding, viscosity, and cohesiveness of the batter. This can be achieved by adding either HPMC or pre-gelatinized starch.

References

Abd Allah, M. A., Mahmoud, R. M., El-Kalyoubi, M. H., and Abou Arab, A. A. (1987). Physical properties of starches isolated from pearl millet, yellow corn, sorghum, sordan and pearl millet. *Starch/Staerke* 39, 9–12.

Almeida-Dominguez, H. D., Suhendro, E. L., and Rooney, L. W. (1997). Corn alkaline cooking properties related to grain characteristics and viscosity (RVA). *J. Food Sci.* 62, 516–523.

Awika, J. M., Gualberto, D., Rooney, L.W., and Rooney, W.L. (2002). Properties of white food sorghums grown in different environments. *AACC 87th Annual Meeting*, October 13–17, Montreal, Quebec Abstract Book, p. 153.

Badi, S. M. and Hoseney, R. C. (1976). Use of sorghum and pearl millet flours in cookies. *Cereal Chem.* 53, 733–738.

Bean, S. R., Chung, O. K., Tuinstra, J. F., and Erpelding, J. (2006). Evaluation of the Single Kernel Characterization System (SKCS) for measurement of sorghum grain attributes. *Cereal Chem.* 83, 108–113.

Bedolla, S., Palacios, M. G., Rooney, L. W., Dielh, K. C., and Khan, M. N. (1983). Cooking characteristics of sorghum and corn for tortilla preparation by several methods. *Cereal Chem.* 60, 263–268.

Belton, P. S., Delgadillo, I., Halford, N. G., and Shewry, P. R. (2006). Kafirin structure and functionality. *J. Cereal Sci.* 44, 272–286.

Bettge, A. D. and Morris, C. F. (2000). Relationships among grain hardness, pentosan fractions, and end-use quality of wheat. *Cereal Chem.* 77, 241–247.

Boyer, C. D. and Shannon, J. C. (2003). Carbohydrates of the kernel. In: White, P. J. and Johnson, L. A. eds. *Corn Chemistry and Technology*, 2nd edn. St. Paul, MN: American Association of Cereal Chemists, pp. 289–312.

Cagampang, G. B. and Kirleis, A. W. (1984). Relationship of sorghum grain hardness to selected physical and chemical measurements of grain quality. *Cereal Chem.* 61, 100–105.

Cagampang, G. B. and Kirleis, A. W. (1985). Properties of starches isolated from sorghum floury and corneous endosperm. *Starch/Staerke* 37, 253–257.

Casier, J. P. J., de Paepe, G., Willems, H., Goffings, G., and Noppen, H. (1977). Bread from starchy tropical crops. II. Bread production from pure millet and sorghum flours, using cereal endosperm—cellwall—pentosan as a universal baking factor. In: Dendy, D. A. V. ed. *Proceedings of a Symposium on Sorghum and Millets for Human Food*. London: Tropical Products Institute, pp. 127–131.

Cauvain, S. P. (1998). Other cereals in breadmaking. In: Cauvain, S. P. and Young, L. S. eds. *Technology of Breadmaking*. London: Blackie Academic & Professional, pp. 330–346.

Chandrashekar, A. and Mazhar, H. (1999). The biochemical basis and implications of grain strength in sorghum and maize. *J. Cereal Sci.* 30, 193–207.

Choto, C. E., Morad, M. M., and Rooney, L. W. (1985). The quality of tortillas containing whole sorghum and pearled sorghum alone and blended with yellow maize. *Cereal Chem.* 62, 51–55.

Dombrink-Kurtzman, M. A. and Bietz, J. A. (1997). Zein composition in hard and soft endosperm of maize. *Cereal Chem.* 70, 105–108.

Duensing, W. J., Roskens, A. B., and Alexandar, R. J. (2003). Corn dry milling: processes, products, and applications. In: White, P. J. and Johnson, L. A. eds. *Corn Chemistry and Technology*, 2nd edn. St. Paul, MN: American Association of Cereal Chemists, pp. 407–447.

Duodu, K. G., Taylor, J. R. N., Belton, P. S., and Hamaker, B. R. (2003). Factors affecting sorghum protein digestibility. *J. Cereal Sci.* 38, 117–131.

Edema, M. O., Sanni, L. O., and Sanni, A. I. (2005). Evaluation of maize-soybean flour blends for sour maize bread production in Nigeria. *Afr. J. Biotechnol.* 4, 911–918.

Esen, A. (1987). A proposed nomenclature for alcohol-soluble proteins (zeins) of maize (*Zea mays* L.) *J. Cereal Sci.* 5, 117–128.

Farnham, D. E., Benson, G. O., and Pearce, R. B. (2003). Corn perspective and culture. In: White, P. J. and Johnson, L. A. eds. *Corn Chemistry and Technology*, 2nd edn. St. Paul, MN: American Association of Cereal Chemists, pp. 1–34.

Hallgren, L. and Murty, D. S. (1983). A screening test for grain hardness in sorghum employing density grading in sodium nitrate solution. *J. Cereal Sci.* 1, 265–274.

Hallgren, L., Rexen, F., Petersen, P. B., and Munck, L. (1992). Industrial utilization of whole crop sorghum for food and industry. In: Gomez, M. I., House, L. R., Rooney, L. W., and Dendy, D. A. V. eds. *Utilization of sorghum and millets. International Crops Research Institute for the Semi-Arid Tropics.* Patancheru, India, pp. 121–130.

Hamaker, B. R. and Bugusu, B. A. (2003). Overview: sorghum proteins and food quality. In: Belton, P. S. and Taylor, J. R. N. eds. *Afripro. Workshop on the Proteins of Sorghum and Millets: Enhancing nutritional and functional properties for Africa.* Pretoria, South Africa, 2–4 April (http://www.afripro.org.uk/papers/Paper08Hamaker.pdf).

Hamaker, B. R., Mohamed, A. A., Habben, J. E., Huang, C. P., and Larkins, B. A. (1995). Efficient procedure for extracting maize and sorghum kernel proteins reveals higher prolamin contents than the conventional method. *Cereal Chem.* 72, 583–588.

Hart, M. R., Graham, R. P., Gee, M., and Morgan Jr., A. I. (1970). Bread from sorghum and barley flours. *J. Food Sci.* 35, 661–665.

Hoseney, R. C. (1994). *Principles of Cereal Science and Technology*, 2nd edn. St. Paul: AACC.

Hugo, L. F., Waniska, R. D., and Rooney, L. W. (1997). Production of bread from composite flours. In: *Harnessing Cereal Science and Technology for Sustainable Development.* CSIR ICC-SA Symposium, Pretoria, South Africa, pp. 100–114.

Johnson, L. A. (2000). Corn: The major cereal of the Americas. In: Kulp, K. and Ponte, J. G. Jr. eds. *Handbook of Cereal Science and Technology*, 2nd edn. New York: Marcel Dekker, pp. 31–80.

Kimber, C. T. (2000). Origins of domesticated sorghum and its early diffusion to China and India. In: Smith, C.W. and Frederiksen, R. A. eds. *Sorghum: Origin, History, Technology, and Production.* New York: John Wiley & Sons, pp. 3–97.

Lawton, J. W. (1992). Viscoelasticity of zein-starch doughs. *Cereal Chem.* 69, 351–355.

Lawton, J. W. (2000). Zein: A history of processing and use. *Cereal Chem.* 19, 1–18.

Lawton, J. W. and Wilson, C. M. (2003). Proteins of the kernel. In: White, P. J. and Johnson, L. A. eds. *Corn Chemistry and Technology*, 2nd edn. St. Paul, MN: American Association of Cereal Chemists, pp. 313–354.

Mazhar, H. and Chandrashekar, A. (1993). Differences in kafirin composition during endosperm development and germination in sorghum cultivars of varying hardness. *J. Cereal Sci.* 70, 667–671.

Mazhar, H. and Chandrashekar, A. (1995). Quantification and distribution of kafirins in the kernels of sorghum cultivars varying in endosperm hardness. *J. Cereal Sci.* 21, 155–162.

Morad, M. M., Doherty, C. A., and Rooney, L. W. (1984). Effect of sorghum variety on baking properties of U.S. conventional bread, Egyptian pita "Balady" bread and cookies. *J. Food Sci.* 49, 1070–1074.

Munck, L. (1995). New milling technologies and products: whole plant utilization by milling and separation of the botanical and chemical components. In: Dendy, D. A. V. ed. *Sorghum and Millets: Chemistry and Technology.* St. Paul, MN: American Association of Cereal Chemists, pp. 223–281.

Murray, J. A. (1999). The widening spectrum of celiac disease. *Am. J. Clin. Nutr.* 69, 354–365.

Murty, D. S. and Kumar, K. A. (1995). Traditional uses of sorghum and millets. In: Dendy, D. A. V. ed. *Sorghum and Millets: Chemistry and Technology.* St. Paul, MN: American Association of Cereal Chemists, pp. 185–222.

Olatunji, O., Koleoso, O. A., and Oniwinde, A. B. (1992a). Recent experience on the milling of sorghum, millet, and maize for making nonwheat bread, cake, and sausage in Nigeria. In: Gomez, M. I., House, L. R., Rooney, L. W., and Dendy, D. A. V. eds. *Utilization of Sorghum and Millets.* Patancheru, India: International Crops Research Institute for the Semi-Arid Tropics, pp. 83–88.

Olatunji, O., Osibanjo, A., Bamiro, E., Ojo, O., and Bureng, P. (1992b). Improvement in the quality of non-wheat composite bread. In: *5th Quadrennial Symposium on Sorghum and Millets.* Schwechat, Austria: International Association for Cereal Science and Technology, pp. 45–54.

Oomah, B. D., Reichert, R. D., and Youngs, C. G. (1981). A novel, multi-sample, tangential abrasive dehulling device (TADD). *Cereal Chem.,* 58, 392–395.

Oria, M. P., Hamaker, B. R., Axell, J. D., and Huang, C. P. (2000). A highly digestible sorghum mutant cultivar exhibits a unique folded structure of endosperm protein. *Proc. Natl Acad. Sci. USA* 97, 5065–5070.

Osborne, B. G. and Anderssen, R. S. (2003). Single-kernel characterization principles and applications. *Cereal Chem.* 80, 613–622.

Oyidi, O. (1976). The vegetative characters and food uses of a mutant sorghum with twin-seeded spikelets, in Northern Nigeria. *Samaru Agric. Newslett.* 18, 44–49.

Pan, Z., Eckhoff, S. R., Paulsen, M. R., and Litchfield, J. B. (1996). Physical properties and dry-milling characteristics of six selected high-oil maize hybrids. *Cereal Chem.* 73, 517–520.

Paulsen, M. R. and Hill, L. D. (1985). Corn quality factors affecting dry milling performance. *J. Agric. Eng. Res.* 31, 255–263.

Pedersen, J. F., Martin, C. R., Felker, F. C., and Steele, J. L. (1996). Application of the single kernel wheat characterization technology to sorghum grain. *Cereal Chem.* 73, 421–423.

Peplinski, A. J., Paulsen, M. R., and Bouzaher, A. (1992). Physical, chemical, and dry-milling properties of corn of varying density and breakage susceptibility. *Cereal Chem.* 397–400.

Pflugfelder, R. L., Rooney, L. W., and Waniska, R. D. (1988). Dry matter losses in commercial corn masa production. *Cereal Chem.* 65, 127–132.

Pomeranz, Z. Y. (1986). Comparison of screening methods for indirect determination of sorghum hardness. *Cereal Chem.* 63, 36–38.

Pratt, R. C., Paulis, J. W., Miller, K., Nesen, T., and Bietz, J. A. (1995). Association of zein classes with maize kernel hardness. *Cereal Chem.* 72, 62–167.

Reichert, R., Mwararu, M., and Mukuru, S. (1988). Characterization of colored grain sorghum lines and identification of high tannin lines with good dehulling characteristics. *Cereal Chem.* 65, 165–170.

Rooney, L. W. and Pflugfelder, R. L. (1986). Factors affecting starch digestibility with special emphasis on sorghum and corn. *J. Anim. Sci.* 63, 1607–1623.

Rooney, L. W. and Serna-Saldivar, S. O. (2000). Sorghum. In: Kulp, K. and Ponte, J. G., Jr. eds. *Handbook of Cereal Science and Technology*, 2nd edn. New York: Marcel Dekker, pp. 149–176.

Rooney, L. W. and Serna-Saldivar, S. O. (2003). Food use of whole corn and dry-milled fractions. In: White, P. J. and Johnson, L. A. eds. *Corn Chemistry and Technology*, 2nd edn. St. Paul, MN: American Association of Cereal Chemists, pp. 495–535.

Rooney, L. W. and Waniska, R. D. (2000). Sorghum food and industrial utilization. In: Smith, C. W. and Frederiksen, R. A. eds. *Sorghum: Origin, History, Technology, and Production*. New York: John Wiley & Sons, pp. 689–750.

Sahai, M. D., Surjewan, I., Mua, J. P., Buendia, M. O., Rowe, M., and Jackson, D. S. (2000). Dry matter loss during nixtamilization of a white corn hybrid, impact of processing parameters. *Cereal Chem.* 77, 254–258.

Sanni, A. I., Onilude, A. A., and Fatungase, M. O. (1998). Production of sour maize bread using starter-cultures, *World J. Microbiol. Biotechnol.* 14, 101–106.

Satin, M. (1988). Bread without wheat. Novel ways of making bread from cassava and sorghum could reduce the Third World's dependence on imported wheat for white bread. *New Sci.* 28 April, 56–59.

Schober, T. J., Messerschmidt, M., Bean, S. R., Park, S. H., and Arendt, E. K. (2005). Gluten-free bread from sorghum: quality differences among hybrids. *Cereal Chem.* 82, 394–404.

Seckinger, H. L. and Wolf, M. J. (1973). Sorghum protein ultrastructure as it relates to composition. *Cereal Chem.* 50, 455–465.

Serna-Saldivar, S. and Rooney, L. W. (1995). Structure and chemistry of sorghum and millets. In: Dendy, D. A. V. ed. *Sorghum and Millets: Chemistry and Technology*. St. Paul, MN: American Association of Cereal Chemists, pp. 69–124.

Serna-Saldivar, S. O., Gomez, M. H., Almeida-Dominguez, H. D., Islas-Rubio, A., and Rooney, L. W. (1993). A method to evaluate the lime cooking properties of corn (*Zea mays*). *Cereal Chem.* 70, 762–764.

Shandera, D. L., Jackson, D. S., and Johnson, B. E. (1997). Quality factors impacting processing of maize dent hybrids. *Maydica* 42, 281–289.

Shull, J. M., Watterson, J. J., and Kirleis, A. W. (1991). Proposed nomenclature for the alcohol-soluble proteins (kafirins) of Sorghum bicolor (*L. Moench*) based on molecular weight, solubility, and structure. *J. Agric. Food Chem.* 39, 83–87.

Smith, C. W. (2000). Sorghum production statistics. In: Smith, C. W. and Frederiksen, R. A. eds. *Sorghum: Origin, History, Technology, and Production*. New York: John Wiley & Sons, pp. 401–408.

Suhendro, E. L., McDonough, C. M., Rooney, L. W., Waniska, R. D., and Yetneberk, S. (1998). Effects of processing conditions and sorghum cultivar on alkaline-processed snacks. *Cereal Chem.* 75, 187–193.

Suhendro, E. L., Kunetz, C. F., McDonough, C. M., Rooney, L. W., and Waniska, R. D. (2000). Cooking characteristics and quality of noodles from food sorghum. *Cereal Chem.* 77, 96–100.

Sweat, V. E., Faubion, J. M., Gonzales-Palacios, L., and Rooney, L. W. (1984). Gelatinization energy and temperature of sorghum and corn starches. *Trans. ASAE* 27, 1960–1984.

Taylor, J. R. N. and Dewar, J. (2001). Developments in sorghum food technologies. In: Taylor, S. ed. *Advances in Food and Nutrition Research*, Vol. 43. San Diego, CA: Academic Press, pp. 217–264.

Taylor, J., Schober, T., and Bean, S. R. (2006). Non-traditional uses of sorghum and pearl millet. *J. Cereal Sci.* 44, 252–271.

Wall, J. S. and Bietz, J. A. (1987). Differences in corn endosperm proteins in developing seeds of normal and opaque-2 corn. *Cereal Chem.* 64, 275–280.

Wallace, J. C., Lopez, M. A., Paiva, E., and Larkins, B. A. (1990). New methods for extraction and quantitaion of zeins reveal a high content of α-zein in modified opaque-2 maize. *J. Plant Physiol.* 92, 191–196.

Waniska, R. (2000). Structure, Phenolic compounds, and anti-fungal proteins of sorghum caryopsis. In: Chandrashekar, A., Bandyopadhyay, R. and Hall, A. J. eds. *Technical and Institutional Options for Sorghum Grain Mold Management: Proceedings of An International Consultation*, 18–19 May 2000. Patancheru, India: ICRISAT, pp. 72–106.

Watson, S. A. (2003). Description, development, structure, and composition of the corn kernel. In: White, P. J. and Johnson, L. A. eds. *Corn Chemistry and Technology*, 2nd edn. St. Paul, MN: American Association of Cereal Chemists, pp. 69–106.

Watterson, J. J., Shull, J. M., and Kirleis, A. W. (1993). Quantitation of a-, b-, and g-kafirins in vitreous and opaque endosperm of *Sorghum bicolor*. *Cereal Chem.* 70, 452–457.

Young, R., Haidara, M., Rooney, L. W., and Waniska, R. D. (1990). Parboiled sorghum: Development of a novel decorticated product. *J. Cereal Sci.* 11, 277–289.

Gluten-free foods and beverages from millets

John R. N. Taylor and M. Naushad Emmambux

Introduction

Millets are not a single species, or even different species within a single genus. They are simply cultivated grasses (cereals) that have small kernels and they are grouped together solely on this basis. The word millet is derived from the French word "mille" meaning thousand, implying that a handful of millet contains thousands of grains. In fact, as can be seen in Table 6.1 there are many different millets, some of which are closely related, like proso millet and little millet, and others which are not, in particular finger millet and teff, which belong to a different tribe to most of the other millets. The study of millet literature is problematical because different common names are used for the same species and even different proper species names are in widespread use. In this account, the English name as given in the table will be used when discussing each species but the list of vernacular names should help when reading the literature.

This chapter will first review each of the more important millet species in respect of their history, production, physical characteristics of the grains, and their nutrient composition. Next, the types of traditional foods and beverages produced from millets will be described, followed by an account of the processing technologies used to make these products. Last, recent and future trends in millet foods and beverages will be examined.

Table 6.1 The different millet species. Information mainly from the USDA Germplasm Resources Information Network (GRIN)

Generally used English name	Other common vernacular names	Taxonomy
Finger millet	Ragi Wimbi	Tribe Eragrostideae *Eleucine coracana* L. Gaertn.
Teff	Tef Teff grass Abyssinian lovegrass	Tribe Eragrostideae *Eragrostis tef* (Zuccagni) Trotter
Job's tears	Adlay Adlay millet	Tribe Andropogoneae *Coix lacryma-jobi* L.
White fonio	Fonio Acha Fonio millet Hungry rice	Tribe Paniceae *Digitaria exilis* (Kippist) Stapf
Black fonio	Black acha Hungry rice	Tribe Paniceae *Digitaria iburua* Stapf
Japanese millet	Japanese barnyard millet	Tribe Paniceae *Echinochloa esculenta* (A. Braun) H. Scholz
Sawa millet	Shama millet Awnless barnyard grass Corn panic grass Deccan grass Jungle ricegrass Jungle rice	Tribe Paniceae *Echinochloa colona* (L.) Link
Proso millet	Common millet Broom millet Hog millet Panic millet	Tribe Paniceae *Panicum miliaceum* L. subsp. *miliaceum*
Little millet	Blue panic Sama	Tribe Paniceae *Panicum sumatrense* Roth.
Kodo millet	Creeping paspalum Ditch millet Indian paspalum Water couch	Tribe Paniceae *Paspalum scrobiculatum* L.
Foxtail millet	Italian millet Foxtail bristle grass German millet Hungarian millet	Tribe Paniceae *Setaria italica* (L.) P. Beauv. subsp. *italica*
Pearl millet	Bulrush millet Cattail millet Babala Bajra/Bajira	Tribe Paniceae *Pennisetum glaucum* (L.) R. Br.
Guinea millet	False signal grass	Tribe Paniceae *Urochloa deflexa* (Schumach.) H. Scholz

Germplasm Resources Information Network (2007).

Review of the more important millet species

The production of millets ranged from 33.6 to 37.3 million tons in 2001 to 2005 (FAO, 2007), slightly higher than the estimated 29.3 million tons in the 1980s (Table 6.2). Quantitatively, the most important millet species, in descending order, are pearl millet, foxtail millet, proso millet, and finger millet. However, mere total world production is not necessarily a guide to the local importance of certain millets. In resource-poor developing countries, millets play a critical role in food security on account of their agronomic characteristics (e.g. in the case of fonio) (Smith, 1996).

Pearl millet

Description, history, and production
Pearl millet is believed to have been domesticated over 5000 years ago in Africa (Andrews and Kumar, 1992) and is widely cultivated across the continent from the Sahelian (Sahara Desert margin) countries in West Africa down to South Africa. It is also widely cultivated in India and was probably introduced into that country some 3000 years ago. The crop is generally cultivated by subsistence farmers but is becoming a commercial crop, for example in Australia. Pearl millet is an annual plant about 2 m tall, with a cylindrical spike, 15–140 cm in length (National Research Council, 1996). Pearl millet is uniquely well-adapted to harsh environmental conditions and can be cultivated in regions with very low annual rainfall, down to 250 mm, and very high temperatures of about 30°C in well-drained loam soils (National Research Council, 1996). It has been reported that pearl millet accounts for over half of the cultivated millets (ICRISAT/FAO, 1996), and the data in Table 6.2 essentially support this statement.

Physical characteristics of the grain
Pearl millet grains are tear-shaped to ovoid and vary greatly in color, from creamy white to gray and purple (Plate 6.1). Their resemblance to seed pearls is the origin of the name. The grains can be up to 2 mm in length and have a 1000 kernel weight of about 8–15 g (Abdelrahman *et al.*, 1984). The structure of the kernel (Figure 6.1)

Table 6.2 Estimated world and regional production (in thousand tons) of different millets in 1981–1985

	Total	Pearl	Foxtail	Proso	Finger	Teff	Fonio	Others
Africa	9 557	7 330	–	–[a]	855	1 063[b]	309	–
Asia	17 048	6 013	5 462	2 279	2 905	–	–	386
World	29 295	13 351	5 489	4 931	3 763	1 063	309	387
World (%)	100	45.6	18.7	16.8	12.8	3.6	1.1	1.3

[a] – no reported values
[b] Only values from Ethiopia
Official and FAO estimates based on country information, as modified by Marathee (1994).

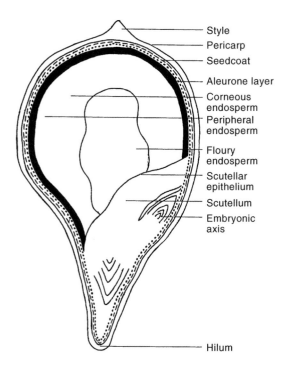

— Style
— Pericarp
— Seedcoat
— Aleurone layer
— Corneous endosperm
— Peripheral endosperm
— Floury endosperm
— Scutellar epithelium
— Scutellum
— Embryonic axis
— Hilum

Figure 6.1 Schematic longitudinal section through a pearl millet grain (Taylor, 2004a). Reprinted with permission of the copyright owner.

is similar to that of sorghum and maize and to most of the other millets, except finger millet. It consists of an outer pericarp, testa, aleurone layer, germ, and starchy endosperm. The kernels are naked (i.e. the hull is removed when they are threshed). What is unique about the pearl millet grain is that it has a relatively large germ, accounting for up to 21% of the whole grain (Abdelrahman *et al.*, 1984). The endosperm can thus be up to about 76% and the pericarp up to about 10% of the grain. The endosperm can be divided into peripheral, corneous, and floury endosperm. The peripheral endosperm has a dense protein matrix with small starch granules (Serna-Saldivar and Rooney, 1995). In the corneous region, the starch granules are more uniformly sized and polygonal, embedded in a protein matrix. The floury endosperm contains large, round-shaped, and loosely packed starch granule in a discontinuous protein matrix. Starch accounts for about 70% of the floury endosperm (Lestienne *et al.*, 2007), and consists of approximately 26% amylose (Muralikrishna *et al.*, 1986).

Nutrients and anti-nutrients

The nutrient composition of pearl millet is given in Table 6.3. It has a relatively high gross energy content of approximately 1475 kJ/100 g. This high energy content is due to the high fat content of the grain, which is related to the large germ size. Similarly, compared with other millets, pearl millet is high in protein (Serna-Saldivar and Rooney, 1995). The prolamin content is about 31–34%, which is lower than that in other millets. This is related to the large germ which is rich in albumins and globulin type proteins. The types of protein in pearl millet affect the amino acid

Table 6.3 Typical nutritional values of grain of the more important millets

	Pearl millet[a]	Foxtail millet[b]	Proso millet[b]	Finger millet[a]	Teff[a]	Fonio[a]	Japanese millet[b]	Kodo millet[c]
Moisture (%)	10	12	12	12	11	10	11	_[f]
Protein (%, N × 6.25)	11.8	9.9	9.9	7.3	9.6	9.0	8.9	11.5
Carbohydrate (%)	70	73	75	74	73	75	76	74
Fat (%)	4.8	1.6	1.7	1.3	2.0	1.8	1.8	1.5
Dietary fiber[d]	8.5[e]	9.4	13.1	11.7			14.3	9.4
Ash (%)	2.3	1.6	0.8	2.6	2.9	3.4	0.9	3.7
Calcium (mg/100 g)	37	–	–	358	159	44	–	–
Iron (mg/100 g)	9.8	–	–	9.9	5.8	8.5	–	–
Energy (kJ/100 g)	1475	–	–	1396	1404	1534	–	–
Vitamin A (microgram retinol equivalents)	22	–	–	6	8	–	–	–
Lysine (g/100 g protein)	3.2	1.7	1.2	2.5	2.3	2.5	1.5	1.7

[a] Values as is basis from National Research Council (1996), unless otherwise specified.
[b] Kasaoka *et al.* (1999), as is basis.
[c] Hulse *et al.* (1980), dry basis.
[d] Malleshi and Hadimani (1993), as is basis.
[e] Taylor (2004), dry basis.
[f] – Values not reported.

composition. Because of a higher globulin and albumin content, the essential amino acid lysine is slightly higher (Table 6.3). Related to this, the true protein digestibility of pearl millet in rats has been reported to be high, at 94–97% (Singh *et al.*, 1987). The lipids in pearl millet are mostly unsaturated fatty acids (about 75%) (Osagie and Kates, 1984). This has an impact on the storability of milled pearl millet due to oxidative rancidity. Starch, the major carbohydrate in pearl millet, has a low *in vitro* digestibility of about 62% (Muralikrishna *et al.*, 1986), which can, however, increase to about 73% when the grain is popped like popped corn. Table 6.3 shows a dietary fiber of 8.5%, but Ragaee *et al.* (2006) reported a value of 15%, and Singh *et al.* (1987) reported values up to 17%. This large discrepancy could be caused by the complexity of dietary fiber determination and varietal differences. Pearl millet also contains about 2% resistant starch. It is a good source of minerals if whole grain is consumed, but decortication can decrease total iron content by 30% (Lestienne *et al.*, 2007).

The major anti-nutrients in pearl millets are phytic acid, goitrogens, and oxalic acid. The phytate content is about 0.7–0.8% (Lestienne *et al.*, 2007). Phytates can decrease the bioavailability of minerals such as calcium, iron, and zinc through binding. The goitrogenic compounds in pearl millet are probably the phenolic flavonoids, C-glycosylflavones and their metabolites (Gaitan *et al.*, 1989). These compounds have also been identified as the cause of off-odors in pearl millet flour (Reddy *et al.*, 1986), characteristically a "mousy" or "mouse-droppings" like flavor (Pelembe,

2001). Decortication, removal of the bran from the grain, can significantly reduce flavonoid-type compounds (Lestienne *et al.*, 2007), suggesting that decortication prior to milling can reduce the goitrogen content. Generally, phenolic type compounds are high in pearl millet (about 1400 µg/g gallic acid equivalents) and this contributes to its high antioxidant activity in comparison with wheat, barley, or rye (Ragaee *et al.*, 2006). Oxalic acid present in pearl millet can significantly reduce calcium availability by forming calcium oxalate complexes (Opoku *et al.*, 1981). However, it was found that malting (sprouting of the grain) can substantially reduce oxalate content from 0.50 to 0.07%.

Foxtail millet

Description, history and production

Foxtail millet is mainly grown in China and other Asian countries. It is one of the most ancient cereals of Euroasia, and China is probably the center of origin (Jiaju and Yuzhi, 1994). It was of great importance in the Neolithic culture in China. Foxtail millet can grow up to 1.5 m high in the single-stem variety, but the types in China are generally 60–70 cm high. It can be highly tillered. Each tiller has a panicle which looks like a fox's tail, hence the name. The panicle is about 10–15 cm long and 1.5–3.0 cm in diameter (House *et al.*, 1995). The crop is well adapted to temperate regions and matures within 70–120 days. The production of foxtail millet in China was over 2.5 million tons in 1988 (Jiaju and Yuzhi, 1994) and the world production is about 5.5 million tons, with Asia the main producer (Table 6.2).

Physical characteristics of the grain

Foxtail millet grains are about 2 mm in length and the glumes can be white, red, yellow, brown, or black. Neither white nor red varieties were found to contain tannins (Hedge and Chandra, 2005). The 1000 kernel weight is about 2.6 g and milling yield is about 77% after removal of the husk and bran (Malleshi and Hadimani, 1994). The starch granules in foxtail millet are generally spherical in the floury endosperm, but polygonal forms have also been found (Kumari and Thayumanavan, 1998). The granule size varies from 0.8 to 9.6 µm. The amylose and amylopectin content of the starch depends on the type of foxtail millet. Foxtail millet can be waxy (high in amylopectin), normal (low amylose) or non-waxy (high amylose) (Nakayama *et al.*, 1998). In normal foxtail millet the amylose content can be up to 33% (Malleshi *et al.*, 1986). The protein bodies are mostly spherical and are 1–2 µm in diameter (Rost, 1971). About 40% of the total extractable nitrogen is prolamin protein and about 20% can only be extracted with a reducing agent (Danno and Natake, 1980). This indicates that foxtail millet, like most other cereals, is rich in prolamin protein but there is a high proportion of disulfide bonds in the protein.

Nutrients and anti-nutrients

The proximate nutrient content of foxtail millet is similar to that of other millets (Table 6.3). *In vitro* protein digestibility of raw and cooked foxtail millet was reported as 77 and 92%, respectively (Ravindran, 1992), a high cooked value. The reported starch content of foxtail millet is about 50–55% (Kumar and Parameswaran, 1998), which is relatively low for cereals. The *in vitro* digestibility of native and popped starch after 3 hours digestion was found to be low, about 47 and 52%, respectively (Muralikrishna *et al.*, 1986). However, Ushakumari *et al.* (2004) reported about 77% starch in decorticated grain and a high starch digestibility of about 96%. This suggests that there is a high proportion of bran in the whole grain, which interferes with starch digestibility. The major fatty acids in decorticated foxtail millet are palmitic acid (C16:0) (46%), stearic (11.5%) (C18:0) and oleic acid (35%) (C18:1) (Ushakumari *et al.*, 2004), which represents an unusually high proportion of saturated fatty acids for cereal grains. The total dietary fiber is around 9.4% (Table 6.3), but Ushakumari *et al.* (2004) reported values of about 8.8% for raw and 11.8% when popped.

Foxtail millet has polyphenols, phytic acid, and oxalate as anti-nutritional factors. These can be decreased by processes such as dehulling (debranning), soaking, and cooking. For example, dehulling increased *in vitro* protein digestibility by 30%, by removing some of the anti-nutritional factors (Pawar and Machewad, 2006). The total phenolic and carotenoid contents of foxtail millet were reported as 47 and 80 μg/100 g, respectively (Choi *et al.*, 2007). A methanolic extract of these compounds was found to have good antioxidant activity. However, in comparison to kodo millet, foxtail millet seems to have a lower free radical quenching potential (Hedge and Chandra, 2005).

Proso millet

Description, history, and production
Proso millet is likely to have originated in Manchuria (House *et al.*, 1995), and it is widely grown in temperature climates across the world. It is an important crop in northwest China and is grown in Kasakhastan, the central and southern states of India and eastern Europe, USA, and Australia. As can be seen in Table 6.2, proso millet is probably the third most important millet, after pearl millet and foxtail millet. It is well adapted to many soil and climatic conditions, and it can be cultivated to altitudes up to 3500 m (Baltensperger, 1996). The plant is short, about 30–100 cm tall and has few tillers.

Physical characteristics of the grain
Proso millet grain varies in color from white cream, yellow, orange, red, brown to black (Plate 6.1). The grains are spherical to oval in shape, about 3 mm long and 2 mm diameter (Baltensperger, 2002). The 1000 kernel weight is about 7.1 g. Milling of the grain yields about 79% flour, with removal of the bran and the husk (Malleshi and Hadimani, 1994). The starch granules in proso millet starchy endosperm are mostly small and spherical rather than large and polygonal, and range from 1.3 to 8.0 μm diameter (Kumari and Thayumanavan, 1998). The endosperm protein bodies are globular in shape and about 2.5 μm in diameter (Jones *et al.*, 1970). Prolamin accounts for up to 80% of the total protein (Kohoma *et al.*, 1999).

Nutrients and anti-nutrients

The proximate nutritional composition of proso millet is similar to that of other millets (Table 6.3). Its starch can vary from 62 to 68% and the amylose content expressed as percentage of the grain is about 17% on a dry basis (Yanez *et al.*, 1991). The rate of starch hydrolysis of proso millet is similar to that of maize (Yanez *et al.*, 1991). Concerning the nutritive value of protein, proso millet has an *in vitro* digestibility of about 80% (Kasaoka *et al.*, 1999). When compared with casein, proso millet protein has been reported to have beneficial effects by suppressing liver injury induced by D-galactosamine (Nishizawa *et al.*, 2002). In terms of the triglycerides, the most common fatty acids are linoleic acid (60%) followed by oleic acid (14%) (Sridhar and Lakshminarayana, 1994). Proso millet has been found to increase the level of the desirable high-density lipoprotein in the blood plasma of mice (Nishizawa *et al.*, 1995). The total polyphenolic and carotenoid contents of proso millet have been reported as 29 and 74 μg/100 g, respectively and a methanolic extract containing these compounds was found to have good antioxidant properties (Choi *et al.*, 2007). With regard to anti-nutrients, proso millet apparently does not have protease inhibitory activity when compared with pearl millet, foxtail millet, and finger millet (Chandrasekher *et al.*, 1982), however, chymotrypsin inhibitors have been detected (Ravindran, 1992).

Finger millet

Description, history, and production

The name finger millet comes from the shape of the head of grains, as it resembles the fingers of the hand. The African native finger millet probably originated in the highlands of Uganda and Ethiopia (National Research Council, 1996). It is estimated that the world production is at least 4.5 million tons, mostly in Asia and Africa, with India being the leading world producer (Table 6.2). The finger millet plant can grow up to about 1.3 m, but is commonly 1.0 m tall. It is widely cultivated in eastern Africa around Lake Victoria, southern Africa, and India. Finger millet requires moderate rainfall (500–1000 mm), an intermediate altitude (500–2400 m), and thrives under hot conditions as high as 35°C in well-drained soils. In Malawi, an average yield of about 4 tons/ha can be obtained for early maturing varieties and the panicle length is about 6 cm (Mnyenyembe, 1994). In India, finger millet apparently can yield up to 5 tons/ha, and it is a highly valued crop as it can be stored for up to 50 years (National Research Council, 1996), which makes it a good reserve against famine.

Physical characteristics of the grain

Finger millet grain is essentially spherical in shape, about 1–2 mm diameter with an average 1000 kernel weight of 2.5 g. The grain can range from white to brown in color (Plate 6.1). White-colored grain is mostly preferred for porridge and the brown-colored varieties are used for traditional opaque beer brewing in southern Africa (Gomez, 1994). Finger millet is unique in its grain characteristics as it is a utricle instead of a true caryopsis like other cereals. The utricle characteristic means that the pericarp is not completely fused with the testa (McDonough *et al.*, 1986). This allows the pericarp to be removed by simply rubbing the dry grain or rubbing it after soaking in water. Finger millet has a five-layered testa which can be red to purple. The color is due to the flavonoids and tannins (Ramachandra *et al.*, 1977). The endosperm contains protein bodies of about 2 μm in diameter and starch granules with a diameter varying from 8 to 21 μm (McDonough *et al.*, 1986). The starch granules in finger millet are compound, with the individual granules being spherical, polygonal as well as rhombic shaped (Malleshi *et al.*, 1986).

Nutrients and anti-nutrients

The general nutritional composition of finger millet is shown in Table 6.3. The protein content is quite low, but it is apparently rich in the essential amino acid methionine (National Research Council, 1996). Average *in vitro* protein digestibilities of raw and cooked finger millet of 71 and 87%, respectively, have been reported (Ravindran, 1992). Carbohydrate (starch) is the major component of the finger millet grain. The starch content is about 60%, with amylose making up about 30% of the starch (Mangala *et al.*, 1999). The *in vitro* starch digestibility of native and popped finger millet starch after 180 minutes hydrolysis have been measured as 66 and 74%, respectively (Muralikrishna *et al.*, 1986). Finger millet grain and malt are a good source of dietary fiber (Rao *et al.*, 2004). The fat content is low (Table 6.2), leading to very good storability of the grains. In terms of the fatty acid composition, unusually almost half is oleic acid (Fernandez *et al.*, 2003). Concerning minerals, finger millet is rich in calcium, iron, magnesium, molybdenum, selenium, and manganese (Fernandez *et al.*, 2003). Because of the low glycemic response, finger millet can be considered in diets for people with non-insulin-dependent diabetes (Kumari and Sumathi, 2002).

With regard to anti-nutrients, the phytate and total oxalate content in finger millet is apparently quite high (0.5 and 0.03%, respectively) (Ravindran, 1991). Tannins and trypsin inhibitors are also anti-nutrients in finger millet. Fermentation significantly reduces the effect of these (Antony and Chandra, 1998). Although tannins can be considered as an anti-nutrient, they can have antioxidant properties (Siwela *et al.*, 2007).

Teff

Description, history, and production

Teff is believed to have originated in Ethiopia and this country is the leading world centre for teff genetic diversity. Teff is a relatively short plant, up to 1.2 m high with slender stems (National Research Council, 1996). This promotes the susceptibility to lodging and thus the grain is harvested manually (Cheverton *et al.*, 1994). Teff is a major cereal in Ethiopia, and its annual production has increased from 1 million tons in the late 1980s (Table 6.2) to 2 million tons in the 1990s (Bultosa and Taylor, 2004a). This represents about 20% of Ethiopia's cereal production. Teff is used as a health grain in the USA, and in South Africa it is used mainly as a forage crop. Teff can be cultivated at a wide range of altitudes, up to 3000 m. However, it is best cultivated at altitudes of 1100–2950 m (National Research Council, 1996). Teff is grown in areas with an average annual rainfall of 1000 mm and it generally yields about 1 ton/ha, but in Ethiopia improved varieties can yield up to 1.8 tons/ha.

Physical characteristics of the grain

The grain can be of several colors, ranging from white to red and brown (Plate 6.1). The white color is mostly preferred for food use. The grain is oval, with a diameter less than 1 mm and a 1000 kernel weight of 2 g. The starchy endosperm contains mainly starch granules and protein bodies. The protein bodies are individual granules and do not coalesce. However, the starch granules, like those in finger millet are compound and polygonal in shape, with individual granules 2–6 μm in diameter (Bultosa *et al.*, 2002).

Nutrients and anti-nutrients

The nutritional composition of teff is similar to that of other millets (Table 6.3). However, unusually, albumin and globulins are in greater proportion than prolamins (Tatham *et al.*, 1996). Because of its lower prolamin content, teff may have better protein digestibility, in comparison to cereals such as maize and sorghum. The amino acid composition can be regarded as well balanced (Bultosa and Taylor, 2004a), except that the lysine content is still relatively low. Teff starch contains about 25–30% amylose (Bultosa *et al.*, 2002). Using porcine pancreatic α-amylase, the *in vitro* starch digestibility of teff starch was found to be similar to that of maize starch (Bultosa and Taylor, 2004b). In terms of minerals, teff is rich in iron, calcium, magnesium, and phosphorus. The iron and calcium contents are reported to be 11–33 and 100–150 mg/100 g, respectively (National Research Council, 1996). Although it has been suggested that teff can contain tannins, Bultosa and Taylor (2004a) found that the testa of neither white nor brown varieties contained tannin.

Fonio

Description, history, and production

Two types of fonio exist, white fonio and black fonio, with the former being the more common (Table 6.1). Fonio is cultivated mostly in the dry savannas of West Africa (the Sahel region) and is probably the oldest African cereal (National Research Council, 1996). The plant grows about 45–50 cm high with a finger-like panicle of up to 15 cm long. The general yield is about 1.0–1.4 tons/ha, but up to 2.0 tons/ha can be harvested under very good agronomic conditions (Ndoye and Nwasike, 1994). Total fonio production in Mali and Senegal is around 10 000 tons (Smith, 1996). Fonio matures very early, and early types can produce grain 6–8 weeks after sowing. Thus, fonio is sometimes referred to as "the grain of life." Fonio can survive in poor soil conditions, such as sandy and acidic soils that are usually considered inferior for sorghum and pearl millet cultivation. It is mostly grown in areas with annual rainfall greater than 400 mm (National Research Council, 1996).

Physical characteristics of the grain

Fonio grains are tiny, with a 1000 kernel weight of only 0.5–0.6 g (Plate 6.1). The grain is about 1.0 mm long and 0.75 mm wide (Irving and Jideani, 1997), with the germ being over a third the length of the grain. The starchy endosperm contains polyhedral starch granules of about 10 μm in diameter. The starch amylose content is normal, about 27% (Jideani et al., 1996). The protein bodies occur in all parts of the grain, but are mostly abundant in the starchy endosperm (Irving and Jideani, 1997). The protein was reported to be mostly glutelin (Jideani et al., 1994), however, this may be misleading as over 55% of the protein was not extracted. The inefficient extraction may be due to the highly crosslinked nature of the protein. Unusually, the aleurone layer and endosperm periphery seem to be abundant in lipids (Irving and Jideani, 1997).

Nutrients and anti-nutrients

Fonio proximate composition is similar to that of other millets (Table 6.3). Fonio is limited in lysine, but rich in methionine (Lumen et al., 1993). According to these authors, the chemical score for the first limiting amino acid lysine is only 50% of that of whole egg for children 2–5 years. To date, the anti-nutrients of fonio have not been investigated.

Japanese millet and Sawa millet

Description, history, and production

Japanese millet is believed to have originated in Japan (House et al., 1995), whereas Sawa millet, another species in the same genus, was domesticated in India. The morphology of both species is similar and so both will be dealt with together, although there is far more information on Japanese millet. Japanese millet is mostly cultivated in Japan, especially in the Tohutu district (Wanatabe, 1999), and it is considered to be an important grain in this area because of its good storage ability. It is also produced in China and Korea. A yield of 3.0 tons/ha can be expected for Japanese millet. The plant generally prefers a warm climate, but is resistant to cold temperatures.

Physical characteristics of the grain

The 1000 kernel weight for Japanese millet is about 3.3 g, with the husk and bran contributing to about 23% of the total weight (Malleshi and Hadimani, 1994). The starch granules in Japanese millet are about 1.2–10.0 μm in diameter, spherical and polygonal in shape (Kumari and Thayumanavan, 1998).

Nutrients and anti-nutrients

The proximate nutritional composition of Japanese millet is similar to that of other millets (Table 6.3). The major protein fraction in Japanese millet is glutelin, followed by prolamin (Suman et al., 1992). The in vitro protein digestibility of native and heated Japanese millet has been found to be quite high, 84 and 89% respectively (Suman et al., 1992). Although the fat content reported in Table 6.3 is 1.8%, Sridhar and Lakshminarayana (1994) reported a total lipids content (including bound and structural lipid) of 8.0%, which seems rather unlikely. More likely was the fact that about 48% of lipids was linoleic acid and 28% oleic acid. Japanese millet appears to have good antioxidant properties. Compounds isolated with antioxidant properties include luteolin and tricin as phenolics and N-(-p-coumaroyl) serotonin as a serotonin derivative (Wanatabe, 1999). To date, the anti-nutrients in Japanese and Sawa millet have not been investigated.

Kodo millet

Description, history, and production
Kodo millet is indigenous to India, and it is believed to have been domesticated some 3000 years ago (House et al., 1995). It well adapted in the tropics and subtropics. Kodo millet is generally cultivated with its weedy species and no distinction is made between the species during harvest. The crop matures in 4–6 months with yields varying from 250 to 1000 kg/ha (Hulse et al., 1980).

Physical characteristics of the grain
Kodo millet has a 1000 kernel weight of 6.7 g. The bran and husk form a large proportion of the grain, about 37% (Malleshi and Hadimani, 1994). The starch granules are large and polygonal, but some small polygonal types can be found (Kumari and Thayumanavan, 1998). The size of the granules varies from 1.2 to 9.5 μm. The amylose content expressed as a proportion of grain weight is about 20% on a dry weight basis.

Nutrients and anti-nutrients
As with other millets, the proximate nutrient composition of kodo millet is not unusual (Table 6.3). The prolamin protein is similar to barnyard and foxtail millets (Parameswaran and Thayumanavan, 1997). It is rich in glutamic acid (mainly glutamine), alanine, leucine, and serine, but deficient in lysine. Sridhar and Lakshminarayana (1994) reported a higher fat content value (3.2%) than that reported in Table 6.2. The fat has similar quantities of linoleic and oleic acid, making up 70% of the total fatty acids of the major lipid fraction. In comparison to finger, barnyard, and pearl millets, kodo millet has the highest free radical quenching potential, indicating possible useful antioxidant activity (Hedge and Chandra, 2005). Not surprisingly, however, the radical quenching activity decreases when the grain is decorticated or heated by roasting and boiling. To date, the anti-nutrients in kodo millet have not been investigated.

Traditional foods and beverages

There are a huge number of traditional millet foods and beverages. They can be categorized as wholegrain foods, foods made from meal/flour, and non-alcoholic and alcoholic beverages. These traditional products are consumed in Africa, the Indian subcontinent, and East Asia. Because of the vast number of different local variations, this account will be limited to describing representative examples of the various categories of foods and beverages.

Wholegrain foods

Many grains, including finger millet, are popped in India. The process involves moistening the grains to about 19% moisture, allowing them to temper for several hours, then agitating the grain in a bed of hot sand (240°C) for a few minutes (Malleshi and Hadimani, 1994). Popping removes the outer pericarp. The popped grain may be consumed as a snack or further processed by milling. Unfortunately, the quality of the products is generally poor due to rancidity and contamination with sand. In Gujarat province, India, wholegrain finger millet may be cooked to produce a rice-like product called kichadi (Subramanian and Jambunathan, 1980). A similar product is also produced from sprouted grain.

Foods made from meal/flour

Not unnaturally, there is a wide range of traditional millet foods produced from meal (coarsely ground grain) or flour. Such foods include flatbreads, couscous, dumplings, and porridges.

Flatbreads

These pancake-like breads are staples in the Horn of Africa (Ethiopia, Eritrea, and Sudan). They are made from a variety of different cereals, especially millets. A feature of many of the flatbreads is that the flour undergoes a mixed lactic acid bacteria and yeast fermentation (Gashe *et al.*, 1982), which gives them a somewhat leavened texture and an acidic flavor. Probably the two most well known of these flatbreads are injera and kisra. Injera from Ethiopia and Eritrea is a large (approx. 50 cm diameter), spongy textured pancake about 5 mm thick. It has a honeycomb-like appearance, very similar to an English crumpet. Teff, followed by finger millet are preferred for making injera. This is because injera made from these millets stales much more slowly than if made from sorghum or other cereals (Yetneberk *et al.*, 2004). Injera is served with just about anything, especially spicy sauces. Kisra from Sudan, in contrast, is a thin, flexible wafer (1–1.5 mm thick) with neither holes nor a spongy texture (Badi *et al.*, 1989). Kisra is served with stews (mullah or tabbikh), relish or sauce, or on its own seasoned with salt and chillies (Ejeta, 1982). In southern India and Sri Lanka, millets may be used to make dosa, a thin, fermented pancake that contains black gram (mung bean) (Murty and Kumar, 1995).

Probably the most well-known unfermented flatbread that may be made from millet (pearl millet and finger millet) is roti, also known as chapatti. This popular staple in India is a very thin (1.3–3.0 mm), 12–25 cm diameter pancake with a soft, flexible puffed texture. Rotis are served with vegetables, meat, fermented milk products, pickles, chutney, or sauce (Murty and Kumar, 1995). In Ethiopia, a similar sweet, unleavened flatbread is called kitta and is preferably made from teff (Bultosa and Taylor, 2004a).

Couscous

In North Africa, couscous (steamed, agglomerated meal) is made from semolina (i.e. particles of wheat endosperm). This is the type of couscous that is widely available in supermarkets across the world. However, in the Sahelian countries of West Africa, such as Mali and Senegal, couscous is produced from pearl millet (Plate 6.2), often mixed with sorghum or maize. This couscous may be steamed. Steamed couscous may be stirred into yoghurt and is called thiakri, thiacry, or tiakri. Agglomerated, but not steamed, couscous is called arraw or karaw. Couscous products seem to be categorized according to the particle size. Fine steamed couscous is known as thiere, whereas thiakri is coarse. Fine couscous is of particle size similar to flaking grits and coarse is about wheat grain size. Couscous is generally served with a spicy aromatic sauce.

Dumplings and other dough products

The Pedi people of northern South Africa traditionally prepared boiled dough dumplings from wholegrain pearl millet meal (Quin, 1959). These dumplings are called dingwa tsa bupi bja leotsa, literally meaning dumplings or breads from pearl millet. They are described as being greenish-brown in color with a firm crumbly texture and a pleasant, slightly bitter, nutty, musty, sweet taste. A similar Indian dumpling product is called mudde (Malleshi and Hadimani, 1994). Also in India, steamed millet dough may be fried to produce a snackfood called ponganum (Subramanian and Jambunathan, 1980). Lin *et al.* (1998) describe a similar type of product from the Shanxi province in the north of China from proso millet called an oil pudding. Strips of steamed millet dough are wrapped around cooked red beans and fried. Apparently, the oil pudding has a sweet aroma and a delicate texture on the inside.

Porridges

There is an almost infinite range of traditional porridges that may be made from millets. The porridges range in consistency from stiff, like mashed potato, to a runny, spoonable gruel. The consistency is primarily related to the solids content of the porridge, which ranges from approximately 30% down to 10%. Serving temperature also plays a role in porridge consistency. Viscous hot porridges will invariable set when allowed to cool. Porridges also vary greatly in flavor. They are frequently soured by lactic acid fermentation or addition of acid such as tamarind juice and today even mayonnaise. Fairly commonly, porridges are made alkaline. Malted cereal flour is often added as an ingredient in porridge making. This imparts a sweeter taste to the porridge and also reduces its viscosity.

Across Africa, the staple food is stiff porridge. In southern Africa, it goes by many names, including pap (South Africa) and sadza (Zimbabwe) and in East Africa it is called ugali, a Kiswahali word. Today, commonly these stiff porridges are made from maize meal. However, pearl millet and finger millet are still used by rural people across the region. In the Sahel region, stiff porridges are commonly called tô and decorticated pearl millet is probably the most popular cereal used. In Mali, tô is often made alkaline by the addition of wood or millet/sorghum stalk leachate or lime (calcium oxide) (Rooney *et al.*, 1986). The pH of the tô is around 8.2 and it is served cool with a sauce. In Shanxi province, China, foxtail millet porridge is a traditional food (Lin *et al.*, 1998).

In northern South Africa and Botswana, a popular intermediate viscosity fermented porridge is called ting. The Pedi people traditionally often made ting from wholegrain pearl millet meal (Quin, 1959). The process involved a one-day fermentation prior to cooking the porridge. The final product had a lactic acid content of around 0.8%.

Probably the most well-known thin porridges are ogi in Nigeria and uji in Kenya. Both these products are commonly soured or made acid. Ogi is often made from pearl millet and uji is very commonly made from finger millet (Plate 6.2).

Non-alcoholic beverages

A very popular pearl millet beverage in Namibia is called oskikundu. It is a lactic acid fermented product made from cooked pearl millet flour with added sorghum malt flour (Taylor, 2004b). Oshikundu is greenish-brown in color with a slightly viscous consistency and a buttery sour taste. Similar products called togwa, made from maize meal and finger millet malt (Oi and Kitabatake, 2003), and kunun zaki (Ayo, 2005) (Plate 6.2), which may be made from pearl millet and white fonio, are produced in Tanzania and Nigeria, respectively. In Zimbabwe, a traditional fermented beverage combines milk and finger millet to give a highly nutritious product (Mugocha *et al.*, 2000).

Alcoholic beverages

Across much of Africa, pearl millet and finger millet are still used widely to make traditional African beers. In southern Africa, traditional beer is often called opaque beer because of its appearance, resulting from semi-suspended particles from the cereal, gelatinized starch, and yeast. Quin (1959) describes how the Pedi people made such beer from 100% pearl millet malt called bjalwa bja leotsa, literally beer from pearl millet. The beer was greenish-brown in color with a milk-like effervescent consistency and a pleasant, musty, bitter-sour taste. The lactic acid content was 1.8%. These opaque beers are effervescent because they are not pasteurized and they are consumed when they are actively fermenting. Opaque beers have a relatively low alcohol content, up to 3%. Today, in Bulawayo, Zimbabwe, pearl millet is malted on a large commercial scale in modern, pneumatic type maltings and used as an ingredient in an industrially brewed opaque beer called Ndlovo, which means elephant in the Ndebele language.

A similar product from the Balkans, Egypt, and Turkey is bosa, also called busa or bouza (Arici and Daglioglu, 2002). The name is derived from buze, the Persian word for millet. Boza can be brewed from various cereals but proso millet is preferred. It is a thick liquid, pale yellow in color with a characteristic acid-alcoholic aroma. The alcohol content is generally low, less than 1% but boza from Egypt can contain up to 7%. In Ethiopia, finger millet and teff are used as ingredients to make a traditional opaque beer called tella and a spirit called katikalla (Bultosa and Taylor, 2004a). In the Himalayas, a traditional beer called chhang or jnard/jaanr is produced from finger millet (Malleshi and Hadimani, 1994). Interestingly, the brewing process does not involve malting the grain (Basappa, 2002).

In contrast, the traditional beers in West Africa, which are made from cereals including pearl millet, are substantially clear. These beers are variously known as burukutu, dolo, pito, sulim, or talla (Taylor and Belton, 2002). They are characterized by being filtered, but remain somewhat cloudy. They are sweetish and generally slightly sour tasting, with a fruity aroma and contain 1–5% alcohol (Demuyakor and Ohta, 1993).

Traditional millet-processing technologies

Milling

In Africa, millets are traditionally milled either using a wooden pestle and mortar or saddle stones comprising the base saddle stone and a roller stone. In India, stone rotary mills are used (Subramanian and Jambunathan, 1980; Murty and Kumar, 1995). These processes are generally two-stage processes (Smith, 1996). The first stage is decortication to remove the bran, which is obtained by moistening the grain. The bran is then removed by winnowing or sieving. The grain is then washed or soaked to remove remaining bran and left to dry (Figure 6.2). During the soaking stage, lactic fermentation can take place which gives a desirable sour flavor to the final flour (Taylor, 2004b). The second stage involves reducing the endosperm into a meal or flour.

Nowadays, these traditional methods have been displaced to a considerable extent by mechanical milling. In Africa, pearl millet is generally milled by first decorticating the grain using an abrasive disk dehuller (Bassey and Schmidt, 1989). A common type is the PRL (Prairie Research Laboratory) Dehuller. Dehullers comprise several carborundum or resinoid disks mounted on a horizontal shaft, within a cylindrical chamber. The shaft is turned at high speed by an electric motor or engine. The bran is rubbed off the grain by the action of the abrasive disks and also the actions of the grains against each other and against the surface of the chamber. The degree of decortication is simply controlled by the duration of decortication. The bran is removed by aspiration or sieving. After decortication, the endosperm is reduced to a meal or flour by hammer milling (Taylor, 2004b). In India, today, millets are generally milled using disk mills, which are called chakki (Subramanian and Jambunathan, 1980). These consist of two vertical stone, cast iron or steel disks, whose inner surfaces are fluted or otherwise raised (Munck, 1995). One rotates and the other is fixed. The grain is fed from the top into the center of the disks and issues out from between the disks. The general principle of operation is the same as a traditional stone mill.

Breadmaking

Millets, like all other cereals, except wheat, do not contain gluten-forming proteins. The wheat gluten proteins are responsible for giving wheat dough its viscoelastic, gas-holding texture, which enables leavened bread to be produced. Despite the absence of gluten, leavened flatbreads, such as injera, can be produced from millets. A key to

Figure 6.2 Processes for making millet foods. Left: Abrasive dehuller for pearl millet (top chamber: dehuller, bottom chamber: sieve separator) (Senegal). Center: Agglomerating pearl millet couscous (Senegal). Top right: Drying decorticated and fermented pearl millet grain (Namibia). Bottom right: Pouring teff injera batter onto the clay griddle (Ethiopia).

this seems to be that in making injera a portion of the fermenting dough, about 20%, is removed, diluted to a paste with water and cooked (Taylor, 2004b). This product is called absit. The cooking gelatinizes the starch and makes the absit viscous. The absit is then added back to dough and the mass is diluted to batter, which is further fermented. The increased viscosity of the batter resulting from cooking the absit seems to enable it to better hold the carbon dioxide produced during fermentation. A second important aspect is how the injera is baked. The fermenting batter is poured onto a hot clay griddle called a mitad (Figure 6.2). The lid of the mitad is then closed over the batter so that the batter is steamed. As the temperature of the batter rises, the carbon dioxide comes out of solution. At the same time, the starch in the batter gelatinizes, increasing its viscosity. The effect is to create gas bubbles in the batter, which turn into cells as the gas escapes and the batter sets. A third important aspect seems to be the nature of the grains themselves. Teff and finger millet make the best-quality injera. In particular, the injera made from these grains is resistant to staling (Yetneberk *et al.*, 2004). The exact reason for the good injera-making quality of these closely related millets is not known, but it seems to be related to their starch and it is probably of great significance that both have compound, and not simple type, starch granules (Bultosa *et al.*, 2002). Compound starch granules consist of many tiny polygonal granules.

Steaming and agglomeration

Couscous has a characteristic particulate texture. Making millet couscous involves first dampening fine meal (particles less than 1 mm) with 30–40% water and agglomerating the meal into particles by hand, then sifting the particles to obtain a uniform size (Figure 6.2) (Galiba *et al.*, 1987). They are then cooked in a steamer. The steamed particles are then agglomerated together by hand, sifted and steam cooked again. This process is repeated a further time. The steaming process gelatinizes the starch, binding the particles together.

Lactic acid fermentation

As mentioned, many traditional millet foods are soured. Traditionally, the process is performed by lactic acid bacteria (LAB) fermentation. The fermentation may be spontaneous (i.e. performed by naturally contained bacteria) or performed by selected starter culture. Another possibility is to use a portion of the fermented food product, or intermediate such as dough, as inoculum for the next fermentation. This process is known as back-slopping (World Health Organization, 1996). Botes *et al.* (2007) measured LAB levels ranging from 9×10^5 to 5×10^7 cfu/mL in three samples of Bulgarian boza. They all belonged to the genus *Lactobacillus* (i.e. *L. brevis*, *L. fermentum*, *L. paracasei*, *L. pentosus*, *L. plantarum*, and *L. rhamnosus*). *Lactobacillus bifermentans* and *Pediococcus pentosaceus* have been isolated from the finger millet fermented beverage jaanr (Saroj and Prakesh, 2004).

Lactic fermentation brings about several nutritional improvements in the grain (Taylor and Belton, 2002). Protein and carbohydrate digestibility are improved, B vitamins increase, and mineral availability is improved. However, the most important health benefit of fermentation is the reduction in pH to below pH 4.0. This inhibits the growth of pathogenic bacteria (Svanberg *et al.*, 1992) and slows down the rate of food spoilage (World Health Organization, 1996). Both these effects are highly desirable in developing countries where many people do not have access to safe water or refrigerated food storage. Because of these benefits, the World Health Organization (1996) sees lactic acid fermentation as an integral part of the fight against food-borne diseases.

Malting and brewing

Malting is the limited germination of cereal grain in moist air, under controlled environmental conditions. Traditionally, in southern Africa, millet and sorghum malting was carried out by first soaking the grain in a grass sack in a river and then allowing the grains to germinate for 2–3 days in the sack or spread out on a floor (Gadaga *et al.*, 1999). The malt was then sun-dried and finely coarsely ground. The major change in the grain brought about by malting is the mobilization of the grain's alpha- and beta-amylases. In porridges where malt is added the action of the malt alpha-amylase reduces the viscosity of the porridge by hydrolyzing the starch. This thinning action is very important when the porridge is used as a weaning food or a food for the infirm. The nutrient value of any food is directly related to its solids content.

Thus, a porridge of 30% solids will have three times the nutrient content of one of 10%, however, such porridges are too stiff to be consumed by infants and the infirm. Reduction of the viscosity can be achieved by adding malted grains. For example, the addition of sorghum malt to a 25% solids pearl millet porridge to give a total solids of 30% reduced the viscosity of the porridge from more than 6000 cP to an acceptable level of only 2500–3000 cP (Thaoge *et al.*, 2003). In addition, the porridge may also be made more palatable by action of the malt beta-amylase (i.e. production of maltose and thus increase of sweetness). It has been shown that the overall levels of amylase activity, referred to as diastatic power, in pearl millet malt (Pelembe *et al.*, 2002) and finger millet malt (Gomez, 1994) are similar to those in sorghum malt. However, pearl millet malt seems to have more beta-amylase activity, approaching the level in barley malt (Pelembe *et al.*, 2004). Malted cereal used to thin porridges is referred to as amylase-rich flour (ARF) or "power flour" (Alnwick *et al.*, 1988). The use of ARF, in conjunction with safe, hygienic practices has been strongly recommended for the preparation of weaning foods in Africa. Malting the grain also has many other effects on cereal grain composition, most of which impact positively on its nutritional value (Taylor and Belton, 2002). The essential amino acid composition is improved, as are protein and starch digestibility. The contents of B vitamins increase and mineral bioavailability is improved through the destruction of phytate.

In beer brewing, the primary function of the malt is to provide amylases which hydrolyze the starch into maltose and other simple sugars that can be fermented by yeasts into ethanol and carbon dioxide. Gadaga *et al.* (1999) described the traditional processes for brewing the opaque beer called "doro" in Zimbabwe using pearl millet or finger millet malt. In these processes, the malt is cooked to gelatinize the starch, allowed to sour, then further malt is added as a source of amylases to hydrolyze the starch. Typically of such traditional beers, alcoholic fermentation is brought about by wild yeasts. This is reflected by the fact that methanol, butanol and other alcohols in addition to ethanol have been detected in a type of this beer. Interestingly, in an analysis of Bulgarian boza, Botes *et al.* (2007) found nine species of yeasts, mainly *Candida* and *Pichia* species, but none of the detected yeasts belonged to the genus *Saccharomyces*, which contains the conventional yeasts responsible for beer alcoholic fermentation. In slight contrast, the finger millet beer jaanr was found to contain *S. cerevisiae*, but also *Saccharomycopsis fibuligera*, as well as *Candida* and *Pichia* species were isolated (Saroj and Prakesh, 2004).

Recent and future trends

Milling

Due to the small size of the grains, a major challenge in millet milling is the efficient separation of clean flour from the bran. This difficulty has stimulated considerable innovation. Perhaps the most remarkable is the fonio dehuller developed in Senegal in the early 1990s by the mechanical engineer Mr Sanoussi Diakité. Apparently, the dehuller comprises pliable abrasive disks which form a kind of spiral propeller with a fixed abrasive tray (Smith, 1996). It can decorticate 2 kg of grain in 6 minutes, compared with the hour or more for manual dehulling. The decorticated grain produced is very clean, with a 95% plus efficiency. In industrial milling of finger millet in Kenya, decortication is achieved using a modified Beall-type degerminator. In the 1970s, a horizontal disk-type debranner, the Palyi-Hansen BR 001-2, which in principle is like a Carter disk grain separator, was developed in Canada for decorticating sorghum and millet (Rasper, 1977). It had a capacity of up to 3 tons of grain per hour and gave a grain yield of about 77% from pearl millet.

Another type of dehuller developed for sorghum and millet is the United Milling Systems machine from Denmark (Munck, 1984). This dehuller aimed to imitate the action of hand pounding. It comprises a screw which transports the pre-moistened grain from below into a decortication chamber. Decortication is achieved by means of a rotor turning in perforated cylindrical screens. The dehuller is operated in conjunction with a vertical disk mill. Flour yields of 80% were claimed for pearl millet. However, in order to achieve this yield, it was necessary to recover by sieving and aspiration up to 60% of the endosperm that had passed through the perforated screen together with the bran.

In Niger, a milling process, SOTRAMIL (Societé de Transformation du Mil) was developed specifically for pearl millet (Goussault and Adrian, 1977; Kent, 1983). Cleaned, washed grain is decorticated using a horizontal millstone type machine. The decorticated grain is then passed through a brush machine to further remove impurities, and then milled to flour using an Ultrafine impact grinder. The air carrying the endosperm material to the grinder is at 85°C. This lowers the moisture content to a safe level, sterilizes the flour, kills insects and eggs and inactivates lipolytic and other deteriorative enzymes in the grain. After grinding, the flour is separated by sieving.

Roller milling as used for wheat milling is not successful by itself for millet milling (Perten, 1977), but is used commercially with additional steps. In Namibia, industrial milling of pearl mill first involves decorticating the grain using a disk dehuller, followed by roller milling the endosperm (Mr S. C. Barrion, Lecturer, University of Namibia, personnel communication). Apparently, in Senegal hammer milling may additionally be used in pearl millet milling to reduce large pieces of endosperm to flour (Smith, 1996).

Foods and beverages

Bread and biscuits (cookies)

For more than 30 years baguette-type bread has been produced in Senegal using pearl millet. The flour is a composite of about 85% wheat flour and 15% millet flour (Perten, 1984). The baguettes are rather smaller and darker than those made from 100% wheat flour, but they still have maintained their popularity.

The absence of gluten requires a change in the breadmaking process to make bread from 100% non-wheat flour. Generally, a batter with some 100–150% water on a flour basis is used instead of a dough (Taylor et al., 2006). Hence, the process is more like cake making. Additionally, hydrocolloids, starch or gum, are normally included in the recipe in combination with the non-wheat flour (Satin, 1988). The added hydrocolloid seems have the same function as the absit in making injera. Around 1990, the Nigerian Federal Institute of Industrial Research carried out pioneering research into these batter breads and achieved notable success. Olatunji *et al.* (1992) made batter breads from 70% maize, sorghum or pearl millet flours, plus 30% cassava starch and the normal ingredients of yeast, salt, sugar, shortening and fungal alpha-amylase. A batter was prepared using 80–100% water on a flour basis. The batter was fermented for 30 minutes, poured into baking pans, fermented for a further 20 minutes and then baked. Of the three cereal flours, pearl millet produced bread with the highest specific volume, 2.33 cm^3/g. Unfortunately, however, the pearl millet gave the bread a grayish crumb color, which was judged unacceptable.

With regard to making biscuits, the major problem when using non-wheat flours is that the biscuits tend to be fragile and crumble. When using pearl millet flour, Badi and Hoseney (1977) found that the only solution was to include wheat flour in the recipe. As the proportion of wheat flour was increased, the biscuits became progressively less fragile and of larger diameter. Interestingly, nearly 30 years later, Indrani *et al.* (2004) were granted a US patent for a process to make finger millet biscuits. The recipe comprises 50–60% finger millet flour and 7–10% wheat gluten powder, plus other ingredients.

Ready-to-eat foods

An excellent puffed proso millet product is manufactured in the USA (Plate 6.2). This ready-to-eat breakfast cereal is produced by gun puffing. In gun puffing, whole grain is put into a pressure vessel (the gun) and live steam at about 1750 kPa is injected. The vessel is heated up to 150°C for 1–2 minutes. A trip valve is released and the grain explodes out of the vessel. The rapid reduction in pressure causes the water in the grain to vaporize instantaneously, gelatinizing the starch and puffing up the grain. The puffing up of the grain also causes the bran to flake off. Research has also been carried out into producing ready-to-eat foxtail millet products by the use of various techniques (Ushakumari *et al.*, 2004). It was found that roller drying gave the highest degree of starch gelatinization, followed by popping, flaking, and extrusion cooking.

Beverages

Malted finger millet powder, which is mixed with hot milk or water to make a beverage, is a popular commercial product in India (Malleshi and Hadimani, 1994). However, apparently, many of the products on the market contain less than 10% millet malt. This is due to problems with powder dispersability. Product development work is required to improve quality. In Nigeria, however, there is an excellent instant pearl millet beverage called kunun tsayima (Plate 6.2), which contains pre-cooked pearl millet flour flavored with tamarind oil.

With regard to lager beer brewing with millets, this has not been researched extensively (Taylor *et al.*, 2006). This is in contrast to the situation with sorghum where commercial sorghum lager beers are now being brewed in many countries. Probably the major reason for this difference is that millets are not generally available at sufficiently low cost to make them a competitive alternative for use in brewing.

Notwithstanding this, the limited research that has been carried out suggests that millets represent useful brewing ingredients. For example, Nzelibe and Nwasike (1995) obtained substantially higher extract (malt solubilization) when laboratory-brewing with pearl millet and fonio malts rather than with sorghum malt.

Healthy foods

Probably the most important health-promoting aspect of millets as foods is that they generally contain substantial amounts of phenolics. Phenolics are notable for their antioxidant activity, which appears to be beneficial in terms of prevention of cardiovascular disease and cancer (Awika and Rooney, 2004). Unfortunately, research on the phenolics of millets is limited. However, it can be stated without doubt that all millets contain phenolic acids (Dykes and Rooney, 2006). In general, ferulic acid, p-coumaric acid, and cinnamic acids are the main types. It appears that the only millet flavonoids are flavones, which are responsible for grain pigmentation. With regard to tannin-type phenolics, it appears that finger millet is unique in that some varieties contain condensed tannins (Ramachandra et al., 1977). The antioxidant activity of the tannin-containing varieties is much higher than that of varieties without tannins, and similar to that of tannin sorghums (Siwela et al., 2007).

Concluding remarks

Despite the fact that millets do not contain gluten-forming proteins, there is a vast number of different types of millet foods and beverages. In the developing world, millets have generally retained their popularity, even as people are becoming increasingly urbanized. This is evident from the popularity of packaged millet-based food products, as shown in Figure 6.2. For example, in the Sahel region of West Africa, the number of processors of sorghum and millets and the quantities they process are increasing rapidly to meet the demands of the growing urban population (Vitale and Sanders, 2005). In developed countries, organic farmers and specialty food companies are turning to millets as niche products. A good example is that of teff farming in the US State of Idaho (National Research Council, 1996; The Teff Company, 2007).

Millets have the potential to add variety to our diet and may have useful health-promoting properties, particularly antioxidant activity. There has been considerable innovation in millet-processing technology and food product development. The area that now seems to require more attention is to improve the cost competitiveness of millets. This can be achieved through development of improved, higher yielding varieties and hybrids that are suitable for mechanized farming. This work is being undertaken by various agencies worldwide, in particular the International Crops Research Institute for the Semi-Arid Tropics (ICRISAT) and the CGIAR (Consultative Group on International Agricultural Research) organization responsible for millets, but more resources are required.

Sources of further information and advice

AACC International (formerly American Association of Cereal Chemists). www.aacc net.org

AFRIPRO (2003). Workshop on the Proteins of Sorghum and Millets: Enhancing Nutritional and Functional Properties in Africa. www. afripro.org.uk

Center for New Crops & Plant Products, Purdue University. www.hort.purdue.edu/ newcrop

Dendy, D. A. V. (1995). *Sorghum and Millets: Chemistry and Technology*. St Paul, MN: American Association of Cereal Chemists.

Food and Agriculture Organization (FAO) (1995). *Sorghum and Millets in Human Nutrition*. Rome: FAO (available at www.fao.org).

Germplasm Resources Information Network (GRIN) (2007). *GRIN Taxonomy for Plants*. www.ars-grin.gov/cgi-bin/npgs

Hard, N. F., Odunfa, S. A., Lee, C-H., Quintero-Ramirez, R., Lorence Quinones, A., and Wacher-Radarte, C. (1999). *Fermented Foods. A Global Perspective*. Rome: FAO (available at www.fao.org).

International Association for Cereal Science and Technology (ICC). www.icc.or.at

International Crops Research Institute for the Semi-Arid Tropics (ICRISAT). www. icrisat.org

International Sorghum and Millet Collaborative Research Support Program (INTSORMIL). www. intsormil.org

National Research Council (1996) *Lost Crops of Africa*, Vol. I: *Grains*. Washington, DC: National Academy Press (available at www.nap.edu/books).

References

Abdelrahman, A., Hoseney, R. C., and Varriano-Marston, E. (1984). The proportions and chemical compositions of hand-dissected anatomical parts of pearl millet. *J. Cereal Sci.* 2, 127–133.

Alnwick, D., Moses, S., and Schmidt, O.G. (1988). *Improving Young Child Feeding in Eastern and Southern Africa: Household-level Food Technology*. Ottawa: International Development Research Centre.

Andrews, D. J. and Kumar, K. A. (1992). Pearl millet for food, feed and forage. *Adv. Agron.* 48, 89–139.

Antony, U. and Chandra, T. S. (1998). Antinutrient reduction and enhancement in protein, starch and mineral availability in fermented flour of finger millet (*Eleusine carocana*). *J. Agric. Food Chem.* 46, 2578–2582.

Arici, M. and Daglioglu, O. (2002). Boza: A lactic fermented cereal beverage as a traditional Turkish food. *Food Rev. Int.* 18, 39–48.

Awika, J. M. and Rooney, L. W. (2004). Review: Sorghum phytochemicals and their potential impact on human health. *Phytochemistry* 65, 1199–1221.

Ayo, J. A. (2005). Effect of acha (*Digitaria exilis* Staph) and millet (*Pennesetum typhodium*) grain on kunun zaki. *Br. Food J.* 106, 512–519.

Badi, S. M. and Hoseney, R. C. (1977). Use of pearl millet and sorghum flours in bread and cookies. In: Dendy, D. A. V. ed. *Proceedings of a Symposium on Sorghum and Millets for Human Food*. London: Tropical Products Institute, pp. 37–39.

Badi, S. M., Bureng, P. L., and Monowar, L. Y. (1989). Commercial production: A breakthrough in kisra technology. In: Dendy, D. A. V. ed. *ICC 4th Quadrennial Symposium on Sorghum and Millets, Lausanne, Switzerland*. Vienna: International Association for Cereal Science and Technology, pp. 31–45.

Baltensperger, D. D. (1996). Foxtail and proso millet. In: Janick, J. ed. *Progress in New Crop*. Alexandria, VA: ASHS Press, pp. 182–190.

Baltensperger, D. D. (2002). Progress in proso, pearl and other millets. In: Janick, J. and Whipkey, A. eds. *Trends in New Crops and New Uses*. Alexandria, VA: ASHS Press, pp. 100–103.

Basappa, S. C. (2002). Investigations on chhang from finger millet (*Eleucine coracana* Gaertn.) and its commercial prospects. *Indian Food Ind* 21, 46–51.

Bassey, M. W. and Schmidt, O. G. (1989). *Abrasive-disk Dehullers in Africa*. Ottawa: International Development Research Centre.

Botes, A., Todorov, S. D., Von Mollendorff, J. W., Botha, A., and Dicks, L. M. T. (2007). Identification of lactic acid bacteria and yeasts from boza. *Process Biochem.* 42, 267–270.

Bultosa, G. and Taylor, J. R. N. (2004a). Teff. In: Wrigley, C., Corke, H., and Walker, C. E. eds. *Encyclopedia of Grain Science*, Vol. 3. Amsterdam: Elsevier, pp. 281–290.

Bultosa, G. and Taylor, J. R. N. (2004b). Functional properties of grain tef (*Eragrostis tef* (Zucc.) Trotter) starch. *Starch/Stärke* 56, 20–28.

Bultosa, G., Hall, A. N., and Taylor, J. R. N. (2002). Physico-chemical characterization of grain tef [*Eragrostis tef* (Zucc.) Trotter] starch. *Starch/Stärke* 54, 461–468.

Chandrasekher, G., Raju, D. S., and Pattabiraman, T. N. (1982). Natural plant enzyme inhibitors, protease inhibitors in millets. *J. Sci. Food Agric.* 33, 447–250.

Cheverton, M., Pullan, M., Didehvar, F., Greig, A., and Chapman, G. (1994). Models for improvement: Genetic advancement of *Eragrostis tef* with particular regard to lodging. In: Riley, K. W., Gupta, S. C., Seetharam, A., and Mushonga, J. N. eds. *Advances in Small Millets*. New York: International Science Publisher, pp. 431–448.

Choi, Y., Jeong, H. S., and Lee, J. (2007). Antioxidant activity of the methanolic extracts from some grains consumed in Korea. *Food Chem.* (in press).

Danno, G. and Natake, M. (1980). Isolation of foxtail proteins and their subunits structure. *Agric. Biol. Chem.* 44, 913–918.

Demuyakor, B. and Ohta, Y. (1993). Characteristics of single and mixed culture fermentation of pito beer. *J. Sci. Food Agric.* 62, 401–408.

Dykes, L. and Rooney, L. W. (2006). Review: Sorghum and millet phenols and antioxidants. *J. Cereal Sci.* 44, 236–251.

Ejeta, G. (1982). Kisra quality: Testing new sorghum varieties and hybrids. In: Mertin, J. V. ed. *Proceedings of the International Symposium on Sorghum Grain Quality*. Patancheru, India: ICRISAT, pp. 67–78.

FAO (2007). http://faostat.fao.org/ (accessed March 2007).

Fernandez, D. R., Vanderjagt, D. J., Millson, M. *et al.* (2003). Fatty acid, amino acid and trace mineral composition of *Eleusine coracana* (Pwana) seeds from northern Nigeria. *Plant Foods Hum. Nutr.* 58, 1–10.

Gadaga, T. H., Mutakumira, A. N., Narvhus, J. A., and Ferusu, S.B. (1999). A review of traditional fermented foods and beverages in Zimbabwe. *Int. J. Food Microbiol.* 53, 1–11.

Gaitan, E., Lindsay, R. H., Reichert, R. D. *et al.* (1989). Antithyroid and goitrogenic effects of millet: role of C-glycosylflavones. *J. Clin. Endocrinol. Metab.* 68, 707–714.

Galiba, M., Rooney, L. W., Waniska, R. D., and Miller, F. R. (1987). The preparation of sorghum and millet couscous in West Africa. *Cereal Foods World* 32, 878–884.

Gashe, B. A., Girma, M., and Bisrat, A. (1982). Tef fermentation. I. The role of microorganisms in fermentation and their effect on the nitrogen content of tef. *SINET: Ethiopian J. Sci.* 5, 69–76.

Germplasm Resources Information Network (GRIN) (2007). *GRIN Taxonomy for Plants.* www.ars-grin.gov/cgi-bin/npgs (accessed February 2007).

Gomez, M. I. (1994). Preliminary studies on grain quality evaluation for finger millet as a food and beverage use in the southern African region. In: Riley, K.W., Gupta, S.C., Seetharman, A., and Mushonga, J. N. eds. *Advances in Small Millets.* New York: International Science Publisher, pp. 289–296.

Goussault, B. and Adrian, J. (1977). The milling of *Pennisetum* millet and the value of protein in the products. In: Dendy, D. A. V. ed. *Proceedings of a Symposium on Sorghum and Millets for Human Food.* London: Tropical Products Institute, pp. 13–17.

Hedge, P. S. and Chandra, T. S. (2005). ESR spectroscopic study reveals higher free radical quenching potential in kodo millet (*Paspalum scrobiculatum*) compared to other millets. *Food Chem.* 92, 177–182.

House, L. R., Osmanzai, M., Gomez, M. I., Monyo, E. S., and Gupta, S. C. (1995). Agronomic principles. In: *Sorghum and Millets: Chemistry and Technology* (Dendy D.A.V. ed.), American Association for Cereal Chemist, St Paul, MN, pp. 27–67.

Hulse, J. H., Laing, E. M., and Pearson, O. E. (1980). *Sorghum and the Millets: their Composition and Nutritive Value.* New York: Academic Press.

ICRISAT/FAO (1996). *The World Sorghum and Millet Economies: Facts, Trends and Outlook.* Patancheru, India: ICRISAT and Rome: FAO.

Indrani, D., Manohar, R. S., Rajiv, J., and Rao, G. V. (2004). Finger millet biscuit and a process for preparing the same. US Patent 2004/0191386 A1.

Irving, D. W. and Jideani, I. A. (1997). Microstructure and composition of *Digitaria exilis* Stapf (acha): a potential crop. *Cereal Chem.* 74, 224–228.

Jiaju, C. and Yuzhi, Q. (1994). Recent developments in foxtail millet cultivation and research in China. In: Riley, K.W., Gupta, S.C., Seetharam, A., and Mushonga, J. N. eds. *Advances in Small Millets.* New York: International Science Publisher, pp. 101–108.

Jideani, I. A., Owusu, R. K., and Muller, H. G. (1994). Proteins of acha (*Gigitaria exilis* Stapf): Solubility fractionation, gel filtration and electrophoresis of protein fractions. *Food Chem.* 51, 51–59.

Jideani, I. A., Takeda, Y., and Hizukuri, S. (1996). Structures and physicochemical properties from acha (*Digitaria exilis*), Iburu (*D. iburua*) and Tamba (*Eleusine carocana*). *Cereal Chem.* 73, 667–685.

Jones, R. W., Beckwith, A. C., Khoo, U., and Inglett, G. E. (1970). Protein composition of Proso millet. *J. Agric. Food Chem.* 18, 37–39.

Kasaoka, S., Oh-Sashi, A., Morita, T., and Kiriyama, S. (1999). Nutritional characterization of millet protein concentrates produced by heat stable α-amylase digestion. *Nutr. Res.* 19, 899–910.

Kent, N. L. (1983). *Technology of Cereals*, 3rd edn. Oxford: Pergamon Press.

Kohoma, K., Ngasawa, T., and Nishizawa, N. (1999). Polypeptide compositions and NH$_2$-terminal amino acid sequences of protein in foxtail and proso millets. *Biosci. Biotechnol. Biochem.* 63, 1921–1926.

Kumar, K. K. and Parameswaran, K. P. (1998). Characterisation of storage from selected varieties of foxtail millet (*Setaria italica* (L) Beauv). *J. Sci. Food Agric.* 77, 535–542.

Kumari, P. L. and Sumathi, S. (2002). Effect of consumption of finger millet on hyperglycemia in non-insulin dependent diabetes mellitus (NIDDM) subjects. *Plant Foods Hum. Nutr.* 57, 205–213.

Kumari, S. K. and Thayumanavan, B. (1998). Characterization of starches of proso, foxtail, barnyard, kodo and little millets. *Plant Foods Hum. Nutr.* 53, 47–56.

Lestienne, I., Buisson, M., Lullien-Pellerin, V., Piq, C., and Treche, S. (2007). Losses of nutrients and anti-nutritional factors during abrasive decortication of two pearl millet cultivars (*Pennisetum glaucum*). *Food Chem.* 100, 1316–1323.

Lin, R., Li, W., and Corke, H. (1998). Spotlight on Shanxi province China: Its minor crops and specialty foods. *Cereal Foods World* 43, 189–192.

Lumen, B. O., Thompson, S., and Odegard, W. J. (1993). Sulfur amino acid rich protein in acha (*Digitaria exilis*), a promising underutilized African cereal. *J. Agric. Food Chem.* 41, 1045–1047.

Malleshi, N. G. and Hadimani, N. A. (1994). Nutritional and technological characteristics of small millets and preparation of value-added products from them. In: Riley, K.W., Gupta, S.C., Seetharman, A., and Mushonga, J.N. eds. *Advances in Small Millets*. New York: International Science Publisher, pp. 271–287.

Malleshi, N. G., Desikachar, H. S. R., and Tharanathan, R. N. (1986). Physico-chemical properties of native and malted finger millet, pearl millet and foxtail millet starches. *Starch/Stärke* 38, 202–205.

Mangala, S. L., Malleshi, N. G., and Tharanathan, M. R. N. (1999). Resistant starch from differently processed rice and ragi (finger millet). *Eur. Food Res. Technol.* 209, 32–37.

Marathee, J. P. (1994). Structure and characteristics of the world millet economy. In: Riley, K. W., Gupta, S. C., Seetharam, A., and Mushonga, J. N. eds. *Advances in Small Millets*. New York: International Science Publisher, pp. 159–180.

McDonough, C. M., Rooney, L. W., and Earp, C. F. (1986). Structural characteristics of *Eleusine coracana* (finger millet) using scanning electron and fluorescence microscopy. *Food Microstructure* 5, 247–256.

Mnyenyembe, P. H. (1994). Past and present research on finger millet in Malawi. In: Riley, K. W., Gupta, S. C., Seetharam, A., and Mushonga, J. N. eds. *Advances in Small Millets*. New York: International Science Publisher, pp. 29–59.

Mugocha, P. T., Taylor, J. R. N., and Bester, B. H. (2000). Fermentation of a composite finger millet-dairy beverage. *World J. Microbiol. Biotechnol.* 16, 341–344.

Munck, L. (1984). Local and industrial milling processes of sorghum and millet as related to process performance and nutritional quality. In: Dendy, D. A. V. ed. *Symposium: The Processing of Sorghum and Millets: Criteria for Quality of Grains and Products for Human Food*. Vienna: International Association for Cereal Science and Technology, pp. 41–47.

Munck, L. (1995). New milling technologies and products: Whole plant utilization by milling and separation of the botanical and chemical components. In: Dendy, D. A. V. ed. *Sorghum and Millets: Chemistry and Technology*. St. Paul, MN: American Association of Cereal Chemists, pp. 223–281.

Muralikrishna, G., Malleshi, N. G., Desikachar, H. S. R., and Tharanathan, R. N. (1986). Effect of popping on the properties of some millet starches. *Starch/Stärke* 38, 48–51.

Murty, D. S. and Kumar, K. A. (1995). Traditional uses of sorghum and millets. In: Dendy, D.A.V. ed. *Sorghum and Millets: Chemistry and Technology*. St. Paul, MN: American Association of Cereal Chemists, pp. 185–222.

Nakayama, H., Afzal, M., and Okuno, K. (1998). Intraspecific differentiation and geographical distribution of Wx alleles for low amylose content in endosperm of foxtail millet, (*Setaria italica* L.) Beauv. *Euphytica* 102, 289–293.

National Research Council (1996). *Lost Crops of Africa*, Vol. 1: *Grains*. Washington DC: National Academy Press.

Ndoye, M. and Nwasike, C. C. (1994). Fonio millet (*Digitaria exilis* stapf) in West Africa. In: Riley, K. W., Gupta, S. C., Seetharam, A., and Mushonga, J. N. eds. *Advances in Small Millets*. New York: International Science Publisher, pp. 65–97.

Nishizawa, N., Fudamoto, Y., and Yoshirahu, Y. (1995). The elevation of plasma concentration of high density lipoprotein cholesterol in mice fed with protein from proso millet. *Biosci. Biotechnol. Biochem.* 59, 333–335.

Nishizawa, N., Sato, D., Ito, Y. *et al.* (2002). Effects of protein of proso millet on liver injury induced by D-galactosamine in rats. *Biosci. Biotechnol. Biochem.* 66, 92–96.

Nzelibe, H. C. and Nwasike, C. C. (1995). The brewing potential of "acha" (*Digitaria exilis*) malt compared with pearl millet (*Pennisetum typhoides*) malts and sorghum (*Sorghum bicolor*) malts. *J. Inst. Brew.* 101, 345–350.

Oi, Y. and Kitabatake, N. (2003). Chemical composition of East African traditional beverage, togwa. *J. Agric. Food Chem.* 51, 7024–7028.

Olatunji, O., Koleoso, O. A., and Oniwinde, A. B. (1992). Recent experience on the milling of sorghum, millet, and maize for making nonwheat bread, cake, and sausage in Nigeria. In: Gomez, M. I., House, L. R., Rooney, L. W., and Dendy, D. A. V. eds. *Utilization of Sorghum and Millets*. Patancheru, India: ICRISAT, pp. 83–88.

Opoku, A. R., Ohenhen, S. O., and Ejiofor, N. (1981). Nutrient composition of nutrient millet (*Pennisetum typhoides*) grains and malt. *J. Agric. Food Chem.* 29, 1247–1248.

Osagie, A. U. and Kates, M. (1984). Lipid composition of millet (*Pannisetum americanum*). *Lipids* 19, 958–965.

Parameswaran, K. P. and Thayumanavan, B. (1997). Isolation and characterization of a 20 kD prolamin from kodo millet (*Paspalum scrobiculatum*) (L.): Homology with other millets and cereals. *Plant Foods Hum. Nutr.* 50, 359–373.

Pawar, V. D. and Machewad, G. M. (2006). Processing of foxtail for improved nutrient availability. *J. Food Process. Preservation* 30, 269–279.

Pelembe, L. A. M. (2001). Pearl millet malting: Factors affecting product quality. PhD thesis, University of Pretoria, Pretoria.

Pelembe, L. A. M., Dewar, J., and Taylor, J. R. N. (2002). Effect of malting conditions on pearl millet malt quality. *J. Inst. Brewing* 108, 7–12.

Pelembe, L. A. M., Dewar, J., and Taylor, J. R. N. (2004). Effect of germination moisture and time of pearl millet malt quality—With respect to its opaque and lager beer brewing potential. *J. Inst. Brewing* 110, 320–325.

Perten, H. (1977). Specific characteristics of millet and sorghum milling. In: Dendy, D. A. V. ed. *Proceedings of a Symposium on Sorghum and Millets for Human Food.* London: Tropical Products Institute, pp. 47–51.

Perten, H. (1984). Industrial processing of millet and sorghum. In: Dendy, D.A.V. ed. *Symposium: The Processing of Sorghum and Millets: Criteria for Quality of Grains and Products for Human Food.* Vienna: International Association for Cereal Science and Technology, pp. 52–55.

Quin, P. J. (1959). *Foods and the Feeding Habits of the Pedi.* Johannesburg: Witwatersrand University Press.

Ragaee, S., Abdel-Aal, A. M., and Noaman, M. (2006). Antioxidant activity and nutrient composition of selected cereals for food use. *Food Chem.* 98, 32–38.

Ramachandra, G., Virupaksha, T. K., and Shadaksharaswamy, M. (1977). Relationship between tannin levels and *in vitro* protein digestibility in finger millet (*Eleusine coracana* Gaertn.). *J. Agric. Food Chem.* 25, 1101–1104.

Rao, M. V. S. S. T., Manohar, R. S., and Muralikrishna, G. (2004). Functional characteristics of non starch polysaccharides (NSP) from native and malted finger millet (ragi, *Eleusine coracana*, indaf-15). *Food Chem.* 88, 453–460.

Rasper, V. F. (1977). Palyi's compact system for debranning of sorghum and millet. In: Dendy, D.A.V. ed. *Proceedings of a Symposium on Sorghum and Millets for Human Food.* London: Tropical Products Institute, pp. 59–72.

Ravindran, G. (1991). Studies on millets: proximate composition, mineral composition, and phytate and oxalate contents. *Food Chem.* 39, 99–107.

Ravindran, G. (1992). Seed protein of millets: amino acid composition, proteinase inhibitors and in-vitro protein digestibility. *Food Chem.* 44, 13–17.

Reddy, V. P., Faubin, J. M., and Hoseney, R. C. (1986). Odor generation in ground, stored pearl millet. *Cereal Chem.* 63, 383–406.

Rooney, L. W., Kirleis, A. W., and Murty, D. S. (1986). Traditional foods from sorghum. In: Pomeranz, Y. ed. *Advances in Cereal Science and Technology*, Vol. VIII. St. Paul, MN: American Association of Cereal Chemists, pp. 317–353.

Rost, T. L. (1971). Fine structure of protein bodies in *Setaria lutescens* (Gramineae). *Protoplasma* 73, 475–479.

Saroj, T. and Prakesh, T. J. (2004). Product characterization of kodo ko jaanr: Fermented finger millet beverage from the Himalayas. *Food Microbiol.* 21, 617–622.

Satin, M. (1988). Bread without wheat: Novel ways of making bread from cassava and sorghum could reduce the Third World's dependence on imported wheat for white bread. *New Sci.* 28 April, 56–59.

Serna-Saldivar, S. and Rooney, L. W. (1995). Structure and chemistry of sorghum and millets. In: Dendy, D. A. V. ed. *Sorghum and Millets, Chemistry and Technology.* St Paul, MN: American Association of Cereal Chemists, pp. 69–124.

Singh, P., Singh, U., Eggum, B. O., Kumar, K. A., and Andrews, D. J. (1987). Nutritional evaluation of high protein genotypes of pearl millet (*Pennisetum americanum* L.) Leeke. *J. Sci. Food Agric.* 38, 41–48.

Siwela, M., Taylor, J. R. N., De Milliano, W. A. J., and Duodu, K. G. (2007). Occurrence and location of tannins in finger millet grain and antioxidant activity of different grain types. *Cereal Chem.* 84, 169–174.

Smith, I. F. (1996). Increasing consumer utilization of minor cereals in West Africa through sustained research and development of appropriate production and processing technologies. In: Dendy, D. A. V. ed. *Sorghum and Millets: Proceedings of the Symposia.* Vienna: International Association for Cereal Science and Technology, pp. 171–197.

Sridhar, R. and Lakshminarayana, G. (1994). Contents of total lipids classes and composition of fatty acids in small millets: Foxtail (*Setaria italica*), proso (*Panicum miliaceum*), and finger (*Eleusine coracana*). *Cereal Chem.* 71, 355.

Subramanian, V. and Jambunathan, R. (1980). Traditional methods of processing sorghum (*Sorghum bicolor*) and pearl millet (*Pennesetum americanum*) grains in India. In: Dendy, D. A. V. ed. *International Association for Cereal Chemistry, 10th Congress, Symposium: Sorghum and Millet Processing.* Vienna: ICC, pp. 115–118.

Suman, C. N., Monteiro, P. V., Ramachandra, G., and Sudharshana, L. (1992). *In vitro* enzymic hydrolysis of storage proteins of Japanese barnyard millet (*Echinochloa frumentacea*). *J. Sci. Food Agric.* 58, 505–509.

Svanberg, U., Sjögren, E., Lorri, W., Svennerholm, A-M., and Kaijser, B. (1992). Inhibited growth of common enteropathogenic bacteria in lactic-fermented cereal gruels. *World J. Microbiol. Biotechnol.* 8, 601–606.

Tatham, A. S., Fido, R. J., Moore, C. M., Kasarda, D. D., Kuzmicky, D. D., Keen, J. N., and Shewry, P. R. (1996). Characterisation of the major prolamins of tef (*Eragrostis tef*) and finger millet (*Eleusine coracana*). *J. Cereal Sci.* 24, 65–71.

Taylor, J. R. N. (2004a). Millet: pearl. In: Wrigley, C., Corke, H., and Walker, C. E. eds. *Encyclopedia of Grain Science*, Vol. 2. Amsterdam: Elsevier, pp. 253–261.

Taylor, J. R. N. (2004b). Foods and nonalcoholic beverages. In: Wrigley, C., Corke, H., and Walker, C. E. eds. *Encyclopedia of Grain Science*, Vol. 1. Amsterdam: Elsevier, pp. 380–390.

Taylor, J. R. N. and Belton, P. S. (2002). Sorghum. In: Belton, P. S. and Taylor, J. R. N. eds. *Pseudocereals and Less Common Cereals*. Berlin: Springer-Verlag, pp. 25–91.

Taylor, J. R. N., Schober, T. J., and Bean, S. R. (2006). Review: Novel and non-food uses for sorghum and millet. *J. Cereal Sci.* 44, 252–271.

Thaoge, M. L., Adams, M. R., Sibara, M. M., Watson, T. G., Taylor, J. R. N., and Goyvaerts, E. M. (2003). Production of improved infant porridges from pearl millet using a lactic acid fermentation step and addition of sorghum malt to reduce viscosity of porridges with high protein, energy and solids (30%) content. *World J. Microbiol. Biotechnol.* 19, 305–310.

The Teff Company (2007). www.teffco.com (accessed March 2007).

Ushakumari, S. R., Latha, S., and Malleshi, N. G. (2004). The functional properties of popped, flaked, extruded and roller-dried foxtail millet (*Setaria italica*). *Int. J. Food Sci. Technol.* 39, 907–915.

Vitale, J. D. and Sanders, J. H. (2005). New markets and technological change in traditional cereals in semiarid sub-Saharan Africa: The Malian case. *Agric. Econ.* 32, 111–129.

Wanatabe, M. (1999). Antioxidative phenolic compounds from Japanese barnyard millet (*Echinochloa utilis*) grains. *J. Agric. Food Chem.* 47, 4500–4505.

World Health Organization (1996). *Fermentation: Assessment and Research*, WHO/FNU/FOS/96.1. Geneva: WHO.

Yanez, G. A., Walker, C. E., and Nelson, L. A. (1991). Some chemical and physical properties of proso millet (*Panicum miliaceum*) starch. *J. Cereal Sci.* 13, 299–305.

Yetneberk, S., De Kock, H. L., Rooney, L. W., and Taylor, J. R. N. (2004). Effects of sorghum cultivar on injera quality. *Cereal Chem.* 81, 314–327.

Pseudocereals

7

Regine Schoenlechner, Susanne Siebenhandl, and Emmerich Berghofer

Introduction

Amaranth and quinoa were major crops for the Pre-Colombian cultures in Latin-America. After the Spanish conquest, however, consumption and cultivation of these crops was suppressed and thereafter only continued in a small scale. Since it has been shown that both grains show good nutritional properties, the interest in them has risen again. The production of quinoa was 25 329 tonnes in Bolivia, 652 tonnes in Ecuador, and 32 590 tonnes in Peru in the year 2006 (FAOSTAT, 2006). Amaranth and quinoa cultivation remain relatively low – amaranth is not even listed in the FAO statistics on production data – although an appreciable commercial cultivation of amaranth for human nutrition does take place. Besides Latin American countries, it is produced in the USA, China, and Europe.

Buckwheat originates from Central Asia and was transferred by nomadic people to Central and Eastern Europe. In the thirteenth century, it reached some importance in Germany, Austria, and Italy, although it then declined due to the cultivation of other cereals. Today, buckwheat is celebrating something of a comeback due to the demand for gluten-fee diets, and it is now grown on 2.5 million hectares, producing 2 million tonnes of grain. In 2005, China produced the most at 800 000 tonnes, followed by Russia (605 640 tonnes), and Ukraine (274 700 tonnes) (FAOSTAT, 2006). In Europe, 72 096 tonnes were produced in Poland, 124 217 tonnes in France and small amounts in Hungary, Slovenia, Latvia, and Lithuania. Japan is the main importer for buckwheat.

From the botanical point of view, amaranth, quinoa, and buckwheat are dicotyledonous plants and thus not cereals (monocotyledonous), but since they produce starch-rich seeds like cereals they are called "pseudocereals." In Plate 7.1 the three plants are shown in the flowering state. According to some phylogenetic classifications the *Amaranthus* and *Chenopodium* genera belong together in the order Caryophyllales, whereas buckwheat (*Fagopyrum*) belongs in the Polygonales.

Gluten-Free Cereal Products and Beverages
ISBN: 9780123737397

Polygonales and Caryophyllales are closely related and are combined together in subclass Caryophyllidae. However, the data obtained by Drzewiecki *et al.* (2003) suggests that there is significant genetic distance between the Polygonales and the Caryophyllales and this has been confirmed recently by Gorinstein *et al.* (2005). It seems that quinoa, buckwheat, and amaranth (as a genus) can be considered as phylogenetically distant taxa, although according to Aphalo *et al.* (2004) the polymerized 11S amaranth globulin (globulin-P) presented cross-reactivity with quinoa globulins, and to a lesser extent with globulins of sunflower and rice.

Over 60 species of amaranth are known worldwide. The main grain amaranth species used today are *Amaranthus caudatus* L. (syn. *edulis* Spegazzini), *Amaranthus cruentus* L. (syn. *paniculatus* L.), and *Amaranthus hypochondriacus*. Among quinoa sweet and bitter varieties exist, dependent on the content of saponins (i.e. if the saponin content is below 0.11% the variety is considered to be a sweet variety) (Koziol, 1991). Two varieties of buckwheat are commonly cultivated: common buckwheat (*Fagopyrum esculentum*) and tartary buckwheat (*Fagopyrum tataricum*).

Amaranth seeds are lentil-shaped and measure about 1 mm in diameter. The 1000 kernel weight is only 0.5–1.4 g. Quinoa seeds are slightly larger than amaranth seeds, the 1000 kernel weight is approximately 1.9–4.3 g. In contrast to cereals, the embryo surrounds the starch-rich tissue (perisperm) in the form of a ring and makes about 25% of the total seed weight. The buckwheat seed is a three-angled achene, 6–9 mm long. The fruit of *F. tataricum* is smaller (4–5 mm) and more rounded at the edges. The 1000 kernel weight (10–20 g) depends mainly on the hull thickness. Structurally and chemically, the endosperm resembles that of a cereal grain consisting of a non-starchy aleurone layer and large cells packed with starch granules constituting the majority of the endosperm.

Chemical composition

Table 7.1 summarizes the chemical composition of amaranth, quinoa, and buckwheat.

Table 7.1 Chemical composition of amaranth, quinoa and buckwheat

Component	Composition (average value in %, range in brackets)		
	Amaranthus spp.	*Chenopodium quinoa*	*Fagopyrum esculentum*
Water	11.1 (9.1–12.5)[a]	12.7	14.1 (13.4–19.4)
Protein ($N \times 5.8$)	14.6 (14.5–14.8)	13.8 (12.2–13.8)	10.9 (10.4–11.0)
Fat	8.81 (6.56–10.3)	5.04 (5.01–5.94)	2.71 (2.40–2.80)
Starch	55.1	67.35[b]	67.2
Crude fiber	3.9–4.4[d]	2.3[c]	–
Dietary fiber	11.14[b]	12.88[b]	8.62[b]
Minerals	3.25	3.33 (2.46–3.36)	1.59 (1.37–1.67)

From Souci *et al.* (2000).
[a] The range given in brackets represents the range from the lowest to the highest value given in the literature.
[b] Own measurements.
[c] Saunders and Becker (1984).
[d] Based on dry matter, Gamel *et al.* (2006a).

Amaranth

Carbohydrates

Starch

Analysis of amaranth starch reveals two major differences in comparison to cereals. First, starch comprises the main component of carbohydrates in amaranth, but is found usually in lower amounts than in cereals (see Table 7.1). Second, amaranth starch is not located in the endosperm but is in the perisperm, where typical compounded starch particles of approximatedly 50–90 μm in diameter are generated in the amyloplasts. When suspended in water, small single starch granules of 1–3 μm can be extracted from the agglomerates (Wilhelm *et al.*, 2002). Compound starch particles consisting of small granules are typical for most starch raw materials. The particles cluster together to minimize the surface and thereby form characteristic compounds. The specific surface area of starches increases remarkably as the granule diameter decreases. Wilhelm *et al.* (2002) gives a value of 5.194 m^2/cm^3 for the surface of the small granule starch of amaranth. Hunjai *et al.* (2004) isolated starch from *A. cruentus*. Using scanning electron microscopy (SEM), starch granules appeared polygonal with a diameter of 0.8–1.0 μm.

X-ray diffraction patterns indicate that amaranth starch shows the typical "A" type crystallinity (Paredes-Lopez *et al.*, 1994; Qian and Kuhn, 1999a; Hunjai *et al.*, 2004). The amylose content of amaranth starch is much lower than that in other cereal starches, with values varying from 0.1 to 11.1% (Stone and Lorenz, 1984; Perez *et al.*, 1993; Hunjai *et al.*, 2004). Amylopectin was found to be composed of short-chain branched glucans with an average molecular weight of 11.8×10^6 g/mol (Praznik *et al.*, 1999). The small size of the starch granule as well as its high amylopectin content explains most of the physical properties of amaranth starch. Compared with other cereals, amaranth starch shows excellent freeze–thaw and retrogradation stability (Baker and Rayas-Duarte, 1998; Wilhelm *et al.*, 2002; Hunjai *et al.*, 2004), a higher gelatinization temperature (Becker *et al.*, 1981) and viscosity (Hunjai *et al.*, 2004; Becker *et al.*, 1981 and others), higher water-binding capacity (Calzetta-Resio *et al.*, 2000; Hunjai *et al.*, 2004), higher sorption capacity at higher water activity values (Paredes-Lopez *et al.*, 1989; Schoenlechner, 1997; Calzetta-Resio *et al.*, 1999), as well as higher solubility, higher swelling power and enzyme susceptibility (Singhal and Kulkarni, 1990; Baker and Rayas-Duarte, 1998; Hunjai *et al.*, 2004). By heating different starch suspensions at 55–95°C, Hunjai *et al.* (2004) could show that amaranth starch has a constant swelling power and shows no increase of swelling power at temperatures higher than 75°C. In addition, solubility did not change after 75°C.

Resistant starch

Resistant starch (RS) is not only naturally present in food but is also formed during processing. Like dietary fiber, RS is not susceptible to human digestive enzymes and thus reaches the colon, where it is fermented by the bacterial biota. RS has beneficial physiological effects, such as lowering blood lipids or lowering the risk of colon cancer. The RS content depends on the characteristics of the starch present in food, type of granule, amylose/amylopectin ratio and crystallinity of starch, as well as on the analytical method used. Food-processing conditions influence the content of RS formed. Gonzalez *et al.* (2007) found a RS content of 0.65%. By extrusion cooking and fluidized bed heating this content was increased, whereas cooking and popping decreased it (Gamel *et al.*, 2005). Lara and Ruales (2002) determined the RS content in popped amaranth, which was approximately 0.5%, showing that the efficiency of utilization of amaranth starch may be very high. Mikulikova and Kraic (2006) determined the RS (enzymatic measurement of RS_3 by AACC Method 32–40) in 18 amaranth genotypes. A wide variation of $1.24 \pm 1.22\%$ of seed was measured, which was explained by the wide variation of amylose content in amaranth species. The RS/total starch proportion was 1.98% in amaranth. Crops containing more than 4.5% RS are considered to be a good source.

Low molecular weight carbohydrates
Mono- and disaccharides can only be found in small amounts in amaranth. According to Gamel *et al.* (2006a), the total sugar content of the two species *A. cruentus* and *A. caudatus* ranges from 1.84 to 2.17 g/100 g. Considering the sugar composition, sucrose was found to be the dominant sugar with values of 0.58–0.75 g/100 g. The values of the other sugars were: galactose plus glucose 0.34–0.42 g/100 g, fructose 0.12–0.17 g/100 g, maltose 0.24–0.28 g/100 g, raffinose 0.39–0.48 g/100 g, stachyose 0.15–0.13 g/100 g, and inositol 0.02–0.04 g/100 g. The values were in good agreement with previously reported data (Becker *et al.*, 1981; Saunders and Becker, 1984).

Fiber
Dietary fibers, soluble and insoluble, are known to have beneficial effects on human health. The fiber content of amaranth lies within the range of other cereals and shows great variation within different species. The fraction of soluble dietary fiber varies between 19.5 and 27.5% in *A. cruentus* and 33.1 and 49.3% in *A. hypochondriacus* (Bressani *et al.*, 1990).

Protein

The nutritional value of pseudocereals is mainly connected to their protein content. Amaranth has a higher protein content than buckwheat or quinoa. Sixty-five percent of the proteins are located in the germ and seed coat, and 35% in the starch-rich endosperm (Saunders and Becker, 1984). The protein content and amino acid pattern depend on genotype and growing conditions.

Storage proteins
While alcohol-soluble prolamins represent the major storage proteins in cereals such as wheat or maize, the storage proteins of dicotyledonous plants are mainly globulins and albumins (Gorinstein *et al.*, 2002; Drzewiecki *et al.*, 2003). According to the Osborne classification, amaranth proteins consist of about 40% albumins, 20% globulins, 25–30% glutelins, and only 2–3% prolamins (Segura-Nieto *et al.*,

1994; Bucaro Segura and Bressani, 2002, and others). Gorinstein *et al.* (1999) found a lower amount of prolamin-like (alcohol-soluble) proteins of about 1.2–1.4%, and even less prolamins (0.48–0.79%) were measured by Muchova *et al.* (2000). According to Gorinstein *et al.* (1999) the protein proportions for amaranth are similar to those of rice. By using SEM and SDS-PAGE, Gorinstein *et al.* (2004) found a close similarity between the protein fractions of amaranth and soybeans. The prolamin showed differences to cereals, whereas the glutelin fraction showed some similarities to maize (Gorinstein *et al.*, 2001, 2004). According to their sedimentation coefficient, two main classes of globulins can be differentiated: 7S and 11S globulins. In amaranth, similar 7S (conamaranthin) and 11S (amaranthin) storage globulins have been found (Marcone *et al.*, 1994; Martinez *et al.*, 1997; Marcone, 1999). Thermal treatment decreased the water-soluble protein fraction (albumins and globulins) and alcohol-soluble fraction (prolamins) (Gamel *et al.*, 2005). It can be concluded that the amaranth proteins are similar to seed proteins in other dicotyledonous crops such as legumes, and have no relationship to the major prolamins of cereals.

Amino acids

The amino acids composition in pseudocereals is outstanding, with a high content of essential amino acids. In particular, methionine, lysine, arginine, tryptophan, and sulfur-containing amino acids are found at higher levels than in other cereals (Matuz *et al.*, 2000a; Gorinstein *et al.*, 2002). For amaranth, the sum of essential amino acids has been reported to be 47.65 g/100 g of protein (Drzewiecki *et al.*, 2003). When compared with soybean, a significantly higher concentration of glutamine, glycine, and methionine was found in amaranth, while tyrosine and cysteine and the essential amino acids (e.g. isoleucine, leucine, and phenylalanine) were significantly lower in amaranth than in soybean. Both amaranth and soybean had comparable or higher amounts of amino acids than whole egg protein. Morales de León Josefina *et al.* (2005) confirmed the high content of sulfur-containing amino acids (4.09–5.34 g/16 g N). Correa *et al.* (1986) found that leucine is the limiting amino acid, however data from the literature concerning the limiting amino acid are controversial. When considering the chemical score, several authors indicate leucine as the limiting amino acid in amaranth (Becker *et al.*, 1981; Saunders and Becker, 1984; Pederson *et al.*, 1987; Abreu *et al.*, 1994; Escudero *et al.*, 2004), whereas when considering the protein efficiency ratio (PER), threonine was recognized to be the limiting amino acid (Bressani *et al.*, 1989). After popping, the loss of the amino acid tyrosine was highest, followed by phenylalanine and methionine (Gamel *et al.*, 2004). Based on the chemical score, lysine was the limiting acid in the popped sample, as previously found by Tovar *et al.* (1989).

Nutritional quality

Protein quality not only depends on the amino acid composition, but also on the bioavailability or digestibility. Protein digestibility, available lysine, net protein utilization (NPU), or PER are widely used as indicators for the nutritional quality of proteins. In this respect, the values for pseudocereal proteins are definitively higher when compared with cereals and are close to those of casein. Bejosano and Corke (1998) measured an average protein digestibility of 74.2% for raw amaranth wholemeal flours, confirming the findings of Guzman-Maldonado and Paredes-Lopez (1994). Slightly higher values were determined by Escudero *et al.* (2004) and by Gamel *et al.* (2004), at 81% and 80–86%, respectively. An increase in protein digestibility by an average of 2.7% was observed after heating. This can be explained by the opening of the carbohydrate–protein complex and/or the inactivation of antinutritional factors such as trypsin inhibitors or polyphenols (tannic acid) (Bejosano and Corke, 1998). In particular, a high correlation was found between the protein digestibility and the presence of polyphenols, whereas only a weak correlation was found with trypsin inhibitors. Fadel *et al.* (1996) demonstrated that heat treatment lowers the activity of trypsin inhibitors, thus improving the nutritive value of amaranth.

Correa *et al.* (1986) calculated a chemical score for amaranth protein of 50–67. The values for calculated PER (C-PER) ranged from 1.39 to 1.80 and those for biological values (BV) from 52 to 68. Similar values have been recently found by Escudero *et al.* (2004). Yanez *et al.* (1994) measured a C-PER value for amaranth of 1.94 compared to 2.77 in casein and 1.64 in wheat. The net protein ratio (NPR) value ranged from 3.04 to 3.20, compared to a NPR of 4.08 in casein. When considering the protein digestibility corrected amino acid score (amino acid score × protein digestibility, PDCAAS), amaranth wholemeal flour has a higher value (0.64) than wheat (0.40) or oat (0.57), whereas the PDCAAS of sodium caseinate is 1.03 (Bejosano and Corke, 1998; Escudero *et al.* 2004).

There is no difference between the *in vivo* protein digestibility of raw and popped seeds, although the *in vitro* digestibility is slightly higher for the popped seeds. Popping reduces the PER value by 14–19%, probably due to a loss of essential amino acids. Feed intake by rats fed amaranth, wheat, or caseinate has been found to be higher when amaranth is used. Furthermore, the growth of the rats fed amaranth was higher than those fed wheat, and similar to those fed caseinate (Gamel *et al.*, 2004).

Allergy and celiac disease

To date, only a few studies have been performed on amaranth allergy or on toxicity of amaranth proteins to people with celiac disease. Recently, a study about allergenic reactions to the prolamin fraction of amaranth was undertaken by Matuz *et al.* (2000b). In contrast to wheat, barley, rye, triticale, and oat, the prolamin fraction of amaranth showed no reactivity against rabbit anti-gliadin (wheat) antibodies. *In vivo* and *in vitro* investigations of general allergic reactions to amaranth revealed that amaranth causes a classical type 1 reaction in sensitized patients (Bossert and Wahl, 2000). On the other hand Hibi *et al.* (2003) found that amaranth grain and its extract inhibited antigen-specific IgE production through augmenting T_h1 cytokine responses *in vivo* and *in vitro*. Genetically modified maize with an amaranth 11S globulin (amarantin) caused no important allergenic reactions to amarantin during *in vitro* investigations (Sinagawa-Garcia *et al.*, 2004). In conclusion, results collected so far indicate that amaranth is not toxic to patients with celiac disease.

Functional properties of proteins

All pseudocereal proteins are highly soluble and thus can be used in functional foods (Segura-Nieto et al., 1999; Bejosano and Corke, 1999; Kovacs et al., 2001; Salcedo-Chavez et al., 2002). Protein concentrates from amaranth exhibited much better solubility, foaming, and emulsification than two commercial soy proteins (Bejosano and Corke, 1999). Recently, it has been suggested that amaranth protein isolate can act as an effective foaming agent (Fidantsi and Doxastakis, 2001). In particular, amaranth globulins have good functional properties (Segura-Nieto et al., 1999). Marcone and Kakuda (1999) found that the functional properties of amaranth globulin isolate are much better than soybean isolate, especially in the vicinity of its electrical point (pH 5–6), showing higher solubility, heat stability, foaming capacity, and stability as well as emulsifying activity. The functional properties of amaranth albumins have been investigated by Silva-Sanchez et al. (2004). The maximum solubility values are above pH 6. When comparing these values to the solubility of egg albumins, amaranth albumins showed excellent foaming capacity and foaming stability at pH 5, suggesting that they could be used as whipping agents like egg albumins. Moreover, the water and oil absorption capacities reached their maximum values at acidic pH. Farinograph and alveograph investigations demonstrated that a 1% albumin addition improves wheat dough properties and bread crumb characteristics.

Depending on protein and thermal conditions, amaranth proteins are able to form self-supporting gels that could be applied in different gel-like foods (Avanza et al., 2005). In addition, Scilingo et al. (2002) found that an amaranth protein isolate hydrolyzed by papain keeps a high solubility after heating, thus indicating that it could be a suitable ingredient in foods submitted to thermal treatments. The solvent (borate or NaOH) used to extract glutelin fractions have been found to influence the physicochemical properties of the proteins (Abugoch et al., 2003), and this in turn may result in different functional properties.

Enzyme inhibitors

Many food plants contain one or more protease inhibitors (e.g. chymotrypsin or trypsin inhibitors) that competitively inhibit the activity of proteolytic enzymes. Protease inhibitors can be anti-carcinogenic, antioxidative, blood glucose regulatory, as well as anti-inflammatory. However, heat treatment can reduce their activity. Compared with other cereals, amaranth contains very low levels of protease inhibitors. Gamel et al. (2006a) found trypsin inhibitor activity (TIU) ranging from 3.05 to 4.34 TIU/mg, chymotrypsin inhibitor activity (CIU) ranging from 0.21 to 0.26 CIU/mg, and amylase inhibitor activity (AIU) ranging from 0.23 to 0.27 AIU/mg. Trypsin, amylase, and, in particular, the chymotrypsin inhibitors decrease after heat treatment or germination.

Lipids

The fat content of amaranth is about 2–3 times higher than that of other cereals (Table 7.1) and it shows again high variation between the species. Amaranth oil contains more than 75% unsaturated fatty acids and is particular rich in linoleic acid

(35–55%). Palmitic acid accounts for 20–23%, palmitoeic acid around 16%, stearic acid 3–4%, and oleic acid for 18–38% (Ayorinde *et al.*, 1989; Becker, 1994; Leon-Camacho *et al.*, 2001; He *et al.*, 2002; Berganza *et al.*, 2003; Escudero *et al.*, 2004). Linolenic acid could not be detected by Escudero *et al.* (2004), while Becker (1994) and Leon-Camacho *et al.* (2001) found an amount of 1%. By applying a multivariate statistical procedure Leon-Camacho *et al.* (2001) demonstrated that the profile of the amaranth fatty acids is similar to oil produced from other cereals such as wheat, barley, maize, rye, oats, or rice.

Amaranth contains high levels of squalene, a highly unsaturated open-chain triterpene, which is usually only found in the livers of deep sea fish and other marine species. Squalenes are widely used in pharmaceutical and cosmetic applications. The content in amaranth ranges from 2 to 8% (Becker *et al.*, 1981; Lyon and Becker, 1987; Qureshi *et al.*, 1996; Leon-Camacho *et al.*, 2001; He *et al.* 2002), whereas in other plant oils it is found in much lower amounts (e.g. olive oil 0.1–0.5% or wheat germ oil 0.1–7%) (Trautwein *et al.*, 1997). Shin *et al.* (2004) found that amaranth squalene exerts a cholesterol-lowering effect by increasing fecal elimination of steroids through interference with cholesterol absorption. The effect was higher than that of shark-liver squalene. In addition, amaranth oil and amaranth grain lowered serum and hepatic cholesterol as well as triglycerides, confirming previous findings (Chaturvedi *et al.*, 1993; Qureshi *et al.*, 1996; Budin *et al.*, 1996; Grajeta, 1999; Gamel *et al.*, 2004). In an *in vivo* study, rats were fed with oat or amaranth containing 1% cholesterol (Czerwiński *et al.*, 2004). Amaranth positively affected the plasma lipid profile, and the effect was directly connected to the contents of bioactive components and antioxidant activities of the amaranth samples. In addition, Kim *et al.* (2006) demonstrated that amaranth grain or amaranth oil significantly decreased the serum glucose and increases serum insulin levels in diabetic rats. In contrast, Berger *et al.* (2003) found no cholesterol-lowering properties of amaranth flakes using hamster feeding experiments, but it was suggested that amaranth grain and oil may be beneficial for the correction of hyperglycemia and in preventing diabetic complications.

Phospholipids constitute about 5% of the oil fraction of amaranth (Becker, 1994). In a previous study, Opute (1979) detected 3.6% phospholipids in amaranth oil, of which the cephalin fraction was 13.3%, the lecithin 16.3%, and the fraction of phosphoinositol 8.2%. Total sterols in amaranth oil are 24.6×10^3 ppm (Leon-Camacho *et al.*, 2001) and almost all sterols of amaranth oil are esterified. In most vegetable oils the percentage of free (non-esterified) sterols is usually much higher. The major sterol present is clerosterol (42%), which has antibacterial activity. The high concentration of sterols makes amaranth oil potentially useful in pharmacological applications (Leon-Camacho *et al.*, 2001).

Minerals

The content of minerals (ash) in amaranth is about twice as high as in other cereals (Table 7.1). Particularly high are the amounts of calcium, magnesium, iron, potassium, and zinc (Saunders and Becker, 1984; Pederson *et al.*, 1987; Bressani,

1994; Yanez *et al.*, 1994; Gamel *et al.*, 2006a). The calcium/phosphorus ratio (Ca:P), which should be around 1:1.5, shows a good value of 1:1.9–2.7 (Bressani, 1994).

Vitamins

Overall, amaranth does not constitute an important source of vitamins. According to Souci *et al.* (2000), the content of thiamine in amaranth is higher than in wheat, in contrast with previous investigations (Bressani, 1994). Amaranth is a good source of riboflavin, vitamin C and in particular of folic acid and vitamin E (Dodok *et al.*, 1994; Gamel *et al.*, 2006a, and others). Folic acid has been found in amounts of 102 μg/100 g, 2.5 times higher than in wheat (40 μg/100 g) (data not published). Vitamin E possesses antioxidative effects and thus increases the stability of the oil. Qureshi *et al.* (1996) and Budin *et al.* (1996) reported a total tocol content of about 45 mg/kg seeds. By using supercritical fluid extraction, Bruni *et al.* (2002) found total tocopherol contents in amounts of 100–129 mg/kg seeds. Among the tocopherols, α-tocopherol, which shows important antioxidant activity, was the most abundant and was found in amounts of 2.97–15.65 mg/kg seed (Lehmann *et al.*, 1994) or 248 mg/kg oil (Leon-Camacho *et al.*, 2001). While no β-tocopherol could be found by Lehmann *et al.* (1994), Leon-Camacho *et al.* (2001) found a high concentration of 546 mg/kg oil. Tocotrienols are important compounds with hypocholesterolemic activity. Contradictory results have been reported about their presence in amaranth. According to Lehmann *et al.* (1994) amaranth grains have significant amounts of β-tocotrienols (5.02–11.47 mg/kg seed) and γ-tocotrienols (0.95–8.69 mg/kg seed), whereas Budin *et al.* (1996) and Leon-Camacho *et al.* (2001) did not detect any tocotrienols in amaranth.

Phytochemicals

In contrast to primary metabolites such as starch, fat, or proteins, phytochemicals are only found in small amounts in plants. Phytochemicals are known to have pharmacological effects and have always been part of the human diet. In the past, plant breeders aimed to remove these substances and food technologists have tried to eliminate them by processing, since they were perceived to be negative for human nutrition (anti-nutrients). However, recent research has shown that phytochemicals have positive effects for human health. Phenolic compounds are the major source of natural antioxidants in plant food. As for vitamins and trace elements, there is a harmful, optimal, essential, or deficient dose which has to be determined.

Total phenolic compounds

Many researchers have measured the polyphenolic compounds expressed as tannic acid or tannins. Tannins are polyphenolic secondary plant metabolites of higher plants and are either galloyl esters and derivatives, or they are oligomeric and polymeric proanthocyanidins. High concentration can be found in the hulls of cereals and legumes and they can negatively influence the digestion and absorption processes by forming complexes with various nutrients or digestive enzymes. Dark amaranth seeds contain more tannins than light ones (104–116 mg/100 g vs. 80–120 mg/100 g) (Bressani, 1994). Becker *et al.* (1981) evaluated 10 different samples and found a range of 80–420 mg/100 g. Breene (1991) gives an average value of 40–120 mg/100 g, whereas higher values (410–520 mg/100 g) were measured by Bejosano and Corke (1998) in various amaranth species using acidified methanol instead of aqueous methanol as extraction medium. Gamel *et al.* (2006b) determined the phenolic compounds (expressed as tannic acid) in amaranth after extraction with acidified methanol and found values ranging from 516 to 524 mg/100 g. Moreover, thermal treatment or germination decreased the content of phenolic compounds.

Total phenolics in amaranth seeds expressed as ferulic acid (an alkali-extractable phenolic compound) were measured by Klimczak *et al.* (2002). Depending on the species considered, values ranging from 39.17 to 56.22 mg/100 g were measured. These values are comparable to other cereals. Free phenolic acids made 27% of the total phenolic acids in *Amaranthus caudatus*. The major compounds were caffeic acid (55.79 µg/g seeds), *p*-hydroxybenzoic acid (20.89 µg/g), and ferulic acid (18.41 µg/g). Low amounts of protocatechuic acid and salicylic acid were detected as well. The amounts of ferulic acid in amaranth insoluble fiber and non-starch polysaccharides were investigated by Bunzel *et al.* (2005). Alkaline hydrolysis released 62 mg/100 g *trans*-ferulic acid and a high content (20.3 mg/100 g) of *cis*-ferulic acid. Three compounds of feruloylated oligosaccharides were identified, indicating that ferulic acid is predominantly bound to pectic arabinans and galactans in amaranth insoluble fiber.

Czerwinski *et al.* (2004) determined total phenolics expressed as gallic acid equivalent (Folin-Ciocalteu reagent), anthocyanins, and flavonoids (spectrophotometrically) in two amaranth samples and compared the results to those from oats. The amounts of polyphenols in the amaranth samples ranged from 14.72 to 14.91 mg/100 g seeds, that of anthocyanins from 59.6 to 62.5 mg/100 g seeds, and that of flavonoids from 13.4 to 14.3 mg/100 g seeds. Overall, these amounts were lower than those obtained from the oat sample.

Antioxidative activity

The antioxidant activity of amaranth and oat extracts has been measured using the β-carotene/linoleate model system and scavenging activity against nitric oxide (NO test) by Czerwinski *et al.* (2004). Amaranth extracts showed less antioxidant activity (23.2–26% β-carotene, 23–25% NO) than the oat extracts. Best correlation was found between the total antioxidant activity and total phenols, and good correlation was also observed for anthocyanins and flavonoids. Recently it has been shown that 0.05% addition of amaranth seeds extract is applicable to inhibit β-carotene degradation in a β-carotene/linoleic acid model system (Klimczak *et al.*, 2002). Unfortunately no information on the concentration and composition of the amaranth extract was provided. Jung *et al.* (2006) determined the antioxidative power (expressed as antioxidative units, corresponding to vitamin C activity) of various seeds by ESR spectroscopy based on the DPPH (1,1-diphenyl-2-picryl-hydrazil) assay, where both the antioxidative capacity and the antioxidant activity were used to characterize the antioxidant. Amaranth seeds showed a rather low antioxidative power of 27, while quinoa seeds had a higher antioxidative power of 458. By inclusion of the kinetic behavior of the reducing process of the antioxidant, the authors drew conclusions on the main antioxidant in the samples. According to this procedure, vitamin C was identified as the main antioxidant in quinoa, whereas polyphenols were found to be the main antioxidants in amaranth.

Saponins

Saponins are strongly bitter tasting, surface active agents (surfactants), which can cause intensive foaming activity in aqueous solutions. They can form complexes with proteins and lipids (e.g. cholesterol) and possess a hemolytic effect. Saponins are only absorbed in small amounts, and their main effect is restricted to the intestinal tract. Saponins can form complexes with zinc and iron, thus limiting their bioavailability (Chauhan *et al.*, 1992). With regard to health-promoting effects, saponins are anti-carcinogenic, anti-microbial, cholesterol decreasing, immune modulating, as well as anti-inflammatory.

Amaranth seeds contain rather low amounts of saponins. Dobos (1992) found contents of on average 0.09% (aescin equivalents) in various amaranth species, and these results have been confirmed by the investigations of Oleszek *et al.* (1999). It was concluded that the low concentration of saponins in amaranth seeds and their relatively low toxicity guarantee that amaranth-derived products create no significant hazard to the consumer.

Phytic acid

Cereals and legumes are particularly rich in phytic acid. Phytic acids can form complexes with the basic protein residues, leading to the inhibition of enzymatic digestive reactions and interference with the adsorption of minerals, in particular with zinc. Amaranth contains phytates in the range 0.2–0.6% (Breene, 1991; Bressani, 1994; Escudero *et al.*, 2004; Gamel *et al.*, 2006a). Recently, it has been shown that cooking reduces the phytate content by approximately 20%, popping by 15% and germination (48 hour) by 22%, indicating that these approaches can be used to reduce the phytate content of amaranth (Gamel *et al.*, 2006a).

Production of flours and their properties

Due to the small size of the amaranth seed and its botanical peculiarities, specific adaptations of the milling procedure are required. While the production of wholemeal flour from amaranth is not complicated, specific demands arise during grinding and separation when producing flour fractions with different chemical compositions and physical properties. In addition to differences in compositions, there are also differences regarding the quality (e.g. uniformity of particle size of the flour fractions). Therefore, the mill and the milling technology used play a key role in determining the quality of the resulting flour fractions. Several research groups have investigated the milling of amaranth to wholemeal flour (Becker *et al.*, 1981; Betschart *et al.*, 1981; Sanchez-Marroquin *et al.*, 1985a, 1985b, 1986). In contrast, very little information can be found on the production of flour fractions. Several different mills (i.e. disk mills with metal mill faces, different hammer mills and a pin mill using several pin configurations) have been tested by Becker *et al.* (1986), however, all only shattered the seeds and produced wholemeal flour. Modification of a stone mill (increased spindle speed of the diametric stones to 3600 rpm and decreased distance between the milling stones) resulted in an intact perisperm separated from the germ and bran.

Some researchers adapted a "Strong-Scott barley pearler" for the production of protein-rich or starch-rich flour fractions (Betschart *et al.*, 1981; Sanchez-Marroquin *et al.*, 1985a, 1985b, 1986). The seed coat and germ could be completely separated and a spherical, intact starch-rich perisperm was left. The bran fraction constituted about 25–26% of the seed weight and contained, similar to cereals, more nutrients than the perisperm. Nitrogen, crude fat, dietary fiber, and ash were found to be 2.3–2.6 times higher than in the whole seed; moreover, the vitamin content was 2.4–3.0 times higher (Betschart *et al.*, 1981). Different milling and fractionating equipments (pin mill and zig-zag sifter, pilot roller mill, vario-technical roller mill, technical scale roller mill in combination with a plansifter) were investigated to obtain different flour fractions (Nanka, 1998; Schoenlechner, 2001). The best results were achieved by applying the technical scale roller mill in combination with a plansifter, as five fractions were obtained which were separated into protein-rich and starch-rich fractions. Interestingly, the starch content in the middlings fraction was higher than in the flour fraction, as a result of the different botanical structure of the amaranth seed. The starch-rich flour and middlings fractions showed higher paste viscosity than the protein-rich semolina fractions.

Gamel *et al.* (2005, 2006a, 2006b) produced on laboratory scale a high-protein amaranth flour (HPF) by milling the seeds to granules passing a 115-mesh sieve, and subjecting the flour to air classification using a zig-zag sifter under the following operation conditions: temperature 23°C, filter pressure 2 mbar, centrifugation at $8000 \times g$, and airflow rate 80 m³/h. The yield of the HPF and the protein content were both 25 g/100 g. Starch content ranged from 30 to 36 g/100 g. Air classification increased the content of minerals by more than 35%, as well as the levels of vitamin B complex, phenolic compounds and phytate, whereas enzyme inhibitors were decreased. In addition, an increased cold paste viscosity and peak viscosity, as well as foam stability were measured in the HPF.

Tosi *et al.* (2001, 2002) applied a differential milling process in order to produce different amaranth flour fractions. In a pilot mill, conditioned amaranth seeds were milled and then classified by sieving and pneumatic separation. A protein-rich fraction containing 40% protein was obtained. The starch-rich semolina and fiber-rich fractions were further improved by additional pneumatic separation until a product with 63.9% insoluble fiber and 6.9% soluble fiber was obtained (Tosi *et al.*, 2001). When 4–12% of wheat flour was replaced by the protein-rich fraction, bread quality did not change. Remarkably, the protein and available lysine contents in the bread were increased by the presence of the protein-rich fraction (Tosi *et al.*, 2002). The starch-rich fraction was modified by extrusion cooking and fluidized bed heating, which allowed the production of pre-cooked amaranth flours with a wide range of hydration and rheological properties. Flours obtained from samples heated by fluidized bed heating gave aqueous dispersions with high consistencies when cooked. They showed low water solubility and preserved some of the starch crystalline structure.

Flours obtained by extrusion cooking gave high water solubility, had lower consistency when cooked and showed a complete loss of the crystalline and granular structure (Gonzalez *et al.*, 2007).

Quinoa

Carbohydrates

Starch

The main component of carbohydrates in quinoa is starch. However, the starch content of quinoa is lower than that in cereals (Table 7.1). As in amaranth, the starch is located in the perisperm, although small amounts occur in the seed coat and embryo (Prego *et al.*, 1998). Quinoa starch consists of single polygonal granules ranging from 0.63 to 1.8 μm (average 1.5 μm) in size (Atwell *et al.*, 1983; Ando *et al.*, 2002). Complexes of starch granules can also be found, in which up to 14 000 single granules are bound together to form one complex (Lorenz, 1990). These complexes, either spheroidal or oblong, are surrounded by a protein matrix and can reach a length of 80 μm. The protein content of quinoa starch is higher than in that of other cereal starches (Atwell *et al.*, 1983).

The two major starch granule-bound proteins have been identified as granule-bound starch synthase I (GBSSI) with molecular masses of 56 and 62 (Lindeboom *et al.*, 2005b). The content of GBSSI correlates positively with the concentration of amylose in the starch. The content of amylose in quinoa starch is much lower than in cereal starches, and can be found in different amounts, ranging from 3 to 22% (Atwell *et al.*, 1983; Qian and Kuhn, 1999a; Tang *et al.*, 2002; Wright *et al.*, 2002; Tari *et al.*, 2003; Lindeboom *et al.*, 2005a, 2005b). The starch has been classified as short-chain branched glucans with an average molecular weight of 11.3×10^6 g/mol (Praznik *et al.*, 1999). Tang *et al.* (2002) found that quinoa amylopectin has a unique chain length distribution similar to a waxy amylopectin as it has a larger ratio of short chains to long chains. X-ray diffraction patterns indicated that the starch shows typical type "A" crystallinity (Qian and Kuhn, 1999a; Wright *et al.*, 2002).

Quinoa starch has higher gelatinization temperatures and higher pasting viscosities than other cereals. These values increase during cooling (Atwell *et al.*, 1983; Lorenz, 1990; Ruales and Nair, 1994; Schoenlechner, 1997). Furthermore, the low amylose content is responsible for a high water-binding capacity, high swelling power, high enzyme susceptibility, and excellent freeze–thaw and retrogradation stabilities (Atwell *et al.*, 1983; Lorenz, 1990; Ahamed *et al.*, 1996; Qian and Kuhn, 1999a). These physical properties have a positive correlation with the amylose content (Lindeboom *et al.*, 2005a). The wide variation within the amylose content is therefore responsible for the differences in physical properties of quinoa starch.

Resistant starch

The resistant starch content (retrograded RS_3 arising during processing) in various cereals and pseudocereals has been measured enzymatically (AACC Method 32–40) by Mikulikova and Kraic (2006). The values for quinoa were 12.6 ± 1.29 g/kg seeds, and thus lower than those for other cereals like wheat ($39.0 \pm 5.7 \, g/kg$) or rye ($49.0 \pm 7.3 \, g/kg$). The proportion of RS/total starch is 2.18% in quinoa, compared to 5.64% and 7.01% for wheat and rye, respectively. The reason for the rather low content of RS in quinoa is the low content of amylose and thus the low formation of RS_3.

Low molecular weight carbohydrates
Conflicting results have been reported on the amount of free sugars in quinoa. Gonzalez et al. (1989) reported a glucose content of 4.55%, fructose of 2.41%, and sucrose of 2.39%. A different ratio of glucose to sucrose was determined by Gross et al. (1989): glucose 0.19%, sucrose 2.79%, raffinose 0.15%, stachyose 0.08%, alpha-galactosides 0.23%, fructose 0%, and verbascose 0%. Also Ogungbenle (2003) found lower amounts of glucose (0.019%) compared to other sugars, i.e. fructose (0.019%), galactose (0.06%), ribose (0.07%), maltose (0.1%), and D-xylose 0.12%. Overall, it can be stated that the amounts of the flatulence-causing raffinose and stachyose are low in quinoa.

Fiber
The dietary fiber content of quinoa (12.88%) is comparable to that of other cereals (Table 7.1), with the embryo containing higher amounts than the perisperm (Hirano and Konishi, 2003). The proportion of soluble fiber is only 13.5% of total dietary fiber (Ranhorta et al., 1993). Cooking and autoclaving decreases the fraction of soluble dietary fiber, while the insoluble fractions do not differ (Ruales and Nair, 1994).

Protein

The protein content of quinoa is higher than that of other cereals, and the quality of the protein is very good (Table 7.1).

Storage proteins
The quinoa proteins are mainly globulins and albumins. The seed protein consists of 31% water, 37% saline, 0.8% alcohol, 11.5% alkali soluble, and 19.7% insoluble protein fractions (Prakash and Pal, 1998; Ando et al., 2002; Watanabe et al., 2003). The amino acid profile of each protein fraction showed a balanced content of essential amino acid, with a high level of lysine (4.5–7.0%) (Watanabe et al., 2003). The two major classes of proteins in quinoa seeds are the 11S (chenopodin) and 2S (high-cysteine) proteins (Brinegar and Goundan, 1993; Brinegar, 1997). The distinctive structural and solubility characteristics of the 11S and 2S proteins suggest that their functional properties differ markedly. The insolubility of chenopodin under acidic conditions is characteristic of other 11S proteins, whereas the quinoa 2S proteins are highly soluble and contain numerous cysteine residues.

Amino acids
Amino acids are present in a concentration of 38.71 g/100 g protein, and this value is only 16% lower than that of whole egg protein (Drzewiecki et al., 2003).

Quinoa protein is close to the FAO recommended pattern in essential amino acids (Prakash and Pal, 1998). The lysine level (6.3%) is comparable to that of soybean, whereas, as in other typical dicotyledonous seed proteins, methionine is deficient (Ranhorta *et al.*, 1993). However, different results have been reported on the limiting amino acids. According to the chemical scores, the aromatic amino acids tyrosine and phenylalanine were found to be limiting, while the content of methionine and cysteine was high (Ruales and Nair, 1992a). Fractionation does not seem to affect the amino acid composition of quinoa (Chauhan *et al.*, 1992).

Nutritional quality

Ranhorta *et al.* (1993) investigated the protein quality of quinoa. The PER values (3.8) and C-PER values (2.7) did not differ significantly from the casein values. The digestibility was 84.3% and was lower than that of casein (88.9%). Quinoa proteins have an NPU value of 75.2, and the biological value (BV) is 82.6 (Ruales and Nair, 1992a). Chauhan *et al.* (1999) found an increased PER, but unchanged BV after saponin extraction. According to Ruales and Nair (1994), the *in vitro* protein digestibility of raw quinoa is lower than that of casein. The removal of the outer seed layers, which contain saponins, improved digestibility by 7%. Increased digestibility was also found after heat treatment. Moreover, heat treatment might also destroy harmful substances (e.g. protease inhibitors). Structural changes might also occur (e.g. reduction of lipid– or starch–protein complexes), which could be responsible for the improvement of the quinoa protein digestibility after heat treatment. However, a prolonged cooking time reduces the level of improvement.

Allergy and celiac disease

No information on the allergenicity of quinoa, or any adverse effects for patients with celiac disease can be found in the literature.

Functional properties of proteins

Protein solubility of the quinoa flour is pH dependent, with minimum solubility at pH 6.0, as found in pearl millet (Oshodi *et al.*, 1999; Ogungbenle, 2003). However, the emulsification capacity and stability are better in quinoa than in pearl millet or wheat, while foaming capacity seems to be lower (Oshodi *et al.*, 1999).

Enzyme inhibitors

Like in amaranth, trypsin inhibitor activity is low in quinoa and can be inactivated by heat treatment (Chauhan *et al.*, 1992). According to Ruales and Nair (1992b) 1.36–5.04 TIU/mL sample can be found, which is lower than in legumes.

Lipids

The content of fat in quinoa is higher than in cereals. On average it ranges from 5 to 6% (Table 7.1), however some varieties even show higher values. For example, Ruales and Nair (1993a) mentioned a fat content of 9.7% and Przybylski *et al.* (1994) measured 7.6%. As for proteins, the fat content is higher in the germ and seed coat than in the perisperm. The fat is characterized by a high content of unsaturated fatty acids, with linoleic acid accounting for more than 50% of the fatty acids. Palmitic acid accounts for around 20%, followed by oleic acid with about 8% and linolenic acid with more than 6%. The degree of unsaturation is over 75% (Przybylski *et al.*, 1994) or, according to Ando *et al.* (2002), even higher than 87%. Due to the high vitamin E content, quinoa lipids have been found to be stable during storage (Ng *et al.*, 2007). Phospholipids constitute 25.2% of total lipids (Przybylski *et al.*, 1994). They are characterized as lysophosphatidyl-ethanolamine (lysocephalin), phosphatidyl-ethanolamine, phosphatidyl-inositol, and phosphatidyl-cholin (lecithin, which represents the 49% of the total phospholipids). Other phospholipids have been detected in trace amounts.

Minerals

The content of minerals (ash) in quinoa seeds is approximately twice as high as in cereals (Table 7.1). Growing conditions seem to have an influence on the mineral composition (Karyotis *et al.*, 2003). Calcium, magnesium, iron, potassium, and zinc can be found in high amounts (Chauhan *et al.*, 1992; Ruales and Nair, 1993a; Ando *et al.*, 2002; Konishi *et al.*, 2004; Ogungbenle, 2003). Quinoa contains higher amounts of potassium than amaranth, and the calcium/phosphorus ratio is 1:4.1 to 1:6 (calculated after Souci *et al.*, 2000; Chauhan *et al.*, 1992). Nutritionists recommend around 1:1.5 (Ca:P). Through removal of saponins (mechanically and/or through washing) the mineral content in quinoa can be strongly reduced (e.g. 46% of the potassium can be lost) (Ruales and Nair, 1993a; Konishi *et al.*, 2004).

Vitamins

The content of vitamins in quinoa is similar to that found in conventional cereals. Ruales and Nair (1993a) described an almost equal content in quinoa and wheat. Quinoa is a good source of thiamine (0.4 mg/100 g), folic acid (78.1 μg/100 g) and vitamin C (16.4 mg/100 g) (Ruales and Nair, 1993a). Like amaranth, quinoa contains more riboflavin (0.2 mg/100 g) than cereals. Furthermore, quinoa is a particularly good source of vitamin E (Coulter and Lorenz, 1990; Ruales and Nair, 1993a; Souci *et al.*, 2000), which contributes to the prolonged stability of the oil. Quinoa seeds contain twice as much γ-tocopherols (5.3 mg/100 g) than α-tocopherols (2.6 mg/100 g). Ruales and Nair (1993a) measured 0.3 mg/100 g β-tocotrienols, but could not detect any α-tocotrienols.

Phytochemicals

Total phenolic compounds

Different tannin contents in quinoa have been reported, with values varying from 0 to 500 mg/100 g (Chauhan *et al.*, 1992; Ruales and Nair, 1993b). This difference can be explained as being due to differences in variety and growing habitats. Anyway, a level of 500 mg/100 g tannins in grain is considered low. Regarding the content of total ferulic acid in insoluble fiber, a value of 635 μg total ferulic acid/g insoluble fiber has been reported (Packert, 1993). This value is comparable to that in amaranth.

Flavonoids

The polyphenols present in quinoa are mainly kaempferol and quercetin glycosides. Two flavonol glycosides have been isolated from seeds (De Simone *et al.*, 1990). Zhu *et al.* (2001) isolated six flavonoids in quinoa, four kaempferol glycosides and two quercetin glycosides. All six compounds exhibited antioxidant activity in a DPPH test. The two quercetin glycosides showed much stronger activity than the four kaempferol 3-glycosides present. In addition, five flavonol glycosides and a vanillic acid glucosyl ester were found by Dini *et al.* (2004).

Antioxidative activity

Quinoa has higher antioxidative power than amaranth (Jung *et al.*, 2006).

Saponins

Quinoa (whole seeds) contains between 0.03 and 2.05% of bitter tasting saponins (Ridout *et al.*, 1991; Chauhan *et al.*, 1992; Gee *et al.*, 1993; Ruales and Nair, 1993b; Cuadrado *et al.*, 1995), but these values are still below those found in soybeans. Saponin present in quinoa seeds comprises oleanic acid and three other sapogenols identified as hederagenin, phytolaccagenic acid, and deoxyphytolaccagenic acid (Cuadrado *et al.*, 1995; Woldemichael and Wink, 2001). Optical and electron microscopy in combination with chemical methods identified the saponin bodies in the pericarp cell of quinoa (Prado *et al.*, 1996). The spherical saponin bodies are about 6.5 μm in diameter and appear to be an aggregate formed of 4 or 5 small granules (2.2 μm in diameter). Thirty-four percent of the saponins are found in the hull (Chauhan *et al.*, 1992). Dehulling and washing decrease the content by up to 72% (Ruales and Nair, 1993b; Gee *et al.*, 1993). Processing can also destroy saponins, but a reduction in the content is not as great as that observed after washing or dehulling (Gee *et al.*, 1993).

Another way to reduce the saponin content in quinoa seeds is by breeding so-called sweet (low saponin content) quinoa species. Mastebroek *et al.* (2000) investigated the saponin content of various species and found only 0.02–0.04% saponins in sweet varieties compared with 0.47–1.13% in bitter varieties. According to Koziol (1991), if the saponin content is less than 0.11% the variety can be considered to be a sweet variety.

Phytic acid

Quinoa contains 0.1–1.0% phytic acid (Chauhan *et al.*, 1992; Ruales and Nair, 1993b). Varriano-Marston and DeFrancisco (1984) found that the phytic acid seems to be concentrated in the embryo, since phosphorus-containing globoid inclusions have been observed in this tissue. However, according to Chauhan *et al.* (1992) and Ruales and Nair (1993b), phytic acid is distributed uniformly in the seed, and is therefore hardly decreased by abrasive dehulling or extraction with water.

Production of flours and their properties

Due to the small size, quinoa is usually milled to wholemeal flour, after removal of the saponins, either by washing or abrasive milling. The production of quinoa flour fractions with different chemical composition has been only poorly investigated. As the saponins are concentrated in the hulls, their content can be minimized by dehulling of the seeds (e.g. tangential abrasive dehulling) (Reichert *et al.*, 1986). Becker and Hanners (1990) milled quinoa using a stone mill and found that 33–40% of the seed can be removed as a bran fraction, indicating a high abrasion (Becker and Hanners, 1990). However, although the bran fraction contains a higher amount of nutrients than the perisperm, it also contains a higher amount of saponins. A laboratory roller mill has been used by Chauhan *et al.* (1992) to separate the bran fraction from the flour fraction. About 40% was removed as a protein- and fat-rich bran fraction, leaving about 50% of a starch-rich flour fraction (mainly perisperm). The production of different flour fractions from quinoa can be obtained using the same equipment as for amaranth (see above; Nanka, 1998; Schoenlechner, 2001). Moreover, in quinoa the combination of technical roller mill and plansifter allows a better separation between starch-rich flour, middlings fractions, and protein-rich semolina fractions. Finally, the middlings fractions show higher starch content than the flour fraction. Caperuto *et al.* (2000) used a Senior Quadrumat Brabender mill to produce quinoa flour. Grain preconditioned at 150 g/kg moisture yielded the highest recovery of break plus reduction flour with an average particle size of 187.7 μm. Unexpectedly, the protein content of the flour fell from 12.5% in the wholemeal to 3.55% in the flour. On the other hand, the protein was not greatly impoverished in lysine, and an increase in methionine and branched-chain amino acid contents was observed. Addition of 100 g/kg quinoa flour to corn flour was sufficient to improve the lysine content of the gluten-free spaghetti produced by a factor of 3.

Buckwheat

Carbohydrates

Starch

Buckwheat has a total carbohydrate content of 67–70% (Li and Zhang, 2001; Steadman *et al.*, 2001a), of which 54.5% is starch (Steadman *et al.*, 2001a). Buckwheat starch granules have a polygonal shape and are often aggregated. The size of starch granules is rather small, with a particle size distribution of 2–14 μm and a mean diameter of 6.5 μm (Acquistucci and Fornal, 1997). Buckwheat starch shows a characteristic fraction composition, in which the ratio between amylose and amylopectin is 1:1. In this respect, buckwheat starch visibly differs from cereal or pulse starch, and is similar to high amylose maize. Amylose contents of buckwheat starch as high as 46% have been found by Qian *et al.* (1998) and Soral-Śmietana *et al.* (1984a), although lower contents have been reported, similar to other cereal starches, ranging from 21.1 to 27.4% (Li *et al.*, 1997; Zheng and Sosulski, 1998; Noda *et al.*, 1998; Qian and Kuhn, 1999b, 1999c; Yoshimoto *et al.*, 2004). In studies of Yoshimoto *et al.* (2004) the actual amylose content accounted for 16–18%, and is thus lower than the apparent amylose content (26–27%), due to the high iodine affinity of buckwheat amylopectin (2.21–2.48%).

Buckwheat starch contains a large amount of long-chain amylopectins (Noda *et al.*, 1998; Praznik *et al.*, 1999; Yoshimoto *et al.*, 2004). The long-chain fraction is abundant (12–13% measured by weight), but the distributions of amylose and short chains of amylopectin (based on molar basis) are similar to those measured for wheat and barley starches (Yoshimoto *et al.*, 2004). Noda *et al.* (1998) found that more than 40% of amylopectin shows a degree of polymerization of 10–12, and that the average weight of buckwheat starch is 94 900, which is closer to values of waxy maize starch than starches isolated from other cereals or pseudocereals. The starch exhibits a typical type "A" pattern of X-ray diffraction and the crystallinity varies from 38.3 to 51.3% (Qian and Kuhn, 1999b; Zheng *et al.*, 1998).

In general, buckwheat starch exhibits a higher gelatinization temperature, peak and set back viscosities (Wei *et al.*, 1995; Zheng *et al.*, 1998) than cereal starches and resembles the pasting behavior of root and tuber starches (Whistle *et al.*, 1984). High viscosity values can be explained by supermolecular glucan structures (Praznik *et al.*, 1999) and by the fact that buckwheat starches exhibit a higher granule swelling and gelling tendency than cereal starches (Pomeranz, 1991; Yoshimoto *et al.*, 2004). Acquistucci and Fornal (1997) suggested that the higher swelling power is a consequence of the weaker but more extensive bonding forces in the granule, whereas Qian *et al.* (1998) suggested that the presence of amylose–lipid complexes could lead to the restriction of swelling power and solubility. The increased susceptibility of buckwheat starch to acid and α-amylase hydrolysis suggests a larger amorphous region in the buckwheat starch granule than in corn and wheat starches (Qian *et al.*, 1998). The water-binding capacity of buckwheat starch is 109.9%, which is higher than that of wheat and corn starch, and is explained by the small size of buckwheat starch granules (Qian *et al.*, 1998).

Resistant starch

Raw buckwheat groats contain 73.5–76.0% starch, of which 33.5–37.8% are resistant starch (RS) (Skrabanja and Kreft, 1998; Skrabanja *et al.* 1998), thus making buckwheat an interesting material for designing low glycemic index foods. Processing affects the distribution of RS. Thermal treatment (e.g. cooking or dry heating to 110°C) decreases the RS to 7.4%, whereas the level of retrograded starch (RS_3) can be increased fourfold by boiling (Skrabanja *et al.*, 1998, 2001). The RS_3 value for buckwheat is 3.79% and the proportion of resistant/total starch has been calculated as 6.51%, which is more than three times the values found in quinoa and amaranth. These results indicate that buckwheat contains more amylose (Mikulikova and Kraic, 2006).

Fiber and D-*chiro*-inositol

Buckwheat bran contains seed coat and embryo tissues, and is the milling fraction most concentrated in proteins (35%), lipids (11%), and dietary fibers (15%). Bonafaccia *et al.* (2003b) reported a total dietary fiber fraction of 27.38% in buckwheat seeds. The soluble fraction was found especially in the bran at levels around 1%, lower than previous findings (Steadman *et al.*, 2001a). Buckwheat bran is also a concentrated source of fagopyritols (2.6%) (indigestible oligosaccharides), galactosyl derivatives of D-*chiro*-inositol that may be useful in the treatment of non-insulin-dependent diabetes mellitus (Steadman *et al.*, 2000). The level of free D-*chiro*-inositol in buckwheat groats ranges from 20.7 to 41.7 mg/100 g (Steadman *et al.*, 2000). D-*chiro*-Inositol is primarily present in the form of fagopyritols (Horbowicz *et al.*, 1998) and is mainly localized in buckwheat embryos.

Protein

Storage proteins

The main components of buckwheat seed proteins are salt-soluble globulins, represented mainly by the 13S legumin-like protein fraction (Aubrecht and Biacs, 1999; Li and Zhang, 2001; Milisavljević *et al.*, 2004). Together with the minor, 8S vicilin-like globulins, storage globulins account for 70% of total seed proteins (Radović *et al.*, 1996, 1999). The 8S globulin contributes to about 7% of total seed proteins. According to Milisavljević *et al.* (2004) 8S globulin is more interesting for biotechnological applications than the 13S buckwheat legumin, which has been reported as the major buckwheat allergen. Remarkably, no cross-reaction has been found for the 8S storage globulin. A significant portion of buckwheat storage proteins is represented by the 2S albumin fraction (18–32% of total protein) (Radović *et al.*, 1999). The contribution of glutelins is minor, and prolamins are found in different amounts, ranging from 0 (Radović *et al.*, 1999) to 1.9% (Aubrecht and Biacs, 2001) or 4.35% of protein (Wei *et al.*, 2003). However, recent results collected from tartary buckwheat are in contrast with the previous studies (Guo and Yao, 2006). The albumin fraction was found to be the predominant protein fraction (43.8%), followed by glutelin (14.6%), prolamin (10.5%), and globulin (7.82%). This discordance can be explained by the use of different extraction methods and/or cultivars in the reported studies.

Amino acids

The amino acid composition depends on the parts of the seed investigated (Li and Zhang, 2001). Buckwheat proteins have a higher or similar content of all amino acids when compared with wheat proteins, except for glutamine and proline, which are found in lower amounts. In particular, the content of the limiting amino acid lysine is 2.5 times higher than that found in wheat flour (Aubrecht and Biacs, 2001). Glutamic acid, followed by aspartic acid, arginine, and lysine are the most represented amino acids. Methionine and cysteine contents have been identified as the less represented. However, different classifications of the limiting amino acids have been reported. Recently, Wei *et al.* (2003) identified leucine as the first limiting amino acid, followed by threonine, methionine, cysteine, phenylamine, and tyrosine. According to these results, the authors suggested that in view of its nutritional value buckwheat is not a suitable material to use as staple food, but it should be used in combination with other cereal grains.

Nutritional quality

The amino acid composition of buckwheat is well balanced and nutritionally superior to that of cereal grains (Pomeranz and Robbins, 1972) in terms of biological value, net protein utilization, and utilizable protein values (Eggum et al., 1980). True digestibility is, however, lower for buckwheat than for wheat (Eggum et al., 1980). Buckwheat proteins suppress gallstone formation and lower the cholesterol level more strongly than soy protein isolates (Kayashita et al., 1995; Tomotake et al., 2000, 2001). It is well accepted that the ratios of lysine/arginine and methionine/glycine are critical factors determining the cholesterol-lowering effects of plant proteins, although the mechanisms are not fully understood (Li and Zhang, 2001). Nevertheless, the cholesterol-lowering effect is also attributed to the low digestibility (<80%) and dietary fiber-like contents of buckwheat (Ikeda et al., 1991; Ikeda and Kishida, 1993; Pandya et al., 1996; Kayashita et al., 1997; Skrabanja et al., 2000). In addition, buckwheat proteins may also retard mammary carcinogenesis by lowering serum estradiol, and suppress colon carcinogenesis by reducing cell proliferation (Kayashita et al., 1999; Liu et al., 2001).

Allergy and celiac disease

Francischi et al. (1994) revealed that buckwheat does not contain toxic prolamins to celiac disease patients (Francischi et al., 1994). However, Radović et al. (1999) stated that buckwheat possesses antinutritive effects and allergen activity for sensitive patients, even though no prolamin could be detected. The estimated prolamin content of buckwheat is in the range 3.8–5.2 mg/100 g seed and all products made of 100% buckwheat flour are well below the permitted limit for gluten-free products (Aubrecht and Biacs, 2001) (10 mg/100 g dry matter, Codex Alimentarius Commission, 2000). Buckwheat has been recognized as a common food allergen in Korea and Japan but not in North America (Park et al., 2000; Taylor and Hefle, 2001; Tanaka et al., 2002).

Buckwheat is known to be a highly potent food allergen related to an IgE-mediated, type I immune reaction. Asthma, allergic rhinitis, urticaria, and angioedema are the main symptoms involved (Li and Zhang, 2001). If patients have a buckwheat-specific IgE antibody level of 1.26 kU$_A$/L or greater, the danger of allergic reactions is already present when only small amounts of buckwheat are ingested or inhaled (Sohn et al., 2003). Bush and Hefle (1996) found four glycoproteins in the molecular weight range of 9–40 kDa as IgE-binding bands. A 24 kDa protein was identified as the major allergen in buckwheat (Kondo et al., 1996). The same 24 kDa protein has been recently isolated from tartary buckwheat seeds (Wang et al., 2004). In another study, allergens of 24, 19, 16, and 9 kDa were identified as strong candidates for allergenicity and the 19 kDa allergen as relatively specific for buckwheat-allergic patients (Park et al., 2000). Allergens with a molecular weight of 67–70 kDa have also been identified (Li and Zhang, 2001). On the other hand, it has been shown that allergens of buckwheat and soy flours can be partially destroyed by high shear forces using twin-screw extrusion cooking (Hayakawa *et al.*, 1996).

Enzyme activities

Several studies have shown that enzymatic activities such as lipoxygenase and per-oxidase (Suzuki *et al.*, 2004b, 2006) or flavonol-3-glucosidase (Suzuki *et al.*, 2002, 2004a) play important roles in the deterioration of buckwheat flour. Inhibitory effects of rutin against *in vitro* lipoxygenase activity were observed by Suzuki *et al.* (2005). Buckwheat not only contains phytic acid but also shows phytase activity (PU) of 2.17 PU/g. The optimum conditions for buckwheat phytase activity are a pH of 5.0 and a temperature of 55°C (Egli *et al.*, 2003).

Lipids

Lipids in buckwheat are concentrated in the embryo, and thus the bran is the most lipid-rich milling fraction. The total lipid content in buckwheat grains amounts to 2.48% (dry weight), of which the free lipids account for 2.41% and bound lipids for 1.09%. Glycolipids and phospholipids make only 1.01 and 0.47%, respectively, of free lipids in buckwheat grains (Soral-Śmietana *et al.*, 1984b). Triacylglycerides are the main component of the neutral lipid fraction. Linoleic acid, oleic acid, and palmitic acid account for 88% of the total fatty acids (Mazza, 1988; Horbowicz and Obendorf, 1992). With typically 80% unsaturated fatty acids and more than 40% of the polyunsaturated essential fatty acid linoleic acid, buckwheat is nutritionally superior to cereal grains (Steadman *et al.*, 2001a). A similar fatty acid composition has been observed in amaranth oil and cotton seed oil (Jahaniaval *et al.*, 2000).

Plant sterols are ubiquitous through the whole buckwheat kernel (Li and Zhang, 2001). In embryo and endosperm tissues, the most abundant sterol is β-sitosterol, which accounts for 70% of the total sterols. The content of sterols in dehulled groats after lipid extraction is about 70 mg/100 g seed for β-sitosterol, about 9.5 mg/100 g for campesterol, and traces of sigmasterol (Horbowicz and Obendorf, 1992).

Minerals

The content of minerals (ash) in buckwheat seeds is lower than in wheat (Table 7.1). However, except for calcium, buckwheat is a richer source of nutritionally important minerals than many cereals such as rice, sorghum, millet, and maize (Adeyeye and Ajewole, 1992). Bonafaccia *et al.* (2003a) reported a 2- to 3-fold higher content of the elements Se, Zn, Fe, Co, and Ni in tartary buckwheat cultivars when compared with common buckwheat. Overall, the minerals were found to be most concentrated in bran. The concentration of P, K, and Mg increased after removal of the hulls, whereas Ca and Zn seem to be accumulated in the hulls (Steadman *et al.*, 2001b). Attempts have been made to increase the Se content in buckwheat seeds: an 8.5-fold increase was obtained after foliar Se fertilization (Na selenate at 1 mg/L) (Stibilj *et al.*, 2004).

Vitamins

The vitamin content of several cereal grain products has recently been investigated (Gujska and Kuncewicz, 2005). Results indicate that buckwheat groats have a higher content of total folate (30 μg/100 g) than rye flour (29 μg/100 g), barley groats (21 μg/100 g), or wheat flours (19–20 μg/100 g). The vitamins B_2 and B_6 are present in buckwheat seeds (Fabjan et al., 2003). Total vitamin B content is higher in tartary buckwheat than in common buckwheat, and, generally, the highest quantity of B vitamins is in the bran. Additionally, tartary buckwheat bran contains about 6% of the daily therapeutic dose of pyridoxine, which is effective (along with folic acid and vitamin B_{12}) in the reduction of blood plasma homocysteine levels and in the decrease of the rate of restenosis after coronary angioplasty (Krkošková and Mrázová, 2005). Watanabe et al. (1998) isolated a thiamine-binding protein with a molecular mass of 42–45 kDa. It has been suggested that, after ingestion, this complex may be cleaved by proteases, thus releasing the thiamine and contributing to its survival in processed foods.

Phytochemicals

Phenolic acids

Total phenolic acids have been determined using first alkaline and then acid hydrolyses (Table 7.2) (Mattila et al., 2005). The content (based on moist mass) of the total ferulic acid was low (1.2 mg/100 g), but the contents of p-hydroxybenzoic acid (11.0 mg/100 g) and caffeic acid (8.5 mg/100 g) were high compared with other grain products. In a recent study, the total phenolic acids content of buckwheat husk and flour were found in the ranges of 30 and 15 mg/100 g, respectively (Gallardo et al.,

Table 7.2 Contents of total phenolic acids (sum of alkaline and acid hydrolysis)

Concentration (mg/kg mm, wholegrain)	Rye flour	Wheat flour	Barley flour	Buckwheat grits
Dry matter	90.0	89.7	90.3	90.9
Caffeic acid	10±2.0	37±1.4	1.7±0.13	85±8.7
Ferulic acid	860±71	890±40	250±32	12±0.69
Sinapic acid	120±12	63±3.6	11±1.7	21±1.0
Protocatechuic acid	9.4±1.6	nd	1.6±0.15	nd
Vanillic acid	22±2.8	15±0.83	7.1±0.83	5.3±0.32
p-Coumaric acid	41±2.8	37±1.2	40±4.9	15±0.89
p-Hydroxybenoic acid	6.8±0.87	7.4±0.06	3.1±0.53	110±14
Syringic acid	6.7±0.33	13±0.007	5.0±0.33	nd
Ferulic acid dehydrodimers	290±25	280±16	130±13	nd
Total	1366	1342	450	248

From Matilla et al. (2005).

2006). In addition, only traces of *p*-coumaric acid and benzoic acid derivatives were measured in buckwheat milling fractions.

Polyphenols

In general, polyphenols are mostly concentrated in bran, whereas flour and grits contain only little amounts. The bran fraction has a high concentration of tannins (0.4 g/100 g non-condensed and 1.7 g/100 g condensed tannins) and other polyphenols (total polyphenols: 1.2 g catechin equivalents/100 g). Remarkably, removal of hull fragments from bran results in a 4-fold reduction of condensed tannins (Steadman *et al.*, 2001b). The content of total flavonoids in the wholemeal fraction of buckwheat has been measured to be 2.42 g/100 g (moist mass) (Liu and Zhu, 2007). The hulls contained 1.53 g/100 g whereas 7.16 g/100 g were found in the fraction containing the crushed embryo, bran, aleuronic layer, and part of the hulls. The molecular weight of the main flavonoid matches that of rutin. Tartary buckwheat is more abundant in flavonoids, containing up to 7 g/100 g (Gu, 1999). Rutin, quercetin, orientin, vitexin, isovitexin, and isoorientin have been identified in the hulls, and rutin and isovitexin in the seeds (Dietrych-Szostak and Oleszek, 1999; Kreft *et al.*, 1999; Gallardo *et al.*, 2006). In addition, Watanabe (1998) isolated catechins from dehulled groats. The flavonolglycosides rutin, quercetin, kaempferol-3-rutinoside, and a trace quantity of a flavonol triglycoside have been isolated from methanol extracts (Tian *et al.*, 2002).

Rutin, a rhamnoglucoside of the flavonol quercetin, is of particular interest, as it is used for medical purposes in many countries. Different values for the amounts of rutin and flavonoids have been reported. Steadman *et al.* (2001b) detected rutin and quercetin mainly in hulls (80–440 mg/100 g) rather than groats (20 mg/100 g). These values are in contrast to a recent investigation of 14 buckwheat varieties, in which a very low content of rutin (0.064–0.390 mg/100 g) was measured (Suzuki *et al.*, 2005). A comparative study revealed higher amounts of rutin (810–1660 mg/100 g) and quercitrin (47–90 mg/100 g) in two varieties of tartary buckwheat than in common buckwheat (Fabjan *et al.*, 2003). However, similar amounts of rutin in tartary and common buckwheat hulls have been reported by Steadman *et al.* (2001b). The bitter taste of tartary buckwheat seeds has been ascribed to these flavonoids (Fabjan *et al.*, 2003). This is further supported by the isolation of a flavonol-3-glucosidase which hydrolyzes rutin and thus causes bitter taste in tartary buckwheat (Suzuki *et al.*, 2004a). Interestingly, Mattila *et al.* (2005) found 4.1 ± 0.41 mg/100 g alkenylresorcinol in buckwheat grits (whole grain). Although the content in buckwheat is the same as that in wheat flour, the presence of alkenylresorcinol adds extra value to this unique crop. This compound is not present in oat products nor in rice, millet, or corn flour.

Antioxidant activity and health benefits

Phenolic compounds (Velioglu *et al.*, 1998) and flavonoids (Watanabe *et al.*, 1997; Watanabe, 1998; Sensoy *et al.*, 2006) isolated from buckwheat hulls have been shown to possess antioxidative activities. Oomah and Mazza (1996) found that the flavonoids content was strongly correlated with rutin but weakly associated with antioxidative activities. Recently, Gallardo *et al.* (2006) found 4.5 and 4.4 total soluble phenolic acids (mg/100 g) in aqueous and 80% methanol extracts of buckwheat flour. Although these contents are only one-seventh of those found in wheat or rye bran extracts, the trolox equivalent antioxidant capacity (TEAC) of the buckwheat fractions was 1.7- to 2-fold that of wheat bran extracts, and even 15-fold the TEAC of rye bran extracts. Sun and Ho (2005) compared the antioxidant activities of buckwheat extracts with butylated hydroxyanisole, butylated hydroxytoluene, and tertiary butylhydrochinone using a β-carotene bleaching assay, a 2,2-diphenyl-β-picrylhydrazyl (DPPH) assay and the Rancimat method. Buckwheat was extracted with solvents of different polarities. The methanol extract showed the highest antioxidant activity coefficient (AAC) when using the β-carotene bleaching method, whereas the longest induction time was observed using the Rancimat method. The acetone extract showed the highest total phenolics of 3.4 ± 0.1 g catechin equivalents/100 g and the highest scavenging activity according to the DPPH method. TPC expressed as gallic acid equivalents indicated that roasting (200°C, 10 minutes) did not significantly affect the phenolic content of either dark (1047 mg/100 g) or white buckwheat flour (180 mg/100 g), whereas the antioxidant activity (DPPH) was decreased.

Phytic acid

Buckwheat seeds generally contain higher amounts of phytic acid than legumes and cereal grains, although the phytic acid content of the flour content is very similar to that of wheat flour (Steadman et al., 2001b). Phytic acid content is highest in bran without hulls (3.5–3.8 g/100 g). In buckwheat, 60–90% of phosphorus is stored as phytic acid.

Production of flours and their properties

Buckwheat seed milling fractions are produced either by roller-milling the intact kernel and sieving the particles into light flour (mainly central endosperm), grits (hard chunks of endosperm) and bran fractions, or by removing the hull through impact dehulling and roller-milling the resulting groats followed by sieving the particles into flour and bran fractions (Steadman et al., 2001a). Milling of buckwheat seeds into fractions concentrates certain components based on the varying proportion of tissues present. Fine flour is mostly endosperm and rich in starch, whereas bran, composed of seed coat and embryo fragments, has low amounts of starch (Skrabanja et al., 2004; Steadman et al., 2001a). Bran contains the pericarp (with seed coat), nuclear remnants, as well as aleurone and subaleurone layers. In mature buckwheat seeds, the outer of the two cotyledons adheres to the seed coat and during milling tears off and separates with bran. Large fragments of embryo from the central endosperm may separate with bran, but some soft embryo tissue is pulverized and separates with the flour (Steadman et al., 2001a). Application of wet milling resulted in 79 and 64% extraction efficiencies for starch and protein, respectively (Zheng et al., 1998). Buckwheat bran is the milling fraction that is of most value in terms of nutritional components, being highly concentrated in proteins (350 g/kg), lipids (110 g/kg), dietary fibers (150 g/kg), and fagopyritols (26 g/kg) (Steadman et al., 2000). Beside starch, proteins are the main endogenous factor responsible for the textural characteristics of buckwheat products (Ikeda *et al.*, 1997). Choosing the appropriate ratio between starch and protein content is thus an important aspect when making and designing different buckwheat products (Skrabanja *et al.*, 2004).

Production and characterization of gluten-free cereal products based on pseudocereals

A worldwide search for "gluten-free" food products based on pseudocereals (i.e. bread, pasta, and cookies) can be performed using the commercial web accessible Productscan® Online Database (www.productscan.com). Using this database, no gluten-free products based on quinoa could be found. Nine gluten-free bread products based on amaranth could be found in North America, and only three gluten-free cookies based on buckwheat were listed for Europe. There were no hits for gluten-free pasta based on any of the three pseudocereals. Most researchers have investigated pasta or bread production of flour blends from pseudocereals combined with wheat. The level of incorporation into wheat dough ranged typically from 10 to 20%. Production of gluten-free bread with amaranth is possible, and the presence of amaranth was shown to increase the nutritional composition (increased protein, fiber, and mineral content) of the final product (Gambus *et al.*, 2002). Kiskini *et al.* (2007) produced amaranth-based gluten-free bread which was fortified with iron. Gluten-free bread containing 8.5% buckwheat flour was produced by Moore *et al.* (2004), but breads were brittle after two days of storage. Di Cagno *et al.* (2004) aimed to produce sourdough bread that is tolerated by celiac disease patients. The results of this study indicate that the combination of selected lactic acid bacteria, non-toxic flours, and a long fermentation time represents a novel tool for decreasing the level of gluten contamination in gluten-free ingredients (e.g. wheat starch or oat). Gluten-free pasta produced with 100% pseudocereal flour blends of amaranth, quinoa, or buckwheat has been investigated (Drausinger, 1999; Wolfrum, 1999; Schoenlechner, 2001; Jurackova, 2005).

Addition of albumen, emulsifier, enzymes, and eventually xanthan increased the quality of noodles produced from 100% pseudocereal flour. Amaranth showed the lowest suitability for noodle production, as the final product was characterized by low texture firmness as well as decreased cooking time and tolerance. Quinoa noodles showed good agglutination, but caused higher cooking loss and a reduced taste. Buckwheat increased texture firmness and decreased cooking loss and therefore was chosen as the best gluten-free raw material suitable to enhance noodle texture. Interestingly, the combination of all three pseudocereal flours seemed most advantageous as the negative effects of using single flours were minimized. The resulting noodles were much better agglutinated, showed good texture firmness and low cooking loss. In addition, cooking stability was highly increased. Addition of emulsifiers improves the quality of gluten-free quinoa pasta. DATEM and sodium stearoyl-2-lactate are the most suitable, while lecithin addition gives only poor improvement (Kovacs *et al.*, 2004). Caperuto *et al.* (2000) investigated the production of gluten-free spaghetti from blends of corn flour and quinoa flour fractions (5–15%), and the resulting spaghetti received good scores by a consumer taste panel. Gluten-free macaroni from blends of quinoa and rice flour produced by extrusion cooking have also been successfully produced (Borges *et al.*, 2003; Ramirez *et al.*, 2003).

Biscuit dough and products has been successfully produced using quinoa and buckwheat as the only starch components (Kuhn *et al.*, 1994). Schober *et al.* (2003) investigated the production of biscuit using 10% buckwheat flour, 50% brown rice flour, 30% potato starch, and 10% millet flakes. The biscuits obtained were non-uniform in structure, softer in dough and thicker in biscuits. In addition, biscuits had high values in moisture and a_w as well as a dark surface color. The physical and chemical properties during cake puffing conditions of buckwheat grit cakes prepared with a rice cake machine were studied by Im *et al.* (2003). Results showed that, in order to obtain cakes with a high specific volume, higher moisture and heating temperature or longer heating time are needed. The production of acceptable biscuits made from common and tartary buckwheat without any addition of wheat flour has been described by Vombergar and Gostenčnik (2005).

Schoenlechner *et al.* (2006) investigated the production of short dough biscuits from amaranth, quinoa, or buckwheat as well as the effect of common bean flour addition. Buckwheat biscuits were found to be crispier than quinoa biscuits, and amaranth biscuits showed the lowest crispness. Addition of bean flour increased the crispness of all biscuits, independent of the pseudocereal used. Partial replacement of amaranth flour by popped amaranth flour increased the textural properties of the resulting biscuits. Granola bars and muesli with good sensory evaluation have been produced using popped or extruded amaranth and quinoa (Wesche-Ebeling *et al.*, 1996; Schoenlechner, 1997). A similar product produced from popped amaranth and honey, called *allegria*, was produced by Latin-American people in pre-Colombian times.

Finally, there is a range of non-traditional gluten-free products from pseudocereals like non-dairy beverages (similar to soy beverage), infant foods, extruded or popped products, and tortillas that can be produced from these pseudocereals. However, the importance of such products in the European market is up to now still insignificant.

Conclusions

The favorable chemical composition of amaranth, quinoa, and buckwheat has been demonstrated in this chapter. In this respect, the excellent protein quality of amaranth and quinoa has to be pointed out, while buckwheat is characterized by a unique concentration of phythochemicals, in particular rutin. However, as amaranth and quinoa have long been neglected within food production and nutrition, mainly on account of wheat, the current knowledge is still very limited. This is one reason why only few food products based on or including pseudocereals are available, in particular Western-type foods like bakery products and pasta. Increased and thorough research should thus be pursued on physico-chemical and functional properties of all three plants in order to enable future product development.

All three pseudocereals do not contain any prolamins toxic to celiac disease and can thus be integrated into gluten-free diets. However, the available research data (in particular for quinoa) is yet not sufficient to clearly state that these three plants can be tolerated by all people with celiac disease. Further research (e.g. animal or clinical studies) is necessary to give detailed recommendations. Celiac disease is often accomplished by malabsorption and subsequent vitamin or mineral deficiencies, which makes high-quality nutrition even more important. As amaranth, quinoa, and buckwheat are highly nutritious, their integration into the gluten-free diet could be a valuable contribution.

References

Abreu, M., Hernandez, M., Castillo, A., Gonzalez, I., Gonzales, J., and Brito, O. (1994). Study on the complementary effect between the proteins of wheat and amaranth. *Die Nahrung* 38, 82–86.

Abugoch, L. E., Martínez, E. N., and Añón, M. C. (2003). Influence of the extracting solvent upon the structural properties of amaranth (*Amaranthus hypochondriacus*) glutelin. *J. Agric. Food Chem.* 51, 4060–4065.

Acquistucci, R. and Fornal, J. (1997). Italian buckwheat (*Fagopyrum esculentum*) starch: Physico-chemical and functional characterization and *in vitro* digestibility. *Food/Nahrung* 41, 281–284.

Adeyeye, A., and Ajewole, K. (1992). Chemical composition and fatty acid profiles of cereals in Nigeria. *Food Chem.* 44, 41–44.

Ahamed, N. T., Singhal, R. S., Kulkarni, P. R., and Pal, M. (1996). Physicochemical and functional properties of *Chenopodium quinoa* starch. *Carbahydr. Polym.* 31, 99–103.

Ando, H., Chen, Y. C., Tang, H., Shimizu, M., Watanabe, K., and Mitsunga, T. (2002). Food components in fractions of quinoa seed. *Food Sci. Technol. Res.* 8, 80–84.

Aphalo, P., Castellani, O. F., Martinez, E. N., and Anon, M. C. (2004). Surface physiochemical properties of globulin-P amaranth protein. *J. Agric. Food Chem.* 52, 616–622.

Atwell, W. A., Patrick, B. M., Johnson, L. A., and Glass, R. W. (1983). Characterization of quinoa starch. *Cereal Chem.* 60, 9–11.

Aubrecht, E. and Biacs, P. Á. (1999). Immunochemical analysis of buckwheat proteins, prolamins and their allergenic character. *Acta Aliment.* 28, 261–268.

Aubrecht, E. and Biacs, P. Á. (2001). Characterization of buckwheat grain proteins and its products. *Acta Aliment.* 30, 71–80.

Avanza, M. V., Puppo, M. C., and Anon, M. C. (2005). Rheological characterisation of amaranth protein gels. *Food Hydrocolloids* 19, 889–898.

Ayorinde, F. O., Ologunde, M. O., Nana, E. Y. *et al.* (1989). Determination of fatty acid composition of amaranthus species. *J. Am. Oil Chem. Soc. JAOCS* 66, 1812–1814.

Baker, L. A. and Rayas-Duarte, P. (1998). Freeze-thaw stability of amaranth starch and the effects of salt and sugars. *Cereal Chem.* 75, 301–307.

Becker, R. (1994). Amaranth oil: composition, processing, and nutritional qualities. In: Paredes-Lopez, O. ed. *Amaranth—Biology, Chemistry, and Technology*. London: CRC Press, pp. 133–142.

Becker, R. and Hanners, G. D. (1990). Compositional and nutritional evaluation of quinoa whole grain flour and mill fractions. *Lebensm.-Wiss. U.-Technol.* 23, 441–444.

Becker, R., Wheeler, E. L., Lorenz, K. *et al.* (1981). A compositional study of amaranth grain. *J. Food Sci.* 46, 1175–1180.

Becker, R., Irving, D. W., and Saunders, R. M. (1986). Production of debranned amaranth flour by stone milling. *Food Sci. Technol.* 19, 372–374.

Bejosano, F. P. and Corke, H. (1998). Protein quality evaluation of *Amaranthus* wholemeal flours and protein concentrates. *J. Sci. Food Agric.* 76, 100–106.

Bejosano, F. P. and Corke, H. (1999). Properties of protein concentrates and hydrolysates from amaranthus and buckwheat. *Sixth Symposium on Renewable Resources for the Chemical Industry*, 23–25 March, 1999 Bonn, p. 53.

Berganza, B. E., Moran, A. W., Rodriguez, M. G., Coto, N. M., Santamaria, M., and Bressani, R. (2003). Effect of variety and location on the total fat, fatty acids and squalene content of amaranth. *Plant Foods Human Nutr.* 58, 1–6.

Berger, A., Monnard, I., Dionisi, F., Gumy, D., Hayes, K. C., and Lambelet, P. (2003). Cholesterol-lowering properties of amaranth flakes and refined oils in hamsters. *Food Chem.* 81, 119–124.

Betschart, A. A., Wood Irving, D., Shepherd, A. D., and Saunders, R. M. (1981). *Amaranthus cruentus:* Milling characteristics, distribution of nutrients within seed components, and the effects of temperature on nutritional quality. *J. Food Sci.* 46, 1181–1187.

Bonafaccia, G., Gambelli, L., Fabjan, N., and Kreft, I. (2003a). Trace elements in flour and bran from common and tartary buckwheat. *Food Chem.* 80, 1–5.

Bonafaccia, G., Marocchini, M., and Kreft, I. (2003b) Composition and technological properties of the flour and bran from common and tartary buckwheat. *Food Chem.* 80, 9–15.

Borges, J. T. da S., Ramirez-Ascheri, J. L., Ramirez-Ascheri, D., Euzebio-do-Nascimento, R., and Silva-Freitas, A. (2003). Cooking properties and physico-chemical character-ization of precooked macaroni of whole quinoa (*Chenopodium quinoa* Willd) flour and rice (*Oryza sativa* L) flour by extrusion cooking. *Bol. Centro Pesquisa Process. Aliment.* 21, 303–322.

Bossert, J. and Wahl, R. (2000). Amaranth—A new allergen in bakeries. *Allergologie* 23, 448–454.

Breene, W. M. (1991). Food uses of grain amaranth. *Cereal Foods World* 36, 426–430.

Bressani, R. (1994). Composition and nutritional properties of amaranth. In: Paredes-Lopez, O. ed. *Amaranth—Biology, Chemistry, and Technology.* London: CRC Press, pp. 185–206.

Bressani, R., Elias, L. G., and Garcia-Soto, A. (1989). Limiting amino acids in raw and processed amaranth grain protein from biological tests. *Plant Foods Hum. Nutr.* 39, 223–234.

Bressani, R., Velásquez, L., and Acevedo, E. (1990). Dietary fiber content in various grain amaranth species and effect of processing. *Amaranth Newsl.* 1, 5–8.

Brinegar, C. (1997). The seed storage proteins of quinoa. *Adv. Exp. Med. Biol.* 415, 109–115.

Brinegar, C. and Goundan, S. (1993). Isolation and characterisation of chenopodin, the 11S seed storage protein of quinoa (*Chenopodium quinoa*). *J. Agric. Food Chem.* 41, 182–185.

Bruni, R., Guerrini, A., Scalia, S., Romagnoli, C., and Sacchetti, G. (2002). Rapid techniques for the extraction of vitamin E isomers from *Amaranthus caudatus* seeds: Ultrasonic and supercritical fluid extraction. *Phytochem. Anal.* 13, 257–261.

Bucaro Segura, M. E. and Bressani, R. (2002). Protein fraction distribution in milling and screened physical fractions of grain amaranth. *Arch. Latinoamericanos Nutr.* 52, 167–171.

Budin, J. T., Breene, W. M., and Putmam, D. H. (1996). Some compositional properties of seeds and oils of eight amaranthus species. *J. Am. Oil Chem. Soc. JAOCS* 73, 475–481.

Bunzel, M., Ralph, J., and Steinhart, H. (2005). Association of non-starch polysaccharides and ferulic acid in grain amaranth (*Amaranthus caudatus* L.) dietary fiber. *Mol. Nutr. Food Res.* 49, 551–559.

Bush, R. K. and Hefle, S. L. (1996). Food allergens. *Crit. Rev. Food Sci. Nutr.* 36, S119–S163.

Calzetta-Resio, A., Aguerre, R. J., and Suarez, C. (1999). Analysis of the sorptional characteristics of amaranth starch. *J. Food Eng.* 42, 51–57.

Calzetta-Resio, T. N., Tolaba, M. P., and Suarez, C. (2000). Some physical and thermal characteristics of amaranth starch. *Food Sci. Technol. Int.* 6, 371–378.

Caperuto, L. C., Amaya-Farfan, J., and Camargo, C. R. O. (2000). Performance of quinoa (*Chenopodium quinoa* Willd) flour in the manufacture of gluten-free spaghetti. *J. Sci. Food Agric.* 81, 95–101.

Chaturvedi, A., Sarojini, G., and Devi, N. L. (1993). Hypocholesterolemic effect of amaranth seeds (*Amaranthus esculantus*). *Plant Foods Hum. Nutr.* 44, 63–70.

Chauhan, G. S., Eskin, N. A. M., and Tkachuk, R. (1992). Nutrients and antinutrients in quinoa seed. *Cereal Chem.*, 69, 85–88.

Chauhan, G. S., Eskin, N. A. M., and Mills, P. A. (1999). Effect of saponin extraction on the nutritional quality of quinoa (*Chenopodium quinoa* Willd) proteins. *J. Food Sci. Technol.* 36, 123–126.

Codex Alimentarius Commission (2000). Draft revised standard for gluten-free foods. CX/NFSDN 00/4, March 2000, pp. 1–4.

Correa, A. D., Jokl, L., and Carlsson, R. (1986). Amino acid composition of some *Amaranthus sp.* grain proteins and of its fractions. *Arch. Latinoam.* 36, 466–476.

Coulter, L. and Lorenz, K. (1990). Quinoa—composition, nutritional value, food applications. *Lebensm.-Wiss. U.-Technol.* 23, 203–207.

Cuadrado, C., Ayet, G., Burbano, C. *et al.* (1995). Occurrence of saponins and sapogenols in Andean crops. *J. Sci. Food Agric.* 67, 169–172.

Czerwinski, J., Bartnikowska, E., Leontowicz, H. *et al.* (2004). Oat (*Avena sativa* L.) and amaranth (*Amaranthus hypochondriacus*) meals positively affect plasma lipid profile in rats fed cholesterol-containing diets. *J. Nutr. Biochem.* 15, 622–629.

Dahlin, K. M. and Lorenz, K. (1993). Protein digestibility of extruded cereal grains. *Food Chem.* 48, 13–18.

De Simone, F., Dini, A., Pizza, C., Saturnino, P., and Schettino, O. (1990). Two flavonol glycosides from *Chenopodium quinoa*. *Phytochemistry* 29, 3690–3692.

Di Cagno, R., Angelis, M. D., Auricchio, S. *et al.* (2004). Sourdough bread made from wheat and nontoxic flours and started with selected lactobacilli is tolerated in celiac sprue patients. *Appl. Environ. Microbiol.* 70, 1088–1096.

Dietrych-Szostak, D. and Oleszek, W. (1999). Effect of processing on the flavonoid content in buckwheat (*Fagopyrum esculentum* Möench) grain. *J. Agric. Food Chem.* 47, 4384–4387.

Dini, I., Tenore, G. C., and Dini, A. (2004). Phenolic constituents of *Kancolla* seeds. *Food Chem.* 84, 163–168.

Dobos, G. (1992). Koerneramaranth als neue Kulturpflanze in Oesterreich. Introduktion und zuechterische Aspekte. PhD thesis, University of Natural Resources and Applied Life Sciences, Vienna, Austria.

Dodok, L., Modhir, A. A., Halasova, G., Polacek, I., and Hozova, B. (1994). Importance and utilisation of amaranth in food industry Part 1. Characteristic of grain and average chemical constitution of whole amaranth flour. *Die Nahrung* 38, 378–381.

Drausinger, J. (1999). Herstellung und Zubereitung von Teigwaren aus Pseudocerealien mittels Transglutaminase. Diplomarbeit, University of Natural Resources and Applied Life Sciences, Vienna, Austria.

Drzewiecki, J., Delgado-Licon, E., Haruenkit, R. *et al.* (2003). Identification and differences of total proteins and their soluble fractions in some pseudocereals based on electrophoretic pattern. *J. Agric. Food Chem.* 51, 7798–7804.

Eggum, B. O., Kreft, I., and Javornik, B. (1980). Chemical composition and protein quality of buckwheat (*Fagopyrum esculentum* Moench). *Qual. Plant. Plant Foods Hum. Nutr.* 30, 175–179.

Egli, I., Davidsson, L., Juillerat, M.-A., Barclay, D., and Hurrell, R. (2003). Phytic acid degradation in complementary foods using phytase naturally occurring in whole grain cereals. *J. Food Sci.* 68, 1855–1859.

Escudero, N. L., De Arellano, M. L., Luco, J. M., Giménez, M. S. and Mucciarelli, S. I. (2004). Comparison of the chemical composition and nutritional value of *Amaranthus Cruentus* flour and its protein concentrate. *Plant Foods Hum. Nutr.* 59, 15–21.

Fabjan, N., Rode, J., Košir, I. J., Wang, Z., Zhang, Z., and Kreft, I. (2003). Tartary buckwheat (*Fagopyrum tataricum* Gaertn.) as a source of dietary rutin and quercitrin. *J. Agric. Food Chem.* 51, 6452–6455.

Fadel, J. G., Pond, W. G., Harrold, R. L., Calvert, C. C., and Lewis, B. A. (1996). Nutritive value of three amaranth grains fed either processed or raw to growing rats. *Can. J. Anim. Sci.* 76, 253–257.

FAOSTAT 2006: FAO Statistics Division (2006). http://faostat.fao.org/, updated: January 9 2007.

Fidantsi, A. and Doxastakis, G. (2001). Emulsifying and foaming properties of amaranth seed protein isolates. *Colloids and Surfaces B: Biointerfaces* 21, 119–124.

Francischi, M. L. P., de, Salgado, J. M., and Leitao, R. F. F. (1994). Chemical, nutritional and technological characteristics of buckwheat and non-prolamine buckwheat flours in comparison of wheat flour. *Plant Foods Hum. Nutr.* 46, 323–329.

Gallardo, C., Jiménez, L., and García-Conesa, M.-T. (2006). Hydroxycinnamic acid composition and in vitro antioxidant activity of selected grain fractions. *Food Chem.* 99, 455–463.

Gambus, H., Gambus, F., and Sabat, R. (2002). The research on quality improvement of gluten-free bread by *Amaranthus* flour addition. *Zywnosc* 9, 99–112.

Gamel, T. H., Linssen, J. P., Alink, G. M., Mossallem, A. S., and Shekib, L. A. (2004). Nutritional study of raw and popped seed proteins of *Amaranthus caudatus* L and *Amaranthus cruentus* L. *J. Sci. Food Agric.* 84, 1153–1158.

Gamel, T. H., Linssen, J. P., Mesallem, A. S., Damir, A. A., and Shekib, L. A. (2005). Effect of seed treatments on the chemical composition and properties of two amaranth species: starch and protein. *J. Sci. Food Agric.* 85, 315–327.

Gamel, T. H., Linssen, J. P., Mesallem, A. S., Damir, A. A., and Shekib, L. A. (2006a). Effect of seed treatments on the chemical composition of two amaranth species: oil, sugars, fibres, minerals and vitamins. *J. Sci. Food Agric.* 86, 82–89.

Gamel, T. H., Linssen, J. P., Mesallem, A. S., Damir, A. A., and Shekib, L. A. (2006b). Seed treatments affect functional and antinutritional properties of amaranth flours. *J. Sci. Food Agric.* 86, 1095–1102.

Gee, J. M., Price, K. R., Ridout, C. L., Wortley, G. M., Hurrell, R. F., and Johnson, I. T. (1993). Saponins of quinoa (*Chenopodium quinoa*): Effects of processing on their abundance in quinoa products and their biological effects on intestinal mucosal tissue. *J. Sci. Food Agric.* 63, 201–209.

Gonzalez, J. A., Roldan, A., Gallardo, M., Escudero, T., and Prado, F. E. (1989). Quantitative determinations of chemical compounds with nutritional value from Inca crops: *Chenopodium quinoa* ('quinoa'). *Plant Foods Hum. Nutr.* 39, 331–337.

Gonzalez, R., Carrara, C., Tosi, E., Anon, M. C., and Pilosof, A. (2007). Amaranth starch-rich fraction properties modified by extrusion and fluidized bed heating. *Food Sci. Technol.* 40, 136–143.

Gorinstein, S., Jaramillo, N. O., Medina, O. J., Rogriques, W. A., Tosello, G. A., and Paredes-Lopez, O. (1999). Evaluation of some cereals, plants and tubers through protein composition. *J. Protein Chem.* 18, 687–693.

Gorinstein, S., Delgado-Licon, E., Pawelzik, E., Permady, H. H., Weisz, M., and Trakhtenberg, S. (2001). Characterisation of soluble amaranth and soybean proteins based on fluorescence, hydrophobicity, electrophoresis, amino acid analysis, circular dichroism, and differential scanning calorimetry measurements. *J. Agric. Food Chem.* 49, 5595–5601.

Gorinstein, S., Pawelzik, E., Delgado-Licon, E., Haruenkit, R., Weisz, M., and Trakhtenberg, S. (2002). Characterisation of pseudocereal and cereal proteins by protein and amino acid analyses. *J. Sci. Food Agric.* 82, 886–891.

Gorinstein, S., Pawelzik, E., Delgado-Licon, E. *et al.* (2004). Use of scanning electron microscopy to indicate the similarities and differences in pseudocereal and cereal proteins. *Int. J. Food Sci. Technol.* 39, 183–189.

Gorinstein, S., Drzewiecki, J., Delgado-Licon, E. *et al.* (2005). Relationship between dicotyledonae-amaranth, quinoa, fagopyrum, soybean and monocots-sorghum and rice based on protein analyses and their use as substitution of each other. *Eur. Food Res. Technol.* 221, 69–77.

Grajeta, H. (1999). Effect of amaranth and oat bran on blood serum and liver lipids in rats depending on the kind of dietary fats. *Die Nahrung* 43, 114–117.

Gross, R., Koch, F., Malaga, I., De Miranda, A. F., Schoeneberger, H., and Trugo, L. C. (1989). Chemical composition and protein quality of some local Andean food sources. *Food Chem.* 34, 25–34.

Gu, Y. (1999). Processing technology for non-staple cereals—buckwheat processing. *Cereal Feed Indust.* 7, 19–26.

Gujska, E. and Kuncewicz, A. (2005). Determination of folate in some cereals and commercial cereal-grain products consumed in Poland using trienzyme extraction and high-performance liquid chromatography methods. *Eur. Food Res. Technol.* 221, 208–213.

Guo, X. and Yao, H. (2006). Fractionation and characterization of tartary buckwheat flour proteins. *Food Chem.* 98, 90–94.

Guzman-Maldonado, H. and Parades-Lopez, O. (1994). Production of high-protein flour and malotdextrins from amaranth grain. *Process Biochem.* 29, 289–293.

Hayakawa, I., Linko, Y. Y., and Linko, P. (1996). Novel mechanical treatments of biomaterials. *Lebensm.-Wiss. U.-Technol.* 29, 395–403.

He, H. P., Cai, Y., Sun, M., and Corke, H. (2002). Extraction and purification of squalene from amaranthus grain. *J. Agric. Food Chem.* 50, 368–372.

Hibi, M., Hachimura, S., Hashizume, S., Obata, T., and Kaminogawa, S. (2003). Amaranth grain inhibits antigen-specific IgE production through augmentation of the IFN-γ response in vivo and in vitro. *Cytotechnology,* 43, 33–40.

Hirano, S. and Konishi, Y. (2003). Nutritional characteristics within structural part of quinoa seeds. *J. Jap. Soc. Nutr. Food Sci.* 56, 283–289.

Horbowicz, M. and Obendorf, R. L. (1992). Changes in sterols and fatty acids of buckwheat endosperm and embryo during seed development. *J. Agric. Food Chem.* 40, 745–750. http://www.productscan.com, updated: December 20, 2006.

Horbowicz, M., Brenac, P., and Obendorf, R. L. (1998). Fagopyritol B1, O-α-D-galactopyranosyl-(1→2)-1 D-*chiro*-inositol, a galactosyl cyclitol in maturing buckwheat seeds associated with desiccation tolerance. *Planta* 205, 1–11.

Hunjai, C., Wansoo, K., and Malshik, S. (2004). Properties of Korean amaranth starch compared to waxy millet and waxy sorghum starches. *Starch/Staerke* 56, 469–477.

Ikeda, K. and Kishida, M. (1993). Digestibility of proteins in buckwheat seed. *Fagopyrum* 13, 21–24.

Ikeda, K., Sakaguchi, T., Kusano, T., and Yasumoto, K. (1991). Endogenous factors affecting protein digestibility in buckwheat. *Cereal Chem.* 68, 424–427.

Ikeda, K., Kishida, M., Kreft, I., and Yasumoto, K. (1997). Endogenous factors responsible for the textural characteristics of buckwheat products. *J. Nutr. Sci. Vitaminol.* 43, 101–111.

Im, J.-S., Huff, H. E., and Hsieh, F.-H. (2003). Effect of processing conditions on the physical and chemical properties of buckwheat grit cakes. *J. Agric. Food Chem.* 51, 659–666.

Jahaniaval, F., Kakuda, Y., and Marcone, M. F. (2000). Fatty acid and triacylglycerol composition of five amaranthus accessions and their comparison to other oils. *J. Am. Oil Chem. Soc.* 77, 847–852.

Jung, K., Richter, J., Kabrodt, K., Luecke, I. M., Schellenberg, I., and Herrling, Th. (2006). The antioxidative power AP—A new quantitative time dependent (2D) parameter for the determination of the antioxidant capacity and reactivity of different plants. *Spectrochim. Acta Part A* 63, 846–850.

Jurackova, K. (2005). Production possibilities of convenience foods from pseudocereals. PhD thesis, University of Natural Resources and Applied Life Sciences, Vienna.

Karyotis, T. H., Iliadis, C., Noulas, C., and Mitsibonas, T. H. (2003). Preliminary Research on Seed Production and nutrient content for certain quinoa varieties in a saline-sodic soil. *J. Agronomy Crop Sci.* 189, 402–408.

Kayashita, J., Shimaoka, I., and Nakajyuh, M. (1995). Hypocholesterolemic effect of buckwheat protein extract in rats fed cholesterol enriched diets. *Nutr. Res.* 15, 691–698.

Kayashita, J., Shimaoka, I., Nakajoh, M., Yamazaki, M., and Kato, N. (1997). Consumption of buckwheat protein lowers plasma cholesterol and raises fecal neutral sterols in cholesterol-fed rats because of its low digestibility. *J. Nutr.* 127, 1395–1400.

Kayashita, J., Shimaoka, I., Nakajoh, M., Kishida, N., and Kato, N. (1999). Consumption of a buckwheat protein extract retards 7,12-dimethylbenz[alpha]anthracene-induced mammary carcinogenesis in rats. *Biosci. Biotechnol. Biochem.* 63, 1837–1839.

Kim, H. K., Kim, M. J., Cho, H. Y., Kim, E. K., and Shin, D. H. (2006). Antioxidative and anti-diabetic effects of amaranth (*Amaranthus esculantus*) in streptozotocin-induced diabetic rats. *Cell Biochem. Funct.* 24, 195–199.

Kiskini, A., Argiri, K., Kalogeropoulos, M. *et al.* (2007). Sensory characteristics and iron dialyzability of gluten-free bread fortified with iron. *Food Chem.* 102, 309–316.

Klimczak, I., Malecka, M., and Pacholek, B. (2002). Antioxidant activity of ethanolic extracts of amaranth seeds. *Nahrung/Food* 46, 184–186.

Kondo, Y., Urisu, A., Wada, E. *et al.* (1996). Allergen analysis of buckwheat by the immunoblotting method. *Jap. J. Allergol.* 42, 142–148.

Konishi, Y., Hirano, S., Tsuboi, H., and Wada, M. (2004). Distribution of minerals in quinoa (*Chenopodium quinoa* Willd) seeds. *Biosci. Biotechnol. Biochem.* 68, 231–234.

Kovacs, E. T., Maraz-Szabo, L., and Varga, J. (2001). Examination of the protein-emulsifier-carbohydrate interactions in amaranth based pasta products. *Acta Aliment.* 30, 173–187.

Kovacs, E. T., Schoenlechner, R., and Berghofer, E. (2004). Use of pseudocereals for pasta production: quinoa and its milling fractions. *Tecnica-Molitoria* 55, 159–169.

Koziol, M. J. (1991). Afrosimetric estimation of threshold saponin concentration for bitterness in quinoa (*Chenopodium quinoa* Willd). *J. Sci. Food Agric.* 54, 211–219.

Kreft, S., Knapp, M., and Kreft, I. (1999). Extraction of rutin from buckwheat (*Fagopyrum esculentum* Moench) seeds and determination by capillary electrophoresis. *J. Agric. Food Chem.* 47, 4649–4652.

Krkošková, B. and Mrázová, Z. (2005). Prophylactic components of buckwheat. *Food Res. Int.* 38, 561–568.

Kuhn, M., Noll, B., and Goetz, H. (1994). Optimierungsversuche mit Biskuitmassen und -gebäcken. *Getreide, Mehl Brot* 48, 56–59.

Lara, N. and Ruales, J. (2002). Popping of amaranth grain (*A. caudatus*) and its effect on the functional, nutritional and sensory properties. *J. Sci. Food Agric.* 82, 797–805.

Lehmann, J. W., Putnam, D. H., and Qureshi, A. A. (1994). Vitamin E isomers in grain amaranths *(Amaranthus spp.)*. *Lipids* 29, 177–181.

Leon-Camacho, M., Garcia-Gonzalez, D. L., and Aparicio, R. (2001). A detailed and comprehensive study of amaranth (*Amaranthus cruentus* L.) oil fatty profile. *Eur. Food Res. Technol.* 213, 349–355.

Li, S. and Zhang, Q. H. (2001). Advances in the development of functional foods from buckwheat. *Crit. Rev. Food Sci. Nutr.* 41, 451–464.

Li, W., Lin, R., and Corke, H. (1997). Physochemical properties of common and tartary buckwheat starch. *Cereal Chem.* 74, 79–82.

Lindeboom, N., Chang, P. R., Falk, K. C., and Tyler, R. T. (2005a). Characteristics of starch from eight quinoa lines. *Cereal Chem.* 82, 216–222.

Lindeboom, N., Chang, P. R., Tyler, R. T., and Chibbar, R. N. (2005b). Granule-bound starch synthase I (GBSSI) in quinoa *(Chenopodium quinoa* Willd*)* and its relationship to amylose content. *Cereal Chem.* 82, 246–250.

Liu, B. and Zhu, Y. (2007). Extraction of flavonoids from flavonoids-rich parts in tartary buckwheat and identification of the main flavonoids. *J. Food Eng.* 78, 584–587.

Liu, Z., Ishikawa, W., Huang, X. et al. (2001). A buckwheat protein product suppresses 1,2-dimethylhydrazine-induced colon carcinogenesis in rats by reducing cell proliferation. *J. Nutr.* 131, 1850–1853.

Lorenz, K. (1990). Quinoa *(Chenopodium quinoa)* starch—physico-chemical properties and functional characteristics. *Starch/Staerke* 42, 81–86.

Lyon, C. K. and Becker, R. (1987). Extraction and refining of oil from amaranth seed. *J. Am. Oil Chem. Soc.* 64, 233–236.

Marcone, M. F. (1999). Evidence confirming the existence of a 7S globulin-like storage protein in *A. hypochondriacus* seed. *Food Chem.* 65, 533–554.

Marcone, M. F. and Kakuda, Y. (1999). A comparative study of the functional properties of amaranth and soybean globulin isolates. *Nahrung/Food* 43, 368–373.

Marcone, M. F., Beniac, D. R., Harauz, G., and Yada, R. Y. (1994). Quaternary structure and model for the oligomeric seed globulin from *Amaranthus hypochondriacus* K343. *J. Agric. Food Chem.* 42, 2675–2678.

Martinez, E. N., Castellani, O. F., and Anon, M. C. (1997). Common molecular features among amaranth storage proteins. *J. Agric. Food Chem.* 45, 3832–3839.

Mastebroek, H. D., Limburg, H., Gilles, T., and Marvin, H. J. P. (2000). Occurrence of sapogenins in leaves and seeds of quinoa (*Chenopodium quinoa* Willd). *J. Sci. Food Agric.* 80, 152–156.

Mattila, P., Pihlava, J.-M., and Hellström, J. (2005). Contents of phenolic acids, alkyl- and alkenylresorcinols and avenathramides in commerical grain products. *J. Agric. Food Chem.* 53, 8290–8295.

Matuz, J., Bartok, T., Morocz-Salomon, K., and Bona, L. (2000a). Structure and potential allergenic character of cereal proteins I. Protein content and amino acid composition. *Cereal Res. Commun.* 28, 263–270.

Matuz, J., Poka, R., Boldizsar, I., Szerdahelyi, E., and Hajos, G. (2000b). Structure and potential allergenic character of cereal proteins II. Potential allergens in cereal samples. *Cereal Res. Commun.* 28, 433–442.

Mazza, G. (1988). Lipid content and fatty acid composition of buckwheat seed. *Cereal Chem.* 62, 27–47.

Mikulikova, D. and Kraic, J. (2006). Natural sources of health-promoting starch. *J. Food Nutr. Res.* 45, 69–76.

Milisavljević, M. D., Timotijević, G. S., Radović, R. S., Brkljačić, Konstantinovic, M. M., and Maksimović, V. R. (2004). Vicilin-like storage globulin from buckwheat (*Fagopyrum esculentum* Moench) seeds. *J. Agric. Food Chem.* 52, 5258–5262.

Moore, M. M., Schober, T. J., Dockery, P., and Arendt, E. K. (2004). Textural comparisons of gluten-free and wheat-based doughs, batters, and bread. *Cereal Chem.* 81, 567–575.

Morales de León Josefina, M. C., Bourges, H., and Camacho, E. (2005). Amino acid composition of some Mexican Foods. *Arch. Latinoamericanos Nutr.* 55, 172–186.

Muchova, Z., Cukova, L., and Mucha, R. (2000). Seed protein fractions of amaranth (*Amaranthus sp.*) *Rostlinna Vyroba* 46, 331–336.

Nanka, G. (1998). Herstellung und Charakterisierung von Mahlprodukten aus Pseudocerealien. Diplomarbeit, University of Natural Resources and Applied Life Sciences, Vienna, Austria.

Ng, S. C., Anderson, A., Coker, J., and Ondrus, M. (2007). Characterisation of lipid oxidation products in quinoa (*Chenopodium quinoa*). *Food Chem.* 101, 185–192.

Noda, T., Takahata, Y., Sato, T. *et al.* (1998). Relationships between chain length distribution of amylopectin and gelatinization properties within the same botanical origin for sweet potato and buckwheat. *Carbohydr. Polym.* 37, 153–158.

Ogungbenle, H. N. (2003). Nutritional evaluation and functional properties of quinoa (*Chenopodium quinoa*) flour. *Int. J. Food Sci. Nutr.* 54, 153–158.

Oleszek, W., Junkuszew, M., and Stochmal, A. (1999). Determination and toxicity of saponins from *Amaranthus cruentus* seeds. *J. Agric. Food Chem.* 47, 3685–3687.

Oomah, B. D. and Mazza, G. (1996). Flavonoids and antioxidative activities in buckwheat. *J. Agric. Food Chem.* 44, 1746–1750.

Opute, F. I. (1979). Seed lipids of the grain amaranths. *J. Exp. Bot.* 30, 601–606.

Oshodi, A. A., Ogungbenle, H. N., and Oladimeji, M. O. (1999). Chemical composition, nutritionally valuable minerals and functional properties of benniseed (*Sesamum radiatum*), pearl millet (*Pennisum typhoides*) and quinoa (*Chenopodium quinoa*) flours. *Int. J. Food Sci. Nutr.* 50, 325–331.

Packert, M. (1993). Analytik und Bedeutung gebundener aromatischer Carbonsäuren der Nahrungsfaser aus Getreide und anderer Nutzpflanzen. PhD thesis, University of Hamburg.

Pandya, M. J., Smith, D. A., Yrwood, A., Gilroy, J., and Richardson, M. (1996). Complete amino acid sequence of two trypsin inhibitors from buckwheat seed. *Phytochemistry* 43, 327–331.

Paredes-Lopez, O., Schevenin, M. L., Hernandez-Lopez, D., and Carabez-Trejo, A. (1989). Amaranth starch-isolation and partial characterization. *Starch/Staerke* 41, 205–207.

Paredes-Lopez, O., Bello-Perez, L. A., and Lopez, M. G. (1994). Amylopectin: Structural, gelatinisation and retrogradation studies. *Food Chem.* 50, 411–417.

Park, J. W., Kang, D. B., Kim, C. W. *et al.* (2000). Identification and characterization of the major allergens of buckwheat. *Allergy* 55, 1035–1041.

Pederson, B., Kalinowski, L. S., and Eggum, B. O. (1987). The nutritive value of amaranth grain (*Amaranthus caudatus*). I. Protein and minerals of raw and processed grain. *Plant Foods Hum. Nutr.* 36, 309–324.

Perez, E., Bahanassey, Y. A., and Breene, W. M. (1993). Some chemical, physical, and functional properties of native and modified starches of *Amaranthus hypochondriacus* and *Amaranthus cruentus*. *Starch/Staerke* 45, 215–220.

Pomeranz, Y. and Robbins, G. S. (1972). Amino acid composition of buckwheat. *J. Agric. Food Chem.* 20, 270–274.

Pomeranz, Y. (1991). Carbohydrates: starch. In: Pomeranz, Y. ed. *Functional Properties of Food Components*. New York: Academic Press, pp. 24–78.

Prado, F. E., Gallardo, M., and Gonzales, J. A. (1996). Presence of saponin-bodies in pericarp cells of *Chenopodium quinoa* Willd (QUINOA). *Biocell* 20, 259–264.

Prakash, D. and Pal, M. (1998). *Chenopodium:* seed protein, fractionation and amino acid composition. *Int. J. Food Sci. Nutr.* 49, 271–275.

Praznik, W., Mundigler, N., Kogler, A., Pelzl, B., and Huber, A. (1999). Molecular background of technological properties of selected starches. *Starch/Staerke* 51, 187–211.

Prego, I., Maldonado, S., and Otegui, M. (1998). Seed structure and localization of reserves in *Chenopodium quinoa*. *Ann. Bot.* 82, 481–488.

Przybylski, R., Chauhan, G. S., and Eskin, N. A. M. (1994). Characterization of quinoa (*Chenopodium quinoa*) lipids. *Food Chem.* 51, 187–192.

Qian, J. Y. and Kuhn, M. (1999a). Characterization of *Amaranthus cruentus* and *Chenopodium quinoa* starch. *Starch/Staerke* 51, 116–120.

Qian, J. Y. and Kuhn, M. (1999b). Physical properties of buckwheat starches from various origins. *Starch/Stärke* 51, 81–85.

Qian, J. Y. and Kuhn, M. (1999c). Evaluation on gelatinization of buckwheat starch: A comparative study of Brabender viscoamylography, rapid visco-analysis, and differential scanning calorimetry. *Eur. Food Res. Technol.* 209, 277–280.

Qian, J. Y., Rayas-Duarte, P., and Grant, L. (1998). Partial characterization of buckwheat (*Fagopyrum esculentum*) starch. *Cereal Chem.* 75, 365–373.

Qureshi, A. A., Lehmann, J. W., and Peterson, D. M. (1996). Amaranth and its oil inhibit cholesterol biosynthesis in 6-week-old female chickens. *J. Nutr.* 126, 1972–1978.

Radović, R. S., Maksimović, R. V., and Varkonji, I. E. (1996). Characterization of buckwheat seed storage proteins. *J. Agric. Food Chem.* 44, 972–974.

Radović, R. S., Maksimović, R. V., Brkljačić, M. J., Varkonji, I. E., and Savić, A. P. (1999). 2S albumin from buckwheat seeds. *J. Agric. Food Chem.* 47, 1467–1470.

Ramirez Ascheri, J. L., Silva Borges, J. T. da, Euzebio do Nascimento, R., and Ramirez Ascheri, D. P. (2003). Functional properties of precooked macaroni of raw quinoa flour (*Chenopodium quinoa* Willd) and rice flour (*Oryza sativa* L.). *Alimentaria* 342, 71–75.

Ranhorta, G. S., Gelroth, J. A., Glaser, B. K., Lorenz, K. J., and Johnson, D. L. (1993). Composition and protein nutritional quality of quinoa. *Cereal Chem.* 70, 303–305.

Reichert, R. D., Tatarynovich, J. T., and Tyler, R. T. (1986). Abrasive dehulling of quinoa (*Chenopodium quinoa*): Effect on saponin content as determined by an adapted hemolytic assay. *Cereal Chem.* 63, 471–475.

Ridout, C. L., Price, K. R., DuPont, M. S., Parker, M. L., and Fenwick, G. R. (1991). Quinoa saponins—analysis and preliminary investigations into the effects of reduction by processing. *J. Sci. Food Agric.* 54, 165–176.

Ruales, J. and Nair, B. M. (1992a). Nutritional quality of the protein in quinoa (*Chenopodium quinoa* Willd) seeds. *Plant Foods Hum. Nutr.* 42, 1–11.

Ruales, J. and Nair, B. M. (1992b). Quinoa (*Chenopodium quinoa* Willd) an important Andean food crop. *Arch. Latinoamericanos Nutr.* 42, 232–241.

Ruales, J. and Nair, B. M. (1993a). Content of fat, vitamins and minerals in quinoa (*Chenopodium quinoa*, Willd) seeds. *Food Chem.* 48, 131–136.

Ruales, J. and Nair, B. M. (1993b). Saponins, phytic acid, tannins and protease inhibitors in quinoa (*Chenopodium quinoa*, Willd) seeds. *Food Chem.* 48, 137–143.

Ruales, J. and Nair, B. M. (1994). Properties of starch and dietary fibre in raw and processed quinoa (*Chenopodium quinoa*, Willd) seeds. *Plant Foods Hum. Nutr.* 45, 223–246.

Salcedo-Chavez, B., Osuna-Castro, J. A., Guevara-Lara, F., Dominguez-Dominguez, J., and Paredes-Lopez, O. (2002). Optimisation of the isoelectric precipitation method to obtain protein isolates from amaranth (*Amaranthus cruentus*) seeds. *J. Agric. Food Chem.* 50, 6515–6520.

Sanchez-Marroquin, A., Domingo, M. V., Maya, S., and Saldana, C. (1985a). Amaranth flour blends and fractions for baking applications. *J. Food Sci.* 50, 789–794.

Sanchez-Marroquin, A., Maya, S., and Domingo, M. V. (1985b). Milling procedures and air classification of amaranth flours. *Arch. Latinoamericanos Nutr.* 35, 621–630.

Sanchez-Marroquin, A., Del Valle, F. R., Escobedo, M., Avitia, R., Maya, S., and Vega, M. (1986). Evaluation of whole amaranth (*Amaranthus cruentus*) flour, its air-classified fractions, and blends of these with wheat and oats as possible components for infant formulas. *J. Food Sci.* 51, 1231–1234.

Saunders, R. M. and Becker, R. (1984). Amaranthus: A potential food and feed resource. *Adv. Cereal Sci. Technol.* 6, 357–396.

Schober, T. J., O'Brien, C. M., McCharthy, D., Darnedde, A., and Arendt, E. K. (2003). Influence of gluten-free flour mixes and fat powders on the quality of gluten-free biscuits. *Eur. Food Res. Technol.* 216, 369–376.

Schoenlechner, R. (1997). Entwicklung und Charakterisierung von Convenience-Produkten aus Amaranth und Quinoa. Diplomarbeit, University of Natural Resources and Applied Life Sciences, Vienna, Austria.

Schoenlechner, R. (2001). Investigation of the processing aspects of the pseudocereals amaranth and quinoa. PhD thesis, University of Natural Resources and Applied Life Sciences, Vienna, Austria.

Schoenlechner, R., Linsberger, G., Kaczyk, L., and Berghofer, E. (2006). Production of gluten-free short dough biscuits from the pseudocereals amaranth, quinoa and buckwheat with common bean. *Ernaehrung/Nutrition* 30, 101–107.

Scilingo, A. A., Molina-Ortiz, S. E., Martinez, E. N., and Anon, M. C. (2002). Amaranth protein isolates modified by hydrolytic and thermal treatments. Relationship between structure and solubility. *Food Res. Int.* 35, 855–862.

Segura-Nieto, M., Barba de la Rosa, A. P., and Paredes-López, O. (1994). Biochemistry of amaranth proteins. In: Paredes-Lopez, O. ed. *Amaranth—Biology, Chemistry, and Technology.* London: CRC Press, pp. 75–101.

Segura-Nieto, M., Shewry, P. R., and Paredes-Lopez, O. (1999). Globulins of the pseudocereals: Amaranth, quinoa, and buckwheat. In: Shewry, P. R. and Casey, R. eds. *Seed Proteins.* Dordrecht: Kluwer Academic Publishers, pp. 453–475.

Şensoy, I., Rosen, R. T., Ho, C.-T., and Karwe, M. V. (2006). Effect of processing on buckwheat phenolics and antioxidant activity. *Food Chem.* 99, 388–393.

Shin, D. H., Heo, H. J., Lee, Y. J., and Kim, H. K. (2004). Amaranth squalene reduces serum and liver lipid levels in rats fed a cholesterol diet. *Br. J. Biomed. Sci.* 61, 11–14.

Silva-Sanchez, C., González-Castaneda, J., De León Rodriguez, A., and Barba de la Rosa, A. P. (2004). Functional and rheological properties of amaranth albumins extracted from two Mexican varieties. *Plant Foods Hum. Nutr.* 59, 169–174.

Sinagawa-García, S. R., Rascón-Cruz, Q., Valdez-Ortiz, A., Medina-Godoy, S., Escobar-Gutiérrez, A., and Paredes-López, O. (2004). Safety assessment by in vitro digestibility and allergenicity of genetically modified maize with an amaranth 11S globulin. *J. Agric. Food Chem.* 52, 2709–2714.

Singhal, R. S. and Kulkarni, P. R. (1990). Some properties of *Amaranthus paniculatas* (Rajgeera) starch pastes. *Starch/Staerke* 42, 5–7.

Skrabanja, V. and Kreft, I. (1998). Resistant starch formation following autoclaving of buckwheat (*Fagopyrum esculentum* Moench) groats. An in vitro study. *J. Agric. Food Chem.* 46, 2020–2023.

Skrabanja, V., Laerke, H. N., and Kreft, I. (1998). Effects of hydrothermal processing of buckwheat (*Fagopyrum esculentum* Moench) groats on starch enzymatic availability *in vitro* and *in vivo* in rats. *J. Cereal Sci.* 28, 209–214.

Skrabanja, V., Nygaard, L. H., and Kreft, I. (2000). Protein-polyphenol interactions and *in vivo* digestibility of buckwheat groat proteins. *Pflug. Arch. Eur. J. Physiol.* 440 (Suppl. 5), R129–R131.

Skrabanja, V., Liljeberg, E. H. G. M., Kreft, I., and Björck, I. M. E. (2001). Nutritional properties of starch in buckwheat products: studies in vitro and in vivo. *J. Agric. Food Chem.* 49, 490–496.

Skrabanja, V., Kreft, I., Golob, T. *et al.* (2004). Nutrient content in buckwheat milling fractions. *Cereal Chem.* 81, 172–176.

Sohn, M. H., Lee, S. Y., and Kim, K.-E. (2003). Prediction of buckwheat allergy using specific IgE concentrations in children. *Allergy* 58, 1308–1310.

Soral-Śmietana, M., Fornal, Ł., and Fornal, J. (1984a). Characteristics of buckwheat grain starch and the effect of hydrothermal processing upon its chemical composition, properties and structure. *Starch/Stärke* 36, 153–158.

Soral-Śmietana, M., Fornal, Ł., and Fornal, J. (1984b). Characteristics of lipids in buckwheat grain and isolated starch and their changes after hydrothermal processing. *Die Nahrung* 28, 483–492.

Souci, S. W., Fachmann, W., and Kraut, H. (2000). *Food Composition and Nutrition Tables*. Stuttgart: Wissenschaft Verlags,

Steadman, K. J., Burgoon, M. S., Schuster, R. L., Lewis, B. A., Edwardson, S. E., and Obendorf, R. L. (2000). Fagopyritols, D-*chiro*-inositol, and other soluble carbohydrates in buckwheat seed milling fractions. *J. Agric. Food Chem.* 48, 2843–2847.

Steadman, K. J., Burgoon, M. S., Lewis, B. A., Edwardson, S. E., and Obendorf, R. L. (2001a). Buckwheat seed milling fractions: description, macronutrient composition and dietary fibre. *J. Cereal Sci.* 33, 271–278.

Steadman, K. J., Burgoon, M. S., Lewis, B. A., Edwardson, S. E., and Obendorf, R. L. (2001b). Minerals, phytic acid, tannin and rutin in buckwheat seed milling fractions. *J. Sci. Food Agric.* 81, 1094–1100.

Stibilj, V., Kreft, I., Smrkolj, P., and Osvald, J. (2004). Enhanced selenium content in buckwheat (*Fagopyrum esculentum* Moench) and pumpkin (*Curcubita pepo* L.) seeds by foliar fertilisation. *Eur. Food Res. Technol.* 219, 142–144.

Stone, L. A. and Lorenz, K. (1984). The starch of *Amaranthus*—physico-chemical properties and functional characteristics. *Starch/Staerke* 36, 232–237.

Sun, T. and Ho, C.-T. (2005). Antioxidant activities of buckwheat extracts. *Food Chem.* 90, 743–749.

Suzuki, T., Honda, Y., Funatsuki, W., and Nakatsuka, K. (2002). Purification and characterization of flavonol-3-glucosidase, and its activity during ripening in tartary buckwheat seeds. *Plant Sci.* 163, 417–423.

Suzuki, T., Honda, Y., Funatsuki, W., and Nakatsuka, K. (2004a). In-gel detection and study of the role of flavonol-3-glucosidase in the bitter taste generation in tartary buckwheat. *J. Sci. Food Agric.* 84, 1691–1694.

Suzuki, T., Honda, Y., and Mukasa, Y. (2004b). Purification and characterisation of lipase in buckwheat seed. *J. Agric. Food Chem.* 52, 7407–7411.

Suzuki, T., Honda, Y., Mukasa, Y., and Kim, S.-J. (2005). Effects of lipase, lipoxygenase, peroxidase, and rutin on quality deteriorations in buckwheat flour. *J. Agric. Food Chem.* 53, 8400–8405.

Suzuki, T., Honda, Y., Mukasa, Y., and Kim, S.-J. (2006). Characterization of peroxidase in buckwheat seeds. *Phytochemistry* 67, 219–224.

Tanaka, K., Matsumoto, K., Akasawa, A. *et al.* (2002). Pepsin resistant 16-kD buckwheat protein is associated with immediate hypersensitivity reaction in patients with buckwheat allergy. *Int. Arch. Allergy Immunol.* 129, 49–56.

Tang, H., Watanabe, K., and Mitsunaga, T. (2002). Characterisation of storage starches from quinoa, barley and adzuki seeds. *Carbohydr. Polym.* 49, 13–22.

Tari, T. A., Annapure, U. S., Singhal, R. S., and Kulkarni, P. R. (2003). Starch-based spherical aggregates: screening of small granule sized starches for entrapment of a model flavouring compound, vanillin. *Carbohydr. Polym.* 53, 45–51.

Taylor, S. L. and Hefle, S. L. (2001). Food allergies and other food sensitivities. *Food Technol.* 55, 68–83.

Tian, Q., Li, D., and Patil, B. S. (2002). Identification and determination of flavonoids in buckwheat (*Fagopyrum esculentum* Moench, Polygonaceae) by high-performance liquid chromatography with electrospray ionisation, mass spectrometry and photodiode array ultraviolet detection. *Phytochem. Anal.* 13, 251–256.

Tomotake, H., Shimaoka, I., Kayashita, J., Yokoyama, F., Nakajoh, M., and Kato, N. (2000). A buckwheat protein product suppresses gallstone formation and plasma cholesterol more strongly than soy protein isolate in hamsters. *J. Nutr.* 130, 1670–1674.

Tomotake, H., Shimaoka, I., Kayashita, J., Yokoyama, F., Nakajoh, M., and Kato, N. (2001). Stronger suppression of plasma cholesterol and enhancement of the fecal excretion of steroids by a buckwheat protein product than by a soy protein isolate in rats fed on a cholesterol-free diet. *Biosci. Biotechnol. Biochem.* 65, 1412–1414.

Tosi, E. A., Ré, E. D., Lucero, H., and Masciarelli, R. (2001). Dietary fibre obtained from amaranth (*A.cruentus*) grain by differential milling. *Food Chem.* 73, 441–443.

Tosi, E. A., Re, E. D., Masciarelli, R., Sanchez, H., Osella, C., and de la Torre, M. A. (2002). Whole and defatted hyperproteic amaranth flours tested as wheat flour supplementation in mold breads. *Lebensm.-Wiss. U.-Technol.* 35, 472–475.

Tovar, L. R., Brito, E., Takahashi, T., Miyazawa, T., Soriano, J., and Fujimoto, K. (1989). Dry heat popping of amaranth seed might damage some of its essential amino acids. *Plant Foods Hum. Nutr.* 39, 299–309.

Trautwein, E. A., Van Leeuwen, A., and Ebersdobler, H. F. (1997). Plant sterol profiles and squalene concentrations in common unrefined and refined vegetable oils. In: Bioactive inositol phosphates and phytosterols in foods. *Proceedings of the 2nd Workshop, COST 916*, Goeteborg, Sweden, pp. 79–81.

Varriano-Marston, E. and DeFrancisco, A. (1984). Ultrastructure of quinoa fruit (*Chenopodium quinoa* Willd). *Food Microstructure* 3, 165–173.

Velioglu, Y. S., Mazza, G., Gao, L., and Oomah, B. D. (1998). Antioxidant activity and total phenolics in selected fruits, vegetables, and grain products. *J. Agric. Food Chem.* 46, 4113–4117.

Vombergar, B. and Gostenčnik, D. (2005). Priprava ajdovih keksov za prehranske poskuse. (Production of buckwheat biscuits for nutritional studies). *Acta agric. Slovenica* 85, 397–409.

Wang, Z., Zhang, Z., Zhao, Z., Wieslander, G., Norbäck, D., and Kreft, I. (2004). Purification and characterization of a 24 kDa protein from tartary buckwheat seeds. *Biosci. Biotechnol. Biochem.* 68, 1409–1413.

Watanabe, M. (1998). Catechins as antioxidants from buckwheat (*Fagopyrum esculentum* Moench) Groats. *J. Agric. Food Chem.* 46, 839–845.

Watanabe, M., Ohshita, Y., and Tsushida, T. (1997). Antioxidant compounds from buckwheat (*Fagopyrum esculentum* Möench) hulls. *J. Agric. Food Chem.* 45, 1039–1044.

Watanabe, K., Shimizu, M., Adachi, T., Yoshida, T., and Mitsunaga, T. (1998). Characterization of thiamine-binding protein from buckwheat seeds. *J. Nutr. Sci. Vitaminol.* 44, 323–328.

Watanabe, K., Ibuki, A., Chen, Y. C., Kawamura, Y., and Mitsuanaga, T. (2003). Compostion of quinoa protein fractions. *Nippon Shokuhin Kagaku Kogaku Kaishi* 50, 546–549.

Wei, Y.-M., Zhang, G.-Q., and Li, Z.-X. (1995). Study on nutritive and physico-chemical properties of buckwheat flour. *Nahrung/Food* 39, 48–54.

Wei, Y.-M., Hu, X.-Z., Zhang, G.-Q., and Ouyang, S.-H. (2003). Studies on the amino acid and mineral content of buckwheat protein fractions. *Nahrung/Food* 47, 114–116.

Wesche-Ebeling, P., Argaíz-Jamet, A., Teutli-Olvera, B., Guerro-Beltrán, J. A., and López-Malo, A. (1996). Development of a high quality granola containing popped amaranth grain varying in fat content, sweetness and degree of toasting. Institute of Food Technologists Annual Meeting: Abstracts, p. 51.

Whistle, R. L., BeMiller, J. N., and Paschall, E. F. (1984). *Starch: Chemistry and Technology*. New York: Academic Press, p. 301.

Wilhelm, E., Aberle, T., Burchard, W., and Landers, R. (2002). Peculiarities of Aqueous Amaranth Starch Suspensions. *Biomacromolecules* 3, 17–26.

Woldemichael, G. M. and Wink, M. (2001). Identification and biological activities of triterpenoid saponins from *Chenopodium quinoa*. *J. Agric. Food Chem.* 49, 2327–2332.

Wolfrum, V. (1999). Herstellung von glutenfreien Teigwaren aus Amaranth und Quinoa. Diplomarbeit, University of Vienna, Vienna, Austria.

Wright, K. H., Huber, K. C., Fairbanks, D. J., and Huber, C. S. (2002). Isolation and characterisation of *Atriplex hortensis* and sweet *Chenopodium quinoa* starches. *Cereal Chem.* 79, 715–719.

Yanez, E., Zacarias, I., Granger, D., Vasquez, M., and Estevez, A. M. (1994). Chemical and nutritional characterisation of amaranthus (*Amaranthus cruentus*). *Arch. Latinoamericanos Nutr.* 44, 57–62.

Yoshimoto, Y., Egashira, T., Hanashiro, I., Ohinata, H., Takase, Y., and Takeda, Y. (2004). Molecular structure and some physicochemical properties of buckwheat starches. *Cereal Chem.* 81, 515–520.

Zheng, G. H. and Sosulski, F. W. (1998). Determination of water separation from cooked starch pastes after refrigeration and freeze-thaw. *J. Food Sci.* 63, 134–139.

Zheng, G. H., Sosulski, F. W., and Tyler, R. T. (1998). Wet-milling, composition and functional properties of starch and protein isolated from buckwheat groats. *Food Res. Int.* 30, 493–502.

Zhu, N., Sheng, S., Li, D., Lavoie, E. J., Karwe, M. V., Rosen, R. T., and Ho, C. T. (2001). Antioxidative flavonoid glycosides from quinoa seeds (*Chenopodium quinoa Willd*). *J. Food Lipids* 8, 37–44.

Oat products and their current status in the celiac diet

8

Tuula Sontag-Strohm, Pekka Lehtinen, and Anu Kaukovirta-Norja

Introduction

The high content of beneficial fibers together with the bioactive co-passengers makes oats an attractive component both for a common and for a gluten-free diet. Oat dietary fiber is nutritionally special due to the high content of soluble, mixed linked $(1{\rightarrow}3),(1{\rightarrow}4)$-β-D-glucan, which comprises 2–7% of the total kernel weight and is the main cell wall component of oat kernel (Wood, 1986). An adequate daily intake of β-glucan is associated with a reduced risk for heart and coronary diseases. In addition to β-glucan, oats, and especially oat bran, contain higher amount of total dietary fiber than most of the other gluten-free flours (Table 8.1). In addition, the protein content in oats is higher than in rice or maize flour.

In addition to dietary fiber and protein, oats deliver substantial amounts of unsaturated fatty acids and bioactive compounds. The total amount of lipids in oats is in the range 3–9% (Brown and Craddock, 1972). The majority of the lipids are unsaturated and the most abundant fatty acids are monounsaturated oleic acid and polyunsaturated linoleic acid. Oats are known to have a high content of antioxidants and the typical tocol content is around 20–30 mg/kg (Lásztity *et al.*, 1980; Peterson and Qureshi, 1993). Other antioxidant compounds present in oats include phenolic acids, avenanthramides, and sterols.

Table 8.1 Typical composition of wholegrain oat flour, oat bran, and common gluten-free flours

	Wholegrain oat flour	Oat bran	Brown rice flour	Soybean	Wholegrain maize flour
Protein	15–17%	15–18%	6–10%	30–40%	5–10%
Starch and sugars	59–70%	10–50%	70–80%	12–17%	75–85%
Fat	4–9%	5–10%	1–4%	10–20%	2–5%
Total dietary fiber	5–13%	10–40%	3–5%	12–17%	1–2%
β-Glucan	2–6%	5–20%	–	–	–

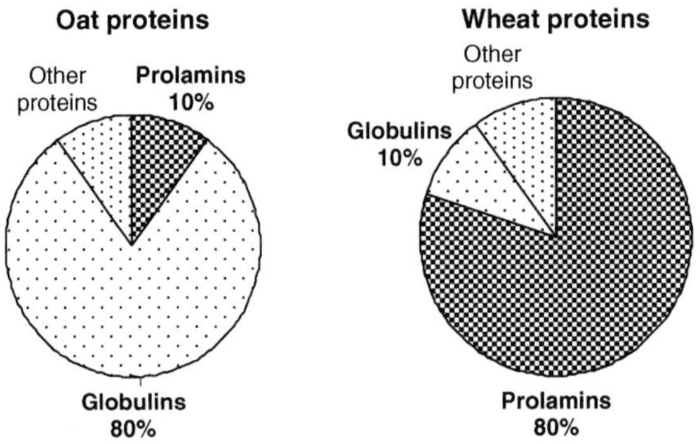

Figure 8.1 Comparison of oat and wheat protein fraction distributions.

The distribution of the different protein classes is unique in oats, with globulins representing the largest group. In contrast, in the gluten-containing cereals wheat, rye, and barley globulins comprise at most 10% of total storage proteins, whereas the prolamins are the largest group, comprising 80% of the total storage proteins (Figure 8.1). Like most of the other cereal seed storage proteins, oat prolamins are rich in glutamine and proline.

Based on numerous clinical studies, in many countries oats are now recommended to be included as a part of the gluten-free diet (Leiss, 2003; Kupper, 2005). At the same time, special oat brands have been introduced in which the cross-contamination of oats with other cereals is minimized by careful control throughout the whole production chain. Consequently, many patients with celiac disease now use oat products to diversify and improve the nutritional quality of their diet.

Gluten-free status of oats

Together with wheat, oats is the most thoroughly clinically studied cereal in connection with celiac disease (Table 8.2). Clinical studies have demonstrated the long-term safety of moderate amounts of oats as part of a gluten-free diet for patients with celiac disease.

The Codex standard from 1981 defined oats as a gluten-containing cereal along with wheat, rye, and barley. However, the standard of gluten-free products is

Table 8.2 Oat challenge in clinical studies

Reference	Oat test and control group[a]	Oat challenge, g/day	Oat challenge, months	Result of the oat challenge	Type of the tests and measurements
Van de Kamer et al., 1953	2 Children	140	1.8	One of the two children had increased fat excretion	Fecal fat excretion
Moulton, 1959	4 Children	56–169	0.8–3.5	No difference, fecal fat excretion in normal limits	Fecal fat excretion
Dissanayake et al., 1974	4 Adults, 1 control	40–60	1	No clinical symptoms	Jejunal (duodenal) biopsies, quantitative histological studies
Baker et al., 1976	11 Adults, 1 child	60	1	No change in xylose excretion for nine, increased excretion for three	Xylose test
Janatuinen et al., 1995 Janatuinen et al., 2000	92 Adults: oat/control 26/26 previously, 19/21 newly diagnosed	44–50	6 / 12	No differences in clinical symptoms, 6 in oat and 5 in control group withdrew from the study	Duodenal biopsies, serological, histological, and morphometrical measurements
Srinivasan et al., 1996	10 Adults	50	3	No serological and no histological changes	Duodenal biopsies, serological, histological, and morphometrical measurements
Hardman et al., 1997	10 Adults with dermatitis herpetiformis	63	3	No deleterious effects on skin or intestine	Duodenal biopsy, skin biopsy, serological, immunohistochemical, and morphometrical measurements
Reunala et al., 1998	23 Adults with dermatitis herpetiformis: 12/11 oat/control	53	6	No changes in small bowel mucosa, three developed rash in the oat and in control group	Duodenal biopsy, skin biopsy, serological, immunohistochemical, and morphometrical measurements
Hardman et al., 1999	2 Adults with dermatitis herpetiformis	2.5 (oat prolamin)	5 days	No toxic effects on skin or intestine by oat prolamin challenge	Duodenal biopsy, skin biopsy, serological, histochemical, and morphometrical measurements

(Continued)

Table 8.2 *Continued*

Reference	Oat test and control group[a]	Oat challenge, g/day	Oat challenge, months	Result of the oat challenge	Type of the tests and measurements
Hoffenberg et al., 2000	10 Children, newly diagnosed	24	6	Improvement in all the tests and clinical measurements in 8 of 10 patients	Duodenal biopsy, serological, histological, and morphometrical measurements
Janatuinen et al., 2002	63 Adults: 35/28 oat/control	34	60 (5 years)	No difference between the oats and control group in 5 year study	Duodenal biopsies, serological, histological, and morphometrical measurements
Lundin et al., 2003	19 Adults	50	3	No clinical sympthoms in 18; one got villous atrophy	Duodenal biopsies, serological, histological, and morphometrical measurements
Störsrud et al., 2003	20 Adults	93	24	No morphological or serological negative effects, five withdraw in the 2 year study	Duodenal biopsies, serological, histological, and morphometrical measurements
Hogberg et al., 2004	116 Children: 57/59 oats/control	15	12	No difference in clinical and small bowel mucosal healing between oats and control group,15 in oat and 7 in control group withdraw the study	Duodenal biopsies, serological, histological, and morphometrical measurements
Peräaho et al., 2004a	39 Adults 23/16 oat/control	50	12	Mucosal integrity was not disturbed but increase in intestinal symptoms in oat group	Serological, histological and morphometrical measurements, psychological general well-being questionnaire, and gastrointestinal symptoms
Holm et al., 2006	32 Children 12/11 oat/control, 9 newly diagnosed	45	24 /7 years follow-up	No difference in clinical and small bowel mucosal healing between oats and control group, with oats better compliance	Serological, histological, and morphometrical measurements

[a]Studies that included a control group of patients with celiac disease are marked.

currently at step 6 of the Codex Alimentarius procedure with the note that the use of oats in gluten-free foods may be determined at national level. The Food and Drug Administration (FDA) in the USA proposed in January 2007 that the definition of the term "gluten-free" does not prohibit oats to be included in gluten-free foods.

Finnish patients with celiac disease and dermatitis herpetiformis have used oat-containing gluten-free diets since 1997. Since 2001, oat foods have been also accepted into the list of the cereals tolerated by children with celiac disease in Finland. In this country, 73% of the adults with celiac disease use oat foods daily in their diet (Peräaho *et al.*, 2004a, 2004b). Patients appreciated the taste, ease of use, and low costs; 94% said that oats diversified the gluten-free diet.

Oat products

Oats are harvested with the hulls, which need to be removed from oat products to be used for human consumption. In addition, the endogenous lipid-modifying enzymes, especially lipase but also lipoxygenase and lipoperoxidase, are typically inactivated by heat treatment before further use. De-hulled and heat-treated oat groats can be processed to various forms of product with different composition, appearance, taste, and technological functionality.

Oat milling fractions

The main technological parts of the oat kernel are hull, cell wall (i.e. bran), and endosperm fractions. Hard outer hull can comprise up to 30–40% of the kernel weight. To clarify the diverse use of different terms used to describe oat products, the oat committee of the American Association of Cereals Chemists (AACC) has started to work on the definitions of oat milling fractions (Table 8.3). The majority of wholegrain oat that is destined for human consumption is either processed to oat flakes by roller mill or to cut grains by cutting the groat into 3–4 pieces by steel cutters. Compared with oat flour, the handling and further processing of these is easier, as oat flour tends to form lumps.

Oat flakes can be used as such for baking processes, as they will disintegrate readily once mixed with water. For applications in which this disintegration is not wanted, flakes with higher thickness are available. In contrast, cut grains will retain part of their structure throughout the baking process and can thus provide a grainy appearance in the final product.

The health effects rely mainly on the total dietary fiber and β-glucan content of oat products, and often oat products are characterized according to their β-glucan content. The β-glucan content of wholegrain de-hulled oat is typically around 4%. Dietary fiber and β-glucan can be enriched into the bran products. Bran is prepared from wholegrain oat flour by removing the starchy endosperm by sieving or air classification. The amount of starchy endosperm remaining in the bran varies from product to product. Regular oat bran contains typically 6–8% β-glucan, whereas novel

Table 8.3 Definition of oat milling fractions

Oat product	Definition
Oat hulls	A product of traditional milling process, is predominantly the fibrous sheat surrounding the oat groat
Oat groat	Whole grain of oats that is portion of oat kernel which remains after removal of the hull
Rolled oats, oat meal	Clean, 100% oat groats that are produced by steaming, cutting if needed, rolling and flaking
Whole oat flour	Products derived without material loss from whole oat groats, by steaming and size reduction
Oat flour	Finely granulated material from which coarce particles or bran fractions have been removed, and is produced by steaming and size reduction of clean oat groat or portions thereof
Steel cut oats (steel cut groats	Clean, 100% oat groats that are produced by steaming and coarsely cutting into two or more pieces
Oat bran	Produced from clean, 100% oat groats or from products derived without material loss from whole groats into fractions with defined beta-glucan, total dietary fiber and soluble dietary fiber contents
Oat fiber	Derived from oat hulls, which is substantially free of groat components and has an insoluble dietary fiber content of at least 85% (dry weight basis)

Adapted from the draft of oat definitions for inclusion in AACC Approved Methods, by AACC Oat Committee, 2003.

oat bran concentrates can contain up to 22% β-glucan. Even higher β-glucan content can be obtained by adapting the extraction process.

Oat milling fractions are also available as extruded products. The extrusion process is applied to modify the flavor by introducing a slightly roasted flavor and to pre-gelatinize the oat starch. Compared with other cereals, oats also have a high lipid content. Even though oat lipids can be considered to be nutritionally beneficial, they have an adverse effect on the processing behavior and storage stability of oat products. Consequently, some oat products are defatted by solvent extraction prior to further processing to improve their stability.

Consumer products containing oats: technology and challenges

Oats can be included into various consumer products to diversify the diet of patients with celiac disease.

Oats alone are unsuitable for a traditional breadmaking process, and most commercial oat breads contain substantial amounts of wheat flour. The protein network in commercial oat breads is often fortified with added gluten. Thus, most of the currently available oat breads are inappropriate for patients with celiac disease. However, by applying a novel baking procedure it is possible to make gluten-free oat bread. Typical gluten-free bread is based on starches. The starch breads often lack cereal flavor and are poor in delivering the palatable crumb structure of gluten-containing breads. When oats are mixed with other gluten-free ingredients, a favorable aroma and taste as well as texture can be accomplished. Although oats are mostly used as wholegrain flakes or flour, they have a mild, nutty flavor. Recently, baking technology for 51–100% oat bread has been developed further (Flander *et al.*, 2007) and the first products have already been commercialized (see www.eho.fi for a 100% oat bread).

In addition to bread, oats are widely used in various other products including snacks and porridge. Fermentation of oat slurry provides a yoghurt-type product that can be used by patients with celiac disease, milk allergenic, or lactose intolerance patients. Several oat-containing drinks have also emerged into the markets (e.g. oat milk) (Lindahl *et al.*, 1997; Önning *et al.*, 1998; Chronakis *et al.*, 2004) and oat-berry beverages. Oat ice cream, oat pancake mix, and meal replacement drinks (Mikola, 2004) all represent totally novel, high-moisture oat products. The studies of Lyly *et al.* (2004) showed that oat β-glucans are technologically feasible thickening agents in soups and have high acceptance among consumers. Due to the high, shear-thinning viscosity of oat extracts, the technology to produce liquid or high-moisture products can be challenging. However, such products have a high market potential beyond just patients with celiac disease (Table 8.4).

Table 8.4 Oat-containing food products: commercial products and future applications

Consumer product type	Typical oat milling fraction/oat ingredient used	Remarks
Bread	All oat products appropriate	Novel breadmaking technology is needed to produce oat bread without the addition of wheat gluten
Porridge	Oat flakes, oat bran	May be combined with other non-gluten flours
Pasta	Oat flour, oat bran	Most oat pasta products contain large proportions of wheat
Snacks	Extruded products	Oat provides flavor and enhance nutritional status
Drinks	Oat flakes	1. Milk-type oat extracts that are free of particulate material 2. Yoghurt-type colloidial products containing particulate material
Muesli	Oat flakes, extruded oat products	
Breakfast cereals	Extruded oat products	Rice or maize starch is usually added to extruded oat products to improve the structure
Confectionery, biscuits	Oat extract, oat flakes	
Processed food	Oat flakes, pre-gelatinized oat products	Oat products can be used to replace wheat-based water binding agents, for example breadcrumbs

How to analyze the gluten-free status of oat products

Oats as well as other cereals or ingredients in gluten-free foods can be contaminated by prohibited cereal species such as wheat, barley, or rye. The contamination can occur in the field, but it can also happen during transportation, storage, milling, or food processing.

The method for analysis of gluten is expected to accurately quantify contaminating prolamins from different food matrices. Two ELISA methods for gluten analysis are widely commercially available. The difference between them is the antibody used to detect prolamins. One is based on a monoclonal ω-gliadin antibody that recognizes the heat-stable ω-fraction from wheat, rye, and barley prolamins, but not from oat avenins (Hill and Skerrit, 1989). Its main disadvantage is that it cannot detect barley and rye prolamins to the same extent as wheat prolamins. In addition, Wieser (2000) showed that the relative amount of ω-gliadin varies between wheat cultivars, thus leading to inaccurate results. The other, more recent, ELISA method is based on a monoclonal antibody R5 raised against rye prolamin (Sorell *et al.*, 1998; Valdés *et al.*, 2003). This method has been given temporary endorsement in the Codex Committee on Methods of Analysis and Sampling (Codex Alimentarius Commission, 2005) promoted by the Codex Committee on Nutrition and Foods for Special Dietary Uses (Codex Alimentarius Commission, 2003). The antibody R5 recognizes a pentapeptide QQPFP (glutamine–glutamine–proline–phenylalanine–proline) present in wheat, rye, and barley prolamins, but not found in oats (Kasarda, 1996). This sequence occurs repeatedly in prolamins, especially in ω-type prolamins (Shewry and Tatham, 1999). Osman *et al.* (2001) have studied different peptides recognized by this antibody and reported that the most important structural requirement within the epitope seems to be a dipeptide FP. Since this peptide is small and occurs in several proteins from different origin, there is a possibility that the antibody R5 recognizes many peptides that are not toxic to people with celiac disease. As a result, there is a risk of false positive results, which could reduce the current variety of gluten-free foods.

Kanerva *et al.* (2006) prepared oat samples with known amounts of barley contamination and measured the prolamin content. The results revealed that the ELISA method based on the R5 antibody multiplied the prolamin contents of the oat-based samples that contained small amounts of barley, making the results much higher than they should be (i.e. 7–30 times higher hordein concentrations in oat samples than actually present). The results were improved when hordein was used as a standard instead of gliadin (Kanerva *et al.*, 2006). This phenomenon makes it difficult to analyze gluten-free products if they contain trace amounts of barley, and may unnecessarily exclude some gluten-free products from the market. Therefore, there is a great need for an analytical method that can differentiate and quantify celiac-toxic polypeptides in food ingredients and processed foods. Since oats have been shown to be suitable as part of the diet of patients with celiac disease, more specific methods are needed to differentiate oats from wheat, rye, and barley.

Future trends and conclusions

According to the present clinical data, it is evident that oats can be included in a gluten-free diet. Holm *et al.* (2006) stated that children with celiac disease can also consume uncontaminated oats as part of their diet for a long time. In addition, oat-containing products were highly acceptable to children. Garsed and Scott (2007) concluded in their recent review that previous conflicting results have at least partly been due to contamination of oats by wheat. In Nordic countries barley is the typical contaminant of oat. It is obvious that an uncontaminated, safe oat production chain from farm to fork is needed before oats can be recommended widely and safely for patients with celiac disease. "Pure oats" production has been developed in recent years and some products have been launched on the market (see, for example, www.creamhillestates.com and www.provena.fi).

Uncontaminated, "pure" oats is a true alternative in a gluten-free diet. Oats deliver a typical cereal character into products and can be used as a raw material in typical cereal applications such as bakery products, porridges, or snacks. The development of whole oat baking technology further improves the quality and enjoyability of traditional-type soft breads. A selection of oat breads for patients with celiac disease is already commercially available (see e.g. www.moilas.fi).

Oats, and especially oat fractions, fulfil the demands of modern consumer for high-fiber products. Oat fractions are suitable for several types of novel health-promoting products such as beverages, yoghurt-type products, and dairy products. As oats are generally widely accepted, it is a choice for the whole family and less complex gluten-free products are needed. This can lead to marked savings for patients with celiac disease, since gluten-free products are typically more expensive than basic products.

The inclusion of oats in the diet will provide several alternatives for a gluten-free diet. At the moment, oats are approved for gluten-free diets in only a few countries. To get broader acceptance for oats, an active collaboration between celiac societies, authorities, scientists, and the whole food production chain is needed.

References

Baker, P. G. and Read, A. E. (1976). Oats and barley toxicity in celiac patients. *Postgrad Med. J.* 52, 264–268.

Brown, C. M. and Craddock, J. C. (1972). Oil content and groat weight of entries in the world oat collection. *Crop Sci.* 12, 514–515.

Chronakis, I. S., Öste Triantafyllou, A., and Öste, R. (2004). Solid-state characteristics and redispersible properties of powders formed by spray-drying and freeze-drying cereal dispersions of varying $(1\rightarrow3,1\rightarrow4)$-β-glucan content. *J. Cereal Sci.* 40, 183–193.

Codex Alimentarius Commission (2003). *Report of the 25th Session of the Codex Committee on Nutrition and Foods for Special Uses.*

Codex Alimentarius Commission (2005). *Report of the 26th Session of the Codex Committee on Methods of Analysis and Sampling.*

Dissanayake, A. S., Truelove, S. C., and Whitehead, R. (1974). Lack of harmful effect of oats on small-intestinal mucosa in celiac disease. *BMJ* 4, 189–191.

Flander, L., Salmenkallio-Marttila, M., Suortti, T., and Autio, K. (2007). Optimization of ingredients and baking process for improved wholemeal oat bread quality. *Lebensm. Wiss. Technol.* 40, 860–870.

Garsed, K. and Scott, B. B. (2007). Can oats be taken in a gluten-free diet? A systematic review. *Scand. J. Gastroenterol.* 42, 171–178.

Hardman, C. M., Garioch, J. J., Leonard, J. N., Thomas, H. J. W., Walker, M. M., and Lortan, J. E. (1997). Absence of toxicity of oats in patients with dermatitis herpetiformis. *N. Engl. J. Med.* 337, 1884–1887.

Hardman, C., Fry, L., Tatham, A., and Thomas, H. J. W. (1999). Absence of toxicity of avenin in patients with dermatitis herpetiformis. *N. Engl. J. Med.* 340, 321.

Hill, A. S. and Skerrit, J. H. (1989). Hybridoma cells and monoclonal antibodies. Patent number GB2207921 (CA1294903).

Hoffenberg, E. J., Haas, J., Drescher, A., Barnburst, R., Osberg, I., and Bao, F. (2000). A trial of oats in children with newly diagnosed celiac disease. *J. Pediatr.* 137, 361–366.

Hogberg, L., Laurin, P., Falth-Magnusson, K., Grant, C., Grodzinsky, E., and Jansson, G. (2004). Oats to children with newly diagnosed coeliac disease: a randomised double blind study. *Gut* 53, 649–654.

Holm, K., Maki, M., Vuolteenaho, N. *et al.* (2006). Oats in the treatment of childhood coeliac disease: a 2-year controlled trial and a long-term clinical follow-up study. *Aliment. Pharmacol. Ther.* 23, 1463–1472.

Janatuinen, E. K., Pikkarainen, P. H., Kemppainen, T. A., Kosma, V.-M., Jarvinen, R. M. K., and Uusitupa, M. I. J. (1995). A comparison of diets with and without oats in adults with celiac disease. *N. Engl. J. Med.* 333, 1033–1037.

Janatuinen, E. K., Kemppainen, T. A., Pikkarainen, P. H., Holm, K. H., Kosma, V.-M., and Uusitupa, M. I. J. (2000). Lack of cellular and humoral immunological responses to oats in adults with coeliac disease. *Gut* 46, 327–331.

Janatuinen, E. K., Kemppainen, T. A., Julkunen, R. J. K., Kosma, V.-M., Maki, M., and Heikkinen, M. (2002). No harm from five-year ingestion of oats in coeliac disease. *Gut* 50, 332–335.

Kanerva, P., Sontag-Strohm, T., Ryöppy, P., Alho-Lehto, P., and ja Salovaara, H. (2006). Analysis of barley contamination in oats using R5 and o-gliadin antibodies. *J. Cereal Sci.* 44, 347–352.

Kasarda, D. D. (1996). Gluten and gliadin: precipitating factors in coeliac disease. In: Mäki, M., Collin, P., and Visakorpi, J. K. eds. *Coeliac Disease, Proceedings of the Seventh International Symposium on Coeliac Disease*, Vammalan kirjapaino, Tampere, Finland, pp. 195–212.

Kupper, C. (2005). Dietary guidelines and implementation for celiac disease. *Gastroenterology* 128, 121–127.

Lásztity, R., Berndorfer-Kraszner, E., and Huszar, M. (1980). On the presence and distribution of some bioactive agents in oat varieties. *Cereals Food Beverages, Recent Prog. Cereal Chem.Technol., [Proc. Int. Conf.]* 429–445.

Leiss, O. (2003). Pitfalls and problems of a gluten-free diet: Is oats harmful to patients with coeliac disease? *Aktuelle Ernahrungsmedizin* 28, 385–397. [In German]

Lindahl, L., Ahldén, I., Öste, R., and Sjöholm, I. (1997). Homogenous and stable cereal suspension and a method of making the same. US Patent 5686123.

Lundin, K. E. A., Nilsen, E. M., Scott, H. G., Loberg, E. M., Gjoen, A., and Bratlie, J. (2003). Oats-induced villous atrophy in celiac disease. *Gut* 52, 1649–1652.

Lyly, M., Salmenkallio-Marttila, M., Suortti, T., Autio, K., Poutanen, K., and Lähteenmäki, L. (2004). The sensory characteristics and rheological properties of soups containing oat and barley β-glucan before and after freezing. *Lebensm. Wiss. Technol.* 37, 749–761.

Mikola, M. (2004). Natural oat bran instant drink. In: Peltonen-Sainio, P. and Topi-Hulmi, M. eds. *Proceedings of the 7th International Oat Conference*, Jokioinen, MTT Agrifood Research, Finland, p. 99.

Moulton, A. L. C. (1959). The place of oats in the coeliac diet. *Arch. Dis. Child.* 34, 51–55.

Önning, G., Åkesson, B., Öste, R., and Lundquist, L. (1998). Effects of consumption of oat milk, soya milk, or cow's milk on plasma lipids and antioxidative capacity in healthy subjects. *Ann. Nutr. Metab.* 42, 211–220.

Osman, A. A., Uhlig, H. H., Valdes, I., Amin, M., Méndez, E., and Mothes, T. (2001). A monoclonal antibody that recognizes a potential celiac-toxic repetitive pentapeptide epitope in gliadins. *Eur. J. Gastroenterol. Hepatol.* 13, 1–5.

Peräaho, M., Collin, P., Kaukinen, K., Kekkonen, L., Miettinen, S., and Maki, M. (2004a). Oats can diversify a gluten-free diet in coeliac disease and dermatitis herpetiformis. *J. Am. Diet. Assoc.* 104, 1148–1150.

Peräaho, M., Kaukinen, K., Mustalahti, N. *et al.* (2004b). Effects of an oats-containing gluten-free diet on symptoms and quality of life in celiac disease. *Scand. J. Gastroenterol.* 39, 27–31.

Peterson, D. M. and Qureshi, A. A. (1993). Genotype and environment effects on tocols in barley and oats. *Cereal Chem.* 70, 157–162.

Reunala, T., Collin, P., Holm, K. *et al.* (1998). Tolerance to oats in dermatitis herpetiformis. *Gut* 43, 490–493.

Shewry, P. R. and Tatham, A. S. (1999). The characteristics, structures and evolutionary relationships of prolamins. In: Shewry, P. R. and Casey, R. eds. *Seed Proteins*. Dordrecht: Kluwer Academic Publishers, pp. 11–33.

Sorell, L., López, J. A., Valdés, I. *et al.* (1998). An innovative sandwich ELISA system based on an antibody cocktail for gluten analysis. *FEBS Lett.* 439, 46–50.

Srinivasan, U., Leonard, N., Jones, E., Kasarda, D. D., Weir, D. G., and O'Farrelly, C. (1996). Absence of oats toxicity in adult celiac disease. *BMJ* 313, 1300–1301.

Störsrud, S., Olsson, M., Arvidsson, L. R., Nilsson, L., Nilsson, O., and Kilander, A. (2003). Adult coeliac disease patients do tolerate large amounts of oats. *Eur. J. Clin. Nutr.* 57, 163–169.

Valdes, I., Garcia, E., Llorente, M., and Mendez, E. (2003). Innovative approach to low-level gluten determination in foods using a novel sandwich enzyme-linked immunosorbent assay protocol. *Eur. J. Gastroenterol. Hepatol.* 15, 465–474.

van de Kamer, J. H., Weijers, H. A., and Dicke, W. K. (1953). Coeliac disease. IV. An investigation into the injurious constituents of wheat in connection with their action on patients with celiac disease. *Acta Paediatr.* 42, 223–231.

Wieser, H. (2000). Comparative investigations of gluten proteins from different wheat species; I. Qualitative and quantitative compositions of gluten protein types. *Eur. Food Res. Techn.* 211, 262–268.

Wood, P. J. (1986). Oat β-glucan: Structure, location and properties. In: Webster, F. H. ed. *Oats: Chemistry and Technology.* St. Paul, MN: American Association of Cereal Chemists, pp. 121–152.

Hydrocolloids

9

James N. BeMiller

Introduction

This chapter on hydrocolloids, also known as food gums, is neither an examination of the properties of gluten and its component proteins (gliadin and glutenin), a detailed explanation of how properties of certain hydrocolloids and mixtures of hydrocolloids might mimic the functionalities of gluten, nor a description of how to use hydrocolloids in preparing gluten-free products. It does describe certain properties, viz. network formation, film formation, thickening, and water-holding capacity, of certain hydrocolloids that might be useful in formulating gluten-free products. More detailed descriptions of the hydrocolloids covered may be found in BeMiller (2007) (a simple overview) and Whistler and BeMiller (1993), Stephen (1995), Imeson (1997), Walter (1998), Dumitriu (1998), Phillips and Williams (2000), and Hoefler (2004) (more extensive presentations).

One way to classify hydrocolloids that is useful in the context of the subject of this book is according to whether or not they can form gels (all hydrocolloids will bind and hold water to different extents, and all will viscosify aqueous systems). Gliadin, glutenin, and hydrocolloids are all biopolymers. Gliadin and glutenin are proteins, whereas most hydrocolloids are polysaccharides. Gelatin, a protein, is often classified as a hydrocolloid, but is not discussed in this chapter. This chapter briefly discusses polysaccharide hydrocolloids as gel formers, thickeners, and water holders in general terms.

Hydrocolloids that can effect gelation

The reason that gel formation might be important in this context is that gelation involves formation of a three-dimensional network. Hydrocolloid gels are viscoelastic, as is dough with developed gluten. However, there are differences. The specific

Gluten-Free Cereal Products and Beverages
ISBN: 9780123737397

rheological properties (viscoelasticity) imparted by gluten are different from those imparted by hydrocolloids, and the rheological properties of gels formed by different hydrocolloids differ from each other. Gelation effected by hydrocolloids (i.e. network formation by hydrocolloids) involves close association of polymer molecules or bundles of polymer molecules that are held together by hydrogen bonds or cross-linking of anionic molecules by multivalent cations (almost always either calcium ions or protein molecules) over portions of their lengths. These associations are called junction zones. The end or ends of molecules or bundles of molecules extending outside the junction zone form junction zones with other molecules or bundles of molecules in another area, forming a three-dimensional network entrapping water (i.e. forming a sponge-like structure). The network formed by any hydrocolloid is, as far as is known, fibrillar in nature. The gluten network that provides the strength to entrap gas bubbles and provide a proper cell structure involves both film/sheet and fibril formation. Certain hydrocolloids will form films, but they are water-soluble films and not known to be involved in the network formation involved in gelation. The bonding involved in hydrocolloid fibril formation involves hydrogen bonding, cationic cross-linking, and in a few cases, hydrophobic interactions. Gluten development involves formation of covalent disulfide bonds and weaker secondary interactions, such as electrostatic interactions, van der Waals interactions, hydrogen bonding, hydrophobic associations, and dipole–dipole interactions. The effects of oxidizing and reducing agents are substantial in the case of gluten development, as the polymers involved are proteins; but there is no known effect of redox reagents on hydrocolloids, which do not contain sulfhydryl groups. Gluten formation is also much more sensitive to the presence of certain anions than are hydrocolloids. However, different gels with different properties, such as modulus, elasticity, hardness (strength), brittleness, cohesiveness, and adhesiveness, can be formed with different hydrocolloids and combinations of hydrocolloids. Extensibility is not something normally considered with hydrocolloid gels.

The hydrocolloids and hydrocolloid systems (Table 9.1) not only have different means of gelation, but also the properties of the gels formed from them (in aqueous systems) (Table 9.2) can vary and thus affect the properties of gluten-free products.

The majority of polysaccharide gels are thermoreversible (i.e. meltable). Thermally reversible gels may not sufficiently retain the network in a dough being heated to hold the gas bubbles and build a proper open cellular (i.e. crumb) structure. Most junction zones (gels) do not reform after being disrupted by shear. Rather the molecules must be redissolved (usually by heating) and then allowed to come partly out of solution (usually by cooling).

How hydrocolloids and their ability to form three-dimensional polymer networks can be utilized to make gluten-free products is not established, but each hydrocolloid family is presented below in a cursory fashion to aid in choosing the one or ones needed. Starches and modified food starches will also form gels but are not classified as hydrocolloids and, therefore, not discussed in this chapter.

Table 9.1 Hydrocolloids that can form gels that might be effective in gluten-free products

Hydrocolloid	Gel conditions
Agar and agarose	Upon cooling of hot solutions
Alginates (algins)	Upon acidification
	With calcium ions
Kappa-type carrageenans	With potassium ions
	With kappa-casein
	With locust bean gum
	With xanthan
Iota-type carrageenans	With calcium ions
Curdlan	Upon heating of warm solutions
Gellans	With any cation
Locust bean gum (LBG)	With kappa-type carrageenans
	With xanthan
Hydroxypropylcelluloses (HPC)	Upon heating of cold solutions
Hydroxypropylmethylcelluloses (HPMC)	Upon heating of cold solutions
Methylcelluloses (MC)	Upon heating of cold solutions
Low-methoxyl pectins (LM-pectins)	With calcium ions
Xanthan	With agarose
	With kappa-type carrageenans
	With locust bean gum

Table 9.2 Characteristics of gels made with hydrocolloid

Characteristic	Types
Texture	Brittle, elastic, plastic, rubbery, tough
Gel strength	Rigid, firm, soft, mushy, spreadable, pourable
Degree of syneresis (an indication of the ability of junction zones to continue to grow after formation)	
Reversibility	Reversible via heating, cooling, or shear
	Irreversible

Thickening and water-binding properties of hydrocolloids

All hydrocolloids can thicken aqueous systems. That includes all those listed above as gel formers, for all are soluble under conditions at which gel formation does not occur. For example, sodium alginate is a thickener of aqueous systems and will not form gels until calcium or hydrogen ions are added. Even the calcium salt form will dissolve if the gel is heated to a sufficient temperature; a gel will reform when the hot solution of calcium alginate is cooled. Also, solutions of xanthan or locust bean gum by themselves will not gel under any condition, but the combination will form firm gels. And solutions of methylcelluloses (MC) and hydroxypropylmethylcelluloses (HPMC) must be heated for reversible gelation to occur. Finally, none will form gels unless at a sufficient concentration.

Not all hydrocolloids are listed in Table 9.1. Some not listed because they are not involved in gel formation and, therefore, thicken aqueous systems without gel formation, might nevertheless be useful in gluten-free products. Such hydrocolloids include carboxymethylcelluloses (CMC), guar gum, and propylene glycol alginates (PGA).

Most hydrocolloids are available in a variety of viscosity grades (viscosity grades represent the viscosity produced when the specific product is dissolved in water at a given concentration). For some, the differences between viscosity grades can be such that the highest viscosity grade produces a solution viscosity that is more than 10 000 times the solution viscosity of the lowest viscosity grade at the same concentration. Higher viscosity grades are used when thickening is the goal. Lower viscosity grades are used when high solid concentrations without high viscosity are desired; an example would be when the hydrocolloid is to be used for film formation or binding. A lower viscosity grade of a gum may produce a firmer gel than a higher viscosity grade of the same gum.

Aqueous systems thickened with hydrocolloids exhibit different rheologies (flow characteristics). Many exhibit shear thinning to some extent. Shear thinning is a reduction in viscosity upon exerting an applied force such as mixing, pumping, chewing, swallowing, etc. There are two kinds of shear thinning rheology: pseudoplasticity and thixotropic. Pseudoplastic flow is instantaneous shear thinning, i.e. as a force is applied the viscosity of the solution/system is reduced instantaneously in proportion to the force applied and increases instantaneously when the force is partly or entirely removed, the resulting viscosity being a function of the remaining force. Solutions/systems with thixotropic rheology thin when a force is applied and thicken when that force is removed or reduced in a time-dependent manner, i.e., it is not instantaneous; rather there is a time lag that can vary from a second or less to hours. These solutions at rest are often weak gels, and as a force is applied then removed, the solutions/systems undergo gel → sol → gel transitions, taking some time to undergo each transition. Other ways in which hydrocolloid solutions/systems differ from each other are in the effects of temperature, pH, and salts on them.

Solutions of all hydrocolloids (save one, viz. xanthan) thin upon heating between 0°C and 100°C. Solutions of MC, HMPC, hydroxypropylcellulose (HPC), and curdlan will gel before 100°C is reached. Hydrocolloids do not denature as do proteins, so while there will be changes in chain mobility and conformation as the temperature increases, the processes are reversible. There are also differences in the effects of salts and pH on hydrocolloid solutions/systems, with the effects being greater on ionic hydrocolloids than on neutral hydrocolloids. As already mentioned, lowering the pH or adding certain cations to solutions of certain anionic hydrocolloids may cause gelation.

Hydrocolloids differ from one another in ease of dissolution. All will bind and hold water, but they differ in their ability to act as humectants. Some can bind as much as 100 times their weight of water. They have been used to keep products moist (especially low-fat bakery products) and to reduce water migration. It should also be noted that using a hydrocolloid as a dough viscosifier can result in a gummy product.

Specific hydrocolloids

Before considering how the properties of each of the hydrocolloids might be beneficial in formulating gluten-free products, it should be noted that not all hydrocolloids or viscosity grades of a given hydrocolloid are used to viscosify or gel aqueous solutions/systems. They are often used to stabilize emulsions, suspensions, foams, and proteins, to inhibit ice and sugar crystal formation and growth, to inhibit syneresis, to encapsulate, as processing aids, and for other reasons, including the already mentioned ability to bind/hold water and to form films. Hydrocolloids differ considerably in their abilities to impart any specific functionality. The anionic hydrocolloids, i.e. those with negative charges, will interact with proteins. The extent and outcome of the interaction is a function of the specific hydrocolloid and the specific protein, including its isoelectric pH (pI) value. Each of the hydrocolloids thought (by the author) to have potential benefit in the formulation of gluten-free products are presented below in alphabetical order.

Agar

Agar is composed of two components—agarose (agaran) and agaropectin. Agarose is the gel-forming component. Usually, agar can only be dissolved in water at 100°C or higher, but preparations that hydrate and dissolve at about 80°C are available. Agar is rather expensive and is little used in food products.

Alginates (algins)

Alginates (or algins) are anionic polymers. They are anionic because each monomer unit in them is an uronic acid unit (either D-mannuronic acid or L-guluronic acid), and uronic acids have carboxyl (-COOH) groups as part of their structure. The carboxyl groups can be in the free acid or any salt form. The most common form is the sodium salt (-COO$^-$Na$^+$) form, followed by the ammonium salt (-COO$^-$NH$_4$$^+$) form.

The predominant characteristic of alginates related to their use in foods is their ability to form gels upon addition of calcium ions. There are three ways this is done: (1) A solution of a soluble calcium salt, such as calcium chloride, can be added to a solution or system containing an alginate, such as sodium alginate. (2) An acidic solution containing a sequestrant is added to a suspension of an insoluble calcium salt, such as dicalcium phosphate or calcium sulfate dehydrate, in a solution of sodium or ammonium alginate. Slow release of calcium ions from the insoluble calcium salt effects gelation. (3) Gelation also occurs when a mixture of an alginate, an insoluble calcium salt, a sequestrant, and a slightly soluble acid is heated and then cooled.

Alginate gels in general are rather heat stable. Lowering the pH of alginate solutions to values of 3 or less will either effect gelation or precipitation, depending on how the acid is added. Alginate solutions themselves are slightly pseudoplastic. Addition of small concentrations of calcium ions makes the solutions thixotropic. Addition of more calcium ions converts the thixotopic solutions to permanent gels. As anionic polymers, alginates can interact with proteins.

In propylene glycol alginates (PGA), 50–85% of the carboxyl groups are esterified with propylene glycol. Solutions of PGA are thixotropic and much less sensitive to acids and calcium ions than are sodium or ammonium alginates. The propylene glycol groups give the molecules a degree of interfacial activity (i.e. foam- and emulsion-stabilizing properties).

Alginates from different sources have different structures (proportions of the two uronic acids making up their structures) and, therefore, different properties, such as their ability to form gels and the type of gel formed.

Carboxymethylcelluloses

The various products that comprise the carboxymethylcellulose (CMC) family of products contain the carboxymethyl ether group in the sodium salt form ($-O-CH_2-COO^-Na^+$) and, therefore, are anionic polymers. They hydrate rapidly, are thickeners but not gel formers, will form water-soluble films, and are compatible with a wide variety of other ingredients. They interact with proteins like soy protein and keep them soluble at their pI, at which they would otherwise precipitate. They are good at water-holding. Most CMC solutions are pseudoplastic, but CMC types are made that make solutions that are thixotropic. Use of CMC in the preparation of low-calorie, yeast-leavened, wheat-free baked products was claimed (Glicksman *et al.*, 1972), but it seems not to have received much attention since the 1972 claim. If it is to be used, it would be important to select the proper type (degree of substitution with carboxymethyl groups, viscosity grade, pseudoplastic or thixotropic type) from among the several types available.

Carrageenans

The variety of products that fall within the family called carrageenans is extensive. There are three basic types known as kappa-, iota- and lambda-type carrageenans. These three types are blended and standardized. Ions, such as potassium ions, may be added. Properties that are controlled by different preparation processes include hydration rate, gel strength, protein interaction, and solution viscosity. The number of products that can be made is almost unlimited (a single supplier may offer more than 100). Only the three major base types are presented here.

Both kappa-type and iota-type carrageenans will form gels, and the sodium salt forms of both are soluble in cold water, but not in cold milk; they are soluble in both hot water and hot milk. The potassium salt form of neither is soluble in cold water. With potassium ions, kappa-type carrageenans form gels that are brittle, undergo synersis, and are not freeze–thaw stable. Iota-type carrageenans will form gels with calcium ions that are soft and elastic, do not undergo syneresis, and are freeze–thaw stable. All salts of lambda-type carrageenans are soluble and its solutions do not gel. Kappa-type carrageenans interact synergistically with kappa-casein and locust bean gum. As a result of their interaction with kappa-casein, they will thicken or gel dairy products. The synergistic interaction with locust bean gum forms firm, rigid, brittle, syneresing gels. The properties of such gels, such as firmness, can be modified (softened) by addition of other ingredients such as guar gum. As anionic polymers, they also interact with other proteins in different ways.

In a study examining the making of gluten-free products using hydrocolloids (one of which was "carrageenan" [type unknown]), it was found that bread quality decreased at hydrocolloid concentrations of greater than 1% (Dluzewska *et al.*, 2001).

Curdlan

Curdlan is a little used, rather expensive, neutral hydrocolloid. It is insoluble in cold water. When aqueous dispersions of curdlan are heated, it first dissolves. When the solution reaches about 55–65°C, then is cooled, a thermoreversible gel forms. When the thermoreversible gel is heated to a temperature above about 80°C, an irreversible gel forms. Heating to higher temperatures results in stronger and stronger gels. Transition temperatures are determined by concentration.

Gellans

The products that make up the family of gellans are known commercially as types of gellan gum. Gels can be made from gellans at concentrations as low as 0.05%; however, firm gels require concentrations of about 0.2%. At concentrations less than about 0.05%, thickening occurs (this is true of all hydrocolloids, i.e., at concentrations below that required to form gels, they thicken aqueous systems). Dispersions of gellans must be heated to 75–85°C (depending on the hardness of the water) to dissolve the gum, which is required before gel formation can take place. Ions increase the dissolution temperature. Sugars reduce gel strength. There are two general types of gellans, viz. native gellan (high-acyl types) and low-acyl (partially deacylated) types that can be blended to form intermediate types. Native (high-acyl) types form thermally irreversible (non-melting) gels with potassium and calcium ions and thermally reversible (meltable) gels with sodium ions. The gels are soft, very elastic, and non-brittle. Gels made from low-acyl types are hard, non-elastic, brittle, and always thermally reversible.

Guar gum

Guar gum is the ground endosperm of the seeds of a legume. The gum powder contains 75–85% polysaccharide (the actual hydrocolloid), 5–6% protein, 8–14% moisture, and other components. The polysaccharide is a neutral polysaccharide and, as such, its solutions are little affected by ions or pH. Different mesh sizes that hydrate and build viscosity at different rates are available. It can produce viscosities that are among the highest of all hydrocolloids at a given concentration and exhibits synergism with agar, kappa-type carrageenan, and xanthan that results in even higher viscosities.

Guar gum was investigated as a water-binding ingredient in gluten-free breads and found to be inferior to other galactomannans (Jud and Bruemmer, 1990). A gluten-free bread made from potato starch and/or rice flour, guar gum, and other ingredients was claimed (Chatelard, 1998). In a study of making gluten-free bread with different hydrocolloids (one of which was guar gum) it was found that bread quality decreased at hydrocolloid concentrations of more than 1% (Dluzewska, 2001).

Gum arabics

Among the hydrocolloids, gum arabic preparations have unique properties. Two of these unique properties are that they produce only low viscosity at high concentrations and that the rheology of gum arabic solutions is not that of shear-thinning rheology over a wide concentration range. Because high concentrations are required to give even modest viscosity, it is not employed as a thickener. However, when a proper amount (as determined by response surface methodology) of gum arabic was included in a formulation for gluten-free, pocket-type flat breads, good attributes were obtained; greater amounts resulted in more cohesive products (Toufeili, 1994). Modified food starch products that mimic the characteristics of gum arabic are available commercially. These products are partially depolymerized starch 1-octenylsuccinate esters.

Hydroxypropylcelluloses

Products within this family of hydrocolloids, like methylcellulose (MC) and hydroxypropylmethylcellulose (HPMC), are soluble in cold water and insoluble in hot water. When solutions of the three types of products are heated, transitions occur within a certain temperature range that are a function of the type of product and the system it is in. MC and HPMC solutions gel as the temperature reaches and exceeds this temperature range. Hydroxypropylcellulose (HPC) will usually precipitate (i.e. become insoluble) as its solutions are. Its food applications are limited, but it does stabilize foams and is used in whipped products.

Hydroxypropylmethylcelluloses

Products in this family of hydrocolloids are often investigated as ingredients in the preparation of gluten-free products. Hydroxypropylmethylcelluloses (HPMCs) are soluble in cold water and undergo reversible thermal gelation, i.e. solutions gel when heated to temperatures above the transition temperature and the gels return to the solution state when cooled. Their solutions exhibit pseudoplastic rheology. They have some interfacial activity, and they can form films.

Use of HPMCs in the preparation of low-calorie, yeast-leavened, wheat-flour-free baked products was claimed (Glicksman et al., 1972). Thermal transitions of gluten-free doughs as affected by HPMCs have been studied (Kobylanski et al.,

2004). It was reported that a formulation containing rice flour, egg and milk proteins, xanthan, and HPMCs created a "bicontinuous matrix with starch fragments, similar to gluten" (Ahlborn *et al.*, 2005). Using response surface methodology, an optimal formulation containing 2.2% HPMC and 79% wheat starch was identified, but wheat starch is not necessarily free of gluten. It was reported that as crumb firmness of the product increased, crust firmness and crumb moisture decreased over 7 days storage (McCarthy *et al.*, 2005). A gluten-free bread prepared with corn starch and a combination of xanthan and HPMC was claimed (Huang *et al.*, 2006).

Konjac glucomannan

Particles in preparations of konjac glucomannan hydrate rapidly, absorbing up to 200 times their weight in water, depending on purity. The glucomannan forms viscous, pseudoplastic dispersions. As with other gums, increasing the temperature and degree of shear during hydration decreases the time required to achieve full viscosity/hydration.

Thermally reversible gels are produced by cooling hot solutions of combinations of konjac glucomannan and kappa-carrageenan or xanthan. Konjac glucomannan is about twice as effective as locust bean gum in its synergistic interaction with kappa-carrageenan. The konjac glucomannan–kappa-carrageenan combination produces strong, elastic, thermally reversible gels, the texture of which can be varied by varying the ratio of the two hydrocolloids and their total concentration. Such gels are more heat stable than are pure kappa-type carrageenan gels.

A combination of xanthan and konjac glucomannan can result in a solution viscosity which is as much as three times the solution viscosity of either gum used alone at the same total concentration. Although neither native konjac glucomannan nor xanthan will form a gel when used alone, hot (85°C) solutions of blends of the two gums (at the optimum ratio) will produce elastic, strong, thermally reversible gels upon cooling. As with konjac glucomannan–kappa-carrageenan gels, heat-stable gels form when konjac glucomannan–xanthan mixtures are heated to higher temperatures. Again, konjac glucomannan is roughly twice as effective in its synergism with xanthan than is locust bean gum.

Konjac glucomannan also interacts with starches and modified starches to produce an increase in viscosity.

Acetyl groups on native konjac glucomannan molecules prevent them from associating with themselves and forming gels. However, when the native gum is de-esterified, thermally stable gels can be formed. Traditionally, the acetyl groups are removed by treatment with calcium hydroxide (lime). However, any food-grade, weak base, such as potassium carbonate, can be used to raise the pH of a konjac glucomannan dispersion to the required pH. Heating the dispersion effects de-esterification. Cooling the deacetylated gum solution produces a gel that will withstand boiling, even retort, temperatures without melting. The gels are strong and somewhat elastic.

Konjac glucomannan is not sensitive to salt. It has the ability to form heat-stable, flexible, high-strength, protective coatings and films.

An attempt to produce gluten-free bread using a xanthan—konjac glucomannan combination was unsuccessful in terms of mimicking the keeping quality of wheat flour bread (Moore *et al.*, 2004).

Locust bean and tara gums

Locust bean gum (LBG), also known as carob gum, is structurally similar to guar gum. It, like guar gum, is a flour made from the endosperm of the seed of a legume. The two gums, however, have important property differences. Most hydrocolloids hydrate faster and produce more viscous solutions if dispersions of them in room-temperature water are heated, then cooled; but LBG is only slightly soluble in room-temperature water. Heating a suspensions to about 85°C is required for good dissolution. Another difference is that, while solutions of LBG by itself do not gel, hot solutions of LBG in combination with agar, kappa-carrageenan, and xanthan will gel when cooled below the gelling temperature. In a study of making gluten-free bread with different hydrocolloids (one of which was LBG), it was found that bread quality decreased at hydrocolloid concentrations of more than 1% (Chatelard, 1998).

Tara gum is similar to guar and locust bean gums with properties somewhat in between them. It is reported that, in the manufacture of gluten-free breads, "a mixture of carob and tara flours (3:1) with a particle size of 75–100 μm gave products with good pore formation and crumb properties, yet without any detrimental effects on taste" (Jud and Bruemmer, 1990).

Methylcelluloses

True methylcellulose (MC) products contain only methyl ether groups rather than both methyl and hydroxypropyl ether groups as does HPMC, although both MC and HPMC are frequently not distinguished from each other and lumped together as methylcellulose products. The properties of MC products are similar to those of HPMC products. It was reported that a formulation based on pre-gelatinized rice flour, pre-gelatinized corn starch, corn flour, methycellulose, egg albumen, and gum arabic optimized by response surface methodology produced gluten-free, pocket-type flat breads with acceptable sensory attributes (Toufeili *et al.*, 1994).

Pectins

The family of pectin products is made up of a variety of hydrocolloids, all of which contain D-galacturonic acid in the sodium salt form as the basic building block. All contain some percentage of the uronic acid units in the form of a methyl ester ($-COOCH_3$); so while all pectin preparations are somewhat anionic, the degree to which they are anionic varies. A variety of pectins are available. Perhaps, the most interesting in the context of gluten-free products is the group of products known as low-methoxyl pectins (LM pectins). Solutions of LM pectins will gel when calcium ions are added to them in a manner similar to gelation of alginate solutions. A special type of LM pectins is the amidated LM pectins which contain $-COO^-Na^+$, $-COOH_3$, and $-CONH_2$ groups. Amidated LM pectins are more sensitive to calcium ions than are conventional LM pectins. As polyanions, proper types of pectin will stabilize certain proteins against heat denaturation or isoelectric precipitation. Stabilization of milk proteins by a proper pectin preparation upon lowering the pH of milk-based products is an example.

Xanthans

Xanthans, known commercially as xanthan gums, form high-viscosity, pseudoplastic solutions that are unaffected by changes in temperature, pH, or salt concentration. Xanthan is a non-gelling hydrocolloid, but it does form gels when combined with agarose, kappa-type carrageenans, konjac glucomannan, or LBG. The gel formed with LBG is rather elastic. At least 10 different categories of xanthans may be available from a single supplier. These include different particle sizes, different viscosity grades, easily dispersible types, rapidly hydrating types, delayed hydrating types, types with reduced pseudoplasticity, and other types. Xanthans from different producers are similar, but differ slightly from each other because of differences in the strain of organism used to produce them and different growth conditions.

A gluten-free bread made with potato starch and/or rice flour plus xanthan and other ingredients is claimed (Chatelard, 1998). In a study of making gluten-free products using hydrocolloids (one of which was xanthan), it was found that bread quality decreased at hydrocolloid concentrations of greater than 1% (Dluzewska et al., 2001). When both xanthan and xanthan + konjac glucomannan were used with corn starch and brown rice, soy, and buckwheat flours to make gluten-free breads, the products became brittle after 2 days of storage (Kobylanski et al., 2004). It was concluded that a continuous protein phase is necessary for sufficient shelf life of gluten-free bread. It was reported that a formulation containing rice flour, egg and milk proteins, xanthan, and HPMC created a "bicontinuous matrix with starch fragments, similar to gluten" (Ahlborn et al., 2005). A formulation used to study the efficacy of transglutaminase in producing a protein network in gluten-free bread contained xanthan (Moore et al., 2006). A gluten-free bread prepared with corn starch and a combination of xanthan and HPMC was claimed (Huang et al., 2006).

Conclusions

To date, hydrocolloids have shown only limited promise for the production of gluten-free bakery products. However, most hydrocolloids are available in a variety (sometimes a large variety) of products, all of which have the same basic name. Each product is designed to have specific properties and to impart specific functionalities in specific products. The different types can be very different from one another; so while, at this time, no hydrocolloid product may be designed specifically for gluten-free products, it is important to choose the type or combinations of types that most closely provides the quality and processing attributes required. This also applies to those combinations of hydrocolloids that are often used in the preparation of food products. For this reason, a complete and thorough investigation of the efficacy of hydrocolloids in preparing gluten-free products is probably yet to be done.

To evaluate the potential of hydrocolloids, the proper hydrocolloid or mixture of hydrocolloids, together with any other indegredient(s), such as calcium ions, required to bring out the desired attribute, need to be evaluated.

References

Ahlborn, G. J., Pike, O. A., Hendrix, S. B., Hess, W. H., and Huber, C. S. (2005). Sensory mechanical and microscopic evaluation of staling in low-protein and gluten-free breads. *Cereal Chem.* 82, 328–335.

BeMiller, J. N. (2007). *Carbohydrate Chemistry for Food Scientists*, 2nd edn. St. Paul, Minnesota: AACC International.

Chatelard, P. (1998). Gluten-free bread and its manufacture. FR Patent 2765076 A1; *Chem. Abstr.* 130, 196095 (1999).

Dluzewska, E., Marciniak, K., and Dojczew, D. (2001). Gluten-free bread concentrates with added selected hydrocolloids. *Zywnosc* 8, 57–67; *Chem. Abstr.* 135, 166234 (2001).

Dumitriu, S., ed. (1998). *Polysaccharides*. New York: Marcel Dekker.

Glicksman, M., Farkas, E. H., and Carter, S. (1972). Low-calorie yeast-leavened baked products. US Patent 3,676,150; *Chem. Abstr.* 77, 138565 (1972).

Hoefler, A. C. (2004). *Hydrocolloids*. St. Paul, Minnesota: American Association of Cereal Chemists.

Huang, W., Yang, X., and Li, X. (2006). Manufacture of gluten free bread from corn starch. CN Patent 1751580; *Chem. Abstr.* 144, 449805 (2006).

Imeson, A., ed. (1997). *Thickening and Gelling Agents for Food*, 2nd edn. London: Blackie Academic and Professional.

Jud, B. and Bruemmer, J. M. (1990). Manufacture of gluten-free breads with special galactomannans. *Getreide Mehl Brot* 44, 178–183.

Kobylanski, J. R., Perez, O. E., and Pilosof, A. M. R. (2004). Thermal transitions of gluten-free doughs as affected by water, egg white and hydroxypropylmethylcellulose. *Thermochim. Acta* 411, 81–89.

McCarthy, D. F., Gallagher, E., Gormley, T. R., Schober, T. J., and Arendt, E. K. (2005). Application of response surface methodology in the development of gluten-free bread. *Cereal Chem.* 82, 609–615.

Moore, M. M., Schober, T. J., Dockery, P., and Arendt, E. K. (2004). Textural comparisons of gluten-free and wheat-based doughs, batters, and breads. *Cereal Chem.* 81, 567–575.

Moore, M. M., Heinbokel, M., Dockery, P., Ulmer, H. M., and Arendt, E. K. (2006). Network formation in gluten-free bread with application of transaminase. *Cereal Chem.* 83, 28–36.

Phillips, G. O. and Williams, P. A., eds (2000). *Handbook of Hydrocolloids*. Boca Ratan, Florida: CRC Press.

Stephen, A. M., ed. (1995). *Food Polysaccharides and Their Applications*. New York: Marcel Dekker.

Toufeili, I., Dagher, S., Shadarevian, S., Noureddine, A., Arakibi, M., and Farran, M. T. (1994). Formulation of gluten-free pocket-type flat breads: optimization of methycellulose, gum arabic, and egg albumen levels by response surface methodology. *Cereal Chem.* 71, 594–601.

Walter, R. H., ed. (1998). *Polysaccharide Association Structures in Food.* New York: Marcel Dekker.

Whistler, R. L. and BeMiller, J. N., eds (1993). *Industrial Gums*, 3rd edn. San Diego, CA: Academic Press.

Dairy-based ingredients 10

Constantinos E. Stathopoulos

Dairy ingredients have long been used in the cereal-processing industry. They have mainly been added for their good functional properties, ease of production, and the nutritional fortification they provide to the final product. These ingredients also find applications in gluten-free products, where substitution of the structural protein complex of gluten is required in order to render the product suitable for consumption by people with celiac disease. An overview of the production of such dairy ingredients is provided here with their properties described, along with their applications and the problems encountered in the production of gluten-free bread.

Introduction

Over the years a number of dairy ingredients have been used in the food industry. Applications have been extensive and variable, especially for dairy proteins. For caseins and caseinates these applications include baked products, cheese and imitation cheese manufacture, coffee creamers, ice creams, pasta products, cultured milk products, whipped toppings, milk-type beverages, non-milk-type beverages, confectionary products, spreads, meat products, and others (Southward, 1989; Mulvihill, 1992; Damodaran, 1997a; Fox and McSweeney, 1998; Mulvihill and Ennis, 2003). For whey protein products, applications include beverages, confectionary, desserts and dressings, meat products, dairy products, and novel dairy products (De Wit, 1989; Mulvihill, 1992; Damodaran, 1997a; Mulvihill and Ennis, 2003). Lactose finds a number of applications in various dairy products, such as sweetened condensed milk, frozen milk products, milk and whey powders, confectionary products, baby foods, and also as flavor enhancer and an anti-caking agent (Morrisey, 1985; Fox and McSweeney, 1998).

Gluten-Free Cereal Products and Beverages
ISBN: 9780123737397

Dairy ingredients are extensively used for their functionality, nutritional value, and ease of production. The most widely used dairy ingredients in gluten-free bread formulations are caseinates (Lazaridou *et al.*, 2007), skim milk powder (Moore *et al.*, 2004; McCarthy *et al.*, 2005), dry milk (Sanchez *et al.*, 2004), whey protein concentrate (Sanchez *et al.*, 2004), and milk protein isolate (Gallagher *et al.*, 2003a).

Production and properties of dairy ingredients: an overview

Production

Caseins

The starting material for the production of caseins, caseinates, and whey proteins is skim milk. The use of skim milk ensures that the fat content of the casein is low enough to minimize flavor defects arising from deterioration of lipids in the dried casein products. Following destabilization the insoluble casein is separated from the soluble whey proteins, lactose, and salts, washed to remove residual soluble solids, and then dried (Mulvihill, 1992; Maubois and Ollivier, 1997). Caseins can be produced in a number of ways, however industrially they are produced either by isoelectric precipitation or proteolytic coagulation. Detailed production methods have been reviewed (Mulvihill, 1989, 1992; Maubois and Ollivier, 1997; Fox and McSweeney, 1998; Mulvihill and Ennis, 2003) and are shown schematically in Figure 10.1. During isoelectric precipitation the pH of the skim milk is reduced to the isoelectric point of casein. The pH decrease can be achieved by addition of a culture of lactic acid bacteria (converting some of the lactose of the milk to lactic acid), or by addition of a dilute mineral or organic acid (producing lactic or acid casein respectively). The pH of skim milk can also be reduced to the isoelectric point of casein by mixing all or part of the milk with an ion-exchange resin at low temperature (Mulvihill, 1989; Maubois and Ollivier, 1997; Lucey and Singh, 2003; Mulvihill and Ennis, 2003).

In the manufacture of acid casein, precipitation is accomplished by under-pressure spraying dilute mineral acids (usually HCl) into preheated milk (25–30°C) flowing in the opposite direction until a pH of 4.6 is reached. Subsequently, steam is injected to heat the acidified milk to the precipitation temperature of about 50°C (Mulvihill, 1992; Maubois and Ollivier, 1997; Mulvihill and Ennis, 2003). During production of lactic casein, pasteurized skim milk is inoculated with one or more defined starters and incubated for 14–16 hours at 22–26°C. Under these conditions, the starters slowly ferment the lactose to lactic acid. As the pH of milk is falling towards the isoelectric point of casein, a casein gel network (the coagulum) is formed, with good water-holding capacity (Mulvihill, 1992; Fox and McSweeney, 1998). The coagulum is then pumped from the coagulation vats and cooked by direct steam injection.

The pH of skim milk can also be reduced to the isoelectric point of casein by mixing skim milk at low temperature (<10°C) with a cation-exchange resin in the hydrogen form in a reaction column. Cations in the milk are replaced by H$^+$ to give a final pH of about 2.2. The deionized acidified milk is then mixed with untreated milk

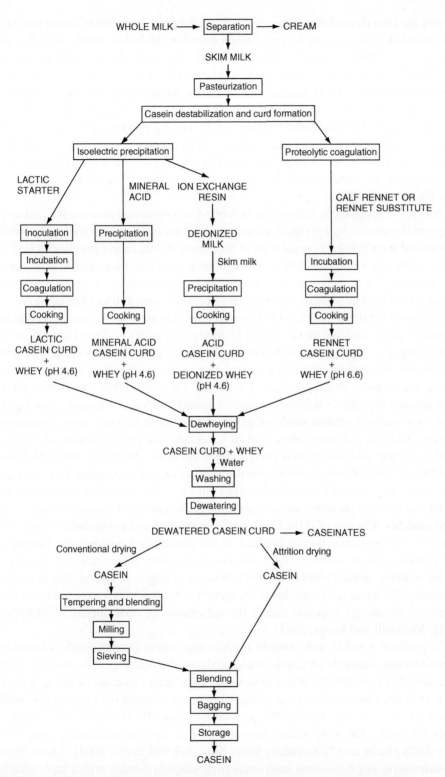

Figure 10.1 Schematic representation of casein production. Reproduced with permission from Mulvihill (1989).

to give the final desired precipitation pH of about 4.6. The mixture is finally heated to the coagulation temperature by direct steam injection (Mulvihill, 1989, 1992; Fox and McSweeney, 1998; Mulvihill and Ennis, 2003). This method is reported to increase the yield by up to 3.5%, giving resulting whey with a lower salt content. However, because of the difficulty of maintaining the ion exchanger under satisfactory bacterial conditions and because of the large volumes of effluents generated, this method is not widely used (Maubois and Ollivier, 1997). During proteolytic coagulation milk is treated with selected proteinases (rennets). The coagulated casein is recovered as rennet casein. However, during this procedure, the κ-casein is hydrolyzed and thus the properties of rennet casein differ fundamentally from those of acid casein (Fox and McSweeney, 1998).

Proteolytic coagulation is therefore described as a two-stage process: the first stage involves the specific hydrolysis of κ-casein to *para*-κ-casein and macropeptides; while the second stage involves coagulation of the rennet-altered casein micelles by Ca^{2+} at temperatures above 20°C. When such coagulum is produced from skim milk, it can be further processed to yield rennet casein (following similar steps as for the production of lactic casein) (Mulvihill, 1989, 1992; Hyslop, 2003; Mulvihill and Ennis, 2003).

Following the destabilization of the casein, the next steps in the casein production are dewheying, washing, dewatering, and drying (Figure 10.1). The efficiency of the dewheying step is of great importance in determining the whey volume recovered for further processing, the efficiency of the washing operation and the quality of the final casein produced (Mulvihill, 1989, 1992; Mulvihill and Ennis, 2003). The equipment used to achieve separation usually includes vibratory, moving, or stationary inclined screens made of nylon or fine mesh stainless steel, or inclined screens made of polyester fabric laid in a cascade-like profile which subjects the curd to turning and rolling as it travels down the slope (Mulvihill, 1989; Mulvihill and Ennis, 2003). Residual whey constituents (lactose, whey proteins, and salts) and free acids are removed from the dewheyed curd to a limited extent by washing the surface of the curd particles; and to a much larger extent by diffusion from within the curd particles. The rate of diffusion depends on the size and permeability of the curd particles, on the concentration gradient of the constituents between the interior of the particles and the washing water, and on the amount, temperature and movement of the washing water (Mulvihill, 1992). When washing is complete, casein curd is mechanically dewatered to minimize the quantity of water to be evaporated, and thus minimize the energy required during the subsequent thermal operation (Mulvihill, 1992; Mulvihill and Ennis, 2003).

To produce a stable and storable product that meets internationally recognized compositional standards for edible grade products, the casein curd is dried to <12% moisture. Traditionally the driers used were of the semi-fluidized vibrating type. In such systems, casein curd passes along vibrating perforated steel conveyors, while warm air is forced up through the perforations, partially fluidizing the curd as it dries. Currently, the most widely used drying technique involves using pneumatic ring driers (Kelly and O'Kennedy, 1986; Mulvihill and Ennis, 2003). These dryers are effectively large, stainless steel ducts (ring-shaped) through which high-velocity heated air and moist disintegrated casein curd are circulated continuously (Mulvihill,

1992; Mulvihill and Ennis, 2003). Dried casein is relatively hot as it emerges from the drier and the moisture content of individual particles varies. Therefore, it is necessary to temper and blend the dried product to achieve a final product of uniform moisture content (Mulvihill, 1989).

A drying process known as "attrition" drying is now widely used in casein manufacture. This process is based on the principle of grinding and drying in a single operation, and it allows the production of a casein product closely resembling spray-dried casein (Kelly and O'Kennedy, 1986; Mulvihill and Ennis, 2003). The drier consists of a fast-revolving, multi-chambered rotor and a stator with serrated surface. In the drier, turbulences, vortices, and cavitation effects result in a highly efficient drying, which produces very small particles with very large surface area. These particles are simultaneously dried in a hot air stream that passes through the drier concurrently with the curd. The dried casein is very fine with an overall average particle size of about 100 μm. The particles have good wettability and dispersability because they are irregular in shape and many contain cavities created by the rapid evaporative process (Kelly and O'Kennedy, 1986; Mulvihill, 1989, 1992; Mulvihill and Ennis, 2003).

Novel methods for casein production include cryoprecipitation, precipitation with ethanol, ultrafiltration, and high-speed centrifugation (Maubois and Ollivier, 1997; Fox and McSweeney, 1998; Mulvihill and Ennis, 2003). During cryoprecipitation, milk is frozen to −10°C. The ionic strength of the liquid phase increases with a concomitant increase in $[Ca^{2+}]$. The pH drops to approximately 5.8, due to the precipitation of calcium phosphates with the release of hydrogen ions. These changes destabilize the casein micelles which precipitate when the milk is thawed. In precipitation with ethanol, the casein in milk coagulates at pH 6.6 following addition of ethanol (to reach a final concentration of about 40%). Stability decreases sharply as the pH is reduced and only 15% ethanol is required at pH 6.

Caseinates

Acid caseins are insoluble in water but will dissolve in alkali under suitable conditions to yield water-soluble caseinates that may be spray- or roller-dried (Kelly and O'Kennedy, 1986; Mulvihill, 1992; Mulvihill and Ennis, 2003). Sodium caseinate, the water-soluble caseinate most commonly used in food, is usually prepared by solubilizing acid casein with NaOH. Towler (1976) has suggested the following protocol for the commercial production of sodium caseinate:

1. Mince casein curd from a dewatering device (about 45% solids) and then mix with water at 40°C to give a solid content of about 25%, before passing through a colloidal mill. The curd particle size must be reduced as much as possible.
2. Sodium caseinate should have a pH in the range of 6.6–6.8. Pumping NaOH into the casein slurry as it emerges from the mill at 45°C will result in the desired final caseinate pH. As the slurry has the consistency of toothpaste, it must be efficiently mixed with the NaOH with a mixer capable of coping with the high viscosity.
3. Transfer the mixture into a vat where solubilization occurs as the mixture is agitated and heated. The slurry is then re-circulated or pumped into a second vat where solubilization is completed as the temperature of the solution is raised to

about 75°C. Use an in-line pH meter to indicate whether the correct amount of NaOH has been added, or if any further regulation is required.

4. Pump the caseinate solution to a balance tank through a heat exchanger (increasing the temperature to about 95°C). Again, an in-line pH meter should be used to control further addition of NaOH, if required.

5. Pump the solution from the balance tank to the spray drier through an in-line viscometer that regulates addition of hot water (to control viscosity) and ensure sufficient atomization of the solution in the drier.

Other caseinates, such as calcium, ammonium, potassium, or citrated caseinates can also be produced. Calcium caseinate is produced by first passing "soft" casein curd through a mixer to get evenly sized particles. The particles are then mixed with water to give about 25% total solids. The mixture is passed through a colloidal mill and the temperature is adjusted to give milled slurry at 35–40°C. This is then mixed with a metered volume of $Ca(OH)_2$ slurry to give the desired final pH. The mixture is agitated and re-circulated in a low-temperature conversion tank until conversion is complete (>10 minutes). Finally the dispersion is heated in a tubular heat exchanger to 70°C and is pumped directly to a spray drier (Roeper, 1977; Mulvihill, 1992; Mulvihill and Ennis, 2003). Ammonium caseinates can be prepared by a method similar to that used for the production of sodium caseinate, but NH_4OH or KOH is used instead of NaOH. Granular ammonium caseinate can be prepared by exposing dry acid casein to ammonia gas and then removing excess ammonia with a stream of air in a fluidized bed degassing system (Mulvihill, 1989). Citrated caseinates can also be produced in a similar fashion by using a mixture of trisodium citrate and tripotassium citrate instead of NaOH (Mulvihill, 1992; Mulvihill and Ennis, 2003).

Whey protein products

About 20% of bovine milk proteins belong to a group of proteins generally referred to as whey proteins. The whey proteins as a group are readily prepared from milk by any of the methods described for casein manufacture (Roberts, 1985; Fox and McSweeney, 1998). Whey and whey protein-enriched solutions are usually pasteurized using minimum temperature and holding times, and maintained at low temperature to minimize microbial and physico-chemical deterioration of the proteins and other whey constituents that would adversely alter functional and organoleptic properties of the resulting products (Mulvihill, 1992). The production of whey and whey protein-enriched fractions have been extensively reviewed (Hugunin, 1985; Kelly and O'Kennedy, 1986; Morr, 1989; Mulvihill, 1992; Maubois and Ollivier, 1997; Mulvihill and Ennis, 2003).

On a commercial scale, whey protein-rich products are prepared by (Fox and McSweeney, 1998):

1. Ultrafiltration/diafiltration of acid or rennet whey to remove varying amounts of lactose, and spray-drying to produce whey protein concentrates (30–80% protein).

2. Ion-exchange chromatography: proteins are adsorbed on an ion exchanger, washed free of lactose and salts, and then eluted by pH adjustment. The eluate is freed

of salts by ultrafiltration, and spray-dried to yield whey protein isolate (about 95% protein).

3. Demineralization by electrodialysis and/or ion exchange, thermal evaporation of water and crystallization of lactose.
4. Thermal denaturation, recovery of precipitated protein by filtration/centrifugation and spray-drying to yield lactalbumin, which has very low solubility and limited functionality.

Coprecipitates

Following precipitation of caseins from milk by acidification or renneting, the whey proteins remain soluble in the whey. However, these can be coprecipitated in combination with the casein by first heating milk to temperatures that denature the whey proteins, thus inducing their complexation with casein. Thereafter, the milk protein complexes are precipitated by acidification to pH 4.6 or by a combination of added $CaCl_2$ and acidification (Mulvihill, 1989, 1992; Fox and McSweeney, 1998; Mulvihill and Ennis, 2003). Products thus obtained are referred to as casein–whey coprecipitates and yields obtained can reach 92–95% of total milk protein, compared with <80% for rennet or acid caseins (Mulvihill, 1992; Mulvihill and Ennis, 2003).

Milk protein concentrate

Skim milk may also be processed directly by ultrafiltration/diafiltration to yield milk protein concentrates that contain a range of protein contents around 80% (Vetter, 1985; Maubois and Ollivier, 1997; Kelly *et al.*, 2003), and in which the casein is in a similar micellar form to that found in milk, while the whey proteins are also reported to be in their native form. These products have a relatively high ash content since protein-bound minerals are retained (Mulvihill, 1992; Kelly *et al.*, 2003).

Properties of dairy ingredients

Dairy proteins are available to food chemists in a wide variety of products. These products are readily available in dry, liquid, or condensed form, depending on the user's needs and handling capabilities. Usually, the selected dairy protein product will reflect both functionality considerations and cost efficiency (Stahel, 1983; Tow, 1985; Mannie, 1999).

Solubility

A typical solubility/pH profile of casein shows that close to its isoelectric pH the acid form of casein is completely insoluble (Roberts, 1985; Fox and McSweeney, 1998; Mulvihill and Ennis, 2003), whereas at pH values >5.5, the casein is converted to a cationic salt (Na, K, or NH_3) and is completely soluble (Mulvihill, 1992). Insolubility in the isoelectric point is clearly advantageous in the production of acid casein and is exploited in the production of two major families of dairy products (i.e. fermented milks and fresh cheeses) (Fox and McSweeney, 1998). Sodium and potassium caseinates exhibit improved solubility and functionality compared with calcium caseinate. This is probably due to larger sized and more strongly interacting calcium caseinate aggregates due to cross-linking by the divalent cations. Sodium and potassium caseinates are completely soluble in water at pH value above 5.5, while calcium caseinate forms stable colloidal dispersions rather than solutions (Kelly and O'Kennedy, 1986). It is well known that the solubility of whey proteins is a function of both pH and ionic strength (De Wit, 1989; Cayot and Lorient, 1997). Due to their native conformation, whey proteins are soluble at low ionic strength over the entire pH range required for food applications (Claypool, 1985). In the undenatured form, whey proteins exhibit little water-binding capacity (Hugunin, 1985; Cayot and Lorient, 1997). However, being globular proteins, salting out at high salt concentrations decreases their solubility (Mulvihill, 1992). The solubility of whey proteins is impaired by heat treatment above 70°C when the pH is between 4.0 and 6.5, and this has serious consequences for the foaming and emulsifying abilities of whey proteins (De Wit, 1989; Cayot and Lorient, 1997).

Solubility of coprecipitates can vary markedly under the influence of pH, agitation, mixing time and power, temperature, particle size and concentration of the casein product as well as the presence of other species (e.g. dissolved salts). The solubility of coprecipitates has generally been considered in the pH range from 6 to 10. However, it has been demonstrated that coprecipitates can also be dissolved in acids at pH 2–3 (Southward and Goldman, 1975). Generally, all grades of coprecipitates are less soluble than sodium caseinate at pH 7, the insoluble fraction consisting mainly of denatured whey protein, and representing from 4 to 15% of the coprecipitates (Southward and Goldman, 1975; Kelly and O'Kennedy, 1986).

Heat stability

Sodium, potassium, and ammonium caseinates are very heat stable. As Mulvihill (1992) states, a 3% (w/v) sodium caseinate solution at pH 7.0 can be heated at 140°C for 60 minutes without coagulation. Calcium caseinate, however, has much lower heat stability and even 1% (w/v) dispersions gel on heating at 50–60°C. Whey proteins are susceptible to denaturation at temperatures higher than 70°C. The susceptibility of whey proteins to heat denaturation is influenced by factors such as pH, Ca^{2+}, protein concentration and the presence of sugars (Mulvihill and Fox, 1989; Mulvihill, 1992; Fox and McSweeney, 1998; Singh and Havea, 2003). When heated, the bonds creating the tertiary structure of the protein globules are destroyed, unfolding of the protein molecules occur and new protein–protein interactions result. Loss of solubility is one functional change which occurs following protein denaturation (Hugunin, 1985; Cayot and Lorient, 1997).

Gelation and coagulation

Milk proteins can undergo gelation, and, in most cases, casein is the component involved (Mulvihill and Fox, 1989; Mulvihill, 1992). Gelation or coagulation occurs when milk is subjected to limited proteolysis by acid proteinases, which hydrolyze the micelle-stabilizing κ-casein, producing *para*-κ-casein-containing micelles which coagulate at the concentration of Ca^{2+} in the milk serum (Mulvihill, 1992; Lucey and

Singh, 2003). Acid gelation is exploited during manufacturing of fermented milks, acid cheeses, and acid caseins. Concentrated calcium caseinate dispersions (>15% protein) gel on heating at 50–60°C. Gelation temperature increases with increased protein concentration (to 20%) and when the pH is in the range 5.2–6.0. The gel liquefies slowly on cooling but reforms on heating. Calcium caseinate is the only milk protein system reported to have reversible thermal gelation properties (Mulvihill and Fox, 1989; Mulvihill, 1992), and κ-casein appears to be the component principally responsible. Hydrophobic bondings are considered to be involved (Mulvihill and Fox, 1989). The caseins are remarkably heat stable and do not undergo thermally induced gelation except under extremely severe conditions.

Thermal sensitivity is undesirable when a soluble whey protein-enriched product is prepared. However, this property can be exploited in the production of thermal gels from whey proteins, which have excellent thermal gelling properties (Mulvihill and Fox, 1989; Mulvihill, 1992; Singh and Havea, 2003). β-Lactoglobulin is considered to be the most important whey protein from a thermal gelation viewpoint, although bovine serum albumin and the immunoglobulins are also known to form stable gels on heating (De Wit, 1989; Mulvihill and Fox, 1989; Cayot and Lorient, 1997; Singh and Havea, 2003). Whey protein concentrates and isolates with a range of gelling properties can be produced by a selection of whey types or variations in processing conditions during manufacturing (Mulvihill and Fox, 1989; Mulvihill, 1992; Carr et al., 2003). Gelation temperature ranges from 50 to 90°C, although it has been shown that it is possible to produce whey protein concentrates which gel at 20°C (Kelly and O'Kennedy, 1986).

Hydration

Many of the functional food applications of dairy proteins depend on their ability to hydrate, and thus bind or entrap water. Under the general property "hydration" one might include solubility, dispersibility, wettability, water absorption, swelling, thickening, gelling, rheological behavior, water-holding capacity, syneresis, and dough formation (Mulvihill and Fox, 1989). The ability to bind and hold water without syneresis is critical in many foods. Although the caseins are relatively hydrophobic, they contain regions of high, medium, or low hydrophobicity and of high negative charge, high positive charge, or low net charge (Carr et al., 2003), and they bind about 2 g water/g, which is typical of proteins (Fox and McSweeney, 1998). The level of hydration of proteins is strongly influenced by the level of available water and it is common to relate the degree of hydration to the relative humidity of the environment to which the protein is exposed (Mulvihill, 1992). Hydration increases with increasing pH and is relatively independent of NaCl concentration, which is especially important in the efficacy of casein in meat-based applications (Fox and McSweeney, 1998). A plot of bound water as a function of relative humidity yields a water sorption isotherm which gives useful information on the water-binding or hydration characteristics of the proteins. Isotherms for sodium caseinate, acid casein, and micellar casein show that the hydration of the acid casein is higher than that of micellar casein, the differences being small for $a_w < 0.6$ and much greater when $a_w > 0.6$. High hydration values for sodium caseinate at high a_w indicate swelling and solubilization (Mulvihill, 1992). The water-holding capacity of sodium caseinate is higher than that of both calcium caseinate and micellar casein (Fox and McSweeney, 1998). The influence of dispersed particles on small and large deformation properties of a concentrated sodium caseinate composite has recently been examined, and the amphiphilic properties of the sodium caseinate were demonstrated (Manski *et al.*, 2007). The water absorption of coprecipitates in flour dough mixtures has been studied and generally insoluble coprecipitates have lower water absorption values than soluble ones (Southward and Goldman, 1975).

Viscosity

Because of hydration, swelling, and polymer–polymer interactions, caseinates form highly viscous solutions at concentrations higher than 15%, and even at high temperatures the viscosity of solutions containing more than 20% protein is so high that they are difficult to process (Mulvihill and Fox, 1989; Mulvihill, 1992). The effects of solution conditions on the viscosity of caseins/caseinates are crucial. The viscosity of sodium caseinate is strongly dependent on pH, with a minimum at pH 7.0 (Mulvihill, 1992). Ammonium caseinate is more viscous than sodium caseinate between pH 6 and 8.5 (Mulvihill and Fox, 1989). When compared with sodium caseinate, calcium caseinate has lower viscosity at the same concentration and pH. The reason lies in the differences between the structure of the calcium caseinate particles in the calcium caseinate dispersions and the sodium caseinate aggregates in the sodium caseinate solutions (Carr *et al.*, 2003). The various manufacturing conditions also affect the viscosity of casein/caseinates. Excessive heating of milk prior to casein manufacture, or of casein curd during drying, leads to increased viscosity of the caseinates prepared from such caseins. Precipitation at lower than normal pH values (e.g. 3.8) and especially at higher pH values (e.g. 5.05) also increases the viscosity of caseinates, while even the viscosity of roller-dried caseinate is higher than that of spray-dried caseinate (Mulvihill and Fox, 1989; Mulvihill, 1992; Carr *et al.*, 2003). Solubilized conventional coprecipitates are more viscous than sodium caseinate and their viscosity increases with increasing calcium concentration. Low calcium coprecipitates were found to have a viscosity similar to that of acid casein, while the viscosity of medium- and high-calcium coprecipitates was relatively high when the pH was above 7 (Southward and Goldman, 1975). Solutions of total milk protein have viscosities between those of sodium caseinate and conventional coprecipitates (Mulvihill, 1992). The viscosity of caseinates can be markedly reduced by treatment with disulfide-reducing and/or sulfhydryl-blocking agents. However, because of the reagents used, those caseinates would be of little interest for the food industry (Mulvihill and Fox, 1989).

Due to their compact globular shapes (Carr *et al.*, 2003), solutions of undenatured whey proteins are much less viscous than caseinate solutions. They exhibit minimum viscosity around the isoelectric point (pH 4.5) and relative to water, their viscosity decreases between 30 and 65°C, but increases thereafter owing to protein denaturation (Mulvihill and Fox, 1989; Mulvihill, 1992; Carr *et al.*, 2003). Above 85°C a further increase in viscosity was observed as a consequence of protein aggregation (De Wit, 1989). Heat-denatured whey proteins, although retaining most of their secondary structure, are linked together and, depending on the environment during denaturation and during measurement, can have a perceived hydration of over 10 g of water/g protein, compared with 0.2 g water/g protein for whey proteins that are in their native globular state (Carr *et al.*, 2003). The viscosity of whey concentrates in the range from 25 to 40% total solids depends strongly on the composition and pre-heat treatment of the whey. Additionally, in a food process, protein solutions are frequently subjected to high degrees of shear and extremes of temperature (Carr *et al.*, 2003). Shear exerted on dispersions of denatured whey proteins may break up large aggregates and this can result in a decrease in viscosity (De Wit, 1989).

Emulsifying—foaming

Soluble caseinates have greater emulsifying capacities than the more aggregated caseins, and aggregated caseins give emulsions with higher viscosity (Mulvihill and Fox, 1989; Dalgleish, 1997). In general, caseinates have superior emulsification properties than whey protein concentrates, presumably due to a more favorable balance between exposed hydrophobic and hydrophilic region which impart surfactant like properties to them (Kelly and O'Kennedy, 1986; Kelly *et al.*, 2003). Caseinates behave quite differently from whey protein concentrate during emulsification. During the formation of emulsions in caseinate protein continues to be absorbed from the bulk phase as new surface is formed, whereas with whey protein concentrate proteins already adsorbed are spread or unfolded over the newly formed surface in preference to further adsorption from the bulk solution (Mulvihill and Fox, 1989; Dalgleish, 1997). The most relevant factors affecting the emulsifying properties of whey proteins are protein concentration, protein solubility, pH, salts, presence of other solutes and temperature (De Wit, 1989; Cayot and Lorient, 1997). Homogenization of oil in whey protein concentrate systems resulted in decreasing droplet size as the whey protein concentration increased 10-fold. The ability of whey proteins to stabilize oil/water emulsions seems to be particularly affected by the ionic strength and the pH of the aqueous phase. Obviously, electrostatic interactions around the isoelectric point are responsible for protein aggregation and, as a consequence, the proteins are less flexible and therefore less prone to form a cohesive interfacial film (De Wit, 1989; Mulvihill and Fox, 1989; Cayot and Lorient, 1997). The presence of salts during the emulsification process (outside the p*I*) may also affect the emulsion activity of whey proteins by influencing their conformation and solubility. Temperature is another factor that affects the emulsifying properties of whey proteins. In particular, the rate of diffusion to the newly formed interface as well as the rates of adsorption and unfolding increase with increasing temperature (De Wit, 1989; Cayot and Lorient, 1997).

Foams can be defined as colloidal systems in which air bubbles are dispersed into an aqueous continuous phase (Damodaran, 1997b). Essential for the formation of protein-based foams is a rapid diffusion of protein to the air–water interface to reduce surface tension, followed by partial unfolding of the protein (De Wit, 1989). The most important foaming characteristics of proteins are foam volume (% overrun) and foam stability. For optimum protein concentrations and whipping times, both sodium and calcium caseinates gave a higher overrun than whey protein concentrate (Mulvihill and Fox, 1989). Southward and Goldman (1978) reported that sodium caseinate gave slightly lower overrun values than egg albumin but the sodium caseinate foams were less stable. Stability increased with the addition of sugar. The whipping properties of industrially prepared whey protein products are affected by several factors. The most relevant of those are: concentration and state of the whey proteins, pH, ionic environment, (pre-) heat treatment and the effect of lipids. As the whey protein concentration is increased the foam becomes denser, with more uniform air bubbles of a finer texture. Generally overrun (foam volume minus initial liquid volume) increases with protein concentration to a maximum value after which it decreases again (De Wit, 1989). In practice, while the caseins are very good emulsifiers and foam readily, the resulting foams are not very stable (Mulvihill, 1992; Fox and McSweeney, 1998). Two macroscopic processes in foams affect the stability of protein stabilized foams, the rate of liquid drainage from the lamellae and the film rupture. The rates of these two processes are dependent on the physical properties of the protein film and the physics of the lamella itself (Damodaran, 1997b). Cayot and Lorient (1997) also suggest that improvement in foam stability is related to the protein's ability to form a cohesive film. Conversely, the flexibility of the protein's polypeptide chain, although essential for foam creation, is detrimental to foam stability. Indeed, for a protein to foam well and stabilize the foam, it should display a proper balance of flexibility and rigidity at the air/water interface (Damodaran, 1997b).

Southward and Goldman (1978) found that soluble high- and medium-calcium coprecipitates both exhibited good emulsion-stabilizing properties. Soluble acid coprecipitates had the lowest stabilizing capacity of those coprecipitates examined, yet it still compared favorably with commercial sodium caseinate. In addition, all coprecipitates when whipped alone or with sugar exhibited greater foam volume and stability than the corresponding sodium caseinate whips (Southward and Goldman, 1978).

Application of dairy ingredients in gluten-free food

The replacement of gluten presents a major technological challenge, as gluten is an essential structure-building protein, contributing to the appearance and crumb structure of many baked products (Gallagher et al., 2004; Lazaridou et al., 2007). A large number of investigative techniques have been used in many studies for understanding the fundamental mechanical/rheological properties of gluten, including small and large deformation tests, temperature and frequency sweeps, bubble inflation, microscopy, and more (Schofield et al., 1984; Weegels et al., 1994; Dobraszczyk and Roberts, 1994; Guerrieri et al., 1996; Janssen et al., 1996; Stathopoulos et al., 2000, 2001; Toufeili et al., 2002; Dobraszczyk and Morgenstern, 2003; Dobraszczyk, 2004; Li et al., 2004; Stathopoulos et al., 2006, 2007). Studies on the rheology of gluten-free baked products have taken place recently (Gallagher et al., 2003a, 2003b, 2004; Schober et al., 2003; Moore et al., 2004, 2006; Sanchez et al., 2004; McCarthy et al., 2005; Lazaridou et al., 2007), and this field is developing rapidly. Over the years a lot of projects have been undertaken in the area of fortification/supplementation of wheat flour with dairy ingredients and products rather than gluten substitution (Stahel, 1983;

Harper and Zadow, 1984; Dubois and Dreese, 1985; Tow, 1985; Gelinas *et al.*, 1995; Erdogdu *et al.*, 1995a, 1995b; Erdogdu-Arnoczky *et al.*, 1996; Mann, 1996; Mannie and Asp, 1999; Kenny *et al.*, 2000, 2001; O'Brien *et al.*, 2000; Crowley *et al.*, 2002; Singh *et al.*, 2003; Gallagher *et al.*, 2005; Esteller *et al.*, 2006). Other ingredients (reviewed in other chapters of this book) have been used in gluten replacement, including starches and gums or hydrocolloids, as well as dietary fiber.

A few years ago it was not believed that, because of their properties, milk proteins can be used to replace gluten in bakery products. However, their use as a nutritional supplement and their functional effects had been accepted (Mulvihill, 1992; Hambraeus and Lonnerdal, 2003). Nutritional benefits include increasing calcium and protein content, as well as supplying essential amino acids (i.e. lysine, methionine, and tryptophan) (Kenny *et al.*, 2000). Recently, the addition of dairy products in gluten-free bread formulas is common practice, for increasing water absorption and therefore enhancing the handling properties of the batter (Gallagher *et al.*, 2004). In addition to the nutritional benefits and the increased water absorption, reduced staling rate and increased crust color are some of the advantages of dairy ingredients in breadmaking (Stahel, 1983; Harper and Zadow, 1984; Dubois and Dreese, 1985; Tow, 1985; Cocup and Sanderson, 1987; Gelinas *et al.*, 1995; Mann, 1996; Mannie and Asp, 1999; Kenny *et al.*, 2000; O'Brien *et al.*, 2000; Crowley *et al.*, 2002; Esteller *et al.*, 2006).

Gallagher *et al.* (2003b) applied seven dairy powders to a gluten-free bread formulation. In general, addition of powders with high protein/low lactose content (i.e. sodium caseinate and milk protein isolate) resulted in breads with an improved overall shape and volume, as well as a firmer crumb texture. The breads also had a better appearance (white crumb and dark crust) and organoleptically they scored well. Depending on the powder and level of addition, differences in loaf volume were observed. Inclusions of dairy powders reduced the loaf volume by about 6%, confirming previous data (Gelinas *et al.*, 1995; Erdogdu-Arnoczky *et al.*, 1996), however, increasing the level of inclusion of sweet whey, sodium caseinate, and milk protein isolate allowed recovery of the loaf volume. The opposite effect was observed when using demineralized whey, fresh milk solids, and skim milk powder. Overall, this work has proved that without a detrimental effect to the loaf volume, application of dairy powders can give products that are more appealing to the panelists than the control formulations (Gallagher *et al.*, 2003a).

Response surface methodology (RSM) has recently been used in order to optimize dry milk and whey protein concentrate fortification of a gluten-free bread formulation. Addition of 7.5% soy flour and 7.8% dry milk to a previously developed formulation increased the protein content from 1 to 7.3% and modified, to a small extent, the sensory quality of the resulting bread (Sanchez *et al.*, 2004). RSM has also been used to optimize water and hydroxypropylmethylcellulose (HPMC) content in a gluten-free bread formulation containing rice flour, potato starch, and skim milk powder (McCarthy *et al.*, 2005).

Another important benefit from using dairy ingredients in bakery and gluten-free products is the extension of shelf-life (Mannie and Asp, 1999; Kenny *et al.*, 2000). Gallagher *et al.* (2003b) assessed the effect of dairy powder addition to the intermediate and long-term shelf-life of gluten-free breads stored under a modified atmosphere by examining the staling profile of the formulations. They found that addition of milk protein isolate resulted in loaves with increased volume and better appearance and acceptability, ultimately there were no changes in the staling rate. Recently, textural studies have been conducted by Moore *et al.* (2004) with two gluten-free bread recipes, one containing 37.5% (dry weight) skim milk powder. Results were compared with those obtained using commercially available gluten-free flour and regular wheat bread. Baking tests showed that wheat bread and the bread made from the commercially available mix yielded significantly higher loaf volumes, while all the gluten-free breads were brittle after 2 days of storage. However, these changes were less pronounced when skim milk powder was present, indicating a positive influence of the dairy product addition. Using confocal laser-scanning microscopy it was shown that the dairy-based gluten-free bread contained network-like structures resembling the gluten network of wheat bread crumb.

The effect of sodium caseinate in combination with different hydrocolloids has been recently investigated (Lazaridou *et al.*, 2007). The type and extent of influence on bread quality was dependent on the specific hydrocolloids used and their supplementation levels.

Problems associated with the incorporation of dairy ingredients in gluten-free cereal products

Currently, many gluten-free breads available in the market are of poor quality and flavor, and many exhibit a dry, crumbly texture (Gallagher *et al.*, 2003a, 2004; McCarthy *et al.*, 2005). Gluten is considered the "structural" protein of breadmaking, and its absence often results in a liquid batter rather than a dough pre-baking. Many gluten-free baked breads exhibit a crumbly texture, poor color, and other quality defects post-baking (Gallagher *et al.*, 2004; McCarthy *et al.*, 2005). Gluten slows the movement of water in the dough by forming an extensible protein network, thus keeping the crumb structure softer (Gan *et al.*, 1995; Gallagher *et al.*, 2003a). In gluten-free batters, the absence of gluten may allow the increase of water movement from crumb to crust, thereby resulting in a firmer crumb and a softer crust (Gallagher *et al.*, 2003a). A further problem of gluten-free bread is the lighter color of the crust (Gallagher and Gormley, 2002). Recently, it was shown that inclusion of dairy powders results in a darkening of the crust, probably due to Maillard browning and caramelization reactions (Gallagher *et al.*, 2003a). However, almost every milk fraction has been described as loaf-volume depressing (Harper and Zadow, 1984; Erdogdu-Arnoczky *et al.*, 1996; Kenny *et al.*, 2000; Singh *et al.*, 2003; Esteller *et al.*, 2006), thus it is important to clearly determine the proper content of dairy ingredients which will allow an increase in color without leading to a reduced volume of the final bread.

An important aspect that has to be considered when developing gluten-free formulations based on dairy products is the lactose-content of the powders. People with celiac disease have been reported to be lactose-intolerant, so products containing high lactose concentration are not suitable for them, because of the absence of the lactase enzyme which is produced by the villi (Ortolani and Pastorello, 1997; Gallagher *et al.*, 2004; Moore *et al.*, 2004). Another problem associated with gluten-free formulations is the selected starch source. Regardless of the type of dairy ingredient involved, the starches used in gluten-free recipes are often wheat starches. Those should, in principle, be free of gluten and gliadins. However, it is very difficult to completely remove the gliadins, therefore trace amounts of the allergenic proteins might be present (McCarthy *et al.*, 2005). Methods for detecting gluten-containing cereals in gluten-free applications have been developed (Olexova *et al.*, 2006), and should be used to investigate the gluten-free status of the selected starch.

Future trends

It is currently believed that, despite the obstacles encountered, it will be possible to completely replace gluten with (one or more) functional dairy ingredients, without negatively affecting the rheological or the organoleptic properties of the gluten-free bread. Presently mixtures of gums, hydrocolloids, and dairy protein products are the most popular approach (McCarthy *et al.*, 2005; Lazaridou *et al.*, 2007). Mixtures of gluten-free flours, soy protein isolate, in combination with gums (locust bean, guar, konjak, xanthan, HPMC) and dietary fiber, are expected to be present in the next generation of gluten-free breads, along with functional dairy ingredients.

Current research aims to completely substitute gluten with a functional casein-based ingredient. The principle behind this approach is that by increasing the calcium concentration to an optimum level in the casein/caseinate ingredient it will be possible, under the correct pH and ionic strength conditions, to replace the highly functional (covalent) S-S bonds in a wheat dough with calcium (coordination) links (Stathopoulos and O'Kennedy, 2007).

Sources of further information and advice

There are a number of resources available regarding celiac disease, e.g. publications, recipe books, and websites. Interested individuals should have no problem locating a wealth of information, while most Western countries have celiac disease societies providing information and support. On the subject of dairy ingredients, their production, and their applications in the food and baking industry, there are a number of reviews available (Mulvihill, 1989, 1992; De Wit, 1989; Southward, 1989; Dalgleish, 1997; Morr, 1989; Mulvihill and Fox, 1989; Maubois and Ollivier, 1997; Cayot and Lorient, 1997; Damodaran, 1997a; Fox and McSweeney, 1998; Fox, 2003; Swaisgood, 2003; De Kruif and Holt, 2003; O'Connell and Fox, 2003; Nieuwenhuijse and van Boekel, 2003; Kelly *et al.*, 2003; Mulvihill and Ennis, 2003; Dickinson, 2003; Singh and Havea, 2003; Carr *et al.*, 2003) and the reader is advised to seek some of those for information in greater detail than provided in this chapter.

References

Carr, A. J., Southward, C. R., and Creamer, L. K. (2003). Protein hydration and viscocity of dairy fluids. In: Fox, P. F. and McSweeney, P. L. H. eds. *Advanced Dairy Chemistry*, Vol. 1: *Proteins Part B*. New York: Kluwer Academic.

Cayot, P. and Lorient, D. (1997). Structure-function relationship of whey proteins. In: Damodaran, S. and Paraf, A. eds. *Food Proteins and their Applications*. New York: Marcel Decker.

Claypool, L. L. (1985). Functional role of components of dairy products in processed cereal products. In: Vetter, J. L. ed. *Dairy Products for the Cereal Processing Industry*. St. Paul, MN: American Association of Cereal Chemists.

Cocup, R. O. and Sanderson, W. B. (1987). Functionality of dairy ingredients in bakery products. *Food Technol.* 41, 86–90.

Crowley, P., O'Brien, C., Slattery, H., Chapman, D., Arendt, E., and Stanton, C. (2002). Functional properties of casein hydrolysates in bakery applications. *Eur. Food Res. Technol.* 215, 131–137.

Dalgleish, D. G. (1997). Structure–function relationships of caseins. In: Damodaran, S. and Paraf, A. ed. *Food Proteins and their Applications*. New York: Marcel Decker.

Damodaran, S. (1997a). Food proteins: an overview. In: Damodaran, S. and Paraf, A. ed. *Food Proteins and their Applications*. New York: Marcel Decker.

Damodaran, S. (1997b). Protein stabilised foams and emulsions. In: Damodaran, S. and Paraf, A. ed. *Food Proteins and their Applications*. New York: Marcel Decker.

De Kruif, C. G. and Holt, C. (2003). Casein micelle structure, functions and interactions. In: Fox, P. F. and McSweeney, P. L. H. eds. *Advanced Dairy Chemistry*, Vol. 1: *Proteins, Part A*, 3rd edn. New York: Kluwer Academic.

De Wit, J. N. (1989). The use of whey protein products. In: Fox, P. F. ed. *Developments in Dairy Chemistry*, Vol. 4: *Functional Milk Proteins*. London: Elsevier.

Dickinson, E. (2003). Interfacial, emulsifying and foaming properties of milk proteins. In: Fox, P. F. and McSweeney, P. L. H. eds. *Advanced Dairy Chemistry*, Vol. 1: *Proteins Part B*. New York: Kluwer Academic.

Dobraszczyk, B. J. (2004). The physics of baking: rheological and polymer molecular structure–function relationships in breadmaking. *J. Non-Newtonian Fluid Mech.* 124, 61–69.

Dobraszczyk, B. J. and Morgenstern, M. P. (2003). Rheology and the breadmaking process. *J. Cereal Sci.* 38, 229–245.

Dobraszczyk, B. J. and Roberts, C. A. (1994). Strain hardening and dough gas cell-wall failure in biaxial extension. *J. Cereal Sci.* 20, 265–274.

Dubois, D. K. and Dreese, P. (1985). Functionality of nonfat dry milk in bakery products. In: Vetter, J. L. ed. *Dairy Products for the Cereal Processing Industry*. St. Paul, MN: American Association of Cereal Chemists.

Erdogdu-Arnoczky, N., Czuchajawoska, Z., and Pomeranz, Y. (1996). Functionality of whey and casein in fermentation and in breadmaking by fixed and optimised procedures. *Cereal Chem.* 73, 309–316.

Erdogdu, N., Czuchajawoska, Z. Y., and Pomeranz, Y. (1995a). Wheat flour and defatted milk fractions characterized by differential scanning calorimetry. II. DSC of interaction products. *Cereal Chem.* 72, 76–79.

Erdogdu, N., Czuchajawoska, Z. Y., and Pomeranz, Y. (1995b). Wheat flour and defatted milk fractions characterized by differential scanning calorimetry. I. DSC of flour and milk fractions. *Cereal Chem.* 72, 70–75.

Esteller, M., Sergio, Zancanaro, O. *et al.* (2006). The effect of kefir addition on microstructure parameters and physical properties of porous white bread. *Eur. Food Res. Technol.* 222, 26–31.

Fox, P. F. (2003). Milk proteins: general and historical aspects. In: Fox, P. F. and McSweeney, P. L. H. eds. *Advanced Dairy Chemistry*, Vol. 1: *Proteins, Part A*, 3rd edn. New York: Kluwer Academic.

Fox, P. F. and McSweeney, P. L. H. (1998). *Dairy Chemistry and Biochemistry.* London: Blackie Academic.

Gallagher, E. and Gormley, T. R. (2002). The quality of gluten free breads produced at retail outlets. *Research Report.* Dublin, Teagasc, The National Food Centre.

Gallagher, E., Gormley, T. R., and Arendt, E. K. (2003a). Crust and crumb characteristics of gluten free breads. *J. Food Eng.* 56, 153–161.

Gallagher, E., Kunkel, A., Gormley, T. R., and Arendt, E. (2003b). The effect of dairy and rice powder addition on loaf and crumb characteristics, and on shelf life (intermediate and long-term) of gluten-free breads stored in a modified atmosphere. *Eur. Food Res. Technol.* 218, 44–48.

Gallagher, E., Gormley, T. R., and Arendt, E. K. (2004). Recent advances in the formulation of gluten-free cereal-based products. *Trends Food Sci. Technol.* 15, 143–152.

Gallagher, E., Kenny, S., and Arendt, E. (2005). Impact of dairy protein powders on biscuit quality. *Eur. Food Res. Technol.* 221, 237–243.

Gan, Z., Ellis, P. R., and Schofield, J. D. (1995). Gas cell stabilisation and gas retention in wheat bread dough. *J. Cereal Sci.* 21, 215–230.

Gelinas, P., Audet, J., Lachance, O., and Vachon, M. (1995). Fermented dairy ingredients for bread: effects on dough rheology and bread characteristics. *Cereal Chem.* 72, 151–154.

Guerrieri, N., Alberti, E., Lavelli, V., and Cerletti, P. (1996). Use of spectroscopic and fluoresence techniques to assess heat-induced molecular modifications of gluten. *Cereal Chem.* 73, 368–374.

Hambraeus, L., and Lonnerdal, B. (2003). Nutritional aspects of milk proteins. In: Fox, P. F. and McSweeney, P. L. H. eds. *Advanced Dairy Chemistry*, Vol. 1. *Proteins Part B.* New York: Kluwer Academic.

Harper, W. J. and Zadow, J. G. (1984). Heat induced changes in whey protein concentrates as related to bread manufacture. *N. Z. J. Dairy Sci. Technol.* 19, 229–237.

Hugunin, A. G. (1985). Functionality of whey protein concentrates in processed cereal products. In: Vetter, J. L. ed. *Dairy Products for the Cereal Processing Industry.* St. Paul, MN: American Association of Cereal Chemists.

Hyslop, D. B. (2003). Enzymatic coagulation of milk. In: Fox, P. F. and McSweeney, P. L. H. eds. *Advanced Dairy Chemistry*, Vol. 1. *Proteins Part B.* New York: Kluwer Academic.

Janssen, A. M., Van Vliet, T., and Vereijken, J. M. (1996). Rheological behaviour of wheat glutens at small and large deformations. Effect of gluten composition. *J. Cereal Sci.* 23, 33–42.

Kelly, A. L., O'Connell, J. E., and Fox, P. F. (2003). Manufacture and properties of milk powders. In: Fox, P. F. and McSweeney, P. L. H. eds. *Advanced Dairy Chemistry*, Vol. 1. *Proteins Part B.* New York: Kluwer Academic.

Kelly, P. M. and O'Kennedy, B. T. (1986). *Production and Functional Properties of Total Milk Protein Isolates.* The Agricultural Institute.

Kenny, S., Wehrle, K., Stanton, C., and Arendt, E. K. (2000). Incorporation of dairy ingredients into wheat bread: effects on dough rheology and bread quality. *Eur. Food Res. Technol.* 210, 391–396.

Kenny, S., Wehrle, K., Auty, M., and Arendt, E. (2001). Influence of sodium caseinate and whey protein on baking properties and rheology of frozen dough. *Cereal Chem.* 78, 458–463.

Lazaridou, A., Duta, D., Papageorgiou, M., Belc, N., and Biliaderis, C. G. (2007). Effects of hydrocolloids on dough rheology and bread quality parameters in gluten-free formulations. *J. Food Eng.* 79, 1033–1047.

Li, W., Dobraszczyk Bogdan, J., and Wilde, P. J. (2004). Surface properties and locations of gluten proteins and lipids revealed using confocal scanning laser microscopy in bread dough. *J. Cereal Sci.* 39, 403–411.

Lucey, J. A. and Singh, H. (2003). Acid coagulation of milk. In: Fox, P. F. and McSweeney, P. L. H. eds. *Advanced Dairy Chemistry*, Vol. 1. *Proteins Part B.* New York: Kluwer Academic.

Mann, E. (1996). Dairy ingredients in Foods – part 2. *Dairy Indust. Int.* 61, 10–11.

Mannie, E. and Asp, E. H. (1999). Dairy ingredients in bread baking. *Cereal Foods World* 44, 143–146.

Manski, J. M., Kretzers, I. M. J., Van Brenk, S., Van Der Goot, A. J., and Boom, R. M. (2007). Influence of dispersed particles on small and large deformation properties of concentrated caseinate composites. *Food Hydrocolloids* 21, 73–84.

Maubois, J. L. and Ollivier, G. (1997). Extraction of milk proteins. In: Damodaran, S. and Paraf, A. eds. *Food Proteins and their Applications.* New York: Marcel Decker.

McCarthy, D. F., Gallagher, E., Gormley, T. R., Schober, T. J., and Arendt, E. K. (2005). Application of response surface methodology in the development of gluten-free bread. *Cereal Chem.* 82, 609–615.

Moore, M. M., Schober, T. J., Dockery, P., and Arendt, E. K. (2004). Textural comparisons of gluten-free breads and wheat-based doughs, batters and breads. *Cereal Chem.* 81, 567–575.

Moore, M. M., Heinbockel, M., Dockery, P., Ulmer, H. M., and Arendt, E. (2006). Network formation in gluten-free bread with application of transglutaminase. *Cereal Chem.* 83, 28–36.

Morr, C. V. (1989). Whey proteins: Manufacture. In: Fox, P. F. ed. *Developments in Dairy Chemistry*, Vol. 4. London: Elsevier.

Morrisey, P. A. (1985). Lactose: chemical and physicochemical properties. In: Fox, P. F. ed. *Developments in Dairy Chemistry*, Vol. 3: *Lactose and Minor Constituents.* London: Elsevier.

Mulvihill, D. M. (1989). Caseins and caseinates: manufacture. In: Fox, P. F. ed. *Developments in Dairy Chemistry*, Vol. 4: *Functional Milk Proteins*. London: Elsevier.

Mulvihill, D. M. (1992). Production, functional properties, and utilisation of milk protein products. In: Fox, P. F. ed. *Advanced Dairy Chemistry*, Vol. 1: *Proteins*. London: Elsevier.

Mulvihill, D. M. and Ennis, M. P. (2003). Functional milk proteins:production and utilisation. In: Fox, P. F. and McSweeney, P. L. H. eds. *Advanced Dairy Chemistry*, Vol. 1: *Proteins Part B*. New York: Kluwer Academic.

Mulvihill, D. M. and Fox, P. F. (1989). Physico-chemical and functional properties of Milk proteins. In: Fox, P. F. ed. *Developments in Dairy Chemistry*, Vol. 4: *Functional Milk Proteins*. London: Elsevier.

Nieuwenhuijse, J. A. and Van Boekel, M. A. J. S. (2003). Protein stability in sterilised milk and milk products. In: Fox, P. F. and McSweeney, P. L. H. eds. *Advanced Dairy Chemistry*, Vol. 1: *Proteins Part B*. New York: Kluwer Academic.

O'Brien, C. M., Grau, H., Neville, D. P., Keogh, M. K. and Arendt, E. (2000). Functionality of microencapsulated high-fat powders in wheat bread. *Eur. Food Res. Technol.* 212: 64–69.

O'Connell, J. E. and Fox, P. F. (2003). Heat-induced coagulation of milk. In: Fox, P. F. and McSweeney, P. L. H. eds. *Advanced Dairy Chemistry*, Vol. 1: *Proteins Part B*. New York: Kluwer Academic.

Olexova, L., Dovicovicova, L., Svec, M., Siekel, P., and Kuchta, T. (2006). Detection of gluten-containing cereals in flours and "gluten-free" bakery products by polymerase chain reaction. *Food Control* 17, 234–237.

Ortolani, C. and Pastorello, E. A. (1997). Symptoms of food allergy and food intolerance. *Study of Nutritional Factors in Food Allergies and Food Intolerance*. Luxemburg: CEC.

Roberts, H. (1985). Processing fluid milk into functional ingredients for the cereal processing industry. In: Vetter, J. L. ed. *Dairy Products for the Cereal Processing Industry*. St. Paul, MN: American Association of Cereal Chemists.

Roeper, J. (1977). Preparation of calcium caseinate from casein curd. *N. Z. J. Dairy Sci. Technol.* 182–189.

Sanchez, H. D., Osella, C. A., and De La Torre, M. A. (2004). Use of response surface methodology to optimize gluten-free bread fortified with soy flour and dry milk. *Food Sci. Technol. Int.* 10, 5–9.

Schober, T. J., O'Brien, C., McCarthy, D. F., Darnedde, A., and Arendt, E. (2003). Influence of gluten-free flour mixes and fat powders on the quality of gluten-free biscuits. *Eur. Food Res. Technol.* 216, 369–376.

Schofield, J. D., Bottomley, R. C., Legrys, G. A., Timms, M. F., and Booth, M. R. (1984). Effects of heat on wheat gluten. In: Graveland, A. and Moonen, J. H. E. eds. *Second International Workshop on Gluten Proteins*. Waageningen: PUDOC.

Singh, H. and Havea, P. (2003). Thermal denaturation, aggregation and gelation of whey proteins. In: Fox, P. F. and McSweeney, P. L. H. eds. *Advanced Dairy Chemistry*, Vol. 1: *Proteins Part B*. New York: Kluwer Academic.

Singh, N., Kaur Bajaj, I., Singh, R. P., and Singh Gujral, H. (2003). Effect of different additives on mixograph and bread making properties of Indian wheat flour. *J. Food Eng.* 56, 89–95.

Southward, C. R. (1989). Use of casein and caseinates. In: Fox, P. F. ed. *Developments in Dairy Chemistry*, Vol. 4: *Functional Milk Proteins*. London: Elsevier.

Southward, C. R. and Goldman, A. (1975). Co-precipitates—A review. *N. Z. J. Dairy Sci. Technol.* 10, 101–112.

Southward, C. R. and Goldman, A. (1978). Co-precipitates and their application in food products II.some properties and applications. *N. Z. J. Dairy Sci. Technol.* 13, 97–105.

Stahel, N. (1983). Dairy proteins for the cereal foods industry: Functions, selection, and usage. *Cereal Foods World* 28, 453–455.

Stathopoulos, C. E., Tsiami, A. A., and Schofield, J. D. (2000). Effect of protein fractions on gluten rheology. In: Shewry, P. R. and Tatham, A. S. eds. *Wheat Gluten*. Cambridge: Royal Society of Chemistry.

Stathopoulos, C. E., Tsiami, A. A., and Schofield, J. D. (2001). Rheological and size characterisation of gluten proteins from cultivars varying in baking quality. In: Gooding, M. J., Barton, S. A., and Smith, G. P. eds. *Aspects of Applied Biology*. Warwick: The Association of Applied Biologists.

Stathopoulos, C. E., Tsiami, A. A., Dobraszczyk, B. J., and Schofield, J. D. (2006). Effect of heat on rheology of gluten fractions from flours with different bread-making quality. *J. Cereal Sci.* 43, 322–330.

Stathopoulos, C. E., Tsiami, A. A., Schofield, J. D., and Dobrasczcyk, B. J. (2007). Effect of heat on rheology, surface hydrophobicity and molecular weight distribution of glutens extracted from flours with different bread-making quality. *J. Cereal Sci.* (in press).

Stathopoulus, C. E. and O'Kennedy, B. T. (2007). The effect of salt on the rheology and texture of a casein based ingredient intended to replace gluten. *Milchwissenschaft* (in press).

Swaisgood, H. E. (2003). Chemistry of the caseins. In: Fox, P. F. and McSweeney, P. L. H. eds. *Advanced Dairy Chemistry*, Vol. 1: *Proteins Part A*. New York: Kluwer Academic.

Toufeili, I., Lambert, I. A., and Kokini, J. L. (2002). Effect of glass transition and cross-linking on rheological properties of gluten: development of a preliminary state diagram. *Cereal Chem.* 79, 138–142.

Tow, G. J. (1985). Some uses of milk in baked products. *Baking Today* 17–18.

Towler, C. (1976). Conversion of casein curd to sodium caseinate. *N. Z. J. Dairy Sci. Technol.* 11, 24–29.

Vetter, J. L. (1985). Utilisation of nonfat dry milk by the baking industry. In: Vetter, J. L. ed. *Dairy Products for the Cereal Processing Industry*. St. Paul, MN: American Association of Cereal Chemists.

Weegels, P. L., Verhoek, J. A., De Groot, A. M. G., and Hamer, R. J. (1994). Effects on gluten of heating at different moisture contents. I. Changes in functional properties. *J. Cereal Sci.* 19, 31–38.

Use of enzymes in the production of cereal-based functional foods and food ingredients

11

Hans Goesaert, Christophe M. Courtin, and Jan A. Delcour

Introduction

There is a growing awareness that the daily diet is an important determinant for a healthy life. Consumers judge food products not only in terms of taste and nutritional needs, but also in terms of the ability to improve their health and well-being. Functional foods and functional food ingredients exert a beneficial influence on body functions to help improve well-being and health and/or reduce the risk of chronic diseases, when consumed at levels that can normally be expected to occur in the diet (Ashwell, 2002) and, hence, meet these new consumer demands. Therefore, it is not surprising that, both from a scientific and a commercial point of view, interest in functional food products and ingredients is high. From a practical point of view, a functional food can be produced by addition of health-promoting component(s), by reducing/removing harmful components, and/or by modifying the nature or the bioavailability of specific components (Ashwell, 2002).

This chapter focuses on the potential use of cereals in the production of functional foods and potential health-promoting ingredients. In particular, in the main part of this chapter, we discuss how the modification of different cereal carbohydrates using enzyme technology may result in functional food components. For this purpose, after giving an overview of the different cereal carbohydrates as well as the main enzymes acting thereupon, a discussion of how enzyme technology can modify the constituent to produce health-related compounds will be provided. Within the concept of cereal carbohydrate-derived functional food ingredients, we consider high molecular weight soluble dietary fiber, prebiotic non-digestible oligosaccharides, and resistant starch.

The second part of this chapter deals with the functional food-related aspects of cereal proteins. Although the proteolytic modification of cereal proteins can lead to potential health-promoting components, such as bioactive peptides (which can affect the cardiovascular, endocrine, immune, and/or nervous systems) (Korhonen *et al.*, 1998; Korhonen and Pihlanto, 2003), we mainly focus on enzyme-related aspects of gluten-free foods. Since the removal of a harmful component (in this case gluten for patients with celiac disease) is a way to produce functional foods (Ashwell, 2002), gluten-free products can be considered as such.

Non-starch polysaccharide-derived functional food ingredients

Cereal non-starch polysaccharides (NSP) are important from a nutritional point of view. They are important dietary fiber components and can be converted to health-promoting food components, such as non-digestible oligosaccharides with prebiotic properties.

Non-starch polysaccharides and NSP-degrading enzymes

Arabinoxylan

In many cereals, but particularly in wheat and rye, arabinoxylans constitute the largest NSP fraction. They are present in a water-extractable (WE-AX) and water-unextractable (WU-AX) form. The latter is strongly cross-linked into the cell wall (Iiyama *et al.*, 1994) and can be solubilized by alkali (AS-AX) and by enzymes (ES-AX).

Both WU-AX and WE-AX are polydisperse polysaccharides with one general structure. Endosperm arabinoxylans contain a backbone of β-1,4-linked D-xylopyranosyl residues, either unsubstituted or substituted at the $C(O)$-3 and/or the $C(O)$-2 position with monomeric α-L-arabinofuranose residues (Perlin, 1951a, 1951b). The $C(O)$-5 of some of the arabinose residues are ester linked to phenolic acids (e.g. ferulic acid) (Fausch *et al.*, 1963). Under oxidizing conditions (e.g. the H_2O_2/peroxidase system), WE-AX can cross-link by covalent coupling of two ferulic acid residues (Vinkx *et al.*, 1991; Figueroa-Espinoza and Rouau, 1998). Bran arabinoxylans contain additional substituents, such as glucuronic acid and its 4-O-methylether, and oligomeric arabinose side-chains (Voragen *et al.*, 1992). The degree of substitution of arabinoxylans is expressed by the arabinose to xylose ratio (A/X), with a typical average value of 0.5–0.6 for the general wheat and rye WE-AX population (Cleemput *et al.*, 1993; Vinkx and Delcour, 1996). However, arabinoxylan subpopulations show a wide range of A/X values of ca. 0.3–1.3 (Cleemput *et al.*, 1995; Dervilly *et al.*, 2000; Trogh *et al.*, 2005a; Verwimp *et al.*, 2007). AS-AX show only small differences in molecular weight (Meuser and Suckow, 1986) and A/X ratios (Gruppen *et al.*, 1993) compared with WE-AX. Furthermore, several arabinoxylan structural models describe a non-random distribution of the arabinose substituents along the arabinoxylan chain with highly branched regions interlinked with lowly substituted, more open regions (Goldschmid and Perlin, 1963; Gruppen *et al.*, 1993).

The structure and related physico-chemical properties of arabinoxylans affect their functionality in cereal-based processes such as breadmaking (Courtin and Delcour, 2002). WE-AX form highly viscous aqueous solutions, whereas WU-AX have strong water-binding capacity.

Arabinoxylan-hydrolyzing enzymes

Due to the heterogenic structure of arabinoxylans, complete arabinoxylan hydrolysis requires the combined action of several hydrolytic enzymes with different specificities (Figure 11.1). Endo-(1,4)-β-D-xylanases, further referred to as endoxylanases (EC 3.2.1.8), are the main arabinoxylan-degrading enzymes as they are able to hydrolyze the xylan backbone of arabinoxylan internally, thereby reducing the molecular weight of arabinoxylan molecules and ultimately forming (arabino)xylooligosaccharides (A)XOS. Hence, they strongly impact arabinoxylan structure and functionality. Endoxylanases are assisted by several types of exo-enzymes. α-L-arabinofuranosidases (EC 3.2.1.55) release the arabinose residues of arabinoxylan (fragments) and create new sites for the endoxylanases to attack. Other substituents, mainly occurring in bran arabinoxylans, are removed by α-D-glucuronidases and

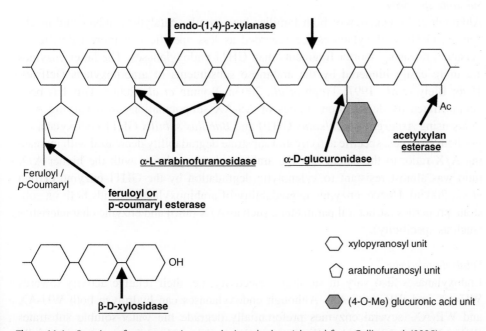

Figure 11.1 Overview of enzymes acting on substituted xylan. Adapted from Collins *et al.* (2005).

acetylxylan esterases. β-D-Xylosidases (EC 3.2.1.37) degrade arabinoxylan fragments from the non-reducing end, releasing xylose. Phenolic acid esterases, such as feruloyl esterases, hydrolyze the ester linkages between the arabinose side-chain residues and the phenolic acids. Furthermore, the activity of different accessory enzymes varies with the properties of the substrate, such as solubility and molecular weight.

Based on amino acid sequences and structural similarities, the majority of endoxylanases are classified into two glycoside hydrolase (GH) families, GH10 and GH11 (Henrissat, 1991; Coutinho and Henrissat, 1999), representing enzymes with different structures and catalytic properties (Jeffries, 1996; Biely *et al.*, 1997; Törrönen and Rouvinen, 1997). Some endoxylanases belonging to GH5, 8, and 43 have also been identified (Coutinho and Henrissat, 1999), but so far few of them have been studied in detail. While each of the above cited GH families contain microbial endoxylanases, all plant endoxylanases thus far identified, including those of cereals, have been classified in GH10 (Simpson *et al.*, 2002). Endoxylanases differ in their mode of action towards substrates. This is evidenced by the variety and size of hydrolysis products obtained. Endoxylanase functionality depends on several parameters such as the biochemical properties of the enzyme (e.g. pH and temperature optima), substrate specificity, substrate selectivity, and its sensitivity to (proteinaceous) endoxylanase inhibitors.

Endoxylanases are routinely used in cereal processing to improve cereal processability and/or product quality. In breadmaking, endoxylanase supplementation increases dough stability, prolongs the oven rise during the initial stage of baking, and results in breads with a higher loaf volume and a finer, softer and more homogeneous bread crumb (Courtin *et al.*, 1999, 2001; Courtin and Delcour, 2002; Goesaert *et al.*, 2006).

Substrate specificity
Although endoxylanases of both families have similar catalytic residues and mechanisms, GH10 endoxylanases are regarded as less specific and more catalytically versatile, releasing shorter fragments than GH11 endoxylanases. The latter enzymes are more easily hindered by the arabinose substituents of arabinoxylans (Jeffries, 1996; Biely *et al.*, 1997; Trogh *et al.*, 2005a; Bonnin *et al.*, 2006). This has been demonstrated by degradation studies of AS-AX subpopulations with varying A/X ratios using *Aspergillus aculeatus* GH10 and *Bacillus subtilis* GH11 endoxylanases. For both enzymes, specific activity and substrate degradability decreased with increasing A/X ratio, to the extent that the arabinoxylan population with the higher A/X ratio was almost resistant to xylanolytic degradation by the GH11 enzyme (Trogh *et al.*, 2005a). Hence, enzymic degradability of arabinoxylans depends both on substrate properties (structural parameters, such as A/X ratio) and enzyme characteristics (such as specificity).

Substrate selectivity
Endoxylanases also vary in substrate selectivity, i.e. their relative activity towards WU-AX and WE- or S-AX. Although endoxylanases can hydrolyze both WU-AX and WE-AX, several enzymes preferentially degrade the water-soluble substrates (WE-AX and S-AX) and have limited activity on WU-AX, while others preferentially

hydrolyze WU-AX (Courtin *et al.*, 2001; Courtin and Delcour, 2002; Moers *et al.*, 2003, 2005; Bonnin *et al.*, 2006). This affects the applicability of endoxylanases in cereal processing. Hydrolysis of WU-AX results in a reduced water-binding capacity of WU-AX and the release of solubilized arabinoxylans, and, consequently, in an increased viscosity of the aqueous phase. In breadmaking, this kind of endoxylanase action is generally regarded as beneficial. Hydrolysis of native WE-AX and S-AX yields arabinoxylan fragments with low molecular weight and a concomitant decrease in viscosity, and this action is considered to be negative for breadmaking (Petit-Benvegnen *et al.*, 1998; Courtin *et al.*, 2001). The relative contribution of each of these actions of endoxylanases in a system containing both WU- and WE-AX, such as in dough, is determined by the substrate selectivity. However, the mechanisms underlying substrate selectivity and its relation to other enzyme properties such as specificity remain unclear. This concept should hence be considered more from a functional and practical point of view rather than a biochemical one (Moers *et al.*, 2005).

Sensitivity to inhibition

The functionality of endoxylanases is also influenced by proteinaceous endoxylanase inhibitors, which are endogenously present in cereal grains. Three types of cereal endoxylanase inhibitors with different structures and specificities have been identified and described (Table 11.1), i.e. TAXI-type (*Triticum aestivum* L. endoxylanase inhibitor) (Debyser *et al.*, 1999; Gebruers *et al.*, 2001, 2004; Goesaert *et al.*, 2003a, 2004), XIP-type (endoxylanase inhibiting protein) (McLauchlan *et al.*, 1999; Goesaert *et al.*, 2003b, 2004; Juge *et al.*, 2004), and TLXI-type (thaumatin-like endoxylanase inhibitors) (Fierens *et al.*, 2007). Recently, molecular engineering of endoxylanases based on several (crystallographic) studies of the interaction between enzyme and inhibitor (Tahir *et al.*, 2002; Sansen *et al.*, 2004a, 2004b; Payan *et al.*, 2004; Fierens *et al.*, 2005) led to the development of endoxylanases which are insensitive to the cereal endoxylanase inhibitors (Sibbesen and Sørensen, 2001; Tahir *et al.*, 2002).

Table 11.1   Comparison of the different cereal endoxylanase inhibitors

	TAXI-type[a]	XIP-type[b]	TLXI-type[c]
Molecular form	Monomer	Monomer	Monomer
Molecular mass	Form A: ~40 kDa Form B: ~30 + 10 kDa	~30 kDa	~18 kDa
pI	>8.8	>8.0	>9.3
Glycosylation	Yes (limited)	Yes	Yes (mostly O-linked)
Specificity	GH 11 of fungal and bacterial origin	GHs 10 and 11, of fungal origin	GH 11 of fungal and bacterial origin

[a] From Gebruers *et al.* (2001, 2004).
[b] From Flatman *et al.* (2002), Juge *et al.* (2004).
[c] From Fierens *et al.* (2007).

β-D-Glucan and β-D-glucan hydrolyzing enzymes

Like arabinoxylans, β-D-glucan is situated in the cereal cell wall, and is present in a water-extractable and water-unextractable form with one general structure. β-D-Glucan is a heterogeneous group of polymers consisting of long, linear chains of (ca. 70%) β-(1,4)- and (ca. 30%) β-(1,3)-linked D-glucopyranosyl residues. More in particular, the β-D-glucan chain is mainly (ca. 90%) made up by blocks of cellotriosyl and cellotetraosyl units, separated by single β-(1,3)-linkages. Approximately 10% of the chain consists of blocks of 4–15 consecutive β-(1,4)-linked glucose residues (Wood *et al.*, 1991, 1994). The β-(1,3)-linkages interrupt the extended, ribbon-like shape of β-(1,4)-linked glucose molecules, inducing kinks in the chain and making the β-D-glucan chains more flexible, more soluble, and less inert than cellulose. A main property of β-D-glucan is its high viscosity-forming potential, which not only depends on the conformation of β-D-glucan, but also on its molecular weight and concentration (Fincher and Stone, 1986).

Several enzymes are able to hydrolyze the internal linkages in β-D-glucan. Endo-β-(1,3)(1,4)-glucanases (EC 3.2.1.73, lichenases, GHs 16 and 17) hydrolyze β-(1,4)-linkages adjacent to the β-(1,3)-linkages. Endo-β-(1,4)-glucanases (EC 3.2.1.4, cellulases) hydrolyze the β-(1,4)-linkages of β-D-glucan primarily at regions with consecutive β-(1,4)-linked glucose units. The latter enzymes can be found in many GH families with many different structures and properties (Coutinho and Henrissat, 1999).

Soluble dietary fiber and enzyme technology

Cereal dietary fiber

Dietary fiber has been defined as the edible parts of plants or analogous carbohydrates that are resistant to digestion and absorption in the human small intestine with partial or complete fermentation [to short chain fatty acids (SCFA) and gasses] in the large intestine (American Association of Cereal Chemists, 2001). Soluble dietary fiber lowers serum cholesterol levels, a risk factor for coronary heart diseases, and reduces post-prandial blood glucose levels in humans, which is potentially beneficial for people with diabetes (Cummings *et al.*, 2004). In general, insoluble dietary fiber has a high water-binding capacity which increases and softens fecal bulk. It also reduces transit time of fecal material through the large intestine (Manthey *et al.*, 1999; American Association of Cereal Chemists, 2001, 2003). For people between 19 and 50 years old, the recommended daily intake of total dietary fiber is 38 g for men and 25 g for women (American Association of Cereal Chemists, 2003).

The cereal NSPs, predominantly arabinoxylans and β-D-glucan, but also cellulose and arabinogalactan-peptides, can all be classified as dietary fiber constituents. To some of them, particularly those belonging to the soluble fraction, health-promoting effects have been ascribed (Lanza *et al.*, 1987). Several studies showed that β-D-glucan can be used as cholesterol and blood glucose-lowering agent, probably because of its highly viscous properties (Klopfenstein, 1988; McIntosh *et al.*, 1993; Yokoyama *et al.*, 1997; Hecker *et al.*, 1998; Cavallero *et al.*, 2002). According to the United States Food and Drug Administration (FDA), consumption of about 3 g/day of β-D-glucan soluble dietary fiber lowers blood cholesterol levels (FDA, 1997). Arabinoxylan is also important from a nutritional point of view. Lu and co-workers (2000a, 2000b) showed that arabinoxylan reduces the post-prandial blood glucose and insulin responses in humans. In addition, there are indications that arabinoxylans can lower blood cholesterol levels because of their highly viscous properties (Bourdon *et al.*, 1999; Rieckhoff *et al.*, 1999) and that both arabinoxylans and β-D-glucan have prebiotic effects (Charalampopoulos *et al.*, 2002; Crittenden *et al.*, 2002; Gråsten *et al.*, 2003).

Production of soluble high molecular weight arabinoxylans

As indicated above, soluble arabinoxylans (of high molecular weight) can be obtained by alkaline treatment or enzyme-assisted conversion of WU-AX. In the latter case, endoxylanase specificity and selectivity, and incubation conditions (i.e. time, dosage) determine to a great extent the yield and properties (molecular weight) of the S-AX.

Production of high molecular weight soluble arabinoxylan as a food ingredient

In *in vitro* experiments, the enzymic hydrolysis of wheat WU-AX by a set of endoxylanases with different substrate selectivity showed the impact of such selectivity on changes in the structural characteristics of the arabinoxylan. For all enzymes tested, incubation of WU-AX with increasing dosages of endoxylanases resulted in an increased solubilization of arabinoxylan polymers and a concomitant decrease in S-AX molecular weight (Moers *et al.*, 2005). However, a gradual decrease in specific solubilizing activity of the endoxylanases was observed in accordance with their substrate selectivity. Furthermore, WU-AX solubilization and subsequent degradation of S-AX fragments by the selected endoxylanases gave rise to widely differing apparent molecular weight profiles, depending on the substrate selectivity of the enzymes. Thus, enzymes with high selectivity towards WU-AX generated higher levels of S-AX, which were of higher molecular weight than those obtained using enzymes with low selectivity. Indeed, a low dosage of GH11 *B. subtilis* XynA (highly selective for WU-AX) solubilized about 58% of total xylose and generated S-AX with a peak degree of polymerization (DP) of about 410 kDa, whereas for a low dosage of a GH10 *A. aculeatus* endoxylanase (highly selective for WE-AX) only 13% of total xylose of the original WU-AX material was solubilized and the apparent peak DP of the solubilized fragments was around 12 kDa (Moers *et al.*, 2005). Although substrate specificity and selectivity are not related (Moers *et al.*, 2005; Bonnin *et al.*, 2006), substrate specificity determines the structural properties of the arabinoxylan fragments. Indeed, highly specific endoxylanases, which are hindered by the presence of substituents, can only hydrolyze a limited number of linkages in arabinoxylans, thus generating larger arabinoxylan fragments (Trogh *et al.*, 2005a).

In addition to enzyme parameters, enzymic arabinoxylan solubilization is also influenced by substrate properties, as well as by the presence of accessory enzymes and endoxylanase inhibitors. Since WU-AX are essentially present in cell wall fragments, cell wall architecture and the interactions between the different cell wall constituents determine to a great extent the accessibility of the arabinoxylan population to the endoxylanases. In general, endosperm cell walls are thin and are rather easily solubilized by endoxylanases. In contrast, enzymic solubilization of bran arabinoxylans is much more limited (Figueroa-Espinoza *et al.*, 2004; Maes *et al.*, 2004). In addition, endoxylanase efficiency differs between the individual bran tissues because of histological and chemical heterogeneity (Beaugrand *et al.*, 2004a). However, in general, enzymes with the higher solubilizing activity on wheat flour WU-AX (e.g. *B. subtilis* XynA) were more efficient in the generation of soluble fiber from bran (Maes *et al.*, 2004). Moreover, xylanolytic solubilization of bran arabinoxylans yielded S-AX with rather low A/X ratios and a highly substituted insoluble residue (Maes *et al.*, 2004). There was no synergistic action on solubilization yield when the GH11 *B. subtilis* XynA and the GH10 *A. aculeatus* enzymes were combined. Solubilization yields can be improved by pretreatment of the bran, such as by extrusion (Figueroa-Espinoza *et al.*, 2004). Furthermore, the action of other hemicellulolytic enzymes, such as β-glucanases, can increase the susceptibility of arabinoxylans in the cell wall to xylanolytic attack. Although the synergistic action of endoxylanases and accessory enzymes, such as arabinofuranosidases and esterases, in the degradation of arabinoxylans is well known (Kormelink and Voragen, 1992), it is still unclear to what extent the latter enzymes can assist the opening of the cell wall structure. In this respect, Petit-Benvegnen and co-workers (1998) reported a limited increase in arabinoxylan solubilization from wheat flour WU-AX when an *A. niger* endoxylanase preparation was combined with either a feruloyl esterase, a cellulase or an endo-β-glucanase. Other researchers reported an increased arabinoxylan solubilization from treated rye bran only for the combined use of an endo-β-glucanase and a *B. subtilis* endoxylanase, and not for the combination of the endoxylanase with an arabinofuranosidase or a feruloyl esterase (Figueroa-Espinoza *et al.*, 2004).

In addition to their effect on endoxylanase activity, endoxylanase inhibitors bind to arabinoxylans as well (Rouau *et al.*, 2006; Fierens, 2007). This may affect the balance between enzymic arabinoxylan solubilization and arabinoxylan depolymerization, as well as having implications for the purity of the generated soluble arabinoxylan material.

Production of high molecular weight soluble fiber during processing

The concept of using endoxylanase enzyme technology for health-related aspects has already been explored in breadmaking. The combined use of hull-less barley flour and enzyme technology has allowed the production of tasty, consumer-acceptable bread products with *in situ* generated increased levels of the health-promoting dietary fiber components arabinoxylans and β-D-glucan (Trogh *et al.*, 2004, 2005b, 2007). Although composite flour breads had a slightly higher arabinoxylan content than the wheat flour breads, their soluble arabinoxylan content (sum of WE-AX and S-AX) was comparable (0.3–0.4 g/100 g bread) (Table 11.2). Endoxylanase addition to the recipe strongly increased the soluble arabinoxylan content because of solubilization of the WU-AX. Thus, endoxylanase not only contributes positively to loaf volume, but it can also be used to significantly increase the soluble arabinoxylan levels. As summarized in Table 11.2, breads made from the combined use of hull-less barley flour and endoxylanase had total arabinoxylan and β-D-glucan levels 1.8 times those of the control wheat flour bread. Moreover, soluble arabinoxylans and β-D-glucan

Table 11.2 Total and soluble AX and β-D-glucan contents (% dry matter) of wheat flour (WF) and composite flour (WF + HBF) (60% wheat flour and 40% hull-less barley flour) breads, with or without endoxylanase[a]

	WF bread (control)	WF bread + endoxylanase	WF + HBF bread	WF + HBF bread + endoxylanase
AX				
Total	1.4	1.4	1.9	1.9
Soluble	0.3	1.0	0.4	0.9
β-D-Glucan				
Total	0.3	0.3	1.2	1.2
Soluble	0.2	0.2	0.5	0.5
AX+β-D-Glucan				
Total	1.7	1.7	3.1	3.1
Soluble	0.5	1.2	0.9	1.4

[a] Data from Trogh et al. (2004).

levels were 2.8 times those of the corresponding wheat flour bread. The endoxylanase-induced increase in soluble dietary fiber levels has a potential nutritional impact. Indeed, a daily consumption of 100 g wheat/hull-less barley flour bread supplemented with endoxylanase would typically result in an intake of 3.1 g total and 1.4 g soluble arabinoxylans and β-D-glucan (compared to 1.7 g and 0.5 g, respectively, for 100 g wheat flour bread), implying a potential positive contribution to the recommended daily total (between 25 and 38 g/day) (American Association of Cereal Chemists, 2003) and soluble dietary fiber levels (about 3 g/day of β-D-glucan soluble dietary fiber) (FDA, 1997).

Prebiotics and enzyme technology

Cereal prebiotic non-digestible oligosaccharides

Prebiotics are food ingredients that resist hydrolysis by the host's salivary and intestinal digestive enzymes, and are subsequently fermented by bacteria in the colon. As such, they fit well within the current definition of dietary fiber. However, in addition, they beneficially affect the host by selectively stimulating the growth and/or activity of one or a limited number of (potentially health-promoting) bacteria in the colon (Gibson and Roberfroid, 1995; Cummings et al., 2004; Swennen et al., 2006a). Non-digestible oligosaccharides (NDOs), such as inulin and fructo-oligosaccharides (FOS), are the best-known prebiotic compounds. Several health-promoting effects of NDOs have been postulated [see Swennen et al. (2006a) and Mussatto and Mancilha (2007) for an overview]. An important aspect of the beneficial effect is their conversion to SCFAs during fermentation in the colon. The acidification of the colon environment is beneficial for the development of bacteria such as bifidobacteria and lactobacilli. In addition, SCFAs may affect mineral absorption, lipid and carbohydrate metabolism, the immune system and the risk of colon cancer (Swennen et al., 2006a; Mussatto and Mancilha, 2007). In the Western world, FOS are among the most used NDOs. They give reproducible prebiotic effects in humans, and particularly increase the bifidobacteria in the gut. However, there are indications that xylo-oligosaccharides (XOS) have stronger prebiotic properties than FOS (Hsu *et al.*, 2004). Furthermore, substituted XOS, i.e. arabinoxylo-oligosaccharides (AXOS), have strong prebiotic potential as well. They can be fermented by some health-promoting bifidobacteria (Van Laere *et al.*, 2000) and a significant increase in bifidobacteria upon supplementation of AXOS to broiler and rat diets has been reported (Swennen, 2007). Furthermore, feruloyl-containing oligosaccharides might have potential as natural antioxidants (Yuan *et al.*, 2005). In general, (A)XOS show very interesting technological properties. Both XOS and AXOS are stable over a wider range of pH and temperatures compared to FOS (Vázquez *et al.*, 2000; Swennen, 2007). AXOS are less sweet than XOS (with DP 2–4) and have a neutral taste. AXOS addition to wheat- or maize-based broiler diets also significantly improved zootechnical performance parameters such as feed conversion rate and body weight, while supplementation with an equal dose of FOS had no such effect (Swennen, 2007). In addition, while XOS products are typically linear molecules of DP 2–4, AXOS differ both in DP and degree of substitution (A/X ratio). This additional structural complexity may have implications for their physiological effects in the colon, e.g. site of fermentation.

Finally, β-gluco-oligosaccharides, as obtained from the enzymic hydrolysis of oat bran β-D-glucan, may be prebiotic as well, as they were found to enhance the growth of lactic acid bacteria (Jaskari *et al.*, 1998; Kontula *et al.*, 1998).

Production of arabinoxylo-oligosaccharides

In general, enzyme processing is an important tool for the production of NDOs. On the one hand, since many NDOs are degradation products of polysaccharides, treatment of the latter, such as xylan and inulin, with specific enzymes yield the preferred NDOs. Currently, synthesis of NDOs from simple sugars using enzymes such as glycosyltransferases, glycosidases, or glycosynthases, has received much attention (Swennen *et al.*, 2006a). Unsubstituted XOS are typically produced by enzymic and/or chemical degradation of xylan-rich raw materials (Vázquez *et al.*, 2000; Mussatto and Mancilha, 2007). Often, enzyme complexes with low exo-xylanase and/or β-xylosidase activity are desired to avoid xylose production (Vázquez *et al.*, 2000).

Similarly, AXOS can be produced by enzymic depolymerization of cereal arabinoxylans. Substrate specificity of the endoxylanase, substrate characteristics, and incubation conditions (i.e. time, dosage) determine to a great extent the yield and properties of the AXOS produced. The use of pure enzymes allows for a better control of the structures of the hydrolysis end-products. Commercial preparations usually contain a combination of different enzymes that may affect the hydrolysis pattern. As mentioned above, arabinoxylan degradation by endoxylanases yields a variety of arabinoxylan fragments, differing in size (DP) and composition (degree of substitution, A/X ratio) (Gruppen *et al.*, 1992; Kormelink *et al.*, 1993; Ordaz-Ortiz *et al.*, 2004; Swennen *et al.*, 2005). In this respect, incubation of wheat flour WU-AX with a GH10 endoxylanase from *A. aculeatus* yielded mixtures of AX poly- and oligosaccharides (Swennen *et al.*, 2005). Higher enzyme dosages and longer incubation times resulted in smaller arabinoxylan fragments, while the average A/X ratio remained the same, demonstrating the importance of the incubation conditions. The AXOS production using a commercial endoxylanase preparation containing the above-mentioned GH10 endoxylanase from *A. aculeatus* was also investigated by Rantanen and co-workers (2007). These authors found that the major AXOS product following enzymic degradation of several AX materials was arabinoxylobiose, with a yield of about 12% of the quantified hydrolysis products in the case of rye WE-AX degradation.

Furthermore, fractionation of the heterogeneous AXOS populations is required in order to investigate AXOS structure–function relationships. Heterogeneous AXOS mixtures have been fractionated by graded ethanol precipitation or ultrafiltration membranes with different molecular mass cut-off. Under these conditions, different AXOS populations were obtained, varying both in DP and A/X ratio (Swennen *et al.*, 2005). Other techniques to isolate specific AXOS molecules include gel filtration chromatography and semi-preparative high-performance anion-exchange chromatography (Gruppen *et al.*, 1992; Kormelink *et al.*, 1993; Ordaz-Ortiz *et al.*, 2004).

Cost-effective production of AXOS requires raw materials with high levels of arabinoxylans, such as (wheat) bran. However, as is the case with the xylanolytic production of soluble fiber from bran, the susceptibility of this substrate to enzymic hydrolysis is rather low. A GH11 endoxylanase was more useful than a GH10 xylanase in the production of (substituted) xylo-oligosaccharides from wheat bran, although the latter enzyme produced smaller arabinoxylan fragments (Beaugrand *et al.*, 2004b). No synergistic action in terms of product yield was found when both enzymes acted simultaneously on wheat bran, but the product mixture was comparable to that produced by the GH10 xylanase alone (Beaugrand *et al.*, 2004b). A large-scale AXOS production based on the enzymic degradation of wheat bran has been developed (Swennen *et al.*, 2006b). Following enzymic removal of starch and proteins from wheat bran, the arabinoxylan-enriched wheat bran fraction was incubated with a GH11 endoxylanase from *B. subtilis*. This allowed kilogram-scale production of AXOS with good purity and relatively good yield. The obtained AXOS had a DP of 15 and A/X ratio of 0.27 and could be further separated using gradual ethanol precipitation, yielding wheat bran AXOS with different structures (i.e. DP ranging between 4 and 59) and A/X ratios ranging between 0.13 and 0.43 (Swennen *et al.*, 2006b).

Starch-derived functional food ingredients

Starch and starch-modifying enzymes

Starch

Starch is the most abundant constituent and most important reserve polysaccharide of cereals. On a molecular level, its major constituents are the glucose polymers amylose and amylopectin. Amylose is an essentially linear molecule, consisting of some 500–6000 α-(1,4)-linked D-glucopyranosyl units. In contrast, amylopectin is a very large and highly branched polysaccharide of up to 3 million glucose units, consisting of linear chains of 10–100 α-(1,4)-linked D-glucopyranosyl units which are connected by α(1,6)-linkages (Manners, 1979; Zobel, 1988). Amylopectin is generally defined in terms of a cluster model (Robin *et al.*, 1974; French, 1984) with polymodal chain length distribution (Hizukuri, 1986) and a non-random nature of branching (Thompson, 2000a). In the cluster model, the short chains, i.e. the unbranched outer A chains and the shortest inner branched chains (B1), form double helices and make up a single cluster, while the longer branched chains (B2–B4) extend into 2 to 4 clusters, respectively.

The amylose/amylopectin ratio varies with the botanical origin of the starch with typical levels of amylose and amylopectin of 20–30% and 70–80%, respectively. Due to deficiency in one or more starch biosynthesis enzymes, such as starch synthases, branching and debranching enzymes, some mutant genotypes of, for example, maize, barley, and rice, contain either an increased amylose content (i.e. high amylose or amylostarch with up to 70% amylose) or an increased amylopectin content (i.e. waxy starch with 99–100% amylopectin), and other starch parameters, such as amylopectin chain length distribution may be altered as well.

Starch occurs as intracellular, water-insoluble semi-crystalline granules of different sizes and shapes, depending on the botanical source. When viewed in polarized light, native starch granules are birefringent and a "Maltese cross" can be observed. This phenomenon results from a degree of order in the starch granule and an orientation of the macromolecules perpendicular to the surface of the granule (Buléon *et al.*, 1998). In addition, native starch is partially crystalline with a degree of crystallinity of 20–40% (Hizukuri, 1996), which is predominantly attributed to structural elements of amylopectin. Different packing of the amylopectin side-chain double helices gives rise to different crystal types. Cereal starches have an A type whereas retrograded starch has a B type X-ray diffraction pattern (Buléon *et al.*, 1998). Several levels of granule organization have been described, including amorphous and semi-crystalline growth rings, blocklets (Gallant *et al.*, 1997), and amorphous and semi-crystalline lamellae (Jenkins *et al.*, 1993; Buléon *et al.*, 1998).

Upon heating starch suspensions in excess water above a critical temperature, starch gelatinizes. During this process, the molecular order of the granule is gradually and irreversibly destroyed. Several events take place during the starch gelatinization process, depending on the conditions and severity of heating: disappearance of the molecular order (and thus the birefringence), granule swelling, water absorption, melting of the starch crystallites, a (limited) leaching of the polymer molecules, mainly amylose, a drastic viscosity increase and a (partial) granule disruption and solubilization (Eerlingen and Delcour, 1995). The gelatinization temperature is characteristic of the starch type and depends on the glass transition of the amorphous fraction of the starch. The total gelatinization event is an endothermic process as demonstrated with differential scanning calorimetry (DSC) (Eerlingen and Delcour, 1995). During cooling and further storage of the gelatinized starch structural transformations occur and new interactions take place in and between the starch polymers. In one definition, these processes are collectively referred to by the term "retrogradation." A new semi-crystalline polymer system is formed, with both amylose (amylose crystallization, in the short term) and amylopectine (amylopectin retrogradation, in the long term) contributing to crystallinity.

Starch-modifying enzymes

Several different enzymes, including endo- and exo-acting amylases, debranching enzymes and transferases, show activity on starch or its derivatives (Figure 11.2). In general, the functionality of amylolytic enzymes comprises several aspects, such as their action on native starch granules, their specificity and the formed degradation products.

GH13, also referred to as the α-amylase family, comprises a variety of amylolytic enzymes, hydrolyzing α-(1,4)- and/or α-(1,6)-linkages between the glucose residues, such as α-amylases and debranching enzymes (Coutinho and Henrissat, 1999; MacGregor *et al.*, 2001; Svensson *et al.*, 2002). α-amylases (EC 3.2.1.1) are typical endo-enzymes, which generate low molecular weight α-dextrins following a more or less random hydrolysis of the α-(1,4)-linkages in the starch polymers (Figure 11.2) (Hoseney, 1994; Bowles, 1996).

The main exo-amylases, i.e. β-amylases (EC 3.2.1.3) and amyloglucosidases (EC 3.2.1.3), are classified in GH14 and 15, respectively. These enzymes are typical inverting exo-amylases, which act on the α-(1,4)-linkages at the non-reducing ends of the starch molecules. β-Amylase releases β-maltose until a branching point is encountered (Figure 11.2). The end-products of β-amylase action on starch are β-maltose and β-limit dextrins. In contrast, amyloglucosidase, also referred to as glucoamylase, has a limited activity on the α-(1,6)-bonds and can hence bypass the side-chains. Theoretically, it can completely convert starch to β-glucose (Figure 11.2) (Bowles, 1996; Hoseney, 1994). The most important debranching enzymes, pullulanase (type I) (EC 3.2.1.41) and isoamylase (EC 3.2.1.68) belong to GH13 as well. These enzymes hydrolyze the α-(1,6)-bonds, thereby liberating the (linear) side-chains (Figure 11.2) (van der Maarel *et al.*, 2002).

Transferases are starch-converting enzymes which cleave α-(1,4)-linkages of the donor molecule and transfer the newly formed reducing end group (donor) to a non-reducing saccharide unit (acceptor) with the formation of a glycosidic bond (van der Maarel *et al.*, 2002). Cyclodextrin glycosyltransferases (CGTase; EC 2.4.1.19; GH13), amylomaltases (EC 2.4.1.25; GHs 13, 57, 77) and amylosucrase

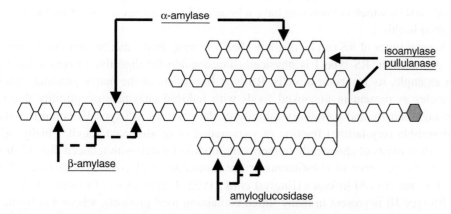

Figure 11.2 Overview of enzymes involved in the enzymic degradation of starch. The gray ring structure represents a reducing glucose residue. Adapted from Goesaert *et al.* (2006).

(EC 2.4.1.4; GH13) form new α-(1,4)-linkages between glucose residues. In general, CGTases have low hydrolytic activity and form cyclic oligosaccharides of 6–8 glucose residues via an intramolecular transglycosylation reaction, while the transglycosylation reaction catalyzed by amylomaltases results in a linear product. Branching enzymes (EC 2.4.1.18, GH13) are transferases which upon cleaving an α-(1,4)-linkage form a new α-(1.6)-glycosidic linkage. They are involved in the biosynthesis of amylopectin and glycogen. Amylosucrase catalyzes the synthesis of amylose-like polymers from sucrose by transferring a glucose residue from sucrose to a glucan polymer (Skov *et al.*, 2001).

Resistant starch and enzyme technology

Resistant starch

It is now recognized that the availability and degradation of dietary starch in the digestive tract differs between different food products, depending on starch and food properties. Indeed, from a nutritional point of view, different starch fractions can be distinguished, i.e. rapidly digestible starch (RDS), slowly digestible starch (SDS) and resistant starch (RS) (Englyst *et al.*, 1992). However, because individuals differ in their ability to digest starch, *in vivo* there is no absolute distinction between these starch fractions (Thompson, 2000b). RDS leads to a rapid glucose release, and hence a rapid increase in blood glucose and insulin levels (which has been related to type 2 diabetes), while RS reduces the glycemic index of foods. In this respect, the reduced starch digestion of the SDS and RS fraction contributes significantly to the control of the glycemic and insulin response (Cummings *et al.*, 2004).

RS is generally defined as starch or starch degradation products not absorbed in the small intestine of healthy individuals, but may be (partially) fermented by the microbiota in the large bowel (Eerlingen and Delcour, 1995). This is in agreement with the definition of dietary fiber and hence RS is a dietary fiber constituent. Although the effect of RS on the increase in stool weight may be modest (Cummings *et al.*, 2004), fermentation of RS gives high yields of butyrate (Martin *et al.*, 2000; Schmiedl *et al.*, 2000; Topping and Clifton, 2001; Cummings *et al.*, 2004; Brouns *et al.*, 2007), which is known to have a beneficial physiological impact on the host's (colon) health.

Several types of RS can be distinguished according to the mechanism that prevents its degradation. RS type I is physically inaccessible for digestive enzymes due to, for example, its inclusion in cells. RS type II consists of the native granular starch granules as present in uncooked foods with reduced enzyme susceptibility due to the high density and partial crystallinity of the granules. RS type III comprises the indigestible (crystalline) fraction of retrograded or reassociated starch. Finally, RS type IV consists of chemically or thermally modified starch with reduced digestibility due to the presence of substituents or the formation of glycoside bonds other than α-(1,4) and α-(1,6)-linkages (Englyst *et al.*, 1992; Eerlingen and Delcour, 1995).

RS type III is present in many starch-containing food products, where it is formed due to starch retrogradation/reassociation following the preparation (heating and cooling) of the food. RS type III is generally believed to consist mainly of associated amylose, with short linear segments of α-(1,4)-glucans arranged in a crystalline structure (Eerlingen and Delcour, 1995). RS levels and characteristics in foods depend on the starch type (waxy, normal, or high amylose), the process conditions (e.g. temperature and storage time), and on the presence of other components (e.g. lipids). Thus, high levels of RS type III may be expected when the starch contains high concentrations of polymer molecules of sufficient length for crystallization/double helices formation (ca. 25 glucose residues), combined with favorable conditions for crystallization (time–temperature; absence of lipids) (Eerlingen and Delcour, 1995). In this respect, high amylose starches are preferred for the production of RS.

Production of resistant starch type III

Usually, the production of RS type III involves the retrogradation/reassociation following the gelatinization/dispersion of the starch. This process can be combined with a partial enzymic degradation.

Hydrolysis after RS formation by α-amylases increases the proportion of RS by removal of non-resistant material (Thompson, 2000b) and allows RS isolation (Garcia-Alonso *et al.*, 1998; Shamai *et al.*, 2003). Partial amylolytic degradation prior to a retrogradation/crystallization step results in an increased polymer mobility for molecular reassociations (Thompson, 2000b). The use of debranching enzymes in the production of RS type III has been well documented in scientific and patent literature (for an overview see Thompson, 2000b). Since amylopectin hinders amylose crystallization, debranching of amylopectin, and hence its removal, allows an enhanced amylose crystallization. Debranching by pullulanase of gelatinized waxy starch and retrogradation/crystallization resulted in RS consisting of linear glucose chains with low DP (DP<40) (Russell *et al.*, 1989). Other researchers have produced crystalline starch fragments (RS type III) by first using a partial amylolytic degradation of starch to produce maltodextrins (with dextrose equivalent below 10), followed by the simultaneous debranching (by isoamylase treatment) and retrogradation/crystallization of the maltodextrin solution (Kettlitz *et al.*, 2000; Pohu *et al.*, 2004). The RS fraction mainly consists of linear α-(1,4)-linked glucose chains of DP 10–35 (Kettlitz *et al.*, 2000).

In addition to their formation by degradation of starch, (insoluble) linear α-(1,4)-glucan of DP<40 can also be synthesized *in vitro* from sucrose using amylosucrase (Schmiedl *et al.*, 2000; Potocki-Veronese *et al.*, 2005). Following heat treatment and a crystallization step at 25°C or 4°C, RS is formed (Schmiedl *et al.*, 2000).

Cereal protein-related functional food aspects

Proteins and protein-modifying enzymes

Proteins of wheat and other cereals

Traditionally, cereal proteins have been classified according to a solubility-based fractionation in albumins (proteins soluble in water), globulins (proteins soluble in dilute salt solutions), prolamins (storage proteins soluble in aqueous alcohol), and glutelins (storage proteins soluble in dilute acid or alkali) (Osborne, 1924).

However, the Osborne fractionation does not provide a clear separation of proteins differing biochemically/genetically or in functionality (Veraverbeke and Delcour, 2002).

Today, particularly in the case of wheat, proteins are preferentially classified from a functional point of view in the non-gluten and the gluten proteins. The former proteins are mostly found in the Osborne albumin and globulin fractions, while the latter are largely insoluble in water and dilute salt solutions. The wheat gluten proteins consist of the gliadins and the functionally distinctly different glutenins. The gliadins form a highly heterogeneous group of monomeric gluten proteins, soluble in aqueous alcohol and, hence are the main components of the Osborne prolamin fraction of wheat. They range in molecular mass from 30 to 80 kDa. The disulfide bonds in gliadins, if present, are intramolecular (Veraverbeke and Delcour, 2002). Glutenins consist of a heterogeneous mixture of polymers with a broad molecular mass range from approximately 80 kDa up into the millions. A large part is soluble in dilute acid conditions and make up the wheat Osborne glutelin fraction. The glutenin polymers are composed of a variety of glutenin subunits (GSs), which are cross-linked by disulfide bonds. Hence, sulfhydryl groups of the GSs are involved in intra- or intermolecular bonds. GSs can be obtained upon reduction of the disulfide bonds and are biochemically related to the gliadins (Veraverbeke and Delcour, 2002; Goesaert et al., 2005).

The properties of gluten proteins allow wheat flour to be transformed into a viscoelastic dough, which is ideally suited for breadmaking and which retains the carbon dioxide produced by the fermenting yeast. These properties are unique and cannot even be found in cereals closely related to wheat such as barley and rye. Due to their large size and the formation of a continuous network, the glutenin polymers provide strength (resistance to deformation) and elasticity to the dough, while the gliadins are believed to act as plasticizers, providing viscosity/plasticity to the dough. Furthermore, in the dough structure, both covalent and non-covalent bonds are involved (Bushuk, 1998; Wrigley et al., 1998). The importance of disulfide cross-links is well established, and oxidative processes are very important during dough development (Wieser, 2003). During baking, gluten proteins undergo a number of complex changes, although the nature of these changes is poorly understood. They are probably a combination of changes in protein surface hydrophobicity, sulfydryl–disulfide interchanges, and formation of new disulfide cross-links (Jeanjean et al., 1980; Schofield et al., 1983; Weegels et al., 1994; Morel et al., 2002).

Protein-modifying enzymes

The functionality of the gluten proteins in wheat-based food products such as bread is strongly determined by the molecular weight of glutenin, the occurrence of covalent and non-covalent bonds between glutenin molecules, and the interactions with other flour components. These different aspects of gluten functionality can be impacted by different enzymes, such as depolymerizing enzymes (proteases) and enzymes enhancing cross-linking reactions (transglutaminases, oxidases).

Proteases

The peptide bonds in the gluten proteins are hydrolyzed by proteases. Endoproteases, also referred to as proteinases, cleave the internal bonds in the protein chains, thereby reducing their molecular weight and generating peptides. Exoproteases and peptidases release amino acids from the chain ends. Classification of most proteolytic enzymes is based on the chemistry of their catalytic mechanism. Serine, thiol or cysteine, metallo, and aspartic proteases can be distinguished, which require a hydroxyl group (serine residue), a sulfhydryl group (cysteine residue), a metal ion (e.g. zinc), and a carboxylic function (aspartic acid residue), respectively, at the active site to function properly (Mathewson, 1998). Protease specificity is related to their preference to cleave peptide bonds involving specific amino acids.

Transglutaminases

Transglutaminase is an acyl transferase catalyzing the transfer of the γ-carboxamide group of protein-bound L-glutamine to primary amines, such as the ε-amino group of protein-bound L-lysine, generating a so-called isopeptide bond (Matheis and Whitaker, 1987). This way, the enzyme builds up new inter- and intramolecular bonds, with the former comprising new covalent non-disulfide cross-links between peptide chains. Side-reactions include amination (i.e. the introduction of free amine groups into proteins) and deamination of glutamine residues (Reinikainen *et al.*, 2003; Gerrard and Sutton, 2005).

Oxidases

Oxidases comprise a family of enzymes that catalyze oxidoreduction reactions with oxygen used as an electron acceptor. Oxidases are classified according to the molecule or functional group that functions as electron donor (http://www.chem.qmul.ac.uk/iubmb/enzyme). Glucose oxidase (EC 1.1.3.4) catalyzes the conversion of glucose and oxygen into gluconolactone (which converts spontaneously into gluconic acid) and hydrogen peroxide (H_2O_2). Hexose oxidase (EC 1.1.3.5) catalyzes a similar reaction but can convert several mono- and oligosaccharides into the corresponding lactones. Tyrosinase is able to oxidize accessible tyrosine residues in proteins to *o*-quinones which are able to condense with each other or with amino and sulfydryl groups of proteins in non-enzymic reactions. It is able to radicalize different aromatic components which can then non-enzymically react with each other (Reinikainen *et al.*, 2003). Glutathione oxidase and sulfydryl oxidase catalyze the formation of disulfide bonds. While the former oxidase is specific for glutathione and releases hydrogen peroxide, the latter may create highly cross-linked gluten fractions (Reinikainen *et al.*, 2003). Other oxidative enzymes, such as laccases (EC 1.10.3.2), can oxidize phenolic compounds. Laccases are copper-containing enzymes that catalyze the oxidation of different phenolic substrates to free radical products, which non-enzymically react with other compounds.

Gluten-free food products and enzyme technology

Gluten intolerance or celiac disease is a frequent, inflammatory, small intestinal disease mainly triggered by the prolamins of wheat, rye, barley, and possibly oats. Gluten proteins are not completely degraded by human gastrointestinal enzymes resulting in toxic peptides, which are mostly derived from glutamine- and proline-rich sections of the gluten proteins.

Enzymes and decreased gluten intolerance

A potential enzyme-induced decrease in gluten intolerance due to proteolytic action has been reported in several studies. Gliadin-derived peptides toxic for patients with celiac disease were rapidly degraded into non-toxic fragments by proteases from germinated wheat, rye, and barley (Hartmann *et al.*, 2006). The proteolytic degradation of gluten proteins has mainly been studied in the context of sourdough fermentations using selected lactic acid bacteria. Sourdough starter cultures were able to degrade gluten proteins under test conditions (Wehrle *et al.*, 1999). Wheat albumin, globulin, and gliadin fractions were hydrolyzed in a sourdough fermentation. Degradation of a toxic peptic-tryptic digest of the gliadin fraction by proteolytic enzymes from selected *Lactobacillus* sp. reduced the toxicity of the digest in *in vitro* experiments (Di Cagno *et al.*, 2002). Likewise, toxicity of rye proteins was decreased following fermentation of rye flour suspensions. This was due to the extensive proteolytic degradation of the rye prolamins. Little bacterial proteolysis of the rye glutelins was observed. Proteolysis of these rye proteins was mainly pH dependent and presumably due to the activation of the rye enzymes (De Angelis *et al.*, 2006). These findings have been confirmed in sourdough breadmaking and pasta-making experiments (Di Cagno *et al.*, 2004, 2005). In breadmaking, after 24 hours of fermentation of a sourdough consisting of a mixture of wheat and non-toxic flours (including buckwheat), the wheat gliadins were almost totally hydrolyzed. This reduced the gluten protein toxicity both in *in vitro* and in *in vivo* experiments (Di Cagno *et al.*, 2004). Likewise, in pasta-making, fermentation of durum wheat semolina by selected lactic acid bacteria resulted in the extensive hydrolysis of the gliadin fraction, which showed a decreased human cell agglutination activity of its peptic-tryptic digest (Di Cagno *et al.*, 2005). These results demonstrate that the level of gluten intolerance in humans can be reduced by a cereal biotechnological approach that uses selected lactobacilli, non-toxic flours, and long (pre)fermentation times (Di Cagno *et al.*, 2004, 2005).

Enzymes in gluten-free breadmaking

In wheat-based breadmaking, enzyme technology is widely used to improve dough handling and end quality of the bread. Many different enzymes have been reported to have a beneficial effect in breadmaking due to their effect on the wheat constituents. This has been extensively discussed by Goesaert and co-workers (2005, 2006). In general, several concepts of enzyme technology in wheat breadmaking, such as the use of amylases for antistaling purposes, are valid for gluten-free breadmaking as well (Gujral *et al.*, 2003).

The classical method of producing gluten-free breads is using a rather complex recipe based on gluten-free cereal ingredients, such as rice and corn flour. One of the main challenging aspects in gluten-free breadmaking is the production of a high-quality bread with good structural properties. Indeed, the proteins of rice and other gluten-free cereals lack the unique protein network and dough-forming properties of the wheat proteins, and are unable to retain the carbon dioxide during fermentation. In this respect, several studies have investigated the functionality of enzymes that induce cross-links, such as transglutaminases and oxidases, in gluten-free breadmaking.

Network formation in rice/corn flour-based gluten-free breads by transglutaminase was reported by Moore and co-workers (2006). The extent of protein network formation was determined by the protein source (with positive effects of e.g. skim milk and egg powder) and enzyme dosage (Moore *et al.*, 2006). In rice breads supplemented with transglutaminase, a decrease in the free amino groups was reported, suggesting enzyme induced cross-linking of the rice proteins (Gujral and Rosell, 2004a). This cross-linking resulted in a dough with improved elastic and viscous properties, which in its turn yielded a rice bread with higher specific volume and crumb strength. Addition of glucose oxidase to rice bread formulations improved rice bread quality as well (Gujral and Rosell, 2004b). Rice proteins were modified as demonstrated by free zone capillary electrophoresis and by the decrease in thiol and amino group content. Overall, these studies demonstrate that the formation of a protein network by transglutaminases or oxidizing enzymes can improve the overall quality, loaf volume and crumb characteristics of gluten-free breads (Moore *et al.*, 2006; Gujral and Rosell, 2004a, 2004b). The functionality of the cross-linking enzymes in gluten-free breadmaking is comparable to that in wheat breadmaking. In wheat breadmaking, transglutaminase can promote the covalent cross-linking of the gluten proteins to larger, insoluble gluten polymers, whereas glucose oxidase functionality is probably related to the generated hydrogen peroxide, which promotes oxidative cross-linking between proteins and/or other components (Goesaert *et al.*, 2006). In contrast to oxidases, the formation of new inter- and intramolecular bonds by transglutaminases is independent of the redox system in dough (Reinikainen *et al.*, 2003).

Final remarks

The specific modification and/or degradation of cereal constituents using enzyme technology is a highly promising way to produce potential health promoting food ingredients, such as soluble (high molecular weight) dietary fiber, prebiotic (A)XOS and resistant starch, and high-quality gluten-free food products.

The *in vitro* production of large quantities of the arabinoxylan- and starch-derived functional food ingredients calls for easily available and cheap raw materials as well as for a careful selection of enzymes with suitable properties. In the case of the production of NSP-derived functional food components, coproducts of cereal processing containing high levels of NSP are abundantly available. However, in many instances they are not readily susceptible to enzyme modification. Also, the endoxylanase-induced solubilization of WU-AX increases viscosity and generates a soluble arabinoxylan population which, in its turn, can be degraded enzymically. Both factors complicate the *in vitro* production of high molecular weight soluble arabinoxylans. Increased understanding of the molecular basis of enzyme properties, such as specificity and selectivity, will give new perspectives for upgrading the waste streams to valuable products consisting of components with potential health effects. In the case of enzyme-assisted resistant starch production, the combined use of specific starches or starch mutants with one or more starch-modifying enzymes, including amylases, debranching enzymes, and/or transferases, shows great potential.

Enzyme technology is specifically well suited for the *in situ* production of the above functional food ingredients. However, this remains a highly challenging research area. Indeed, the functionality of the enzyme-modified cereal constituents often differs from that of the unmodified polymers and this has to be taken into account when producing high-quality food products. Furthermore, sufficient levels of the functional food components should be produced for the end-product to have a potential health effect. In this respect, the combined use of hull-less barley flour and enzyme technology has allowed to produce tasty and consumer acceptable bread products with *in situ* generated increased levels of the health-promoting soluble dietary fiber components WE-AX and β-glucan.

As in the case of the *in situ* generation of functional food components, enzyme technology (in combination with fermentation technology) also holds promise for the *in situ* removal of gluten proteins, which are toxic for patients with celiac disease. This way, food products with reduced gluten intolerance properties may become available. However, the different functionality of the modified cereal constituent gluten proteins has to be taken into account. Finally, enzyme supplementation can be an efficient way to obtain high-quality food products from gluten-free ingredients. Research in this area will benefit from more insight into the properties of the constituents (protein, starch, and NSP) of the gluten-free cereals, combined with an increased understanding of the functionality of the different enzymes.

References

American Association of Cereal Chemists (2001). Report of the Dietary Fiber Definition Committee. The definition of dietary fiber. *Cereal Foods World* 46, 112–126.

American Association of Cereal Chemists (2003). Report by the AACC Dietary Fiber Technical Committee. All dietary fiber is fundamentally functional. *Cereal Foods World* 48, 128–132.

Ashwell, M. (2002). *Concepts of Functional Foods*. Washington DC: International Life Science Institute Press, 40pp.

Beaugrand, J., Reis, D., Guillon, F., Debeire, P., and Chabbert, B. (2004a). Xylanase-mediated hydrolysis of wheat bran: evidence for subcellular heterogeneity of cell walls. *Int. J. Plant Sci.* 165, 553–563.

Beaugrand, J., Chambat, G., Wong, V. W. K. *et al.* (2004b). Impact and efficiency of GH10 and GH11 thermostable endoxylanases on wheat bran and alkaliextractable arabinoxylans. *Carbohydr. Res.* 339, 2529–2540.

Biely, P., Vrsanskà, M., Tenkanen, M., and Kluepfel, D. (1997). Endo-beta-1,4-xylanase families: differences in catalytic properties. *J. Biotechnol.* 57, 151–166.

Bonnin, E., Daviet, S., Sørensen, J. F. *et al.* (2006). Behaviour of family 10 and 11 xylanases towards arabinoxylans with varying structure. *J. Sci. Food. Agric.* 86, 1618–1622.

Bourdon, I., Yokoyama, W., Davis, P. *et al.* (1999). Postprandial lipid, glucose, insulin, and cholecystokinin responses in men fed barley pasta enriched with β-glucan. *Am. J. Clin. Nutr.* 69, 55–63.

Bowles, L. K. (1996). Amylolytic enzymes. In: Hebeda, R. E. and Zobel, H. F. eds. *Baked Goods Freshness: Technology, Evaluation, and Inhibition of Staling.* New York: Marcel Dekker, pp. 105–129.

Brouns, F., Arrigoni, E., Langkilde, A. M. *et al.* (2007). Physiological and metabolic properties of a digestion-resistant maltodextrin, classified as type 3 retrograded resistant starch. *J. Agric. Food Chem.* 55, 1574–1581.

Buléon, A., Colonna, P., Planchot, V., and Ball, S. (1998). Starch granules: structure and biosynthesis. *Int. J. Biol. Macromol.* 23, 85–112.

Bushuk, W. (1998). Interactions in wheat doughs. In: Hamer, R. J. and Hoseney, R. C. eds. *Interactions: The Keys to Cereal Quality.* St. Paul, MN: American Association of Cereal Chemists, pp. 1–16.

Cavallero, A., Empilli, S., Brighenti, F., and Stanca, A. M. (2002). High (1→3,1→4)-β-glucan barley fractions in bread making and their effects on human glycemic response. *J. Cereal Sci.* 36, 59–66.

Charalampopoulos, D., Wang, R., Pandiella, S. S., and Webb, C. (2002). Application of cereals and cereal components in functional foods: a review. *Int. J. Food Microbiol.* 79, 131–141.

Cleemput, G., Roels, S. P., Van Oort, M., Grobet, P. J., and Delcour, J. A. (1993). Heterogeneity in the structure of water-soluble arabinoxylans in European wheat flours of variable bread-making quality. *Cereal Chem.* 70, 324–329.

Cleemput, G., van Oort, M., Hessing, M. *et al.* (1995). Variation in the degree of D-xylose substitution in arabinoxylans extracted from a European wheat flour. *J. Cereal Sci.* 22, 73–84.

Collins, T., Gerday, C., and Feller, G. (2005). Xylanases, xylanase families and extremophilic xylanases. *FEMS Microbiol. Rev.* 29, 3–23.

Courtin, C. M. and Delcour, J. A. (2002). Arabinoxylans and endoxylanases in wheat flour bread-making. *J. Cereal Sci.* 35, 225–243.

Courtin, C. M., Roelants, A., and Delcour, J. A. (1999). Fractionation-reconstitution experiments provide insight into the role of endoxylanases in bread-making. *J. Agric. Food Chem.* 47, 1870–1877.

Courtin, C. M., Gelders, G. G., and Delcour, J. A. (2001). The use of two endoxylanases with different substrate selectivity provides insight into the functionality of arabinoxylans in wheat flour breadmaking. *Cereal Chem.* 78, 564–571.

Coutinho, P. M. and Henrissat, B. (1999). Carbohydrate-active enzymes server. http://www.cazy.org/.

Crittenden, R., Karppinen, S., Ojanen, S. *et al.* (2002). *In vitro* fermentation of cereal dietary fibre carbohydrates by probiotic and intestinal bacteria. *J. Sci. Food. Agric.* 82, 781–789.

Cummings, J.H., Edmond, L.M., and Magee, E.A. (2004). Dietary carbohydrates and health: do we still need the fibre concept? *Clin. Nutr.* Suppl. 1, 5–17.

De Angelis, M., Coda, T., Silano, M. *et al.* (2006). Fermentation by selected sourdough lactic acid bacteria to decrease coeliac intolerance to rye flour. *J. Cereal Sci.* 43, 301–314.

Debyser, W., Peumans, W. J., Van Damme, E. J. M., and Delcour, J. A. (1999). *Triticum aestivum* xylanase inhibitor (TAXI), a new class of enzyme inhibitor affecting bread-making performance. *J. Cereal Sci.* 30, 39–43.

Dervilly, G., Saulnier, L., Roger, P., and Thibault, J.-F. (2000). Isolation of homogeneous fractions from wheat water-soluble arabinoxylans. Influence of the structure on their macromolecular characteristics. *J. Agric. Food Chem.* 48, 270–278.

Di Cagno, R., De Angelis, M., Lavermicocca, P. *et al.* (2002). Proteolysis by sourdough lactic acid bacteria: effects on wheat flour protein fractions and gliadin peptides involved in human cereal intolerance. *Appl. Environ. Microbiol.* 68, 623–633.

Di Cagno, R., De Angelis, M., Auricchio, S. *et al.* (2004). Sourdough bread made from wheat and nontoxic flours and started with selected lactobacilli is tolerated in celiac sprue patients. *Appl. Environ. Microbiol.* 70, 1088–1096.

Di Cagno, R., De Angelis, M., Alfonsi, G. *et al.* (2005). Pasta made from durum wheat semolina fermented with selected lactobacilli as a tool for a potential decrease of the gluten intolerance. *J. Agric. Food Chem.* 53, 4393–4402.

Eerlingen, R. C. and Delcour, J. A. (1995). Formation, analysis, structure and properties of type III enzyme resistant starch. *J. Cereal Sci.* 22, 129–138.

Englyst, H. N., Kingman, S. M., and Cummings, J. H. (1992). Classification and measurements of nutritionally important starch fractions. *Eur. J. Clin. Nutr.* 46, S33–S50.

Fausch, H., Kündig, W., and Neukom, H. (1963). Ferulic acid as a component of a glycoprotein from wheat flour. *Nature* 199, 28.

FDA (Food and Drug Administration) (1997). Food labeling: Health claims; Oats and coronary heart disease; Rules and Regulations. *Fed. Reg.* 62, 3584–3601.

Fierens, E., Rombouts, S., Gebruers, K. *et al.* (2007). TLXI, a novel type of xylanase inhibitor from wheat (*Triticum aestivum*) belonging to the thaumatin family. *Biochem. J.* 403, 583–591.

Fierens, K., Gils, A., Sansen, S. *et al.* (2005). His374 of wheat endoxylanase inhibitor TAXI-I stabilises complex formation with glycoside hydrolase family 11 endoxylanases. *FEBS J* 272, 5872–5882.

Fierens, E. (2007). TLXI, a thaumatin-like xylanase inhibitor: isolation, characterisation and comparison with other wheat (Triticum aestivum L.) xylanase inhibiting proteins. PhD dissertation, Katholieke Universiteit Leuven, Leuven, Belgium.

Figueroa-Espinoza, M. C. and Rouau, X. (1998). Oxidative crosslinking of pentosans by a fungal laccase and horseradish peroxidase, mechanism of linkage between feruloylated arabinoxylans. *Cereal Chem.* 75, 259–265.

Figueroa-Espinoza, M. C., Poulsen, C., Soe, J. B., Zargahi, M. R., and Rouau, X. (2004). Enzymatic solubilization of arabinoxylans from native, extruded and high-shear–treated rye bran by different endo-xylanases and other hydrolyzing enzymes. *J. Agric. Food Chem.* 52, 4240–4249.

Fincher, G. B. and Stone, B. A. (1986). Cell walls and their components in cereal grain technology. In: Pomeranz, Y. ed. *Advances in Cereal Science and Technology*, Vol. VIII. St. Paul, MN: American Association of Cereal Chemists, pp. 207–295.

Flatman, R., McLauchlan, W. R., Juge, N. *et al.* (2002). Interactions defining the specificity between fungal xylanases and the xylanase-inhibiting protein XIP-I from wheat. *Biochem. J.* 365, 773–781.

French, D. (1984). Organization of starch granules. In: Whistler, R. L., BeMiller, J. N., and Paschal, E. F. eds. *Starch Chemistry and Technology*, 2nd edn. New York: Academic Press, pp. 183–212.

Gallant, D. J., Bouchet, B., and Baldwin, P. M. (1997). Microscopy of starch: evidence of a new level of granule organization. *Carbohydr. Polym.* 32, 177–191.

Garcia-Alonso, A., Calixto, F. S., and Delcour, J. A. (1998). Influence of botanical source and processing on formation of resistant starch type III. *Cereal Chem.* 75, 802–804.

Gebruers, K., Debyser, W., Goesaert, H., Proost, P., Van Damme, J., and Delcour, J. A. (2001). *Triticum aestivum* L. endoxylanase inhibitor (TAXI) consists of two inhibitors, TAXI I and TAXI II, with different specificities. *Biochem. J.* 353, 239–244.

Gebruers, K., Brijs, K., Courtin, C. M. *et al.* (2004). Properties of TAXI-type endoxylanase inhibitors. *Biochim. Biophys. Acta* 1696, 213–221.

Gerrard, J. A. and Sutton, K. H. (2005). Addition of transglutaminase to cereal products may genereate the epitope responsable for coeliac disease. *Trends Food Sci. Technol.* 16, 510–512.

Gibson, G. R. and Roberfroid, M. B. (1995). Dietary modulation of the human colonic microbiota: introducing the concept of prebiotics. *J. Nutr.* 125, 1401–1412.

Goesaert, H., Gebruers, K., Brijs, K., Courtin, C. M., and Delcour, J. A. (2003a). TAXI type endoxylanase inhibitors in different cereals. *J. Agric. Food Chem.* 51, 3770–3775.

Goesaert, H., Gebruers, K., Brijs, K., Courtin, C. M., and Delcour, J. A. (2003b). XIP-type endoxylanase inhibitors in different cereals. *J. Cereal Sci.* 38, 317–324.

Goesaert, H., Elliott, G., Kroon, P. A. *et al.* (2004). Occurrence of proteinaceous endoxylanase inhibitors in cereals. *Biochim. Biophys. Acta* 1696, 193–202.

Goesaert, H., Brijs, K., Veraverbeke, W. S., Courtin, C. M., Gebruers, K., and Delcour, J. A. (2005). Wheat flour constituents: how they impact bread quality, and how to impact their functionality. *Trends Food Sci. Technol.* 16, 12–30.

Goesaert, H., Gebruers, K., Courtin, C. M., Brijs, K., and Delcour, J. A. (2006). Enzymes in breadmaking. In: Hui, Y. H., Corke, H., De Leyn, I., Nip, W.-K., and Cross, N. eds. *Bakery Products: Science and Technology*. Ames, Iowa: Blackwell Publishing Company, Chapter 19, pp. 337–364.

Goldschmid, H. R. and Perlin, A. S. (1963). Interbranch sequences in the wheat arabinoxylans. Selective enzymolysis studies. *Can. J. Chem.* 41, 2272–2277.

Gråsten, S., Liukkonen, K.-H., Chrevatidis, A., El-Nezami, H., Poutanen, K., and Mykkänen, H. (2003). Effects of wheat pentosan and inulin on the metabolic activity of fecal microbiota and on bowel function in healthy humans. *Nutr. Res.* 23, 1503–1514.

Gruppen, H., Hoffmann, R. A., Kormelink, F. J. M., Voragen, A. G. J., Kamerling, J. P., and Vliegenthart, J. F. G. (1992). Characterization by H-1-NMR spectroscopy of enzymatically derived oligosaccharides from alkali-extractable wheat flour arabinoxylan. *Carbohydr. Res.* 233, 45–64.

Gruppen, H., Kormelink, F. J. M., and Voragen, A. G. J. (1993). Water-unextractable cell wall material from wheat flour. III. A structural model for arabinoxylans. *J. Cereal Sci.* 19, 111–128.

Gujral, H. S. and Rosell, C. M. (2004a). Functionality of rice flour modified with a microbial transglutaminase. *J. Cereal Sci.* 39, 225–230.

Gujral, H. S. and Rosell, C. M. (2004b). Improvement of the breadmaking quality of rice flour by glucose oxidase. *Food Res. Int.* 37, 75–81.

Gujral, H. S., Haros, M., and Rosell, C. M. (2003). Starch hydrolysing enzymes for retarding the staling of rice bread. *Cereal Chem.* 80, 750–754.

Hartmann, G., Koehler, P., and Wieser, H. (2006). Rapid degradation of gliadin peptides toxic for coeliac disease patients by proteases from germinating cereals. *J. Cereal Sci.* 44, 368–371.

Hecker, K. D., Meier, M. L., Newman, R. K., and Newman, C. W. (1998). Barley β-glucan is effective as a hypocholesterolaemic ingredient in foods. *J. Sci. Food. Agric.* 77, 179–183.

Henrissat, B. (1991). A classification of glycosyl hydrolases based on amino acid sequence similarities. *Biochem. J.* 280, 309–316.

Hizukuri, S. (1986). Polymodal distribution of the chain lengths of amylopectins, and its significance. *Carbohydr. Res.* 147, 342–347.

Hizukuri, S. (1996). Starch: analytical aspects. In: Eliasson, A.-C. ed. *Carbohydrates in Food*. New York: Marcel Dekker, pp. 347–429.

Hoseney, R. C. (1994). *Principles of Cereal Science and Technology*, 2nd edn. St. Paul, MN: Association of Cereal Chemists, pp. 81–101, 229–273.

Hsu, C. K., Liao, J. W., Chung, Y. C., Hsieh, C. P., and Chan, Y. C. (2004). Xylooligosaccharides and fructooligosaccharides affect the intestinal microbiota and precancerous colonic lesion development in rats. *J. Nutr.* 134, 1523–1528.

Iiyama, K., Lam, T. B. T., and Stone, B. A. (1994). Covalent cross-links in the cell wall. *Plant Physiol.* 104, 315–320.

Jaskari, J., Kontula, P., Siitonen, A., Jousimies-Somer, H., Mattila-Sandholm, T., and Poutanen, K. (1998). Oat beta-glucan and xylan hydrolysates as selective substrates for *Bifidobacterium* and *Lactobacillus* strains. *Appl. Microbiol. Biotechnol.* 49, 175–181.

Jeanjean, M. F., Damidaux, R., and Feillet, P. (1980). Effect of heat treatment on protein solubility and viscoelastic properties of wheat gluten. *Cereal Chem.* 57, 325–331.

Jeffries, T. W. (1996). Biochemistry and genetics of microbial xylanases. *Curr. Opin. Biotechnol.* 7, 337–342.

Jenkins, P. J., Cameron, R. E., and Donald, A. M. (1993). A universal feature in the structure of starch granules from different botanical sources. *Starch/Stärke* 45, 417–420.

Juge, N., Payan, F., and Williamson, G. (2004). XIP-I, a xylanase inhibitor protein from wheat: a novel protein function. *Biochim. Biophys. Acta* 1696, 203–211.

Kettlitz, B. W., Coppin, J. V. J.-M., Roeper, H. W. W., and Bornet, F. (2000). Highly fermentible resistant starch. US Patent 6,043,229.

Klopfenstein, C. F. (1988). The role of cereal beta-glucans in nutrition and health. *Cereal Foods World* 33, 865–869.

Kontula, P., von Wright, A., and Mattila-Sandholm, T. (1998). Oat bran β-gluco- and xylooligosaccharides as fermentative substrates for lactic acid bacteria. *Int. J. Food Microbiol.* 45, 163–169.

Korhonen, H. and Pihlanto, A. (2003). Food-derived bioactive peptides—opportunities for designing future foods. *Curr. Pharmac. Design* 9, 1297–1308.

Korhonen, H., Pihlanto-Leppälä, A., Rantamäki, P., and Tupasela, T. (1998). Impact of processing on bioactive proteins and peptides. *Trends Food Sci. Technol.* 9, 307–319.

Kormelink, F. J. M. and Voragen, A. G. J. (1992). Combined action of xylandegrading and accessory enzymes on different ([glucurono]-arabino)xylans. In: Visser, J., Beldman, G., Kusters-van Someren, M. A., and Voragen, A. G. J. eds. *Xylans and Xylanases, Progress in Biotechnology.* Amsterdam: Elsevier Science Publishers, pp. 415–418.

Kormelink, F. J. M., Hoffmann, R. A., Gruppen, H., Voragen, A. G. J., Kamerling, J. P., and Vliegenthart, J. F. G. (1993). Characterization by H-1-NMR spectroscopy of oligosaccharides derived from alkali-extractable wheat-flour arabinoxylan by digestion with endo-(1,4)-beta-D-xylanase III from *Aspergillus awamori. Carbohydr. Res.* 249, 369–382.

Lanza, E., Jone, D. Y., Block, G., and Kessler, L. (1987). Dietary fiber intake in the U.S. population. *Am. J. Clin. Nutr.* 46, 790–797.

Lu, Z. X., Walker, K. Z., Muir, J. G., Mascara, T., and O'Dea, K. (2000a). Arabinoxylan fiber, a byproduct of wheat flour processing, reduces the postprandial glucose response in normoglycemic subjects. *American Journal of Clinical Nutrition*, 71, 1123–1128.

Lu, Z. X., Gibson, P. R., Muir, J. G., Fielding, M., and O'Dea, K. (2000b). Arabinoxylan fiber from a by-product of wheat flour processing behaves physiologically like a soluble, fermentable fiber in the large bowel of rats. *J. Nutr.* 130, 1984–1990.

MacGregor, E. A., Janecek, Š., and Svensson, B. (2001). Relationship of sequence and structure to specificity in the α-amylase family of enzymes. *Biochim. Biophys. Acta* 1546, 1–20.

Maes, C., Vangeneugden, B., and Delcour, J. A. (2004). Relative activity of two endoxylanases towards water-unextractable arabinoxylans in wheat bran. *J. Cereal Sci.* 39, 181–186.

Manners, D. J. (1979). The enzymic degradation of starches. In: Blanshard, J. M. V. and Mitchell, J. R., eds. *Polysaccharides in Food.* London: Butterworths, pp. 75–91.

Manthey, F. A., Hareland, G. A., and Huseby, D. J. (1999). Soluble and insoluble dietary fiber content and composition in oat. *Cereal Chem.* 76, 417–420.

Martin, L. J. M., Dumon, H. J. W., Lecannu, G., and Champ, M. (2000). Potato and high amylose maize starches are not equivalent producers of butyrate for the colonic mucosa. *Br. J. Nutr.* 84, 689–696.

Matheis, G. and Whitaker, J. R. (1987). A review: Enzymatic cross-linking of proteins applicable to foods. *J. Food Biochem.* 11, 309–327.

Mathewson, P. R. (1998). Common enzyme reactions. *Cereal Foods World* 43, 798–803.

McIntosh, G. H., Le Leu, R. K., Kerry, A., and Goldring, M. (1993). Barley grain for human food use. *Food Australia* 45, 392–394.

McLauchlan, W. R., Garcia-Conesa, M. T., Williamson, G., Roza, M., Ravestein, P., and Maat, J. (1999). A novel class of protein from wheat which inhibits xylanases. *Biochem. J.* 338, 441–446.

Meuser, F. and Suckow, P. (1986). Non-starch polysaccharides. In: Blanshard, J. M. V., Frazier, P. J., and Galliard, T. eds. *Chemistry and Physics of Baking*. London: The Royal Society of Chemistry, pp. 42–61.

Moers, K., Courtin, C. M., Brijs, K., Delcour, J. A. (2003). A screening method for endo-β-1,4-xylanase substrate selectivity. *Analyt. Biochem.* 319, 73–77.

Moers, K., Celus, I., Brijs, K., Courtin, C. M., and Delcour, J. A. (2005). Endoxylanase substrate selectivity determines degradation of wheat water-extractable and waterunextractable arabinoxylan. *Carbohydr. Res.* 340, 1319–1327.

Moore, M. M., Heinbockel, M., Dockery, P., Ulmer, H. M., and Arendt, E. K. (2006). Network formation in gluten-free bread with application of transglutaminase. *Cereal Chem.* 83, 28–36.

Morel, M. H., Redl, A., and Guilbert, S. (2002). Mechanism of heat and sheat mediated aggregation of wheat gluten protein upon mixing. *Biomacromolecules* 3, 488–497.

Mussatto, S. I. and Mancilha, I. M. (2007). Non-digestible oligosaccharides: a review. *Carbohydr. Polym.* 68, 587–597.

Ordaz-Ortiz, J. J., Guillon, F., Tranquet, O., Dervilly-Pinel, G., Tran, V., and Saulnier, L. (2004). Specificity of monoclonal antibodies generated against arabinoxylans of cereal grains. *Carbohydr. Polym.* 57, 425–433.

Osborne, T. B. (1924). *The Vegetable Proteins*. London: Longmans Green and Co.

Payan, F., Leone, P., Porciero, S. *et al.* (2004). The dual nature of the wheat xylanase protein inhibitor XIP-I. Structural basis for the inhibition of family 10 and family 11 xylanases. *J. Biol. Chem.* 279, 36029–36037.

Perlin, A. S. (1951a). Isolation and composition of the soluble pentosans of wheat flour. *Cereal Chem.* 28, 370–381.

Perlin, A. S. (1951b). Structure of the soluble pentosans of wheat flours. *Cereal Chem.* 28, 282–393.

Petit-Benvegnen, M. D., Saulnier, L., and Rouau, X. (1998). Solubilization of arabinoxylans from isolated water-unextractable pentosans and wheat flour doughs by cell-wall-degrading enzymes. *Cereal Chem.* 75, 551–556.

Pohu, A., Planchot, V., Putaux, J. L., Colonna, P., and Buléon, A. (2004). Split crystallization during debranching of maltodextrins at high concentration by isoamylase. *Biomacromolecules* 5, 1792–1798.

Potocki-Veronese, G., Putaux, J. L., Dupeyre, D. *et al.* (2005). Amylose synthesized in vitro by amylosucrase: morphology, structure and properties. *Biomacromolecules* 6, 1000–1011.

Rantanen, H., Virkki, L., Tuomainen, P., Kabel, M., Schols, H., and Tenkanen, M. (2007). Preparation of arabinoxylobiose from rye xylan using family 10 *Aspergillus aculeatus* endo-1,4-β-D-xylanase. *Carbohydrate Polym.* 68, 350–359.

Reinikainen, T., Lantto, R., Niku-Paavola, M.-L., and Buchert, J. (2003). Enzymes for cross-linking of cereal polymers. In: Courtin, C. M., Veraverbeke, W. S., and Delcour, J. A. ed. *Recent Advances in Enzymes in Grain Processing*. Leuven, Belgium: Laboratory of Food Chemistry (K.U. Leuven), pp. 91–99.

Rieckhoff, D., Trautwein, E. A., Mälkki, Y., and Erbersdobler, H. F. (1999). Effects of different cereal fibers on cholesterol and bile acid metabolism in the Syrian golden hamster. *Cereal Chem.* 76, 788–795.

Robin, J. P., Mercier, C., Charbonnière, R., and Guilbot, A. (1974). Lintnerized starches. Gel filtration and enzymatic studies of insoluble residues from prolonged acid treatment of potato starch. *Cereal Chem.* 51, 389–406.

Rouau, X., Daviet, S., Tahir, T., Cherel, B., and Saulnier, L. (2006). Effect of the proteinaceous wheat xylanase inhibitor XIP-I on the performances of an *Aspergillus niger* xylanase in breadmaking. *J. Sci. Food. Agric.* 86, 1604–1609.

Russell, P. L., Berry, C. S., and Greenwell, P. (1989). Characterisation of resistant starch from wheat and maize. *J. Cereal Sci.* 9, 1–5.

Sansen, S., De Ranter, C. J., Gebruers, K. *et al.* (2004a). Crystallization and preliminary X-ray diffraction study of two complexes of a TAXI-type xylanase inhibitor with glycoside hydrolase family 11 xylanases from *A. niger* and *B. subtilis*. *Acta Crystallogr. D* 60, 555–557.

Sansen, S., De Ranter, C. J., Gebruers, K. *et al.* (2004b). Structural basis for inhibition of *Aspergillus niger* xylanase by *Triticum aestivum* xylanase inhibitor-I. *J. Biol. Chem.* 279, 36022–36028.

Schmiedl, D., Bäuerlein, M., Bengs, H., and Jacobasch, G. (2000). Production of heat-stable, butyrogenic resistant starch. *Carbohydr. Polym.* 43, 183–193.

Schofield, J. D., Bottomley, R. C., Timms, M. F., and Booth, M. R. (1983). The effect of heat on wheat gluten and the involvement of sulphydryl-disulphide interchange reactions. *J. Cereal Sci.* 1, 241–253.

Shamai, K., Bianco-Peled, H., and Shimoni, E. (2003). Polymorphism of resistant starch type III. *Carbohydr. Polym.* 54, 363–369.

Sibbesen, O. and Sørensen, J.F. (2001). Enzyme. Patent application WO 01/66711 A1.

Simpson, D. J., Fincher, G. B., Huang, A. H. C., and Cameron-Mills, V. (2002). Structure and function of cereal and related higher plant (1→4)-β-xylan endohydrolases. *J. Cereal Sci.* 37, 111–127.

Skov, L. K., Mirza, O., Henriksen, A. *et al.* (2001). Amylosucrase, a glucan-synthesizing enzyme from the alpha-amylase family. *J. Biol. Chem.* 276, 25273–25278.

Svensson, B., Tovborg Jensen, M., Mori, H. *et al.* (2002). Fascinating facets of function and structure of amylolytic enzymes of glycoside hydrolase family 13. *Biologia* 57, Suppl. 11, 5–19.

Swennen, K. (2007). Production, characterisation and functionality of arabinoxylooligosaccharides with different structures. PhD dissertation, Katholieke Universiteit Leuven, Leuven, Belgium.

Swennen, K., Courtin, C. M., Van der Bruggen, B., Vandecasteele, C., and Delcour, J. A. (2005). Ultrafiltration and ethanol precipitation for isolation of arabinoxylooligosaccharides with different structures. *Carbohydr. Polym.* 62, 283–292.

Swennen, K., Courtin, C. M., and Delcour, J. A. (2006a). Non-digestible oligosaccharides with prebiotic properties. *Crit. Rev. Food Sci. Nutr.* 46, 459–471.

Swennen, K., Courtin, C. M., Lindemans, G. C. J. E., and Delcour, J. A. (2006b). Large scale production and characterisation of wheat bran arabinoxylooligosaccharides. *J. Sci. Food. Agric.* 86, 1722–1731.

Tahir, T. A., Berrin, J.-G., Flatman, R. *et al.* (2002). Specific characterisation of substrate and inhibitor binding sites of a glycosyl hydrolase family 11 xylanase from *Aspergillus niger. J. Biol. Chem.* 277, 44035–44043.

Thompson, D. B. (2000a). On the non-random nature of amylopectin branching. *Carbohydr. Polym.* 43, 223–239.

Thompson, D. B. (2000b). Strategies for the manufacture of resistant starch. *Trends Food Sci. Technol.* 11, 245–253.

Topping, D. L. and Clifton, P. M. (2001). Short chain fatty acids and human colonic function: roles of resistant starch and non-starch polysaccharides. *Physiol. Rev.* 81, 1031–1064.

Törrönen, A. and Rouvinen, J. (1997). Structural and functional properties of low molecular weight endo-1,4-beta-xylanases. *J. Biotechnol.* 57, 137–149.

Trogh, I., Courtin, C. M., Andersson, A. A. M., Åman, P., Sørensen, J. F., and Delcour, J. A. (2004). The combined use of hull-less barley flour and xylanase as a strategy for wheat/hull-less barley flour breads with increased arabinoxylan and (1–3)(1–4)-β-D-glucan levels. *J. Cereal Sci.* 40, 257–267.

Trogh, I., Croes, E., Courtin, C. M., and Delcour, J. A. (2005a). Enzymic degradability of hull-less barley flour alkali-solubilised arabinoxylan fractions by endoxylanases. *J. Agric. Food Chem.* 53, 7243–7250.

Trogh, I., Courtin, C. M., Goesaert, H. *et al.* (2005b). From hull-less barley and wheat to soluble dietary fiber-enriched bread. *Cereal Foods World* 50, 253–260.

Trogh, I., Courtin, C. M., and Delcour, J. A. (2007). Barley β-glucan and wheat arabinoxylan soluble fiber technologies for health-promoting bread products. In: Marquart, L., Jacobs, D. R., McIntosh, G. H., Poutanen, K., and Reicks, M. eds. *Whole Grains and Health*. Ames, Iowa: Blackwell Publishing, Chapter 13, pp. 157–176.

van der Maarel, M. J. E. C., van der Veen, B., Uitdehaag, J. C. M., Leemhuis, H., and Dijkhuizen, L. (2002). Properties and applications of starch-converting enzymes of the α-amylase family. *J. Biotechnol.* 94, 137–155.

Van Laere, K. M. J., Hartemink, R., Bosveld, M., Schols, H. A., and Voragen, A. G. J. (2000). Fermentation of plant cell wall derived polysaccharides and their corresponding oligosaccharides by intestinal bacteria. *J. Agric. Food Chem.* 48, 1644–1652.

Vázquez, M. J., Alonso, J. L., Domínguez, H., and Parajó, J. C. (2000). Xylooligosaccharides: manufacture and applications. *Trends Food Sci. Technol.* 11, 387–393.

Veraverbeke, W. S. and Delcour, J. A. (2002). Wheat protein composition and properties of wheat glutenin in relation to breadmaking functionality. *CRC Crit. Rev. Food Sci. Nutr.* 42, 179–208.

Verwimp, T., Van Craeyveld, V., Courtin, C. M., and Delcour, J. A. (2007). Variability in the structure of rye flour alkali-extractable arabinoxylans. *J. Agric. Food Chem.* 55, 1985–1992.

Vinkx, C. J. A., Van Nieuwenhove, C. G., and Delcour, J. A. (1991). Physicochemical and functional properties of rye nonstarch polysaccharides. III. Oxidative gelation of a fraction containing water-soluble pentosans and proteins. *Cereal Chem.* 68, 617–622.

Vinkx, C. J. A. and Delcour, J. A. (1996). Rye (Secale cereale L.) arabinoxylans. *J. Cereal Sci.* 24, 1–14.

Voragen, A. G. J., Gruppen, H., Verbruggen, M. A., and Viëtor, R. J. (1992). Characterization of cereal arabinoxylans. In: Visser, J., Beldman, G., Kusters-van Someren, M. A., and Voragen, A. G. J. eds. *Xylans and Xylanases, Progress in Biotechnology.* Amsterdam: Elsevier Science, pp. 51–67.

Weegels, P. L., de Groot, A. M. G., Verhoek, J. A., and Hamer, R. J. (1994). Effects on gluten of heating at different moisture contents. II. Changes in physico-chemical properties and secondary structure. *J. Cereal Sci.* 19, 39–47.

Wehrle, K., Crowe, N., Van Boeijen, I., and Arendt, E. K. (1999). Screening methods for the proteolytic breakdown of gluten by lactic acid bacteria and enzyme preparations. *Eur. Food Res. Technol.* 209, 428–433.

Wieser, H. (2003). The use of redox agents. In: Cauvain, S. P. ed. *Bread Making: Improving Quality.* Cambridge: Woodhead Publishing, pp. 424–446.

Wood, P. J., Weisz, J., and Blackwell, B. A. (1991). Molecular characterization of cereal β-D-glucans. Structural analysis of oat β-D-glucan and rapid structural evaluation of β-D-glucans from different sources by high-performance liquid chromatography of oligosaccharides released by lichenase. *Cereal Chem.* 68, 31–39.

Wood, P. J., Weisz, J., and Blackwell, B. A. (1994). Structural studies of $(1 \rightarrow 3),(1 \rightarrow 4)$-β-D-glucans by ^{13}C-nuclear magnetic resonance spectroscopy and by rapid analysis of cellulose-like regions using high-performance anion-exchange chromatography of oligosaccharides released by lichenase. *Cereal Chem.* 71, 301–307.

Wrigley, C. W., Andrews, J. L., Békés, F. *et al.* (1998). Protein-protein interactions essential to dough rheology. In: Hamer, R.J. and Hoseney, R.C. eds. *Interactions: the Keys to Cereal Quality.* St. Paul, MN: American Association of Cereal Chemists, pp. 17–46.

Yokoyama, W. H., Hudson, C. A., Knuckles, B. E. *et al.* (1997). Effect of barley β-glucan in durum wheat pasta on human glycemic response. *Cereal Chem.* 74, 293–296.

Yuan, X., Wang, J., Yao, H., and Chen, F. (2005). Free radical-scavenging capacity and inhibitory activity on rat erythrocyte hemolysis of feruloyl oligosaccharides from wheat bran insoluble fibre. *Lebensm. Wiss. Technol.* 38, 877–883.

Zobel, H. F. (1988). Starch crystal transformations and their industrial importance. *Starch/Stärcke* 40, 1–7.

Sourdough/lactic acid bacteria

12

Marco Gobbetti, Maria De Angelis, Raffaella Di Cagno, and Carlo Giuseppe Rizzello

The sourdough

The use of sourdough as the natural starter for leavening is one of the oldest biotechnological processes in food fermentation (Röcken and Voysey, 1995). Sourdough is a mixture of flour (e.g. wheat, rye), water, and other ingredients (e.g. NaCl) that is fermented by naturally occurring lactic acid bacteria and yeasts. Although these microorganisms originate mainly from flours and process equipment, the resulting composition of the sourdough microbiota is determined by endogenous (e.g. chemical and enzyme composition of the flour) and exogenous (e.g. temperature, redox potential, dough yield and time of the fermentation process) factors (Hammes and Gänzle, 1998). In the mature sourdoughs, lactic acid bacteria dominate, occurring in numbers $>10^8$ cfu/g, whereas the number of yeasts is orders of magnitude lower (Ehrmann and Vogel, 2005). Overall, three standard protocols are distinguished for sourdough fermentation (Böcker *et al.*, 1995; De Vuyst and Neysens, 2005). Type I sourdough is manufactured with a traditional technique and is characterized by continuous, daily refreshments to maintain the microorganisms in an active state, as indicated by their high metabolic activity. The process is carried out at room temperature (20–30°C) and the final pH of the sourdough is ca. 4.0. Type II sourdough is mainly used as dough acidifier. The fermentation lasts 2–5 days at >30°C to speed up the process and the pH is <3.5 after 24 hours of fermentation (Hammes and Gänzle, 1998).

The microorganisms are in the late stationary phase of growth and exhibit restricted metabolic activity. Type III is a dried sourdough in powder form, that is fermented by defined starter cultures. It is used as acidifier supplement and aroma carrier during breadmaking. In contrast to type I, type II and III sourdoughs often require the addition of baker's yeast (*Saccharomyces cerevisiae*) for leavening (De Vuyst and Neysens, 2005). Beyond this classification, artisan and industrial technologies largely use other traditional and tailored protocols (Gobbetti *et al.*, 2005).

Sourdough lactic acid bacteria

Microbiological studies have revealed that more than 50 species of lactic acid bacteria and more than 25 species of yeasts, especially belonging to the genera *Saccharomyces* and *Candida*, occur in mature sourdoughs. Sourdough is considered as a unique food ecosystem in that it (i) selects for lactic acid bacteria strains that are adapted to their environment and (ii) harbors lactic acid bacteria communities specific for each sourdough (Gobbetti, 1998; De Vuyst *et al.*, 2002; Gobbetti *et al.*, 2005; De Vuyst and Neysens, 2005). Representatives genera of sourdough lactic acid bacteria are *Lactobacillus*, *Leuconostoc*, *Pediococcus*, and *Weissella* (Table 12.1) (De Vuyst and Vancanneyt, 2007). The largest biodiversity was found within the genus *Lactobacillus* and a relatively high number of species was discovered recently (Hammes and Gänzle, 1998; De Vuyst and Neysens, 2005; Valcheva *et al.*, 2005, 2006; Vancanneyt *et al.*, 2005; Aslam *et al.*, 2006; Scheirlink *et al.*, 2007). Except for the *Lactobacillus salivarius* group, representative sourdough isolates are found for each phylogenetic group currently distinguished within the genus *Lactobacillus* (http://141.150.157.117:8080/prokPUB/index.htm). Depending on the protocols used for sourdough fermentation various microbial consortia of mainly obligate and facultative hetero-fermentative lactic acid bacteria are found. *Lactobacillus brevis*, *Lactobacillus fermentum*, *Lactobacillus paralimentarius*, *Lactobacillus plantarum*, *Lactobacillus pontis*, and, especially, *Lactobacillus sanfranciscensis*, considered to be a key sourdough bacterium (Gobbetti and Corsetti, 1997), are commonly isolated from traditional type I sourdoughs. Type II sourdoughs are mainly characterized by the occurrence of *L. brevis*, *L. fermentum*, *Lactobacillus frumenti*, *L. pontis*, *Lactobacillus panis*, and *Lactobacillus reuteri* (De Vuyst and Vancanneyt, 2007).

Sourdough properties and functions

Beyond its natural and additive-free image, it is generally accepted that sourdough has various positive effects when used for making baked goods. Compared with other leavening agents (e.g. baker's yeast), it improves the texture, flavor, nutritional value, and shelf-life (Table 12.2). Notwithstanding the role of sourdough yeasts, the main metabolic properties of lactic acid bacteria determining the above effects are described briefly below.

Table 12.1 Species of lactic acid bacteria isolated from sourdough

Species	Type of sourdough	Obligate hetero-fermentative	Facultative heterofermentative	Obligate homofermentative	*Lactobacillus* Filogenetic group[a]	Reference
L.[b] *brevis*	I/II/III	x			*L. buchneri*	Vancanneyt et al., 2005
L. *fermentum*	I/II	x			*L. reuteri*	Hammes and Gänzle, 1998
L. *paralimentarius*	I	x			*L. plantarum*	Cay et al., 1999
L. *plantarum*	I/II		x		*L. plantarum*	De Vuyst and Vancanneyt, 2007
L. *pontis*	I/II	x			*L. reuteri*	Vogel et al., 1994
L. *sanfranciscensis*	I/I	x			*L. buchneri*	Weiss and Schillinger, 1984
L. *panis*	II	x			*L. reuteri*	Wiese et al., 1996
L. *reuteri*	I/II	x			*L. reuteri*	De Vuyst and Neysens, 2005; Corsetti et al., 2004
L. *mindensis*	I	x			*L. plantarum*	Ehrmann et al., 2003
L. *spicheri*	IPRWS-I[c]		x		*L. buchneri*	Meroth et al., 2004
L. *rossiae*	TS[d]	x			*L. reuteri*	Corsetti et al., 2005
L. *zymae*	TS	x			*L. buchneri*	Vancanneyt et al., 2005
L. *acidifarinae*	TS	x			*L. buchneri*	Vancanneyt et al., 2005
L. *hammesii*	TS	x			*L. buchneri*	Valcheva et al., 2005
L. *nantensis*	TS			x	*L. plantarum*	Valcheva et al., 2006
L. *buchneri*	I	x			*L. buchneri*	Hammes and Gänzle, 1998; Vogel et al., 1999
L. *fructivorans*	I	x			*L. buchneri*	Hammes and Gänzle, 1998; Vogel et al., 1999
W.[e] *cibaria*	I	x			NI[f]	Hammes and Gänzle, 1998; Vogel et al., 1999
W. *confusa*	II	x			NI	Müller et al., 2001; Vogel et al., 1999
L. *alimentarius*	I		x		*L. plantarum*	Hammes and Gänzle, 1998; Vogel et al., 1999
L. *casei*	I		x		*L. casei*	De Vuyst and Neysens, 2005
L. *acidophilus*	I/II			x	*L. delbruecki*	Hammes and Gänzle, 1998; Vogel et al., 1999

(*Continued*)

Table 12.1 *Continued*

Species	Type of sourdough	Obligate hetero-fermentative	Facultative heterofermentative	Obligate homofermentative	*Lactobacillus* Filogenetic group[a]	Reference
L. delbruecki	I/II			x	*L. delbruecki*	Hammes and Gänzle, 1998; Vogel et al., 1999
L. amylovorus	I/II	x			*L. delbruecki*	Müller et al., 2001
L. farciminis	I/II			x	*L. plantarum*	Hammes and Gänzle, 1998; Vogel et al., 1999
L. frumenti	II	x			*L. reuteri*	Müller et al., 2000
L. johnsonii	II			x	*L. delbruecki*	Müller et al., 2001; Vogel et al., 1999
P.[g]*pentosaceus*	III		x		NI	De Vuyst and Neysens, 2005
L. siliginis	TS		x		*L. reuteri*	Aslam et al., 2006
L. namurensis	TS	x			*L. buchneri*	Scheirlinck et al., 2007
Lc.[h]*mesenteroides*	I	x			NI	Arendt et al., 2007
Lc. citreum	TS	x			NI	De Vuyst and Neysens, 2005

[a] http://141.150.157.117:8080/prokPUB/index.htm.
[b] *L, Lactobacillus.*
[c] IPRWS-I, Industrial processed rice and wheat sourdough type I.
[d] TS, Traditional sourdough.
[e] *W, Weissella.*
[f] NI, Not included.
[g] *P, Pediococcus.*
[h] *Lc., Leuconostoc.*

Table 12.2 Sourdough properties and functions

Properties	Effect	Mechanism of action	Reference
Texture	Increase of the bread volume	Lactic acidification	Corsetti et al., 2000; Crowley et al., 2002; Clarke et al., 2002
	Decrease of the dough resistance to extension	Lactic acidification	Di Cagno et al., 2002
	Increase of the dough extensibility and softening	Lactic acidification	Di Cagno et al., 2002
	Improvement of the dough gas retention	Lactic acidification	Hammes and Gänzle, 1998; Clarke et al., 2002
Flavor	Increase of the synthesis of acetic acid	Embden-Meyerhof-Parnas and phosphogluconate energy routes	Gänzle et al. 2007; Vermeulen et al., 2006
		Use of external acceptors of electrons	
		Recycling of NADH co-factors	
		Hierarchical and simultaneous use of various energy sources	Gobbetti et al., 2000; Gobbetti and Corsetti, 1996
		Interactions with endogenous and exogenous enzymes	Di Cagno et al., 2003
	Liberation of free amino acids and their derivatives	Secondary proteolysis	Gobbetti, 1998; Thiele et al., 2002; Gobbetti et al., 2005; Kieronczyk et al., 2001; De Angelis et al., 2002; Schieberle, 1996
		General catabolism of free amino acids	
		Arginine deiminase pathway	
Nutrition	Improvement of the texture and palatability of whole grain and fiber-rich bread	Lactic acidification	Katina et al., 2005
	Stabilization or increase of the levels of various bioactive compounds	Lactic acidification	Liukkonen et al., 2003
	Improvement of the mineral bioavailability	Degradation of phytate Lactic acidification	De Angelis et al., 2003; Lopez et al., 2001; Katina et al., 2005
	Retard of the starch bioavailability and decrease of the glycemic index	Lactic acidification and unknown mechanisms	Ostman et al., 2002; De Angelis et al., 2007a
Shelf-life	Decrease of the rate of bread staling	Lactic acidification Probable slight degradation of starch molecules	Corsetti et al., 2000; Crowley et al., 2002
	Anti-ropeness activity Anti-bacterial activity Anti-fungal activity	Lactic acidification Synthesis of bacteriocins, bacteriocin-like inhibitory substances (BLIS) Synthesis of low-molecular mass antibiotic reutericyclin Synthesis of antifungal metabolites (e.g. acetic, caproic and formic acids and phenyllactic and 4-hydroxy-phenyllactic acids)	Kirchener and Von Holy 1989; Schnürer and Magnusson, 2005; Corsetti et al., 2004; Holtzel et al., 2000; Gobbetti et al., 2005; Corsetti et al., 1998; Lavermicocca et al., 2003

Texture

Depending on the level of lactic acidification, the use of sourdough leads to an increase in bread volume (Corsetti *et al.*, 2000; Crowley *et al.*, 2002; Clarke *et al.*, 2002). A decrease in dough resistance to extension and an increase in both extensibility and degree of softening were also shown before baking (Di Cagno *et al.*, 2002). Overall, sourdough fermentation improves the gas retention in bread dough (Hammes and Gänzle, 1998; Clarke *et al.*, 2002). Acidification impacts on the solubility of structure-forming components such as gluten, starch, and arabinoxylans, and positively interferes with the activity of cereal endogenous enzymes (Korakli *et al.*, 2001). Lactic acidification also influences the mixing behavior of the dough whereby, when low pH are reached, a shorter mixing time and less stability than normal dough are achieved (Hoseney, 1994).

Flavor

The fermentation of soluble carbohydrates (e.g. maltose, glucose, and fructose), metabolism of nitrogenous compounds and generation of volatile compounds by sourdough lactic acid bacteria directly or indirectly influence the flavor of baked goods. Beyond the Embden–Meyerhof–Parnas (EMP, facultative hetero-fermentative strains) and phosphogluconate (obligate hetero-fermentative strains) energy routes, (i) the use of external acceptors of electrons (e.g. fructose) and recycling of NADH co-factors (Vermeulen *et al.*, 2006; Gänzle *et al.*, 2007), (ii) the hierarchical (e.g. pentoses instead of hexoses) (Gobbetti *et al.*, 2000) and simultaneous (e.g. co-fermentation of citrate and maltose) use of various energy sources (Gobbetti and Corsetti, 1996), and (iii) the interactions with endogenous and exogenous enzymes (Di Cagno *et al.*, 2003) lead to different quotients of fermentation (molar ratio between lactic and acetic acids) that differently impact on the flavor of baked goods.

Overall, sourdough fermentation with lactic acid bacteria results in an increase of free amino acids (FAAs), whereas dough fermentation with yeasts alone reduces the concentration of FAAs (Gobbetti, 1998). Proteolysis during sourdough fermentation may be categorized as: (i) primary, regarding the hydrolysis of proteins to intermediate sized polypeptides through, especially, flour endogenous proteinases; and (ii) secondary, regarding the liberation of FAAs from intermediated sized polypeptides through, especially, the lactic acid bacteria peptidase system (Thiele *et al.*, 2002; Gobbetti *et al.*, 2005). Once liberated, FAAs contribute directly to flavor or are further subjected to chemical conversion during baking or enzymatic catabolism (Kieronczyk *et al.*, 2001) thus leading to the synthesis of flavor volatile compounds. Within the catabolism of FAAs, the expression of the arginine deiminase (ADI) pathway in sourdough lactic acid bacteria has a marked practical significance. The expression of this pathway in *L. sanfranciscensis* CB1 (De Angelis *et al.*, 2002) enhances the growth and tolerance to acid environmental stress, and, especially, increases the synthesis of ornithine, which is the precursor of the 2-acetyl-pyrroline, responsible for the roasty note of the wheat bread crust (Gobbetti *et al.*, 2005).

Alcohols, aldehydes, ketones, acids, esters, ether derivates, furan derivates, hydrocarbons, lactones, pyrazines, pyrrol derivates, and sulfur compounds are the flavor stimuli in baked goods (Schieberle, 1996). Chemically acidified dough with levels of amino acids corresponding to those of sourdough improves the bread flavor only slightly (Thiele *et al.*, 2002), indicating the significant role of sourdough in directly originating volatile components. Overall, homo-fermentative lactic acid bacteria mainly synthesized diacetyl, acetaldehyde, and hexanal, and hetero-fermentative strains are characterized by the production of ethyl-acetate, alcohols, and aldehydes. Iso-alcohols (2-methyl-1-propanol, 2,3-methyl-1-butanol), with their respective aldehydes and ethyl-acetate, are characteristics volatile compounds of yeast fermentation (Damiani *et al.*, 1996).

Nutrition

Sourdough fermentation modifies the healthiness of cereals in a number of ways: it may (i) improve texture and palatability of whole grain and fiber-rich cereals; (ii) stabilize or increase levels of various bioactive compounds; (iii) retard starch bioavailability (low glycemic index products); and (iv) improve mineral bioavailability (Katina *et al.*, 2005). Lactic acidification increases the levels of bioactive compounds (e.g. phenolic compounds) or decreases the levels of thiamine, ferulic acid dehydrodimers, and tocopherols (Liukkonen *et al.*, 2003). The degradation of phytate in sourdough processes leads to an increased mineral bioavailability (Lopez *et al.*, 2001; De Angelis *et al.*, 2003). Furthermore, lactic acidification also increases the magnesium and phosphorus solubility (Katina *et al.*, 2005) and has been found to be a protective factor for β-glucan in breads. Organic acids such as those produced during sourdough fermentation have also been shown to play a role in the post-prandial glycemic responses. The presence of lactic acid during heat treatment promotes interactions between starch and gluten, reducing starch bioavailability and, consequently, the glycemic index of baked goods (Östman *et al.*, 2002). Finally, the effect of the biological acidification seems to be more pronounced with respect to that of chemical acidification (De Angelis *et al.*, 2007a).

Shelf-life

The improvement of the loaf-specific volume and crumb softness by sourdough fermentation have been associated with the decrease of the rate of bread staling (Corsetti *et al.*, 2000; Crowley *et al.*, 2002). The anti-staling effect is dependent on the particular strain performing the fermentation and involves dynamics other than those associated with the degree of acidification. Besides, starch molecules may be affected by enzymes synthesized by lactic acid bacteria, causing a variation in the retrogradation properties of the starch which, in turn, slows the rate of staling.

Acidification through sourdough fermentation has been shown to inhibit the endospore germination and growth of *Bacillus* spp. responsible for rope spoilage (Kirschner and Von Holy, 1989). Besides various compounds (e.g. organic acids, hydrogen peroxide, diacetyl), sourdough lactic acid bacteria may inhibit the growth of other related microorganisms by synthesizing bacteriocins, bacteriocin-like inhibitory substances (BLIS) (Corsetti *et al.*, 2004; Gobbetti *et al.*, 2005) and low-molecular mass antibiotics such as the reutericyclin of *L. reuteri* LTH2584 (Höltzel *et al.*, 2000). A number of antifungal metabolites, e.g. cyclic dipeptides, phenyllactic acid, proteinaceous compounds, and 3-hydroxylated fatty acids, are potentially synthesized by lactic acid bacteria (Schnürer and Magnusson, 2005; Dal Bello *et al.*, 2007). Mixtures of organic acids (e.g. acetic, caproic and formic acids) acting in a synergistic way were responsible for the *in vitro* inhibitory activity of *L. sanfranciscensis* CB1 against molds responsible for bread spoilage (Corsetti *et al.*, 1998). Phenyllactic and 4-hydroxy-phenyllactic acids were the antifungal compounds synthesized by *L. plantarum* 20B, showing inhibitory activity against *Aspergillus*, *Penicillium*, *Eurotium*, and *Monilia* (Lavermicocca *et al.*, 2003).

Applications of sourdough in gluten-free products

If the use of sourdough in the manufacture of conventional baked goods promotes a number of positive effects, it would be natural to consider its application for gluten-free products. Overall, the gluten-free products available on the market are of low quality, exhibiting poor mouth-feel and flavor (Gallagher *et al.*, 2004). Since they do not contain gluten, and are mainly starch based, and the onset of staling is more rapid than in gluten-containing breads (Moore, 2005). Besides, when limiting the use of gluten-free flours to the most common sources (e.g. rice, corn, and starches), nutrient deficiencies may occur due to very low dietary fiber content and excess calories (Diowksz *et al.*, 2006). Nevertheless, the current literature contains a very limited number of papers dealing with the use of sourdough in gluten-free goods. The few available results indicate that sourdough has a positive effect on the baking quality, particularly regarding volume, texture, and flavor. The influence of sourdoughs fermented by different lactic acid bacteria strains on the textural quality of gluten-free bread was evaluated during storage and compared to that of chemically acidified or non-acidified doughs (Clarke *et al.*, 2002; Crowley *et al.*, 2002). The growth of selected lactic acid bacteria in gluten-free batters was similar to that reported for wheat sourdoughs (Clarke *et al.*, 2002). Sourdough fermentation caused an increase in the dough elasticity and staling was delayed (Ryan *et al.*, 2006). These effects were mainly attributed to the breakdown of non-gluten proteins and starch components by sourdough lactic acid bacteria. Based on triangle tests, gluten-free sourdough bread was discriminated from the control breads and clearly preferred. In a recent patent (Giuliani *et al.*, 2006) *L. sanfranciscensis* LS40 and LS41, and *L. plantarum* CF1, previously isolated from traditional sourdoughs, were selected. This microbial mixture was used to ferment gluten-free ingredients (e.g. corn starch, rice, buckwheat, and millet flours) and compared with baker's yeast fermentation. The sourdough fermentation allowed the researchers to: (i) completely degrade ca. 300 ppm of gluten, eventually present as contaminant; (ii) increase by ca. 10-fold the concentration of FAAs; (iii) increase by ca. 10-fold the phytase activity during fermentation; and (iv) improve the sensory characteristics of the resulting bread as evaluated by descriptive analysis. A recent patent (Sikken and Lousche, 2003) describes the use of *L. fermentum* for the manufacture of high-quality gluten-free baked goods. In view of the sensory, texture, and nutritional improvements, the commercial use of sourdough for the manufacture of gluten-free goods should be recommended.

Sourdough lactic acid bacteria as a tool for detoxifying gluten

Beyond genetic predisposition, several environmental factors influence the prevalence of celiac disease. Recent epidemiological studies show that, besides being frequently found in countries where individuals are mostly of European origin, celiac disease is a common disorder in many areas of the developing world where agriculture started over 10 000 years ago. More recently, cereal food biotechnology has changed dramatically by influencing the dietary habits of entire populations previously naïve to gluten exposure. Cereal baked goods are currently manufactured using accelerated processes and the long fermentations of sourdoughs have been replaced too often by the indiscriminate use of chemical and/or baker's yeast leavening agents (Gobbetti, 1998). Under these circumstances, the traditional biotechnology of sourdough breadmaking has recently been exploited for its capacity to degrade toxic epitopes during food processing. Extensive research in this field is ongoing at the authors' laboratory in a joint project with medical specialists to show the potential of proteolytic enzymes of sourdough lactic acid bacteria as has been demonstrated for prolyl endopeptidases (PePs) of *Flavobacterium meningosepticum* (Pyle *et al.*, 2005), *Myxococcus xanthus* (Shan *et al.*, 2004), and *Aspergillus niger* (Stepniak *et al.*, 2006).

Selection of sourdough lactic acid bacteria

During endoluminal proteolytic digestion, mainly prolamins of wheat (α-, β-, γ-, and ω-gliadin subgroups), rye (e.g. secalin), and barley (e.g. hordein) release a family of proline- and glutamine-rich polypeptides that are responsible for the inappropriate T cell-mediated immune response (Sollid and Khosla, 2005). Although still debated and updated monthly, several fragments (f) are defined as indubitably toxic; e.g. f31–43 of α2-gliadin (Picarelli *et al.*, 1999), f62–75 of α2-gliadin (Shan *et al.*, 2002), 33-mer epitope, corresponding to f57–89 of α2-gliadin (Shan *et al.*, 2002), f134–153 of γ-gliadin (Aleanzi *et al.*, 2001), and f57–89 of α2-gliadin (Arentz-Hansen *et al.*, 2000). Recently, it has also been shown that glutenins contain cryptic regions which produce toxic epitopes (Molberg *et al.*, 2003). Overall, the large proportion and location of proline residues in the amino acid sequences of these toxic peptides make them extremely resistant to further hydrolysis (Hausch *et al.*, 2002). To adequately deal with such peptides, a group of specific peptidases is necessary to hydrolyze peptide bonds in which a proline residue occurs as a potential substrate. Sourdough lactic acid bacteria have been considered as cell factories for multiple and complementary enzyme activities to be exploited over a long fermentation period. Since a single unique strain may not possess the entire portfolio of peptidases needed to hydrolyze all the potential peptides involving proline, four sourdough strains—*Lactobacillus alimentarius* 15M, *L. brevis* 14G, *L. sanfranciscensis* 7A, and *Lactobacillus hilgardii* 51B—were selected based on their large enzyme substrate specificity and capacity to hydrolyze the 33-mer peptide, the most potent inducer of gut-derived human T cell lines in patients with celiac disease (Table 12.3) (Di Cagno *et al.*, 2004). Later, similar results were achieved using a mixture of probiotic strains such as VSL#3 (De Angelis *et al.*, 2005). Nevertheless, the hydrolyzing capacity was lost when the individual strains comprising VSL#3 were tested, confirming that no single strain contains the entire portfolio of peptidases necessary to degrade proline-rich polypeptides (Figure 12.1). Furthermore, VSL#3 completely hydrolyzed the f62–75 of α2-gliadin, previously reported as immunomodulatory peptides involved in the pathogenesis of celiac disease (Shan *et al.*, 2002). The need for a complex system of enzymes was further demonstrated by a study dealing with the purification and characterization of an X-prolyl dipeptidyl aminopeptidase (PepX) from *L. sanfranciscensis* (Gallo *et al.*, 2005). No hydrolysis of the proline-rich 33-mer epitope was found when it was treated with PepX alone. When the general aminopeptidase type N from the same bacterium was combined with PepX, the hydrolysis of the 33-mer peptide (0.2 mmol/L) was complete after 24 hours of incubation at 30°C. In conclusion, these studies (Di Cagno *et al.*, 2004; De Angelis *et al.*, 2005; Gallo *et al.*, 2005) show that selected sourdough or probiotic lactic acid bacteria may possess complementary

Table 12.3 Enzyme activity[a] of selected sourdough lactobacilli (*Lactobacillus alimentarius* 15M, *Lactobacillus brevis* 14G, *Lactobacillus sanfranciscensis* 7A, and *Lactobacillus hilgardii* 51B) on various substrates containing proline residues

Source of enzyme activity	Substrate (concentration [mM])	Average activity (U) ± SD[a]
Cells[b]	Pro-*p*-NA (2)	0.3 ± 0.01[d]
Cells	Gly-Pro-*p*-NA (2)	5.2 ± 0.03
Cells	Z-Gly-Pro-NH-trifluoromethylcoumarin (2)	12.3 ± 0.4[e]
Cells	Val-Pro (2.3)	2.1 ± 0.03[f]
Cells	Pro-Gly (3)	1.9 ± 0.04
Cells	Gly-Pro-Ala (2)	2.2 ± 0.02
Cells	Bradykinin (0.3)	5.5 ± 0.3
Cells	Fragment 62–75 of A-gliadin (0.45)	9.7 ± 0.5
Pooled cells and CE[c]	Fragment 62–75 of A-gliadin	15.0 ± 0.5
Cells	33-mer (0.200)	0.08 ± 0.002
Pooled cells and CE	33-mer	0.2 ± 0.01

[a] Each value is the average of three enzyme assays, and standard deviations were calculated.
[b] Aliquots (25 μL) of each cell suspension were used in the enzyme assays.
[c] Aliquots (12.5 μL) of the pooled cells and cytoplasmic extracts (CE) of each species were used in the enzyme assays.
[d] A unit of enzyme activity on *p*-NA substrates was defined as the amount of enzyme that produced an increase in absorbance at 410 nm of 0.01/min.
[e] A unit of enzyme activity on Z-Gly-Pro-NH-trifluoromethylcoumarin was the amount of enzyme that produced an increase in fluorescence of 0.1/min.
[f] A unit of enzyme activity on di-, tri-, and polypeptides was the amount of enzyme that liberates 1 μmol of substrate/min. From Di Cagno *et al.* (2004).

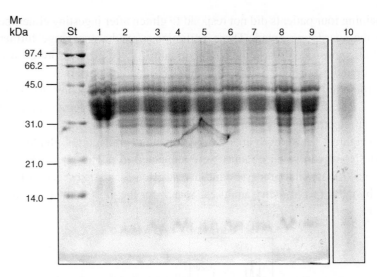

Figure 12.1 SDS-PAGE analysis of gliadins polypeptides from wheat flour doughs incubated for 24 hours with the different cell preparations (10^9 cfu/mL) which composed the VSL#3 preparation. Protein standard (St). Chemically acidified dough (1); doughs incubated with cells of *Bifidobacterium longum* (2); *Lactobacillus delbrueckii* subsp. *bulgaricus* (3); *L. plantarum* (4); *L. casei* (5); *B. infantis* (6); *L. acidophilus* (7); *Streptococcus thermophilus* (8); *B. breve* (9); and VSL#3 preparation (10). From De Angelis *et al.* (2005).

peptidase activities which will allow gluten epitopes to be managed to some extent during baked good processing.

Sourdough wheat bread

A sourdough made from a mixture of wheat (30%) and the non-toxic oat, millet, and buckwheat flours was started with the selected sourdough lactobacilli mentioned above (*L. alimentarius* 15M, *L. brevis* 14G, *L. sanfranciscensis* 7A, and *L. hilgardii* 51B; ca. 10^9 cfu/g dough) and subjected to long-term (24 hours) fermentation (Di Cagno *et al.*, 2004). The semi-liquid pre-fermentation of wheat flour was essential to fully exploit the potential of sourdough lactic acid bacteria enzymes. An almost complete hydrolysis of gliadins was achieved, while prolamins from oats, millet, and buckwheat were not affected during fermentation. A comparison with a chemically acidified dough or with a dough started with baker's yeast alone showed that the hydrolysis was due to the proteolytic activity of the sourdough lactic acid bacteria and that gliadins were not affected during dough fermentation with yeasts. A wheat sourdough started with selected lactic acid bacteria was allowed to ferment for 24 hours at 30°C, mixed with non-toxic flours at an optimal ratio (3:7), further fermented for 2 hours at 30°C with baker's yeast and baked at 220°C for 20 minutes. This bread and a baker's yeast started bread, containing ca. 2 g of gluten, were used for an *in vivo* double-blind acute challenge of patients with celiac disease. Thirteen of the 17 patients showed a marked alteration of intestinal permeability after ingestion of baker's yeast bread. When given the sourdough bread to eat, the same 13 patients had values for intestinal permeability that did not differ significantly from baseline values.

The remaining four patients did not respond to gluten after ingesting either the baker's yeast or the sourdough bread. These preliminary results encourage further studies since they demonstrated that a moderate amount (2 g) of pre-hydrolyzed wheat flour can be tolerated by patients with celiac disease in an acute trial.

The probiotic VSL#3 preparation also showed the capacity to decrease the toxicity of wheat flour during long fermentation (De Angelis *et al.*, 2005). Two-dimensional electrophoresis, immunological (R5 antibody), and mass spectrometry analyses (Figure 12.2) showed an almost complete degradation of gliadins during fermentation. Non-hydrolyzed gliadins were subjected to peptic–tryptic (PT) digestion to mimic gastrointestinal processes and investigated for the presence of known toxic epitopes by mass spectrometry analyses. Search for the most known epitopes showed

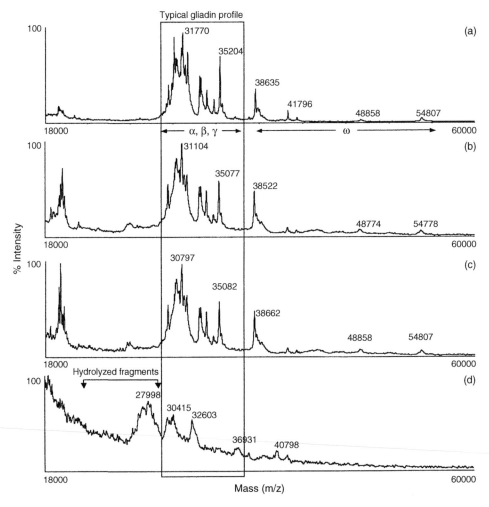

Figure 12.2 MALDI-TOF mass spectra of aqueous ethanol extract of wheat gliadin: (a) European gliadin standard showing the α-, β-, γ-, and ω-gliadin ranges; (b) chemically acidified dough (control) incubated for 24 hours at 37°C; (c) chemically acidified dough with heat inactivated VSL#3 cells incubated for 24 hours at 37°C; and (d) fermented dough incubated with VSL#3 for 24 hours at 37°C. The typical α-, β-, γ-gliadin profile is displayed in a box. From De Angelis *et al.* (2005).

the presence of α2-gliadin f62–75 at a very low concentration (sub-parts per million range). With respect to the previous study on sourdough lactobacilli (Di Cagno *et al.*, 2004), new *in vitro* and *ex vivo* analyses were carried out. Compared with rat intestinal epithelial cells IEC-6 exposed to intact gliadins from the control, VSL#3 pre-digested gliadins caused a less pronounced reorganization of the F-actin, which was mirrored by an attenuated effect on intestinal mucosa permeability. The release of zonulin, a molecule which sustains the increase of the intestinal permeability as a mechanism of response to toxic peptides (Clemente *et al.*, 2003; Drago *et al.*, 2003), from intestinal epithelial cells treated with gliadins was considerably lower when digested with VSL#3. Wheat proteins were extracted from doughs and subjected to PT digestion. Compared with the PT digest from chemically acidified dough, celiac jejunal biopsies exposed to the PT digest from the dough fermented by VSL#3 did not show an increase in the infiltration of $CD3^+$ intraepithelial lymphocytes. Overall, $CD3^+$ intraepithelial lymphocytes increased after challenge of small intestine mucosa from patients with celiac disease with gluten (Troncone *et al.*, 1998; Mazzarella *et al.*, 2005). Based on the above results, it seemed that the probiotic preparation VSL#3 also has the capacity to decrease the toxicity of gluten epitopes.

Pasta from fermented durum wheat semolina

The same approach as that described for sourdough wheat bread (Di Cagno *et al.*, 2004; De Angelis *et al.*, 2005) was adapted for pasta-making. A pool of selected lactic acid bacteria was used to ferment durum wheat semolina under liquid conditions (Di Cagno *et al.*, 2005). After fermentation, the dough was freeze-dried, mixed with buckwheat flour at a ratio of 3:7, and used to produce "fusilli" type Italian pasta. Pasta without pre-fermentation was used as the control. The two types of pasta were subjected to sensory analysis. The scores for stickiness and firmness were slightly higher for the pasta control. Odor and flavor did not differ between the two types of pasta. Two-dimensional electrophoresis and mass spectrometry MALDI-TOF analyses showed an almost complete hydrolysis of the gliadin fraction (Figure 12.3). As shown by immunological analysis by R5-Western blot, the concentration of gluten decreased from 6280 ppm in the control pasta to 1045 ppm in the fermented pasta. Although this type of pasta still contained 1045 ppm of gluten, which may trigger celiac disease, the use of a mixture which includes 20% of fermented durum wheat semolina in the pasta formulas may theoretically lead to a novel pasta product within the safe threshold for celiac disease. Gliadins were extracted from fermented and non-fermented durum wheat semolina dough, and used to produce the PT digests for an *in vitro* agglutination test on K 563(S) subclone cells of human myelogenous leukemia origin (Auricchio *et al.*, 1984). The whole PT digests did not cause agglutination. In contrast to bread wheat, rye, and barley, it seemed that durum wheat contained a decapeptide which had the capacity to prevent the agglutination by PT digest and which may have a protective effect on celiac disease (De Vincenzi *et al.*, 1998). Affinity chromatography separated the PT digests into three fractions. Only the smallest of these separated fractions contained agglutination activity. The minimal agglutinating activity of this fraction from the PT digest of fermented durum wheat

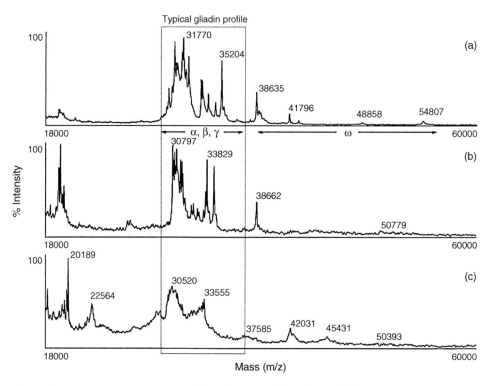

Figure 12.3 MALDI-TOF mass spectra of ethanol extract of wheat durum gliadin: (a) European gliadin standard showing the α-, β-, γ-, and ω-gliadin ranges; (b) chemically acidified dough incubated for 24 hours at 37°C; and (c) durum wheat semolina fermented with the mixture of selected lactic acid bacteria for 24 hours at 37°C. The typical α-, β-, and γ-gliadin profiles are displayed in the box. From Di Cagno *et al.* (2005).

semolina was ca. 80 times higher than that of durum wheat semolina, indicating a decreased toxicity. Based on these results, it seems that the use of selected sourdough lactobacilli could be also adaptable to the manufacture of pasta from durum wheat semolina with decreased gluten toxicity.

Rye fermentation

The pool of selected sourdough lactic acid bacteria also showed the capacity to degrade prolamins contained in rye flour (De Angelis *et al.*, 2006). Prolamins were extracted from rye flour and used to produce a PT digest for *in vitro* tests with Caco-2/TC7 cells of human origin (De Angelis *et al.*, 1998; Giovannini *et al.*, 2003). Hydrolysis of rye PT digest by selected sourdough lactic acid bacteria decreased the toxicity of PT digest itself towards Caco-2/TC7 cells as estimated by cell viability, caspase-3 activity, and release of nitric oxide. On the other hand, prolamins and glutelins were extracted from fermented rye sourdough and subjected to PT digestion. Compared with PT digest from chemically acidified dough, celiac jejunal biopsies exposed to the PT digest from the dough fermented by lactic acid bacteria did not show an increase in the infiltration of CD3[+] intraepithelial lymphocytes (Figure 12.4), as shown by Fas expression, which is a measure of cell apoptosis (Maiuri *et al.*, 2001).

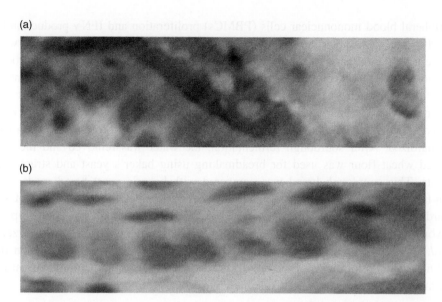

Figure 12.4 Fas expression in a celiac disease specimen treated with PT digests from secalin and glutelin proteins extracted from doughs chemically acidified (a) or fermented with selected lactic acid bacteria for 24 hours at 37°C (b). Original magnification 200×; immunohistochemistry; Red Fuchsin APAAP staining technique. From De Angelis *et al.* (2006).

Highly efficient gluten degradation by lactobacilli and fungal proteases

Although a number of *in vitro* (e.g. agglutination and Caco-2/TC assays), *ex vivo* (biopsy-derived T cells) and acute *in vivo* (intestinal permeability) tests were carried out, the above results (Di Cagno *et al.*, 2004, 2005; De Angelis *et al.*, 2005, 2006) only showed a marked decrease of the gliadin fraction. This route might be helpful to eliminate the risk of cross-contamination of gluten-free products but not to completely eliminate the toxicity of wheat flour. Consequently, further efforts were done to increase the hydrolyzing capacity of sourdough lactic acid bacteria. Together with fungal proteases, routinely used in breadmaking, other lactobacilli strains, characterized by a marked peptidase activity towards proline-rich peptides (De Angelis *et al.*, 2007b), were used during long fermentation of semi-liquid wheat flour doughs. As determined by R5-sandwich and competitive ELISA, the residual concentration of gluten in the fermented sourdough was <20 ppm, as required by the standard of the Codex Alimentarius Commission for gluten-free products. Two-dimensional electrophoresis and MALDI-TOF mass spectrometry analyses showed the complete hydrolysis of albumins/globulins and gliadins. Only ca. 20% of glutenins persisted. After hydrolysis, the spray-dried flour from fermented sourdough was mainly a mixture of water/salt-soluble low molecular weight peptides and amino acids. Low molecular weight epitopes were not detectable as determined by strong cation–exchange liquid chromatography (SCX-LC) and capillary liquid chromatography–electrospray ionization (CapLC-ESI)-q-ToF-MS and R5-Western blot analyses. PT digests of all protein fractions extracted from fermented sourdough were used for *in vitro* assays on peripheral blood mononuclear cells (PBMCs) proliferation and IFNγ production by PBMCs, and intestinal T cell lines (iTCLs) of 12 CD patients. PT digests underwent tissue transglutaminase (tTG)-mediated deamidation before iTCL assay. All protein fractions gave an activation of PBMCs and induced IFNγ as the negative control. None of iTCLs demonstrated immunoreactivity towards PT digests. Because it was uncertain whether such wheat flour would be suitable to ensure bread quality after the complete degradation of gluten, a biotechnological protocol for breadmaking was standardized. After sourdough fermentation, the water was removed and the pre-treated wheat flour was used for breadmaking using baker's yeast and structuring agents. This sourdough bread was compared with baker's yeast bread made with non-treated flour and without structuring agents. The specific loaf volume of the sourdough bread was similar to that of the baker's yeast bread and showed the typical flavor of the sourdough wheat bread as judged by an internal panel test (Rizzello et al., 2007). These results may potentially encourage further investigation on the technological performance of the pre-digested wheat flour as the main ingredient for the manufacture of baked goods tolerated by patients with celiac disease.

Future trends

Compliance with a gluten-free diet is an extremely challenging task, given a number of problems related to cross-contamination, lack of clear food labeling policies and poor quality of gluten-free products compared with their gluten-rich counterpart. Even if the exploitation of sourdough in gluten-free systems is still in its infancy, the literature data available strongly indicate that sourdough may undoubtedly be considered as a technological tool for improving the texture and flavor characteristics of gluten-free products. Commercial application of this traditional biotechnology should be warranted. On the other hand, the role of the sourdough lactic acid bacteria in the management of gastrointestinal diseases has been defined as "emerging and intriguing" (Yan and Polk, 2004). Based on the above studies, the use of sourdough lactic acid bacteria would certainly eliminate any traces of gluten epitopes in processed foods and will minimize the long-term risk to the multitude of individuals affected by celiac disease worldwide. Furthermore, long-term in vivo trials have been recently started to evaluate the effective tolerance of patients with celiac disease to breads made of wheat flour alone which has been pre-digested to a concentration of gluten <20 ppm by sourdough lactic acid bacteria and fungal proteases.

Sources of further information and advice

De Vuyst, L. and Gänzle, M. eds (2005). Second International Symposium on Sourdough—From Fundamental to Applications. *Trends Food Sci. Technol.* Special Issue 16, 1–124.

Gobbetti, M. and Gänzle, M. eds (2007). Third International Symposium on Sourdough—From Traditional to Innovation. *Food Microbiol*, Special Issue 24, 113–196.

Hammes, W. P. and Gänzle, M. G. (1998). Sourdough breads and related products. In: Woods, B. J. B. eds. *Microbiology of Fermented Foods*, Vol. 1. London: Blackie Academic/Professional, pp. 199–216.

References

Aleanzi, M., Demonte, A. M., Esper, C., Garcilazo, S., and Waggener, M. (2001). Celiac disease: antibody recognition against native and selectively deaminated gliadin peptides. *Clinical Chemistry* 47, 2023–2028.

Arendt, E. K., Liam, A., Ryan, M., and Dal Bello, F. (2007). Impact of sourdough on the texture of bread. *Food Microbiology* 24, 165–174.

Arentz-Hansen, H., Korner, R., Molberg, O. *et al.* (2000). The intestinal T cell response to alpha-gliadin in adult celiac disease is focused on a single deaminated glutamine targeted by tissue transglutaminase. *J. Exp. Med.* 191, 603–612.

Aslam, Z., Im, W. T., Ten, L. N., Lee, M. J., Kim, K. H., and Lee, S. T. (2006). *Lactobacillus siliginis* sp. nov., isolated from wheat sourdough in South Korea. *Int. J. Syst. Evol. Microbiol.* 56, 2209–2213.

Auricchio, S., De Ritis, G., De Vincenzi, M., Minetti, M., Sapora, O., and Silano, V. (1984). Agglutination activity of gliadin-derived peptides from bread wheat: implications for coeliac disease pathogenesis. *Biochem. Biophys. Res. Commun.* 21, 428–433.

Böcker, G., Stolz, P., and Hammes, W. P. (1995). Neue Erkenntnisse zum Ökosystem Sauerteig und zur Physiologie des sauerteig-typischen Stämme *Lactobacillus sanfrancisco* und *Lactobacillus pontis*. *Getreide Mehl Brot* 49, 370–374.

Cai, Y., Okada, H., Mori, H., Benno, Y., and Nakase, T. (1999). *Lactobacillus paralimentarius* sp. nov., isolated from sourdough. *Int. J. Syst. Bacteriol.* 49, 1451–1455.

Clarke, C. I., Schober, T. J., and Arendt, E. K. (2002). The effect of single strain and traditional mixed strain starter cultures on rheological properties of wheat dough and bread quality. *Cereal Chem.* 79, 640–647.

Clemente, M. G., De Virgiliis, S., Kang, J. S. *et al.* (2003). Early effects of gliadin on enterocyte intracellular signalling involved in intestinal barrier function. *Gut* 52, 218–223.

Corsetti, A., Gobbetti, M., Rossi, J., and Damiani, P. (1998). Antimould activity of sourdough lactic acid bacteria: identification of a mixture of organic acids produced by *Lactobacillus sanfrancisco* CB1. *Appl. Microbiol. Biotechnol.* 50, 253–256.

Corsetti, A., Gobbetti, M., De Marco, B., Balestrieri, F., Paletti, F., and Rossi, J. (2000). Combined effect of sourdough lactic acid bacteria and additives on bread firmness and staling. *J. Agric. Food Chem.* 48, 3044–3051.

Corsetti, A., Settanni, L., and Van Sinderen, D. (2004). Characterization of bacteriocin-like inhibitory substances (BLIS) from sourdough lactic acid bacteria and evaluation of their in vitro and in situ activity. *J. Appl. Microbiol.* 96, 521–534.

Corsetti, A., Settanni, L., van Sinderen, D., Felis, G. E., Dellaglio, F., and Gobbetti, M. (2005). *Lactobacillus rossiae* sp. nov., isolated from wheat sourdough. *Int. J. Syst. Evol. Microbiol.* 56, 587–591.

Crowley, P., Schober, T., Clarke, C., and Arendt, E. (2002). The effect of storage time on textural and crumb grain characteristics of sourdough wheat bread. *Eur. Food Res. Technol.* 214, 489–496.

Dal Bello, F., Dal Bello, C. I., Clarke, L. A. *et al.* (2007). Improvement of the quality and shelf life of wheat bread by fermentation with the antifungal strain *Lactobacillus plantarum* FST 1.7. *J. Cereal Sci.* 45, 309–318.

Damiani, P., Gobbetti, M., Cossignani, L., Corsetti, A., Simonetti, M. S., and Rossi, J. (1996). The sourdough microflora. Characterization of hetero- and homofermentative lactic acid bacteria, yeasts and their interactions on the basis of the volatile compounds produced. *Lebensm.-Wiss. U.-Technol.* 29, 63–70.

De Angelis, I., Vicentini, O., Brambilla, G., Stammati, A., and Zucco, F. (1998). Characterization of furazolidone apical-related effects to human polarized intestinal cells. *Toxicol. Appl. Pharmacol.* 152, 119–127.

De Angelis, M., Mariotti, L., Rossi, J. *et al.* (2002). Arginine catabolism by sourdough lactic acid bacteria: purification and characterization of the arginine deiminase pathway enzymes from *Lactobacillus sanfranciscensis* CB1. *Appl. Environ. Microbiol.* 68, 6193–6201.

De Angelis, M., Gallo, G., Corbo, M. R. *et al.* (2003). Phytase activity in sourdough lactic acid bacteria: purification and characterization of a phytase from *Lactobacillus sanfranciscensis* CB1. *Int. J. Food Microbiol.* 87, 259–270.

De Angelis, M., Rizzello, C. G., Scala, E. *et al.* (2005). VSL#3 probiotic preparation has the capacity to hydrolyze gliadin polypeptides responsible for celiac sprue. *Biochim. Biophys. Acta* 1762, 80–93.

De Angelis, M., Coda, R., Silano, M. *et al.* (2006). Fermentation by selected sourdough lactic acid bacteria to decrease the intolerance to rye and barley flours. *J. Cereal Sci.* 43, 301–314.

De Angelis, M., Rizzello, C. G., Alfonsi, G. *et al.* (2007a). Use of sourdough lactobacilli and oat fibre to decrease the glycemic index of white wheat bread. *Br. J. Nutr.* doi:10.1017/S0007114507772689.

De Angelis, M., Di Cagno, R., Gallo, G. *et al.* (2007b). Molecular and functional characterization of *Lactobacillus sanfranciscensis* strains isolated from sourdoughs. *Int. J. Food Microbiol.* 114, 69–82.

De Vincenzi, M., Stammati, A., Luchetti, R., Silano, M., Gasbarrini, G., and Silano, V. (1998). Structural specifities and significance for coeliac disease of wheat gliadin peptides able to agglutinate or to prevent agglutination of K562(S) cells. *Toxicology* 127, 97–106.

De Vuyst, L. and Gänzle, M. G. (2005). *Trends in Food Science & Technology Special Issue—Second International Symposium on Sourdough: from Fundamentals to Applications*, Vol. 16. Amsterdam: Elsevier.

De Vuyst, L. and Neysens, P. (2005). The sourdough microflora: biodiversity and metabolic interactions. *Trends Food Sci. Technol.* 16, 43–56.

De Vuyst, L. and Vancanneyt, M. (2007). Biodiversity and identification of sourdough lactic acid bacteria. *Food Microbiol.* 24, 120–127.

De Vuyst, L., Schrijvers, V., Paramithiotis Hoste, B. *et al.* (2002). The biodiversity of lactic acid bacteria in greek traditional wheat sourdoughs is reflected in both composition and metabolite formation. *Appl. Environ. Microbiol.* 68, 6059–6069.

Di Cagno, R., De Angelis, M., Lavermicocca, P. *et al.* (2002). Proteolysis by sourdough lactic acid bacteria: effects on wheat flour protein fractions and gliadin peptides involved in human cereal intolerance. *Appl. Environ. Microbiol.* 68, 623–633.

Di Cagno, R., De Angelis, M., Corsetti, A. *et al.* (2003). Interaction between sourdough lactic acid bacteria and exogenous enzymes: effects on the microbial kinetics of acidification and dough textural properties. *Food Microbiol.* 20, 67–75.

Di Cagno, R., De Angelis, M., Auricchio, S. *et al.* (2004). A sourdough bread made from wheat and non-toxic flours and started with selected lactobacilli is tolerated in Celiac Sprue. *Appl. Environ. Microbiol.* 70, 1088–1096.

Di Cagno, R., De Angelis, M., Alfonsi, G. *et al.* (2005). Pasta made from durum wheat semolina fermented with selected lactobacilli as a tool for a potential decrease of the gluten intolerance. *J. Agric. Food Chem.* 53, 4379–4402.

Diowksz, A., Koziol, G., Kordialik-Bogacka, E., Ambroziak, W., and Sucharzewska, D. (2006). β-Glucan content in gluten free sourdough breads supplemented with soya sprouts. *Proceedings of the 3rd International Symposium on Sourdough,* October 25–28, Bari, Italy.

Drago, S., Asmar, R. E., D'Agate, C. *et al.* (2003). Gliadin induces increased intestinal permeability, zonulin release, and occluding down-regulation in an ex-vivo human intestinal model of celiac disease. *Gastroenterology* 124, A658.

Ehrmann, M. A. and Vogel, R. F. (2005). Molecular taxonomy and genetics of sourdough lactic acid bacteria. *Trends Food Sci. Technol.* 16, 31–42.

Ehrmann, M. A., Müller, M. R. A., and Vogel, R. F. (2003). Molecular analysis of sourdough reveals *Lactobacillus mindensis* sp. nov. *Int. J. Syst. Evolutionary Microbiol.* 53, 7–13.

Gallagher, E., Gormleya, T. R., and Arendt, E. K. (2004). Recent advances in the formulation of gluten-free cereal-based products. *Trends Food Sci. Technol.* 15, 143–152.

Gallo, G., De Angelis, M., McSweeney, P. L. H., Corbo, M. R., and Gobbetti, M. (2005). Partial purification and characterization of an X-prolyl dipeptidyl aminopeptidase from *Lactobacillus sanfranciscensis* CB1. *Food Chem.* 9, 535–544.

Gänzle, M. G., Vermeulen, N., and Vogel, R. F. (2007). Carbohydrate, peptide and lipid metabolism of lactic acid bacteria in sourdough. *Food Microbiol.* 24, 128–138.

Giovannini, C., Matarrese, P., Scazzocchio, B., Varì, R., D'Archivio, M., and Straface, E. (2003). Wheat gliadin induces apoptosis of intestinal cells via an autocrine mechanism involving Fas-Fas ligand pathway. *FEBS Lett.* 540, 117–124.

Giuliani, G. M., Benedusi, A., Di Cagno, R., De Angelis, M., Luisi, A., and Gobbetti, M. (2006). Miscela di batteri lattici per la preparazione di prodotti da forno senza glutine. RM2006A000369.

Gobbetti, M. (1998). The sourdough microflora: interactions of lactic acid bacteria and yeasts. *Trends Food Sci. Technol.* 9, 267–274.

Gobbetti, M. and Corsetti, A. (1996). Co-metabolism of citrate and maltose by *Lactobacillus brevis* subsp. *lindneri* CB1 citrate-negative strains: effect on growth, end-products and sourdough fermentation. *Z. Lebensm. Untersuch. Forsch.* 203, 82–87.

Gobbetti, M. and Corsetti, A. (1997). *Lactobacillus sanfrancisco* a key sourdough lactic acid bacterium: a review. *Food Microbiol.* 14, 175–187.

Gobbetti, M., Lavermicocca, P., Minervini, F., De Angelis, M., and Corsetti, A. (2000). Arabinose fermentation by *Lactobacillus plantarum* in sourdough added of pentosans and α-L-arabinofuranosidase: a tool to increase the production of acetic acid. *J. Appl. Microbiol.* 88, 317–324.

Gobbetti, M., De Angelis, M., Corsetti, A., and Di Cagno, R. (2005). Biochemistry and physiology of sourdough lactic acid bacteria. *Trends Food Sci. Technol.* 16, 57–69.

Hammes, W. P. and Gänzle, M. G. (1998). Sourdough breads and related products. In: Woods, B. J. B. ed. *Microbiology of Fermented Foods*, Vol. 1. London: Blackie Academic/Professional, pp. 199–216.

Hausch, F., Shan, L., Santiago, N. A., Gray, G. M., and Khosla, C. (2002). Intestinal digestive resistance of immunodominant gliadin peptides. *Am. J. Physiol. Gastrointest. Liver Physiol.* 283, 996–1003.

Höltzel, A., Gänzle, M. G., Nicholson, G. J., Hammes, W. P., and Jung, G. (2000). The first low-molecular-weight antibiotic from lactic acid bacteria: reutericyclin, a new tetramic acid. *Angew. Chem. Int. Ed.* 39, 2766–2768.

Hoseney, C. (1994). *Principles of Cereals Science and Technology*, 2nd edn. St. Paul MN: American Association of Cereal Chemists.

Katina, K., Arendt, E. H., Liukkonen, K. H., Autio, K., Flander, L., and Poutanen, K. (2005). Potential of sourdough for cereal products. *Trends Food Sci. Technol.* 16, 104–112.

Kieronczyk, A., Skeie, S., Olsen, K., and Langsrud, T. (2001). Metabolism of amino acids by resting cells of non-starter lactobacilli in relation to flavour development in cheese. *Int. Dairy J.* 11, 217–224.

Kirschner, L. and Von Holy, A. (1989). Rope spoilage of bread. *S. Afr. J. Sci.* 85, 425–427.

Korakli, M., Rossmann, A., Gänzle, G., and Vogel, R. F. (2001). Sucrose metabolism and exopolysaccharide production in wheat and rye sourdoughs by *Lactobacillus sanfranciscensis*. *J. Agric. Food Chem.* 49, 5194–5200.

Lavermicocca, P., Valerio, F., and Visconti, A. (2003). Antifungal activity of phenyllactic acid against moulds isolated from bakery products. *Appl. Environ. Microbiol.* 69, 634–640.

Liukkonen, K. H., Katina, K., Wilhelmson, A. *et al.* (2003). Process-induced changes on bioactive compounds in whole grain rye. *Proc. Nutr. Soc.* 62, 117–122.

Lopez, H., Krspine, V., Guy, C., Messager, A., Demigne, C., and Remesy, C. (2001). Prolonged fermentation of whole wheat magnesium. *J. Agric. Food Chem.* 49, 2657–2662.

Maiuri, L., Ciacci, C., Vacca, L. *et al.* (2001). IL-15 drives the specific migration of CD94+ and TCR-gammadelta+intraepithelial lymphocytes in organ cultures of treated celiac patients. *Am. J. Gastroenterol.* 96, 150–156.

Mazzarella, G., Maglio, M., Paparo, F. *et al.* (2005). An immunodominant DQ8 restricted gliadin peptide activates small intestinal immune response in in vitro cultured mucosa from HLA-DQ8 positive but not HLA-DQ8 negative celiac patients. *Gut* 52, 57–62.

Meroth, C. B., Hammes, W. P., and Hertel, C. (2004). Characterization of the microbiota of rice sourdoughs and description of *Lactobacillus spicheri* sp. nov. *Syst. Appl. Microbiol.* 27, 151–159.

Molberg, O., Solheim Flaete, N., Jensen, T., Lundin, K. E., Arentz-Hansen, H., and Anderson, O. D. (2003). Intestinal T-cell responses to high-molecular-weight glutenins in celiac disease. *Gastroenterology* 125, 337–344.

Moore, M. M. (2005). Novel approaches in the structural development of gluten free bread. Doctoral dissertation, University College Cork Ireland.

Müller, M. R. A., Ehrmann, M. A., and Vogel, R. F. (2000). *Lactobacillus frumenti* sp. nov., a new lactic acid bacterium isolated from rye-bran fermentations with a long fermentation period. *Int. J. Syst. Evol. Microbiol.* 50, 2127–2133.

Müller, M. R. A., Wolfrum, G., Stolz, P., Ehrmann, M. A., and Vogel, R. F. (2001). Monitoring the growth of *Lactobacillus* species during rye flour fermentation. *Food Microbiol.* 27, 217–227.

Östman, E., Nilsson, M., Liljeberg-Elmståhl, H., Molin, G., and Björck, I. (2002). On the effect of lactic acid on blood glucose and insulin responses to cereal products: mechanistic studies in healthy subjects and in vitro. *J. Cereal Sci.* 36, 339–346.

Picarelli, A., Di Tola, L., Sabbatella, M., Greco, L., Silano, R., and De Vincenzi, M. (1999). 31–43 amino acid sequence of the α-gliadin induces endomysial antibody production during in vitro challenge. *Scand. J. Gastroenterol.* 34, 1099–1102.

Pyle, G. G., Paaso, B., Anderson, B. E. *et al.* (2005). Effect of pretreatment of food gluten with prolyl endopeptidase on gluten-induced malabsorption in celiac sprue. *Clin. Gastroenterol. Hepatol.* 3, 687–694.

Rizzello, C. G., De Angelis, M., Di Cagno, R. *et al.* (2007). Highly efficient gluten degradation by lactobacilli and fungal proteases during food processing: new perspectives for celiac disease. *Appl. Environ. Microbiol.* 73, 4499–4507.

Röcken, W. and Voysey, P. A. (1995). Sour-dough fermentation in bread making. *J. Appl. Bacteriol.* 79, 38S–48S.

Ryan, L. A., Dal Bello, F., Renzetti, S., and Arendt, E. K. (2006). The use of selected lactic acid bacteria to improve the baking and rheological quality of gluten-free batter and bread. Meeting Abstract, San Francisco.

Scheirlink, I., Van der Meulen, R., Van Schoor, A. *et al.* (2007). *Lactobacillus namurensis* sp. nov., isolated from a traditional Belgian sourdough. *Int. J. Syst. Evol. Microbiol.* 57, 223–227.

Schieberle, P. (1996). Intense aroma compounds—useful tools to monitor the influence of processing and storage on bread aroma. *Adv. Food Sci.* 18, 237–244.

Schnürer, J. and Magnusson, J. (2005). Antifungal lactic acid bacteria as bio-preservatives. *Trends Food Sci. Technol.* 16, 70–78.

Shan, L., Molberg, O., Parrot, I. *et al.* (2002). Structural basis for gluten intolerance in coeliac sprue. *Science* 297, 2275–2279.

Shan, L., Marti, T., Sollid, L. M., Gray, G. M., and Khosla, C. (2004). Comparative biochemical analysis of three bacterial prolyl endopeptidases: implications for celiac sprue. *Biochem. J.* 383, 311–318.

Sikken, D. and Lousche, K. (2003). Starter preparation for producing bakery products. European Patent EP13611796.

Sollid, L. M. and Khosla, C. (2005). Future therapeutic options for celiac disease. *Nat. Clin. Pract. Gastroenterol. Hepatol.* 2, 140–147.

Stepniak, D., Spaenij-Dekking, L., Mitea, C. *et al.* (2006). Highly efficient gluten degradation with a newly identified prolyl endoprotease: implications for celiac disease. *Am. J. Physiol. Gastrointest. Liver Physiol.* 291, 460–468.

Thiele, C., Gänzle, M. G. and Vogel, R. F. (2002). Contribution of sourdough lactobacilli, yeast and cereal enzymes to the generation of amino acids in dough relevant for bread flavour. *Cereal Chem.* 79, 45–51.

Troncone, R., Mazzarella, G., Leone, N. *et al.* (1998). Gliadin activates mucosal cell mediated immunity in cultured rectal mucosa from coeliac patients and a subset of their siblings. *Gut* 43, 484–489.

Valcheva, R., Korakli, M., Onno, B. *et al.* (2005). *Lactobacillus hammesii* sp. nov., isolated from French sourdough. *Int. J. Syst. Evol. Microbiol.* 55, 763–767.

Valcheva, R., Ferchichi, M., Korakli, M. *et al.* (2006). *Lactobacillus nantentis* sp. nov. isolated from French wheat sourdough. *Int. J. Syst. Evol. Microbiol.* 56, 587–591.

Vancanneyt, M., Neysens, P., Dewachter, M. *et al.* (2005). *Lactobacillus acidifarinae* sp. nov., from wheat sourdoughs. *Int. J. Syst. Evol. Microbiol.* 55, 615–620.

Vermeulen, N., Kretzer, J., Machalitza, H., Vogel, R. F., and Gänzle, M. G. (2006). Influence of redox-reactions catalysed by homo-and hetero-fermentative lactobacilli on gluten in wheat sourdoughs. *J. Cereal Sci.* 43, 137–143.

Vogel, R. F., Bocker, G., Stolz, P. *et al.* (1994). Identification of lactobacilli from sourdough and description of *Lactobacillus pontis* sp. nov. *Int. J. Syst. Bacteriol.* 44, 223–229.

Vogel, R. F., Knorr, R., Müller, M. R. A., Steudel, U., Gänzle, M. G., and Ehrmann, M. A. (1999). Non-dairy lactic fermentations: the cereal world. *Antonie van Leeuwenhoek* 76, 403–411.

Weiss, N. and Schillinger, U. (1984). *Lactobacillus sanfrancisco* sp. nov., nom. rev. *Syst. Appl. Microbiol.* 4, 507–511.

Wiese, B. G., Strohmar, W., Rainey, F. A., and Diekmann, H. (1996). *Lactobacillus panis* sp. Nov., from sourdough with a long fermentation period. *Int. J. Syst. Bacteriol.* 46, 449–453.

Yan, F. and Polk, D. B. (2004). Commensal bacteria in the gut: learning who our friends are. *Curr. Opin. Gastroenterol.* 20, 565–571.

Gluten-free breads

13

Elke K. Arendt, Andrew Morrissey, Michelle M. Moore, and Fabio Dal Bello

Introduction

Permanent lifelong withdrawal of gluten from the diet is the only effective treatment for celiac disease (Chapter 1). However, removal of gluten from bread formulations often results in a liquid batter, rather than a dough system during the pre-baking phase, and can result in baked bread with crumbling texture, poor color and other quality defects (Gallagher *et al.*, 2004a). Indeed, gluten is the main structure-forming protein present in wheat flour, and plays a major role in breadmaking functionality of wheat flours by providing viscoelasticity to the dough, good gas-holding properties, and good crumb structure of many baked products (Gallagher *et al.*, 2004a; Moore *et al.*, 2004). Currently many of the gluten-free baked products that are available on the market are of low quality, exhibiting poor mouth-feel and flavor (Arendt *et al.*, 2002). These problems present major technological challenges to both the cereal technologist and the baker, and have led to the search for alternatives to gluten in the manufacture of gluten-free baked products. Gluten-free bread requires polymeric substances that mimic the viscoelastic properties of gluten in bread dough (Toufeili *et al.*, 1994). The production of gluten-free breads mainly involves the incorporation of starches, protein-based ingredients like dairy proteins (Chapter 12), and hydro-colloids (Chapter 9) into a gluten-free base flour that could mimic the viscoelastic properties of gluten and result in improved texture, mouth-feel, acceptability, and shelf-life of these products (Ylimaki *et al.*, 1991; Haque and Morris, 1994; Gujral *et al.*, 2003a, 2003b; Gallagher *et al.*, 2003, 2004a; Moore *et al.*, 2004, 2006, 2007a; Sivaramakrishnan *et al.*, 2004; McCarthy *et al.*, 2005; Ahlborn *et al.*, 2005; Lazaridou *et al.*, 2007).

The gluten-free diet

Total lifelong avoidance of gluten ingestion remains the cornerstone treatment for celiac disease. The gluten-free diet is sometimes called the "drug of choice" by patients, because nutrition therapy is a strict gluten-free diet for life (Kupper, 2005). The overall goal of the gluten-free diet is to achieve healing and maintain health through the adoption of a well-balanced diet that avoids gluten. The diet requires ongoing education of patients and their families by both doctors and dieticians. Compliance with a strict gluten-free diet is not easy, because a strict diet could lead to a form of social isolation to patients with celiac disease, and to nutritional deficiencies in B vitamins, calcium, vitamin D, iron, zinc, magnesium and fiber.

Worldwide, there is a major debate regarding the accepted definition of what constitutes "gluten-free." Products labeled "gluten-free" in Canada must meet standards of less than 20 mg gluten per kg, whereas other countries use 200 mg/kg, and still others prefer a double standard for those products rendered gluten-free and those naturally gluten-free. The current Codex Alimentarius Standard for "Gluten-free Foods" was adopted by the Codex Alimentarius Commission in 1976, and amended in 1983. In this document, gluten is defined as those storage proteins commonly found in wheat, triticale, rye, barley, or oats. The definition of gluten-free came under review in the 1990s and the definition of gluten-free continues at step 7 while the Codex Committee awaits research on the scientific basis for the establishment of a tolerance level and a method of detection is clarified (Codex Alimentarius Commission, 2003). Gluten-free foods are described as foods:

> (a) consisting of, or made only from ingredients which do not contain any prolamins from wheat or all *Triticum* species such as spelt, kamut or durum wheat, rye, barley, oats or their crossbred varieties with a gluten level not exceeding 20 mg/kg; or (b) consisting of ingredients from wheat, rye, barley, oats, spelt or their crossbred varieties, which have been rendered gluten-free; with a gluten level not exceeding 200 mg/kg; or (c) any mixture of two ingredients as in (a) and (b) mentioned above with a level not exceeding 200 mg/kg.

In this context standard gluten was defined as a protein fraction from wheat, rye, barley, oats, or their crossbred varieties and derivatives thereof, that are soluble in water and 0.5 mol/L NaCl and to which some individuals are intolerant.

There are problems around the world on the issue of labeling foods as gluten-free because the exact amount of toxic prolamins that individuals with celiac disease may consume without damaging the mucosa of the small intestine has not been adequately defined (Thompson, 2000). Acceptance of the ELISA method by the Codex Commission, and results of ongoing research on tolerance levels will allow the Commission to move towards adopting a revised definition of "gluten-free."

The role of gluten in bakery products

It has long been known that the breadmaking quality of wheat flour depends on both the quantity and quality of its gluten proteins. The gluten proteins contribute 80–85% of the total wheat protein and are the major storage proteins of wheat (Figure 13.1). They belong to the prolamin class of seed storage proteins (Shewry and Halford, 2002). Gluten proteins are largely insoluble in water or dilute salt solutions. Two functionally distinct groups of gluten proteins can be distinguished: monomeric gliadins and polymeric (extractable and unextractable) glutenins (Lindsay and Skerritt, 1999). Gliadins and glutenins are usually found in more or less equal amounts in wheat. Gluten has a unique amino acid structure, with Glu/Gln and Pro accounting for more than 50% of the amino acid residues (Eliasson and Larsson, 1993; Lasztity, 1995). The low water solubility of gluten is attributable to its low content of Lys, Arg, and Asp residues, which together amount to less than 10% of the total amino acid residues. About 30% of the amino acid residues in gluten are hydrophobic, and the residues contribute greatly to its ability to form protein aggregates by hydrophobic interactions and to bind lipids and other non-polar substances. The high glutamine and hydroxyl amino acids ($\sim$10%) content of gluten are responsible for its water-binding properties. In addition, hydrogen bonding between glutamine and hydroxyl residues of gluten polypeptides contributes to its cohesion–adhesion properties. Cysteine and cystine residues account for 2–3% of the total amino acid residues, and during formation of dough, these residues undergo sulfhydryl–disulfide interchange reactions, resulting in extensive polymerization of gluten proteins. It is generally accepted that the breadmaking quality of wheat is related to the presence and properties of gluten proteins. The gliadin fraction has been reported to contribute to the viscous properties and dough extensibility of wheat dough (Pomeranz, 1988; Don *et al.*, 2003a, 2003b). The glutenin fraction of wheat gluten has long been considered to have a prominent role in the elastic and strengthening of dough (MacRitchie, 1980; Xu *et al.*, 2007). The relative proportions of gliadin and glutenin found in dough affect the physical

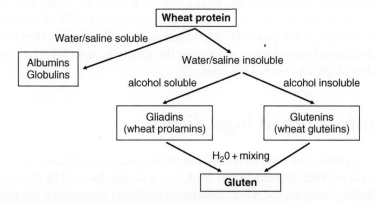

Figure 13.1 Illustration of the component of wheat protein.

properties of dough, with higher relative proportions of glutenin imparting greater dough strength (MacRitchie, 1987).

In addition, wheat gluten makes possible many non-food applications, such as gluten-based films and molded biodegradable plastics. Because of their ability to polymerize extensively via sulfhydryl–disulfide interchange reactions, which occur during dough formation, glutenins contribute greatly to the elasticity of dough. Also, because of their unique structure and functional properties, it is technologically extremely challenging to find alternative ingredients that mimic these properties in breadmaking.

It is clear that gluten protein functionality is central to bread quality. Fractionation and reconstitution experiments show clearly that variations in breadmaking performance are determined by the gluten proteins (Veraverbeke and Delcour, 2002). Wheat flour breadmaking performance is linearly related to flour protein content, and thus with gluten protein content, because this protein fraction increases much more than the non-gluten protein fraction with increasing grain protein content (Hoseney, 1994). When kneading/mixing flour with water, gluten proteins enable the formation of cohesive viscoelastic dough that is capable of holding gas produced during fermentation and oven-rise resulting in the typical fixed open foam structure of bread after baking. Although the dough's rheological properties, essential for breadmaking, are largely determined by the wheat gluten proteins, interactions of gluten protein matrix with other flour components [e.g. flour lipids (Eliasson and Larsson, 1993), arabinoxylans (Goesaert et al., 2005), non-gluten proteins (Veraverbeke and Delcour, 2002)] may affect its rheological properties. Wheat gluten rheological properties can be modified by the addition of oxidants, reducing agents, or by the addition of lipids/emulsifiers or hemicelluloses that can modify gluten protein interactions (Veraverbeke and Delcour, 2002; Goesaert et al., 2005).

The single most important factor in the acceptance of a food product is its sensory character, which is the integrated response to the chemical and physical stimuli imparted by the food through its texture, taste, color, aroma, and irritant components (Forde and Delahunty, 2004). Generally foods that are deemed pleasant are selected over those that are regarded as unpleasant. The keystone treatment of patients with celiac disease is a lifelong diet in which food products containing gluten are avoided. The technological approach to the production of gluten-free foods that meet the unique nutritional and sensory requirements of patients with celiac disease, includes the use of starches, dairy products, gums, and hydrocolloids, and other non-gluten proteins, as alternatives to gluten, to improve the structure, mouth-feel, acceptability, and shelf-life of gluten-free bakery products.

Gluten-free bread ingredients

The different gluten-free flours available for the production of gluten-free bread have been the topic of other chapters in this book, and will therefore not be discussed here. In the following sections, the most important ingredients constituting the gluten-free formulation will be introduced and discussed singularly.

Starch

Physiological properties of starch

Starch is the most important reserve polysaccharide and the most abundant constituent in many plants. Starch is unique among carbohydrates because it occurs naturally as discrete semi-crystalline granules that are relatively dense and insoluble, and hydrate only slowly in cold water. Starch has some unique properties that determine its functionality in many food applications, in particular bakery products where it contributes to texture, appearance, and overall acceptability of cereal-based foods (Ward and Andon, 2002). Its structure and physico-chemical properties have been extensively reviewed (Parker and Ring, 2001; Eliasson and Gudmundsson, 2006).

Most starch granules are composed of a mixture of amylose and amylopectin. Amylose is an essentially linear molecule, consisting of α (1,4)-linked D-glucopyranosyl units with a degree of polymerization in the range of 500–600 glucose residues. In contrast, amylopectin is a very large, highly branched polysaccharide with a degree of polymerization ranging from 3×10^5 to 3×10^6 glucose units. It is composed of chains of α (1,4)-linked D-glucopyranosyl residues, which are interlinked by α (1,6)-bonds. The amylose/amylopectin ratio differs between starches, but typical levels of amylose and amylopectin are 25–28% and 72–75%, respectively. However, the starches of some mutant genotypes of maize, barley, and rice contain either an increased amylose content (high amylose or amylostarch with up to 70% amylose) or an increased amylopectin content (waxy starch with 99–100% amylopectin content) (Goesaert et al., 2005).

Starch is present as intracellular water-insoluble granules of different sizes and shapes, depending on the botanical source (Moon and Giddings, 1993). A significant fraction of the starch granule (about 8%) is damaged during milling. This mechanical damage to the granule structure greatly affects starch properties (Hoseney, 1994). Damaged starch has a higher water absorption capacity and is more susceptible to enzymatic hydrolysis. At room temperature and in sufficient water, starch granules can imbibe up to 50% of their dry weight of water, and then return to their original size on drying (BeMiller and Whistler, 1996). When the starch suspension is heated in water above a specific temperature, it undergoes disruption of the molecular order, loss of crystallinity, and irreversible granule swelling. This process is termed gelatinization. Heating and hydration of the non-crystalline regions facilitate molecular mobility in these regions and dissociation of the amylopectin double helices and melting of crystallites (Tester and Debon, 2000). The gelatinization process is also associated with limited starch solubilization (mainly amylose leaching), which increases the viscosity of the starch suspension. During further heating, and above the gelatinization temperature, swelling and leaching continue and a continuous phase of solubilized macromolecules (mainly amylose) and a discontinuous phase of swollen, amorphous starch granules or remnants are formed (BeMiller and Whistler, 1996; Tester and Debon, 2000; Eliasson and Gudmundsson, 2006).

When the starch paste is cooled, the starch polysaccharides re-associate to a more ordered or crystalline state. This process is termed retrogradation. The kinetics of starch retrogradation of the two starch polymers amylose and amylopectin differ considerably (Hug-Iten et al., 2003). At starch concentrations above 6%, double helices are formed between amylose molecules that were solubilized during gelatinization and pasting, and a continuous network develops. After some hours, these double helices form very stable crystalline gel structures. The re-crystallization of the short amylopectin side-chains is a much slower process (several days or weeks) and occurs in the gelatinized granules or remnants (Miles *et al.*, 1985). Therefore, amylose retrogradation determines to a great extent the initial firmness of a starch gel, while amylopectin retrogradation determines the long-term development of gel structures and crystallinity in starch systems (Miles *et al.*, 1985). Starch retrogradation is influenced by a number of conditions and substances, including pH and the presence of salts, sugars, and lipids (Eliasson and Gudmundsson, 2006). An important characteristic of amylose is its ability to form helical inclusion complexes with a number of substances, in particular polar lipids. Amylose forms a left-handed single helix and the hydrocarbon chain of the lipid is situated in the central cavity. The presence of polar lipids affect starch properties to a large extent, in particular its gelatinization and retrogradation characteristics (BeMiller and Whistler, 1996; Eliasson and Gudmundsson, 2006).

The role of starch in breadmaking

During dough preparation, starch absorbs up to about 45% water and is considered to act as an inert filler in the continuous matrix of the dough (Bloksma, 1990). On the other hand, Eliasson and Larsson (1993) described dough as a bicontinuous network of protein and starch. In a later study, Larsson and Eliasson (1997) reported that the rheological behavior of wheat dough is influenced by the specific properties of the starch granule surface. Due to the combination of heat, moisture, and time during baking, starch granules gelatinize (i.e. they swell and are partially solubilized), but still maintain their granular identity (Hug-Iten *et al.*, 2001). The two starch polymers amylose and amylopectin tend to demix and a small amount of amylose is leached into the inter-granular phase. Part of this amylose forms inclusion complexes with both added and endogenous polar lipids of wheat. Due to phase separation, amylose and amylopectin are not homogeneously distributed in the granule (Hug-Iten *et al.*, 2001). On cooling, the solubilized amylose forms a continuous network, in which swollen and deformed starch granules are embedded and interlinked. Because of its rapid retrogradation, amylose is an essential structural element of bread and is a determining factor for initial loaf volume (Eliasson and Larsson, 1993). During storage, bread loses its freshness and stales, and the crust toughens, the crumb becomes more firm and less elastic, and moisture and flavor is lost (Hoseney, 1994). On staling, water migrates from crumb to crust and leads to a glass to rubber transition of the two components, and as a consequence, the crust becomes soft and leathery (Eliasson and Larsson, 1993). An increase in firmness and crumbliness is observed as typical changes in the crumb during aging. Reorganization of the starch fractions, amylopectin and amylose, and the increase of starch network rigidity due to increase of polymer order are important changes during aging. Migration of water and amylopectin retrogradation, in particular the formation of double helical structures and crystalline regions, are considered to be primarily responsible for the staling of bread during ageing (Zobel and Kulp, 1996; Gray and BeMiller, 2003;

Hug-Iten *et al.*, 2003). However, formation of ordered amylose structures in the centre of the granules may also contribute to granular rigidity (Hug-Iten *et al.*, 2003). Therefore, both molecular reorganization of the amylopectin-rich and amylose-rich regions in the starch granules, results in an increased granular rigidity, and the formation of a structured network consisting of interlinked crystallites, contribute to crumb firming. Hug-Iten *et al.* (2003) also proposed that formation of cross-links by hydrogen bonds and entanglements between the two polymers have an impact on the mechanical properties of bread. It has been proposed that crumb firming during aging can be attributed to some extent to gluten–starch interaction (Every *et al.*, 1998). However, Ottenhof and Farhat (2004) concluded that there was no evidence of any significant effect of the presence of gluten on the kinetics and extent of amylopectin retrogradation. Interaction is unlikely because of the thermodynamic incompatibility of the gluten and amylopectin (Tolstoguzov, 1997).

Dairy ingredients

The replacement of gluten with other protein sources such as dairy proteins (Chapter 10) is another approach used to improve the quality of gluten-free products. However, the supplementation of gluten-free breads with high lactose-content dairy ingredients is not suitable for people with celiac disease who have encountered significant damage to their intestinal villi, since they lack the enzyme lactase, which is normally generated by the villi (Ortolani and Pastorello, 1997). According to Murray (1999), approximately 50% of people with celiac disease have lactose intolerance. Nonetheless, several studies have addressed the inclusion of dairy proteins in gluten-free systems, showing that they have functional properties similar to gluten. Dairy proteins are capable of forming networks and have good swelling properties; they are also regarded as highly functional ingredients and, due to their versatility, they can be readily used in gluten-free food products (Gallagher *et al.*, 2003). Dairy ingredients are currently applied in bakery products for both nutritional and functional benefits such as the improvement of flavor and texture, and the reduction of bread staling (Cocup and Sanderson, 1987; Mannie and Asp, 1999; Kenny *et al.*, 2001; Gallagher *et al.*, 2003). Recently, improvement in the volume, appearance, and sensory properties of gluten-free bread was obtained upon addition of the dairy ingredients molkin, demineralized whey powder, skim milk powder replacer, skim milk powder, sodium caseinate, or milk protein isolate (Gallagher *et al.*, 2003). In another study, Moore *et al.* (2004) investigated gluten-free breads produced from commercial gluten-free flour and mixtures of gluten-free ingredients with and without the addition of dairy ingredients. The study clearly showed that the breads containing the dairy ingredients had the best quality and resembled a wheat bread most closely. The high quality of the dairy ingredients-containing breads was attributed to the ability of the dairy ingredients to form a network similar to gluten (Figure 13.2).

Overall, the type of dairy ingredient added is a key factor determining the quality of gluten-free products (Nunes *et al.*, 2007). As shown in Plate 13.1, the addition

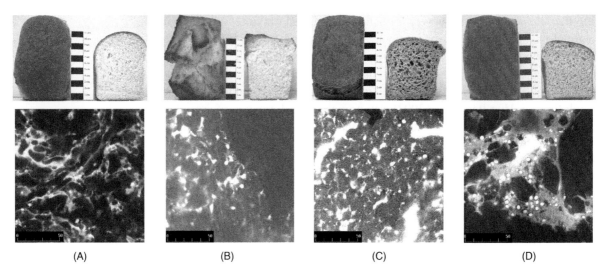

(A)   (B)   (C)   (D)

Figure 13.2 Wheat bread control (A) and gluten-free breads [bread from commercial gluten-free flour (B), non-dairy recipe (C), dairy recipe (D)]. Outer appearance of the breads and microscopical structure as detected by confocal laser-scanning microscopy (magnification bar corresponds to 50 μm).

of low lactose powders, sodium caseinate, milk protein isolate, whey protein isolate, or whey protein concentrate leads to breads with distinctly different appearances.

From the studies carried out so far it can be concluded that dairy ingredients can certainly improve the overall quality of gluten-free foods as well as their nutritional value. However, it is essential that the dairy ingredients used in gluten-free products are lactose-free or at least contain low levels of lactose. In addition, it should be noted that there are large differences in the properties of the various dairy ingredients, not only due to their composition but also in the way they have been processed (Kenny *et al.*, 2000, 2001).

Soya

Soyabeans belong to the plant family Fabaceae, also known as "legumes" or "pulses." Members of this plant family are characteristically rich in protein, but deficient in S-containing amino acids (Belitz and Grosch, 1987). Soya has been reported to have a number of properties that have made it an attractive ingredient for functional foods. Soya isoflavones have been shown to have positive effects on bone tissue and therefore they could be used in functional foods that target a reduction in the risks of osteoporosis (Brouns, 2002). Furthermore, soya isoflavones are reported to reduce the risk of cardiovascular disease, reduce the oxidation of low-density lipoproteins, and prevent breast cancer. Soya flour and soya products have been used to increase the protein content as well as to improve the structural properties of gluten-free products. For example, Sanchez *et al.* (2002) found that the inclusion of 0.5% soya in a gluten-free formulation enhanced the crumb grain score, bread volume, and overall bread score. Moore *et al.* (2004) also found that soya has a positive impact on gluten-free bread quality (i.e. the inclusion of soya increased the nutritional and water absorption properties of gluten-free breads). However, according to most recent European Union directives soya is listed as a high allergenic ingredient (Fernández-Rivas and Ballmer-Weber, 2007), and therefore its use for gluten-free products needs to be carefully evaluated.

Egg

Eggs are not only added to food products to increase their nutritional value, but also to improve color, flavor, and to enhance the product's emulsifying, whipping/foaming, and/or coagulation/gelation properties (Mine, 2002). The yolk is rich in the fat-soluble vitamins A, D, E, and K as well as in phospholipids, including the emulsifier lecithin. Nutritionally, eggs are a good source of fat, protein, vitamins, and minerals, especially iron. However, eggs contain about 240 mg of cholesterol, which is present in the yolk. For this reason, people who must restrict their cholesterol intake usually consume fewer whole eggs. In certain baked products such as cakes, egg constituents may perform multiple functions, including emulsification and stabilization of fat in the batter. Research by Jonagh *et al.* (1968) showed that proteins such as egg albumen are able to link starch granules together.

Kato *et al.* (1990) suggested that egg proteins form strong cohesive viscoelastic films, which are essential for stable foaming (Figure 13.2). In a gluten-free bread system egg proteins form viscous solutions, a film-like continuous protein structure similar to wheat gluten as can be observed using confocal laser-scanning microscopy (Moore *et al.*, 2004). This film-like protein structure can be further enhanced by using transglutaminase (Moore *et al.*, 2006), an enzyme which catalyzes protein cross-linking and which has been used to increase the quality of gluten-free breads (Renzetti *et al.*, 2007). Confirmation of the positive effects of egg powder is given by the fact that, without the addition of transglutaminase, in a gluten-free bread system film formation can be observed only when egg powder is present (Figure 13.3).

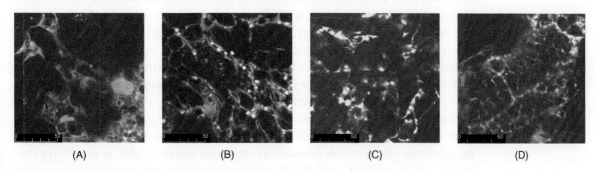

(A)   (B)   (C)   (D)

Figure 13.3 Confocal laser-scanning micrographs of gluten-free breads: Egg control bread (A); Egg with 1U transglutaminase/g protein (B); skim milk powder (SMP) control bread (C), and SMP with 10U transglutaminase/g protein (D). The magnification bar corresponds to 50 μm for the objective × 63.

Although studied by very few groups, the limited data available suggest that egg is a valuable ingredient in the formulation of gluten-free products. The only problems are the costs involved as well as the possible complication associated with processing of gluten-free formulations containing eggs.

Hydrocolloids

Function of hydrocolloids in breadmaking

Hydrocolloids are hydrophilic polymers of vegetable, animal, microbial, or synthetic material that generally contain many hydroxyl groups, and may be polyelectrolytes. They are widely used to control the functional properties of foodstuffs (Williams and Phillips, 2000; Chapter 9). They are usually added to starch-containing products due to their desirable effect on the acceptability of food systems. Hydrocolloids have been widely used as additives to: (i) improve food texture and viscoelastic characteristics (Armero and Collar, 1996a, 1996b); (ii) slow down the retrogradation of starch (Davidou *et al.*, 1996); (iii) act as water binders; (iv) function as fat replacers; (v) extend the overall quality of products during storage; and (vi) also function as gluten substitutes in the formation of gluten-free breads (Toufeili *et al.*, 1994; Gurkin, 2002). Hydrocolloids, due to their high water retention capacity, give stability to products that undergo successive freeze–thaw cycles (Lee *et al.*, 2002). They also show good properties as substitutes for fats in different products (Albert and Mittal, 2002). Even though hydrocolloids are often present at concentrations less than 1% they can have a significant influence on the textural and organoleptic properties of foods.

In the baking industry, hydrocolloids are of increasing importance as breadmaking improvers, and several studies have been carried out showing their potential in this area. Carboxymethylcellulose (CMC) and guar gum have been added to rye flour to improve the bread quality (Mettler and Seibel, 1995). In addition, Rosell *et al.* (2001) showed that an improvement in wheat dough stability during proofing can be obtained by the addition of the hydrocolloids sodium alginate, κ-carrageenan, xanthan gum, or hydroxypropylmethylcellulose (HPMC). Furthermore, the hydrocolloids increased the specific volume, with the exception of alginate, as well as both moisture retention and water activity. In addition, textural studies revealed that addition of κ-carrageenan or HPMC reduced the firmness of the bread crumb. The authors concluded that κ-carrageenan and HPMC were effective improvers in breadmaking performance. An earlier study by Davidou *et al.* (1996) using locus bean gum, xanthan gum, and alginate also revealed a softening effect by those colloids, which was attributed primarily to the high water-retention capacity in the case of locus bean gum. In the case of the xanthan gum and alginates, the softening effect was caused by hindering gluten–starch interactions. In wheat breads, Collar *et al.* (2001) studied the effects of CMC and HPMC addition on dough and bread performance, as well as their interactions with α-amylose and emulsifiers. In this case, and mainly with HPMC, the softening effect was attributed to their water-retention capacity, and a possible inhibition of amylopectin retrogradation. The authors concluded that HPMC preferentially binds to starch, and as a consequence starch–gluten interactions were prevented. Guar gum also has a softening effect, which is probably caused by inhibition of amylopectin retrogradation, since guar gum preferentially binds to starch (Collar *et al.*, 2001).

The hydrophilic character of hydrocolloids also prevents water release and polymer aggregation during refrigeration. Recently, a protective effect by HPMC was observed in partially baked bread stored at frozen temperatures, which resulted in improved loaf volume and softer crumb in the fully baked bread (Barcenas *et al.*, 2004; Barcenas and Rosell, 2005, 2006). HPMC decreased hardening rates of the bread, and also retarded amylopectin retrogradation. Cryo-scanning electron microscopy revealed an intimate interaction between HPMC chains and the constituents of the bread crumb (Barcenas and Rosell, 2006). Bread loaves prepared with locust bean gum retained moisture to a greater extent, and the loaves were softer when compared with the controls (Sharadanant and Khan, 2003).

Guarda *et al.* (2004) evaluated the effects of a range of hydrocolloids (sodium alginate, xanthan, κ-carrageenan, and HPMC) on fresh bread quality and bread staling. Different effects were associated with different hydrocolloids. HPMC improved specific volume index, width/height ratio, and crumb hardness. In addition, visual appearance, aroma, flavor, crunchiness, and overall acceptability were enhanced. All the hydrocolloids reduced moisture loss during bread storage, and alginate and HPMC had an anti-staling effect. Some studies (Sharadanant and Khan, 2003; Barcenas *et al.*, 2004) found that κ-carrageenan was not an effective improver in partially baked frozen breads.

Hydrocolloids have been used extensively to modify the gelatinization–gelation process, namely the pasting properties of starch. The influence of selected hydrocolloids (guar gum, pectin, alginate, κ-carrageenan, xanthan, and HPMC) on the pasting properties and gelling behavior of wheat flour were evaluated by Rojas *et al.* (1999). The greatest effect on pasting temperature was observed when alginate was added, which implied an earlier beginning of starch gelatinization and, in turn, an increase in availability of starch as enzyme substrate during the baking period. Xanthan and pectin increased the cooking stability, while κ-carrageenan, and alginate did not modify stability. The formation of amylose–lipid complexes was favored by κ-carrageenan, alginate, and pectin, and only slightly affected by xanthan and HPMC. Thus, when looking for reduction in staling, κ-carrageenan was considered the best hydrocolloid due to both its softening, and retardation of firmness during storage.

Hydrocolloids and gluten-free bread

The replacement of gluten in gluten-free bread presents a major technological challenge, since gluten is an essential structure-building protein, contributing to appearance and crumb structure of many baked products. In recent years, there has been increasing interest in the incorporation of starches, dairy ingredients and/or hydrocolloids into a gluten-free flour base (rice and corn flour) that could mimic the viscoelastic properties of gluten in bread dough, and thus result in improved structure, mouth-feel, acceptability, and shelf-life of these products (Toufeili *et al.*, 1994; Gallagher *et al.*, 2003, 2004a, 2004b; Moore *et al.*, 2004, 2006; Ahlborn *et al.*, 2005;

McCarthy *et al.*, 2005). In these studies, a number of hydrocolloids were investigated for the production of high-quality gluten-free breads, including HPMC, CMC, methylcellulose, β-glucan, psyllium gum, locust bean gum, guar gum, and xanthan.

An improving effect of several hydrocolloids, such as HPMC, CMC, locust bean gum, guar gum, κ-carrageenan, xanthan, β-glucan, and psyllium, was reported by Haque and Morris (1994), Gallagher *et al.* (2004a) and Moore *et al.* (2004, 2006). According to Rosell *et al.* (2001) hydrocolloids improve dough development and gas retention by increasing dough viscosity, thereby increasing loaf volume. However, McCarthy *et al.* (2005) reported a slight decrease in loaf volume with increasing levels of HPMC in a gluten-free bread base of rice flour, potato starch and milk protein. The optimized formula contained 2.2% HPMC and 79% water. Haque and Morris (1994) produced rice bread with good loaf volume when combinations of HPMC and psyllium were used in gluten-free formulations, but when the polymers were added alone, the increase in volume was reduced. In addition, the volume increased up to a certain hydrocolloid concentration, but further increases in polymer concentration resulted in a decrease in loaf volume. In a recent study, Lazaridou *et al.* (2007) evaluated the effects of pectin, CMC, agarose, xanthan and oat β-glucan on dough rheology of gluten-free bread. The elasticity and resistance to deformation of dough followed the order of xanthan > CMC > pectin > agarose > β-glucan. The type and extent of influence on bread quality was also dependent on the specific hydrocolloid used and its concentration. They observed that the volume of breads increased with addition of hydrocolloids at 1% supplementation level with the exception of xanthan and pectin, compared with the respective control samples. However, when the hydrocolloid concentrations were increased from 1 to 2%, a reduction in bread volume was observed with the exception of pectin. The loaf volume of the pectin formulation increased significantly compared to the control. Incorporation of xanthan at 1% in gluten-free breads did not change loaf volume, and at 2% loaf volume decreased (Lazaridou *et al.*, 2007). This formulation produced the lowest volume of all the preparations used. Similarly, Haque and Morris (1994) observed no influence of xanthan incorporation in rice flour bread, and Schober *et al.* (2005) noted a decrease in loaf volume of gluten-free breads from sorghum with increasing xanthan levels.

Important physico-chemical parameters of bread quality are porosity and elasticity (Lazaridou *et al.*, 2007). High porosity was found in gluten-free breads supplemented with CMC and β-glucan at 1% concentration, and pectin at 2%. On the other hand, porosity of breads containing 2% xanthan was the lowest. Wang *et al.* (1998) observed that incorporation of β-glucan into wheat bread improved crumb grain by stabilizing air cells in the dough and preventing coalescence of cells. On the other hand, the porosity of bread containing 2% xanthan was the lowest. Lazaridou *et al.* (2007) pointed out that, in addition to porosity, the uniform size distribution of gas cells is also important for bread quality. For some formulations, such as those containing agarose (1%) or β-glucan (1%), despite large differences in porosity values, the visual appearance and internal structure of the loaf indicated that both samples exhibited a considerable number of non-uniform large gas cells, which adversely affected the uniformity of the crumb structure, and subsequent loaf quality (Lazaridou *et al.*,

2007). High values for crumb elasticity were observed in breads supplemented with CMC, pectin, and xanthan at the 2% level. An increase in lightness (L values) of crust was observed with the addition of β-glucan (1%), whereas whiteness of crumb was improved when xanthan was included in the formulation.

Sensory evaluation by a consumer (untrained) panel showed that those gluten-free breads containing 2% CMC were highly acceptable. Compared with control formulations, crumb firmness (compression test) was not increased significantly on addition of pectin (2%), CMC (2%), agarose (1 and 2%), or β-glucan (1%). On the other hand, addition of xanthan (1 and 2%) and β-glucan (2%) resulted in crumb hardening. Gluten-free breads supplemented with xanthan exhibited the greatest increase in firmness, which was consistent with the large decrease in a_w values during storage. Schober et al. (2005) also reported an increase in crumb hardness when xanthan gum was added to gluten-free breads made from sorghum. According to Biliaderis et al. (1997) the effects of hydrocolloids on starch structure and mechanical properties result from two opposite phenomena: an increase in rigidity as a consequence of a decrease in swelling of starch granules and reduced amylose leaching from the granules; and a weakening effect on the composite starch network structure due to inhibition of inter-particle contacts among swollen granules. Lazaridou et al. (2007) concluded that a combination of these factors determine the overall effect on mechanical properties of the bread structure. This effect is dependent on each specific hydrocolloid used in the formulation of gluten-free bread.

From the above, it is evident that the properties and functionality of the hydrocolloids vary to a great extent, depending on their origin and chemical structure. A high variation in hydrocolloid functionality occurs due to their origin and processing procedures (Rojas et al., 1999; Guarda et al., 2004). The cellulose derivatives (methylcellulose, CMC, and HPMC) are obtained by chemical modification of cellulose, which ensures their uniform properties (Guarda et al., 2004). These hydrocolloids have high water retention properties because of their hydrophilic groups, which induce additional properties, including interfacial activity within the system during proofing, and forming gel networks during the breadmaking process.

The addition of methyl and hydroxypropyl groups to the cellulose chain leads to a polymer with a high surface activity, and unique hydration–dehydration characteristics in the solution state and during temperature changes. The network structures of methylcellulose, CMC, or HPMC formed during baking, serve to increase viscosity and to strengthen the boundaries of the expanding cells in the dough, thus increasing gas retention during baking, and consequently leads to increased loaf volume (Bell, 1990). The hydrophobic–hydrophilic balance in HPMC allows it to act as an emulsifier, strengthen crumb grain and increases the moisture content of the crumb (Bell, 1990). CMC has a preferred interaction with proteins, while HPMC preferentially binds to starch (Collar et al., 2001). CMC and pectin appear to significantly increase loaf volume, porosity and elasticity in gluten-free breads. The pasting properties of wheat starch are largely modified by hydrocolloid addition, and the extent of the effect depends on the chemical structure of the hydrocolloid. Xanthan and pectin increase the cooking stability, while κ-carageenan mainly affects the formation of amylase–lipid complex (Rojas et al., 1999).

The overall effect of a hydrocolloid on wheat bread or gluten-free bread undoubtedly depends on the source of the hydrocolloid, its chemical structure, extraction process, chemical modification, the dosage of hydrocolloid into dough formulations, and interaction with wheat bread and gluten-free bread constituents.

Water as a constituent of bread

Water is an essential ingredient in dough formation: it is necessary for solubilizing other ingredients, for hydrating proteins and carbohydrates, and for the development of gluten networks (Maache-Rezzoug et al., 1998). Water has a complex function, since it determines the conformational state of biopolymers, it affects the nature of interactions between the various constituents of the formula, and contributes to dough structuring (Eliasson and Larsson, 1993). It is an essential factor in the rheological behavior of flour dough (Bloksma and Bushuk, 1988). Adding water to flour reduces the viscosity and increases dough extensibility. On the other hand, if the proportion of water is too low, the dough becomes brittle, not consistent, and exhibits a marked "crust" effect due to rapid dehydration at the surface. In general, dough stiffness changes between 5 and 15% when the water content is changed by 1% of flour mass (Bloksma and Bushuk, 1988). While the various constituents absorb water in dough, native starch is the only constituent whose water content in dough can be estimated with some precision. In equilibrium with water, native starch absorbs approximately 0.45 kg water per kg dry matter (Bloksma and Bushuk, 1988). Water content and its distribution govern textural properties such as softness of crumb, crispness of the crust and shelf-life (Wagner et al., 2007). Water also plays an important role in the major changes (e.g. starch gelatinization) that take place during bread-baking and that contribute to the structure and eating quality of the baked product.

When flour is added to water, the outer layers of the flour particles are hydrated and a sticky mass is obtained. As mixing continues, the hydrated outer surface layers are stripped away, exposing new layers of the flour particles that are then hydrated. This continues until all flour particles are hydrated and disappear (Hoseney and Rogers, 1990). Several physical and chemical transformations occur during mixing and kneading of a mix of flour and water (Damodaran, 1996). Under the applied shear and tensile forces, gluten proteins absorb water and partially unfold. The partial unfolding of protein molecules facilitates hydrophobic interactions and sulfhydryl–disulfide interchange reactions, which result in formation of thread-like polymers. These linear polymers, in turn, are believed to interact with each other, probably via hydrogen bonding, hydrophobic associations, and disulfide cross-linking, to form a sheet-like film capable of trapping gas.

Changes in starch structure, such as melting, gelatinization, or fragmentation, are affected by water/starch ratios, temperature, rate of heating, amylose/amylopectin ratio, shear, granule size distribution, addition of sugars, salt, protein, lipids, and other factors (Kokini et al., 1992). During dough preparation, starch absorbs up to about 46% water (Goesaert et al., 2005). When heated in water, starch granules undergo gelatinization, which involves the disruption of molecular order within granules (Morris, 1994). Leaching of amylose out of the granules also occurs during gelatinization. Total gelatinization usually occurs over a temperature range (BeMiller and Whistler, 1996). Continued heating of starch granules in excess water results in further granule swelling, additional leaching of soluble components (primarily amylose), and eventually, total disruption of granules. This phenomenon results in the formation of starch paste (BeMiller and Whistler, 1996). Water acts as a plasticizer during gelatinization. The mobility-enhancing effect of water takes place initially in the amorphous region, which have the nature of glass. When starch granules are heated in the presence of sufficient water (at least 60%), and a specific temperature (the glass transition temperature) is reached, the plasticized amorphous regions of the granule undergo a phase transition from a glassy state to a rubbery state (BeMiller and Whistler, 1996). During the above process, water molecules enter between chains, break inter-chain bonds, and establish hydration layers around the separated molecules. This lubricates the chains so they become more fully separated and solvated.

As a result of the combination of heat, moisture, and time during baking, the starch granules gelatinize and swell, and a small amount of starch (mainly amylose) is leached into the inter-granular phase (Goesaert et al., 2005). Davidou et al. (1996) stressed the importance of starch swelling and the plasticizing effect of water during breadmaking. Upon cooling, the solubilized amylose forms a continuous network, in which swollen and deformed starch granules are embedded and interlinked. Because of its rapid retrogradation, amylose is an essential structural element of bread and is a determining factor for initial loaf volume (Eliasson and Larsson, 1993). Moisture content controls the level and rate of starch retrogradation (Davidou et al., 1996). In bread crumb, a maximum in the melting energy of retrograded starch was observed in the 35–45% moisture range. Analytical data showed that the viscoelastic behavior of the crumb was similar to that of synthetic polymers, and the crumb rigidity decreased as the water content increased. According to Biliaderis (1992, 1998) water is the most important plasticizer in foods. Increased absorption of water during baking can enhance initial softness and decrease firming of bread.

During storage, bread gradually loses its freshness and stales. The staling process comprises several aspects: the crust becomes tougher, the crumb becomes more firm and less elastic, soluble starch decreases and moisture and flavor is lost (Hoseney, 1994). Rogers et al. (1988) and Davidou et al. (1996) reported that bread moisture content influenced the firming rate and starch retrogradation during storage of bread. Rogers et al. (1988) also observed that the firming rate in wheat bread was retarded when the moisture content was high. It is generally concluded that water migration and transformation in the starch fraction are the important factors in the staling process (Goesaert et al., 2005).

As already discussed, hydrocolloids are extensively used in wheat bread and gluten-free bread to improve structure, mouth-feel, acceptability and shelf-life of these products. Used in small quantities (<1%, w/w) they are expected to increase loaf volume and to decrease firmness (Davidou et al., 1996). Water absorption is increased by hydrocolloid addition and the extent of this increase depends on the structure of the hydrocolloid added (Rosell et al., 2001; Lazaridou et al., 2007). The presence of hydrocolloids influences melting, gelatinization, fragmentation, and retrogradation of starch (Fanta and Christianson, 1996). These effects influence pasting properties, dough rheological behavior (Rojas *et al.*, 1999) and bread staling (Davidou *et al.*, 1996).

In baking studies, results have shown that the moisture contents of bread samples containing hydrocolloids are significantly higher than those of the controls (Friend *et al.*, 1993; Rosell *et al.*, 2001; Guarda *et al.*, 2004; Barcenas and Rosell, 2005). Lower crumb hardness has also been reported in wheat breads and gluten-free breads containing hydrocolloids (Rosell *et al.*, 2001; Gallagher *et al.*, 2003; Sharadanant and Khan, 2003; Guarda *et al.*, 2004; Barcenas and Rosell, 2005; Lazaridou *et al.*, 2007). Gallagher *et al.* (2003) observed that added water (10 or 20%) to gluten-free flours resulted in higher loaf volume and a much softer crust and crumb texture. McCarthy *et al.* (2005) also observed that increasing water content of gluten-free bread significantly decreased crumb firmness. HPMC and water showed significant interactions on crumb grain structure, and the optimized levels of 2.2% HPMC and 79% water yielded good-quality gluten-free bread.

Barcenas and Rosell (2005) observed that the presence of HPMC in wheat bread decreased the hardening rate and also retarded amylopectin retrogradation, and they concluded that the reduction in amylopectin retrogradation, and the delay in bread staling may be due to the water content of the bread containing HPMC. Kobylanski *et al.* (2004), using differential scanning calorimetry, observed that the level of water and HPMC in dough greatly influence the glass transition temperature (i.e. the transition from a glassy state to a rubbery state) and that HPMC–water interaction mainly controlled the onset temperature of starch gelatinization. According to Davidou *et al.* (1996), the hydrocolloids affect the retrogradation level in breads by limiting both the diffusion and the loss of water from bread crumb. Thus, the control of water content and its mobility may be key factors controlling loaf volume and crumb firmness in breads.

Nutritional improvement of gluten-free breads

Cereals are an important source of dietary fiber, contributing to about 50% of the fiber intake in Western countries (Nyman *et al.*, 1989). The role of dietary fiber in providing roughage and bulk, and in contributing to a healthy intestine has long been recognized. Diets that contain moderate quantities of cereal grains, fruits, and vegetables are likely to provide sufficient fiber. Due to the fact that gluten-free products generally are not enriched or fortified, and are frequently made from refined flour or starch, they may not contain the same levels of nutrients as the gluten-containing counterparts they are intended to replace. Therefore, uncertainty still exists as to whether patients with celiac disease living on a gluten-free diet are ensured a nutritionally balanced diet, especially regarding the dietary fiber intake.

Grehn *et al.* (2002) screened the intake of nutrients and foods of 49 adults diagnosed with celiac disease and following a gluten-free diet. They were found to have a lower intake of fiber when compared with a control group of people on a normal diet. Similarly, Lohiniemi *et al.* (2000) found that the average fiber consumption amongst patients with celiac disease in Sweden was lower than recommended. In their studies with adolescents with celiac disease, Mariani *et al.* (1998) concluded that adherence to a strict gluten-free diet worsens the already nutritionally unbalanced diet of adolescents (dietary levels of nutrients and fiber were found to be low). Similar findings were also revealed by Thompson (2000).

The enrichment of gluten-free baked products with dietary fibers has, therefore, been a topic of research for various teams of technologists. Studies have shown that the addition of high-fiber ingredients can give texture, gelling, thickening, emulsifying, and stabilizing properties to gluten-free foods (Sharma, 1981; Dreher, 1987). Inulin is one of the ingredients used to increase the dietary fiber content of gluten-free foods. Inulin is a storage polysaccharide consisting of a chain of β $(2 \rightarrow 1)$-linked fructose units with a terminal glucose molecule (Leite-Toneli *et al.*, 2007). It is present in more than 30 000 vegetable products. In particular, chicory roots (*Cichorium intybus*) are considered suitable for industrial applications. Inulin is not digested or absorbed in the small intestine but is instead fermented in the colon by the beneficial bacteria (Lopez-Molina *et al.*, 2005). The difference observed between inulin extracted from various plants relates to the degree of polymerization.

Due to its functional properties, inulin has been used in a wide range of products, where it is promoted for its prebiotics properties. It has also been used as fat replacer in a wide range of food products. Silva (1996) for example reported on the interactions between inulin and certain hydrocolloids. The authors found that inulin and hydrocolloids show a synergistic effect, which will increase the viscosity of the system significantly. It was found that when inulin is mixed with an aqueous solution, the inulin particles form a gel-like network, which results in a product with a creamy texture and spreadable attributes. The mixture can easily be applied in food systems and has the potential to replace up to 100% fat (Lopez-Molina *et al.*, 2005).

There are only very few studies investigating the influence of inulin on the quality and nutritional properties of bakery products. Recently, Korus *et al.* (2006) determined the influence of a range of prebiotics (inulin, oligosaccharide syrup, and bitter-free chicory flour) on the quality of gluten-free breads. These authors used addition levels of 3, 5, or 8% and stored the breads for 48 hours. The authors found that 5% inulin resulted in bread with the highest sensory values. Five and 8% inulin, oligosaccharide syrup, and chicory flour reduced the rate of staling during the 3-day storage period. Overall, the authors concluded that it is possible to produce good-quality gluten-free breads supplemented with prebiotics. Among the applied additives, the most beneficial effect on bread quality was found to be 5% addition of inulin, which led to increased loaf volume, reduced rate of crumb hardening, and a positive sensory level evaluation.

A further study by Gallagher *et al.* (2004a) investigated the impact of different levels of inulin on the quality as well as the nutritional value of gluten-free bread. These authors found that the addition of inulin did not only improve the quality of the gluten-free bread, but it also increased the dietary fiber content of the product significantly. From the limited studies performed as well as based on the nutritional analysis of existing gluten-free cereal products it is essential to increase the dietary fiber content of gluten-free products. Inulin seems to be a good candidate, but further studies are required to evaluate the suitability of other sources of dietary fiber for the use in gluten-free products.

Gluten-free bread production

The production of gluten-free breads differs significantly to that of standard wheat breads (Figure 13.4). Traditionally, wheat dough is mixed, bulk fermented, divided/molded, proofed, and finally baked. Most gluten-free breads tend to contain higher water levels and have a more fluid-like structure. In addition, they require shorter mixing, proofing and baking times than their wheat counterparts. A new method to produce high-quality gluten-free bread was developed by Moore *et al.* (2004), which consists of mixing, proofing and baking. This method was successfully applied in further studies on gluten-free bread (Moore *et al.*, 2006, 2007a, 2007b; Moore and Arendt, 2007; Schober *et al.*, 2005; Renzetti *et al.*, 2007). In the same work, Moore *et al.* (2004) produced and compared the quality of gluten-free bread with or without dairy ingredients to that of wheat bread or a gluten-free bread made from a commercial mix based on wheat starch. The non-dairy gluten-free bread contained corn starch, brown rice flour, soya, buckwheat flour, and xanthan gum. The dairy gluten-free bread was based on brown rice flour, skim milk powder, whole eggs, potato and corn starch, soya flour, xanthan and konjac gum. The commercial

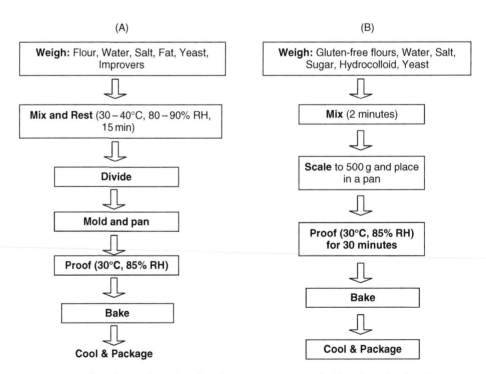

Figure 13.4 Outline of a standard wheat bread process (A) compared with a gluten-free bread development process (B).

and non-dairy breads achieved a high volume, but at the expense of quick staling. Increased water levels in combination with wholemeal cereals could not notably delay the staling. However, the addition of proteins in sufficient amounts improved the keeping quality and the formation of a continuous phase and film-like structure was observed (Figure 13.2). The continuous phase and film-like structure can mask changes caused by starch retrogradation and is therefore a key factor determining the quality of gluten-free bread.

Techniques such as response surface methodology (RSM) are useful tools when developing a new food, such as gluten-free bread. RSM may be used to identify combinations of levels of ingredients or processing parameters (e.g. time, temperature) to be tested for subsequent measurement of appropriate responses (e.g. color, volume, acceptability). It is also used for building of models using these data to identify local maxima and minima (i.e. select optimal conditions to be used in the process). RSM allows multiple parameters to be varied simultaneously in a manner which nonetheless yields reliable data to the manufacturer with a minimum number of trials (and hence minimized cost and time input) (Chapter 19). Successful application of RSM in the production of different types of wheat bread has been reported (Lee and Hoseney, 1982; Malcolmson *et al.*, 1993; Clarke *et al.*, 2002; Gallagher *et al.*, 2004a; Clarke *et al.*, 2004).

Ylimaki *et al.* (1991) used RSM to produce and objectively measure gluten-free breads based on three types of rice flour (varying in grain size and grinding method). Optimal loaves were obtained when using medium grain, finely ground white rice flour, low levels of HPMC and low levels of CMC, and the resulting bread was the most similar to wheat breads. The same three rice flours were used in a second trial, where gluten-free yeast breads were produced based on the rice flours (80%) and potato starch (20%). Using sensory evaluation based on a trained panel, RSM was applied to find optimal CMC, HPMC, and water levels and combinations of these ingredients for the different rice flours. With respect to moistness, cohesiveness, flavor, color, and cell structure, it was found that gluten-free loaves made with medium grain rice flours were of a higher standard than those made from long grain rice flour (Ylimaki *et al.*, 1991).

Subsequently, Toufeili *et al.* (1994) applied RSM to analyze the effects of methylcellulose, gum arabic, and egg albumen on the sensory properties of gluten-free flat breads baked from formulae based on pre-gelatinized corn starch with corn flour. Methylcellulose, and egg albumen were identified as the major ingredients improving the sensory attributes of the breads. When 3% gum arabic, and 2–4% methylcellulose and egg albumen were used, gluten-free breads comparable to wheat breads were produced. However, the breads staled more rapidly over a 2-day period than regular wheat bread. RSM was also used by Demiate *et al.* (2000) for optimization of the processing as well as formulation of gluten-free breads and biscuits based on cassava starch. In another study, Sanchez *et al.* (2002) successfully used RSM to improve the texture of gluten-free bread derived from corn starch, cassava starch, and rice flour with 0 and 0.5% soy flour addition. Finally, Schober *et al.* (2005) used RSM to study the influence of different sorghum hybrids on the quality of sorghum-based gluten-free breads. The different sorghum hybrids led to differences in the quality of the gluten-free bread, and RSM optimized the performance of a certain variety by optimizing the product formulation.

In conclusion, the studies performed so far have clearly showed that mathematical modeling is an excellent tool to optimize the production of gluten-free products, and that RSM can be applied to optimize product formulations as well as processing conditions with a minimum amount of trials.

Improvement of gluten-free bread quality

A review by Arendt *et al.* (2002) pointed out that most commercial gluten-free breads are of poor quality due to the rapid onset of staling, dry crumbly texture, and potent off-flavors. Gluten-free breads tend to have a rapid onset of staling, mainly because of the high amount (almost 100% of the flour base) of isolated starches present. Furthermore, due to the absence of gluten, more water is available and hence causes an increase in crumb firmness and softer crust (Gallagher *et al.*, 2004a). Several studies have been carried out whereby a range of gluten-free cereals in combination with enzymes, proteins, hydrocolloids, and/or lactic acid bacteria have been used to increase the quality of gluten-free breads (Sanchez *et al.*, 2002, 2004; Gallagher *et al.*, 2003, 2004a, 2004b; Moore *et al.*, 2004, 2007a, 2007b; Moore and Arendt, 2007; McCarthy *et al.*, 2005; Schober *et al.*, 2005; Renzetti *et al.*, 2007).

Enzymes

The functions of enzymes are widespread throughout the baking industry, for example in decolorizing (bleaching) of doughs, improving the volume and texture of doughs, or increasing the shelf-life (Gélinas and Lachance, 1995; Sahlström and Brathen, 1997; Grossman and De Barber, 1997; Vemulappali and Hoseney, 1998; Delcros *et al.*, 1998; Corsetti *et al.*, 2000; Rosell *et al.*, 2001; Chapter 12). Enzymes can either be naturally present in the raw materials or can be added from external sources. Amylases, proteases, hemicellulases, lipases, and oxidases have been reported to influence all aspects of the baking process as well as the quality of baked goods (Hozová *et al.*, 2002).

To date, there are few published reports on the application of amylases and their impact on gluten-free foods. However, Gujral *et al.* (2003a, 2003b) investigated the effectiveness of two starch-hydrolyzing enzymes from the *Bacillus* species (α-amylase of intermediate thermostability and cyclodextrin glycosyl transferase (CGTase)) in retarding rice bread staling. The presence of CGTase, in particular, decreased amylopectin retrogradation and showed a significant anti-staling effect. A favorable increase in loaf-specific volume following addition of this enzyme was also noted. Therefore, this study indicates that amylolytic enzymes might be helpful in preventing gluten-free bread staling.

Transglutaminase is a relatively new tool used in the manufacture of baked goods (Diez Poza, 2002). It can modify proteins by amine incorporation, cross-linking, or deamination. Cross-linking occurs when the ϵ-amino groups of lysine residues in proteins act as an acyl-receptor, ϵ-(γ-Glu)Lys bonds (isopeptide bonds) are formed both intra-and intermolecularly (Ando *et al.*, 1989). In the absence of primary amines in the reaction system, water is used as an acyl acceptor leading to a deamination of glutamine residues (Motoki and Kumazawa, 2000). Transglutaminase can also catalyze incorporation of primary amines into proteins (Folk and Chung, 1973; Folk, 1980). Transglutaminase has the ability to link proteins of different origin: casein and albumin from milk, animal protein from eggs and meat, soya protein, and wheat protein. The enzyme can be obtained from a range of different sources, such as animal tissue, fish, plant, or microorganisms (Kuraishi *et al.*, 1996). The transglutaminases used in baking applications are obtained from microbial cultures. The enzyme is active against wheat gluten (Larré *et al.*, 1998, 2000; Bauer *et al.*, 2003) and has a positive effect on the specific volume of wheat-based croissants (Gerrard *et al.*, 2000).

Moore *et al.* (2006) evaluated the impact of transglutaminase (at different levels) in gluten-free bread in conjunction with the protein sources soya, skim milk, or egg powder. The most pronounced effect was the reduction in volume due to network formation. Bread containing skim milk powder and 10 units of enzyme showed the most compact structure (Plate 13.1), and the authors concluded that network formation in gluten-free bread depends on the level of transglutaminase and type of protein used. Renzetti *et al.* (2007) evaluated in depth the impact of transglutaminase on a range of gluten-free cereals. A significant increase in the pseudoplastic behavior of buckwheat and brown rice batters was observed when 10 units of transglutaminase were used. The resulting buckwheat and brown rice breads showed improved baking characteristics as well as overall macroscopic appearance. Three-dimensional CLSM image elaborations confirmed the formation of protein complexes by transglutaminase action. However, transglutaminase showed negative effects on corn flour as its application was detrimental for the elastic properties of the batters. Nevertheless, the resulting breads showed significant improvements in terms of increased specific volume and decreased crumb hardness and chewiness. No effects of transglutaminase could be observed on breads from oat, sorghum, or tef. The authors concluded that transglutaminase can be successfully applied to gluten-free flours to improve their breadmaking potential by promoting network formation. However, the protein source is a key element determining the impact of the enzyme.

Sourdough and its role in improving gluten-free bread quality

The use of sourdough represents an attractive alternative to increase the quality of gluten-free breads (Chapter 12). Sourdough addition has a well-established role in improving the quality of gluten-containing bread. There is considerable consensus with regard to the positive effects, including improvements in bread volume and crumb structure (Corsetti *et al.*, 2000; Clarke *et al.*, 2002; Crowley *et al.*, 2002), flavor (Thiele *et al.*, 2002), nutritional value (Salovaara and Göransson, 1983; Larsson and Sandberg, 1991; Liljeberg and Björck, 1994; Liljeberg *et al.*, 1995), and mold-free shelf-life (Lavermicocca *et al.*, 2000, 2003; Magnusson and Schnürer, 2001; Dal Bello *et al.*, 2006). The improvement in flavor due to sourdough addition is of particular interest for gluten-free bread production. The flavor of the bread can be influenced by the type of starter cultures used and characteristic flavors are obtained from organic acids and amino acids released during fermentation (Barber *et al.*, 1992). Gas-holding properties are mainly influenced by the swelling capacity of flour although starch granules are relatively water insoluble and hydrate only slightly in cold water. Acidification of flour by sourdough fermentation can replace the function of gluten in a way and enhance the swelling properties of polysaccharides (pentosans in rye). This property may be beneficial to the structure of gluten-free breads.

The influence of sourdough on the quality of gluten-free bread was recently investigated (Moore *et al.*, 2007a). During fermentation, protein degradation occurred, however this process was far less obvious than that occurring in gluten-containing sourdoughs. Incorporation of 20% sourdough has remarkable effects on the final quality of gluten-containing breads. However, when gluten-free sourdoughs were incorporated at a 20% level into the gluten-free batters, no significant differences were observed in the structure. Nonetheless, the onset of staling was delayed. Remarkably, Moore *et al.* (2007b) recently showed that addition of sourdough can effectively retard the growth of spoilage organisms on gluten-free bread, thereby increasing the shelf-life of these products.

Research on the use of sourdough for the production of high-quality gluten-free bread is still at its infancy, but data available so far clearly indicate that sourdough represents an attractive tool to increase the quality (e.g. flavor and shelf-life) of gluten-free bread.

Conclusions

The keystone treatment for patients with celiac disease is a lifelong elimination diet in which food products containing gluten are avoided. However, gluten is an essential structure-building protein, contributing to the appearance, crumb structure, and consumer acceptability of many baked products. Therefore, the biggest challenge for food scientists and bakers in the area of gluten-free products is probably the production of high-quality gluten-free bread. Extensive market research showed that the majority of breads currently on the market are of very poor quality. In wheat bread, gluten has such a wide range of functions that it is not possible to replace wheat flour with one single ingredient. Good-quality gluten-free bread can only be produced if a range of flours and polymeric substances, which mimic the viscoelastic properties of gluten, are included in the gluten-free formulation. It is recommended to use a range of gluten-free flours rather than just one flour to achieve products with good sensory and textural properties. The addition of a certain percentage of starch to a gluten-free formulation does certainly improve the overall quality of the gluten-free bread. Naturally gluten-free starches such as that from rice, potatoes, or tapioca, rather than wheat starch, should be used for this purpose.

Hydrocolloids are an essential ingredient for gluten-free bread production, since they are able to mimic the viscoelastic properties of gluten to a certain extent. They are also known to reduce staling, improve water binding, and improve the overall structure of the bread. Research performed so far suggest that xanthan gum and HPMC are the most suitable hydrocolloids for gluten-free bread formulations, but further research is needed to optimize the application of these or other hydrocolloids in gluten-free systems. Protein-based ingredients are also essential in the improvement of gluten-free bread, and the most promising are probably the dairy-based ingredients; however, it is essential that only low lactose dairy ingredients are used. One of the most important ingredients in any gluten-free formulation is water, and therefore it is essential to optimize the water level for every formulation in order to achieve optimal results.

Recently, research has also focused on the application of enzymes to improve the texture of gluten-free bread. Among other enzymes, transglutaminase has been shown to improve the texture of gluten-free bread, but showed a dependency on the raw material taken into consideration. Lactic acid bacteria/gluten-free sourdough are also one possibility to improve gluten-free bread quality, particularly its sensory properties. Even if the research on gluten-free products is still in its infancy, researchers have been able to create products that are superior to the ones currently on the market, and which patients with celiac disease might soon be able to see available in the stores.

References

Ahlborn, G. J., Pike, O. A., Hendrix, S. B., Hess, W. H., and Huber, C. S. (2005). Sensory, mechanical, and microscopic evaluation of staling in low-protein and gluten-free bread. *Cereal Chem.* 83, 328–335.

Albert, S. and Mittal, G.S. (2002). Comparative evaluation of edible coatings to reduce the uptake in a deep fried cereal product. *Food Res. Int.* 35, 445–458.

Ando, H., Adachi, M., Umeda, K. *et al.* (1989). Purification and Characteristics of novel transglutaminase derived from microorganism. *Agric. Biol. Chem.* 53, 2613–2617.

Arendt, E. K., O'Brien, C. M., Schober, T. J., Gallagher, E., and Gormley, T. R. (2002). Development of gluten-free cereal products. *Farm Food* 21–27.

Armero, E. and Collar, C. (1996a). Anti-staling additives, flour type and sourdough process effects on functionality of wheat doughs. *J. Food Sci.* 61, 299–303.

Armero, E. and Collar, C. (1996b). Anti-staling additive effects on fresh wheat bread quality. *Food Sci. Technol. Int.* 2, 323–333.

Barber, B., Ortolá, C., Barber, S., and Fernández, F. (1992). Storage of packaged white bread. III. Effects of sour dough and addition of acids on bread characteristics. *Z Lebensm Untersuch Forsch* 149, 442–449.

Barcenas, M. E. and Rosell, C. M. (2006). Different approach for improving the quality and extending the shelf-life of the partially baked bread: low temperature and HPMC addition. *J. Food Eng.* 72, 92–99.

Barcenas, M. E. and Rosell, C. M. (2005). Effect of HPMC addition on the microstructure, quality and aging of wheat bread. *Food Hydrocolloids* 19, 1037–1043.

Barcenas, M. E., Benedito, C., and Rosell, C. M. (2004). Use of hydrocolloids as bread improvers in interrupted baking process with frozen storage. *Food Hydrocolloids* 19, 769–774.

Bauer, N., Koehler, P., Wieser, H., and Schieberle, P. (2003). Studies on the effects of microbial transglutaminase on gluten proteins of wheat. *Recent Adv. Enzymes Grain Process.* 107–113.

Belitz, H. D. and Grosch, W. (1987). *Food Chemistry*. Berlin: Springer Verlag, pp. 379–388; 395–396; 403–412; 536–538.

Bell, D. A. (1990). Methylcellulose as a structure enhancer in bread baking. *Cereal Food World* 35, 1001–1006.

BeMiller, J. N. and Whistler, R. L. (1996). Carbohydrates. In: Fennema, O. R. ed. *Food Chemistry*, 3rd edn. New York: Marcel Dekker, pp. 158–223.

Biliaderis, C. G. (1998). Structure and phase transition of starch polymers. In: Walker, R. H. ed. *Polysaccharide Association of Structures in Foods*. New York: Marcel Dekker, New York, pp. 57–168.

Biliaderis, C. G., Arvanitoyannis, T. S., Ijydorczyk, M. S., and Prokopowich, D. J. (1997). Effect of hydrocolloids on gelatinization and structure formation in concentrated waxy maize and wheat starch gels. *Starch/Staerke* 49, 278–283.

Biliaderis, C. G. (1992). Characterization of starch networks by small strain dynamic rheometry. In: Alexander, R. J. and Zobel, H. F. eds. *Developments in Carbohydrate Chemistry*. St. Paul, MN: American Association of Cereal Chemists, pp. 87–135.

Bloksma, A. W. (1990). Dough structure, dough rheology, and baking quality. *Cereal Foods World* 35, 237–243.

Bloksma, A. H. and Bushuk, W. (1988). Rheology and chemistry of dough. In: Pomeranz, Y. ed. *Wheat, Chemistry and Technology*, Vol. II, 3rd edn. St. Paul, MN: American Association of Cereal Chemists, pp. 131–217.

Brouns, F. (2002). Soya isoflavones: a new and promising ingredient for the health foods sector. *Food Res. Int.* 35, 187–193.

Clarke, C. I., Schober, T. J., Dockery, P., O'Sullivan, K., and Arendt, E. K. (2004). Wheat sourdough fermentation: Effects of time and acidification on fundamental rheological properties. *Cereal Chem.* 81, 409–417.

Clarke, C. I., Schober, T. J., and Arendt, E. K. (2002). Effect of single strain and traditional mixed strain starter cultures on rheological properties of wheat dough and on bread quality. *Cereal Chem.* 79, 640–647.

Cocup, R. O. and Sanderson, W. B. (1987). Functionality of dairy ingredients in bakery products. *Food Technol.* 41, 86–90.

Codex Alimentarius Commission (2003). Draft revised standard for gluten free foods. Report of the Joint FAO/WHO Food Standards Programme Codex Alimentarius Commission, Twenty-fifth Session. Paragraphs 9 and 10 Rome.

Collar, C., Martinez, J. C., and Rosell, C. M. (2001). Lipid binding of fresh and stored formulated wheat breads. Relationship with dough and bread technological performance. *Food Sci. Technol. Int.* 7, 501–510.

Corsetti, A., Gobbetti, M., De Marco, B. *et al.* (2000). Combined effects of sourdough lactic acid bacteria and additives on bread firmness and staling. *J. Agric. Food Chem.* 48, 3044–3051.

Crowley, P., Schober, T. J., Clarke, C. I., and Arendt, E. K. (2002). The effect of storage time on textural and crumb grain characteristics of sourdough wheat bread. *Eur. Food Res. Technol.* 214, 489–496.

Dal Bello, F., Clarke, C. I., Ryan, L. A. M. *et al.* (2006). Improvement of the quality and shelf life of wheat bread by using the antifungal strain *Lactobacillus plantarum* FST 1.7. *J Cereal Sci.* 45, 309–318.

Damodaran, S. (1996). Amino acids, peptides, and proteins. In: Fennema, O. R. ed. *Food Chemistry*, 3rd edn. New York: Marcel Dekker, pp. 321–429.

Davidou, S., Le Meste, M., Debever, E., and Bekaert, D. (1996). A contribution to the study of staling of white bread: effect of water and hydrocolloid. *Food Hydrocolloids* 10, 375–383.

Delcros, J. F., Rakotozafy, L., Boussard, A. (1998). Effect of mixing conditions on the behaviour of lipoxygenase, peroxidase, and catalase in wheat flour. *Cereal Chem.* 75, 85–93.

Demiate, I. M., Dupuy, N., Huvenne, J. P., Cereda, M. P., and Wosiacki, G. (2000). Relationship between baking behaviour of modified cassava starches and starch chemical structure by FTIR spectroscopy. *Carbohydr. Polym.* 42, 149–158.

Diez Poza, O. (2002). Transglutaminase in baking applications. *Cereal Foods World* 47, 93–95.

Don, C., Lichtendonk, W. J., Pfijter, J. J., and Hamer, R. J. (2003a). Glutenin macropolymer: a gel formed by glutenin particles. *J. Cereal Sci.* 37, 1–7.

Don, C., Lichtendonk, W. J., Plijter, J. J., and Hamer, R. J. (2003b). Understanding the link between GMP and dough: from glutenin particles in flour towards developed dough. *J. Cereal Sci.* 38, 157–165.

Dreher, M. L. (1987). *Handbook of Dietary Fiber: An Applied Approach*. New York: Marcel Dekker.

Eliasson, A.-C. and Larsson, K. (1993). *Cereals in Breadmaking: A Molecular Colloidal Approach*. New York: Marcel Dekker.

Eliasson, A.-C. and Gudmundsson, M. (2006). Starch: physicochemical and functional aspects. In: Eliasson, A.-C. ed. *Carbohydrates in Foods*. Boco Raton, FL: Taylor and Francis, pp. 391–470.

Every, D., Gerrard, J. A., Gilpin, M. J., Ross, M., and Newberry, M. P. (1998). Staling in starch bread; the effect of gluten additions on specific loaf volume and firming rates. *Starch/Starke* 50, 443–446.

Fanta, G. F. and Christianson, D. D. (1996). Starch-hydrocolloid composites prepared by steam jet cooking. *Food Hydrocolloids* 10, 173–178.

Fernández-Rivas, M. and Ballmer-Weber, B. (2007). Food allergy:current diagnosis and management. In: Mills, C., Wichers, H., and Hoffmann-Sommergruber, K. eds. *Managing Allergens in Food*. Woodhead Publish, *Food Sci. Technol. Nutr.* 1, 3–28.

Folk, J. E. (1980). Transglutaminases. *Annu. Rev. Biochem.* 49, 517–531.

Folk, J. E. and Chung, I. S. (1973). In: Meister, A. ed. *Advances in Enzymology*, Vol. 38. New York: John Wiley and Sons, p. 109.

Forde, C. G., and Delahunty, C. M. (2004). Understanding the role cross-model sensory interactions play in food acceptability in younger and older consumers. *Food Qual. Pref.* 15, 715–727.

Friend, C. P., Waniska, R. D., and Rooney, L. W. (1993). Effects of hydrocolloids on processing and qualities of wheat tortillas. *Cereal Chem.* 70, 252–256.

Gallagher, E., Gormley, T. R., and Arendt, E. K. (2004a). Crust and crumb characteristics of gluten-free breads. *J. Food Eng.* 56, 153–161.

Gallagher, E., Gormley, T. R., and Arendt, E. K. (2004b). Recent advances in the formulation of gluten-free cereal-based products. *Trends Food Sci. Technol.* 15, 143–152.

Gallagher, E., Kunkel, A., Gormley, T. R., and Arendt, E. K. (2003). The effect of dairy and rice powder addition on loaf and crumb characteristics, and on shelf life (intermediate and long term) of gluten-free breads stored in a modified atmosphere. *Eur. J. Food Res.* 218, 44–48.

Gelinas, P. and Lachance, O. (1995). Development of fermented dairy ingredients as flavour enhancers for bread. *Cereal Chem.* 72, 17–21.

Gerrard, J. A., Newberry, M. P., Ross, M., Wilson, A. J., Fayle, S. E., and Kavale, S. (2000). Pastry lift and croissant volume as affected by microbial transglutaminase. *J. Food Sci.* 65, 312–314.

Goesaert, H., Brijs, K., Veraverbeke, W. S., Courtin, C. H., Gebruers, K., and Delcour, J. A. (2005). Wheat flour constituents: how they impact bread quality, and how to impact their functionality. *Trends Food Sci. Technol.* 16, 12–30.

Gray, J. A. and BeMiller, J. N. (2003). Bread staling: molecular basis and control. *Rev. Food Sci. Safety* 2, 1–20.

Grehn, S., Fridell, K., Lilliecreutz, M., and Hallert, C. (2002). Dietary habits of Swedish adult celiac patients treated by a gluten-free diet for 10 years. *Scand. J. Nutr.* 45, 178–182.

Grossman, M. V. and De Barber, B. C. (1997). Bread staling. Simultaneous effect of bacterial alpha-amylase and emulsifier on firmness and pasting properties of bread crumbs. *Arch. Latinoam. Nutr.* 47, 229–233.

Guarda, A., Rosell, C.M., Benedito, C., and Galotto, M. J. (2004). Different hydrocolloids as bread improvers and antistaling agents. *Food Hydrocolloids* 18, 241–247.

Gujral, H. S. and Rosell, M. C. (2004). Improvement of the breadmaking quality of rice flour by glucose oxidase. *Food Res. Int.* 37, 75–81.

Gujral, H. S., Haros, M., and Rosell, C. M. (2003a). Starch hydrolysing enzymes for retarding the staling for rice bread. *Cereal Chem.* 80, 750–754.

Gujral, H. S., Guardiola, I., Carbonell, J. V., and Rosell, C. A. (2003b). Effect of cyclodextrinase on dough rheology and bread quality from rice flour. *J. Agric. Food Chem.* 51, 3814–3818.

Gurkin, S. (2002). Hydrocolloids—Ingredients that add flexibility to tortilla processing. *Cereal Foods World* 47, 41–43.

Haque, A. and Morris, E. R. (1994). Combined use of ispaghula and HPMC to replace or augment gluten in breadmaking. *Food Res. Int.* 27, 379–393.

Hoseney, R. C. (1994). *Principles of Cereal Science and Technology*, 2nd edn. St. Paul, MN: American Association of Cereal Chemists, pp. 40, 147–148, 194–195, 197, 342–343.

Hoseney, R. C. and Rogers, D. E. (1990). The formation and properties of wheat flour doughts. *Crit. Rev. Food Sci. Nutr.* 29, 73–93.

Hozová, B., Jančovičová, J., Dodok, L., Buchtová, V., and Staruch, L. (2002). Use of transglutaminase for improvement of quality of pastry produced by frozen-dough technology. *Czech. J. Food Sci.* 20, 215–222.

Hug-Iten, S., Conde-Petit, B., and Echer, F. (2001). Structural properties of starch in bread and bread model systems—influence of an antistaling α-amylose. *Cereal Chem.* 78, 421–428.

Hug-Iten, S., Echer, F., and Conde-Petit, B. (2003). Staling of bread, role of amylose and amylopectin and influence of starch-degrading enzymes. *Cereal Chem.* 80, 651–661.

Jonagh, G., Slim, T., and Greve, H. (1968). Bread without gluten. *Baker's Digest* 6, 24–29.

Kato, A., Ibrahim, H., Watanabe, H., Honma, K., and Kobayashi, K. (1990). Enthalpy of denaturation and surface functional properties of heated egg white proteins in the dry state. *J. Food Sci.* 55, 1280–1282.

Kenny, S., Wehrle, K., Auty, M., and Arendt, E. K. (2001). Influence of sodium caseinate and whey protein on baking properties and rheology of frozen dough. *Cereal Chem.* 78, 458–463.

Kenny, S., Wehrle, K., Stanton, C., and Arendt, E. K. (2000). Incorporation of dairy ingredients into wheat bread: effects on dough rheology and bread quality. *Eur. Food Res. Technol.* 210, 391–396.

Kobylanski, J. R., Perez, O. E., and Pilosof, A. M. R. (2004). Thermal transitions of gluten-free doughs as affected by water, egg white and hydroxypropylmethylcellulose. *Thermochim. Acta* 411, 81–89.

Kokini, J. L., Lai, L.-S., and Chedid, L. L. (1992). Effect of starch structure on starch rheological properties. *Food Technol.* 41, 124–139.

Korus, J., Grzelak, K., Achremowicz, K., and Sabat, R. (2006). Influence of prebiotic additions on the qulaity of gluten-free bread and on the content of inulin and fructooligosaccharides. *Food Sci. Technol.* 12, 489–495.

Kupper, C. (2005). Dietary guidelines and implementation for celiac disease. *Gastroenterology* 128, S121–S127.

Kuraishi, C., Sakamoto, J., and Soeda, T. (1996). The usefulness of transglutaminase for food processing. *Biotechnology for Improved Foods and Flavours*, pp. 29–38.

Larré, C., Denery-Papini, S., Popineau, Y., Deshayes, G., Desserme, C., and Lefebvre, J. (2000). Biochemical analysis and rheological properties of gluten modified by transglutaminase. *Cereal Chem.* 77, 121–127.

Larré, C., Deshayes, G., Lefebvre, J., and Popineau, Y. (1998). Hydrated gluten modified by transglutaminase. *Nahrung* 42, 155–157.

Larsson, H. and Eliasson, A.-C. (1997). Influence of the starch granule surface on the rheological behaviour of wheat flour dough. *J. Texture Studies* 28, 487–501.

Larsson, M. and Sandberg, A. S. (1991). Phytate reduction in bread containing oat flour, oat bran or rye bran. *J. Cereal Sci.* 14, 141–149.

Lasztity, R. (1995). *The Chemistry of Cereal Proteins*, 2nd edn. Boca Raton, FL: CRC Press.

Lavermicocca, P., Valerio, F., and Visconti, A. (2003). Antifungal activity of phenyllactic acid against moulds isolated from bakery products. *Appl. Environ. Microbiol.* 69, 634–640.

Lavermicocca, P., Valerio, F., Evidente, A., Lazzaroni, S., Corsetti, A., and Gobetti, M. (2000). Purification and characterization of novel antifungal compounds from the sourdough *Lactobacillus plantarum* strain 21B. *Appl. Environ. Microbiol.* 66, 4084–4090.

Lazaridou, A., Duta, D., Papageorgiou, M., Belc, N., and Biliaderis, C. G. (2007). Effects of hydrocolloids on dough rheology and bread quality parameters in gluten-free formulations. *J. Food Eng.* 79, 1033–1047.

Lee, C. C. and Hoseney, R. C. (1982). Optimization of the fat-emulsifier system and the gum-egg white-water system for a laboratory scale single stage cake mix. *Cereal Chem.* 59, 392–395.

Lee, M. H., Baek, M. H., Cha, D. S., Park, H. J., and Lim, S. T. (2002). Freeze-thaw stabilization of sweet potato starch gel by polysaccharide gums. *Food Hydrocolloids* 16, 345–352.

Leite-Toneli, J. T. C., Mürr, F. E. X., Martinelli, P., Dal Fabbro, I. M., and Park, K. J. (2007). Optimization of a physical concentration process for inulin. *J. Food Eng.* 80, 832–838.

Liljeberg, H. G. M., Lönner, C. H., and Björck, I. M. E. (1995). Sourdough fermentation of addition of organic acids or corresponding salts to bread improves nutritional properties of starch in healthy humans. *J. Nutr.* 125, 1503–1511.

Liljeberg, H. and Björck, I. (1994). Bioavailability of starch in bread products. Postprandial glucose and insulin responses in healthy subjects and in vitroresistant starch content. *Eur. J. Clin. Nutr.* 48, 151–163.

Lindsay, M. P. and Skerritt, J. H. (1999). The glutenin macropolymer of wheat flour doughs: structure-function perspective. *Trends Food Sci. Technol.* 10, 247–253.

Lohiniemi, S., Maki, M., Kaukinen, K., Laippala, P., and Collin, P. (2000). Gastrointestinal symptoms rating scale in coeliac patients on wheat starch-based gluten-free diets. *Scand. J. Gastroenterol.* 35, 947–949.

Lopez-Molina, D., Navarro-Martinez, M. D., Melgaejo, F. R., Hiner, A. N. P., Chazarro, S., and Rodriguez-Lopez, J. N. (2005). Molecular properties and prebiotic effect of inulin obtained from artichoke (*Cyanar scolymus* L.). *Phytochemistry* 66, 1476–1484.

Maache-Rezzoug, Z., Bouvier, J.-M., Allaf, K., and Patras, C. (1998). Effect of principal ingredients on rheological behaviour of biscuit dough and on quality of biscuits. *J. Food Eng.* 35, 23–42.

MacRitchie, F. (1980). Studies of gluten protein from wheat flours. *Cereal Foods World* 25, 382–385.

MacRitchie, F. (1987). Evaluation of contributions from wheat protein fractions to dough mixing and breadmaking. *J. Cereal Sci.* 6, 257–268.

Magnusson, J. and Schnürer, J. (2001). *Lactobacillus coryniformis* subsp. *coryniformis* strain Si3 produces a broad-spectrum proteinaceous antifungal compound. *Appl. Eviron. Microbiol.* 67, 1–5.

Malcolmson, L. J., Matsuo, R. R., and Balshaw, R. (1993). Textural optimisation of spaghetti using response surface methodology: effects of drying temperature and durum protein level. *Cereal Chem.* 70, 417–423.

Mannie, E. and Asp, E. H. (1999). Dairy ingredients in baking. *Cereal Foods World* 44, 143–146.

Mariani, P., Grazia, V. M., Montouri, M. *et al.* (1998). The gluten-free diet: a nutritional risk factor for adolescents with coeliac disease. *J. Pediatr. Gastroenterol. Nutr.* 27, 519–523.

McCarthy, D. F., Gallagher, E., Gormley, T. R., Schober, T. J., and Arendt, E. K. (2005). Application of response surface methodology in the development of gluten-free bread. *Cereal Chem.* 82, 609–615.

Mettler, E. and Seibel, W. (1995). Optimizing of rye bread recipes containing mono-diglycerides, guar gum, and carboxymethylcellulose using a maturograph and an oven rise recorder. *Cereal Chem.* 72, 109–115.

Miles, M. J., Morris, V. J., Orford, P. D., and Ring, S. G. (1985). The role of amylase and amylopectin in the gelation and retrogradation of starch. *Carbohydr. Res.* 135, 271–281.

Mine, Y. (2002). Recent advances in egg protein functionality in the food system. *World's Poultry Sci. J.* 58, 31–39.

Moon, M. H. and Giddings, J. C. (1993). Rapid separation and measurement of particle size distribution of starch granules by sedimentation/steric field-flow fractionation. *J. Food Sci.* 58, 1166–1171.

Moore, M. M., Schober, T. J., Juga, B. *et al.* (2007a). Effect of lactic acid bacteria on the properties of gluten-free sourdoughs, batters and the quality and ultrastructure of gluten-free bread. *Cereal Chem.* 84, 357–364.

Moore, M. M., Dal Bello, F., and Arendt, E. K. (2007b). Sourdough fermented by Lactobacillus plantarum FST 1.7 improves the quality and shelf life of gluten-free bread. *Eur. Food Res. J.* Online.

Moore, M. M. and Arendt, E. K. (2007). Fundamental study on gluten-free flours and their potential use in bread systems. Paper read at Cereal and Europe Spring Meeting, Montpellier, France.

Moore, M. M., Heinbockel, M., Dockery, P., Ulmer, H. M., and Arendt, E. K. (2006). Network formation in gluten-free bread with the application of transglutaminase. *Cereal Chem.* 83, 28–36.

Moore, M. M., Schober, T. J., Dockery, P., and Arendt, E. K. (2004). Textural comparison of gluten-free and wheat based doughs, batters and breads. *Cereal Chem.* 81, 567–575.

Morris, V. J. (1994). Starch gelation and retrogradation. *Trends Food Sci. Technol.* 1, 2–6.

Motoki, M. and Kumazawa, Y. (2000). Recent research trends in transglutaminase technology for food processing. *Food Sci. Technol. Res.* 6, 151–160.

Murray, J. A. (1999). The widening spectrum of celiac disease. *Am. J. Clin. Nutr.* 69, 354–363.

Nunes, M. H. B., Moore, M. M., and Arendt, E. K. (2007). Fundamental studies on the impact of emulsifiers and dough improvers on gluten-free bread quality. Paper read at Young Cereal Scientist workshop, Montpellier, France.

Nyman, M. I., Björck, I., Siljeström, M., and Asp, N. G. (1989). Dietary fibre in cereals-composition, fermentation and effect of processing. In: *Proceedings from an International Symposium on Cereal Science and Technology*, Lund University, pp. 40–54.

Ortolani, C. and Pastorello, E. A. (1997). Symptoms of food allergy and food intolerance. In: *Study of Nutritional Factors in Food Allergies and Food Intolerance*. Luxembourg: CEC, pp. 26–45.

Ottenhof, M.-A. and Farhat, I.A. (2004). The effect of gluten on the retrogradation of wheat starch. *J. Cereal Sci.* 40, 269–274.

Parker, R. and Ring, S. G. (2001). Aspects of the physical chemistry of starch. *J. Cereal Sci.* 34, 1–17.

Pomeranz, Y. (1988). Composition and functionality of wheat flour components. In: Pomeranz, Y. ed. *Wheat Chemistry and Technology II*, 3rd edn. St. Paul, MN: American Association of Cereal Chemists, pp. 219–370.

Renzetti, S., Dal Bello, F., and Arendt, E. K. (2007). Impact of transglutaminase on the microstructure, fundamental rheology and baking characteristics of batters and breads made from different gluten-free flours. *J. Cereal Sci.* (in press).

Rogers, D. E., Zeleznak, K. L., Lai, C. S., and Hoseney, R. C. (1988). Effect of native lipid, shortening, and bread moisture on bread firming. *Cereal Chem.* 65, 398–401.

Rojas, J. A., Rosell, C. M., and Benedito de Barber, C. (1999). Pasting properties of different flour-hydrocolloid systems. *Food Hydrocolloids* 13, 27–33.

Rosell, C. M., Haros, M., Escriva, C., and Benedito De Barber, C. (2001). Experimental approach to optimise the use of alpha-amylases in breadmaking. *J. Agric. Food Chem.* 49, 2973–2977.

Sahlström, S. and Brathen, E. (1997). Effects of enzyme preparations for baking, mixing time and resting time on bread quality and bread staling. *Food Chem.* 58, 75–80.

Salovaara, H. and Göransson, M. (1983). Nedbrytning av fytinsyra vid franställning av surt och osyrat råggbröd. *Näringsforskning* 27, 97–101.

Sanchez, H. D., Osella, C. A., and de la Torre, M. A. G. (2002). Optimisation of gluten-free bread prepared from cornstarch, rice flour and cassaca starch. *J. Food Sci.* 67, 416–419.

Sanchez, H. D., Osella, C. A., and de la Torre, M. A. G. (2004). Use of response surface methodology to optimize gluten-free bread fortified with soy flour and dry milk. *Int. J. Food Sci. Technol.* 10, 5–9.

Schober, J. T., Messerschmidt, M., Bean, S. R., Park, S. H., and Arendt, E. K. (2005). Gluten-free bread from sorghum:quality differences among hybrids. *Cereal Chem.* 82, 394–404.

Sharadanant, R. and Khan, K. (2003). Effect of hydrophilic gums on the quality of frozen dough: II. Bread characteristics. *Cereal Chem.* 80, 773–780.

Sharma, S. C. (1981). Gums and hydrocolloids in oil-water emulsion. *Food Technol.* 35, 59–67.

Shewry, P. R. and Halford, N. G. (2002). Cereal seed storage proteins: Structures, properties and role in grain utilization. *J. Exp. Bot.* 53, 947–958.

Silva, R. F. (1996). Use of inulin as a natural texture modifier. *Cereal Foods World* 41, 792–795.

Sivaramakrishnan, P. H., Senge, B., and Chattopadhyay, K. P. (2004). Rheological properties of rice dough for making rice bread. *J. Food Eng.* 62, 37–45.

Tester, R. F. and Debon, S. J. J. (2000). Annealing of starch—a review. *Int. J. Biol. Macromol.* 27, 1–12.

Thiele, C., Gänzle, M. G., and Vogel, R. F. (2002). Contribution of sourdough lactobacilli, yeast and cereal enzymes to the generation of amino acids in dough relevant for bread flavour. *Cereal Chem.* 79, 45–51.

Thompson, T. (2000). Folate, iron and dietary fibre contents of the gluten-free diet. *J. Am. Diet. Assoc.* 1000, 1389–1396.

Tolstoguzov, V. (1997). Thermodynamic aspects of dough formation and functionality. *Food Hydrocolloids* 11, 181–193.

Toufeili, I., Dagher, S., Shadarevian, S., Noureddine, A., Sarakbi, M., and Farran, T. M. (1994). Formulation of gluten-free pocket-type flat breads: Optimization of methyl-cellulose, gum arabic, and egg albumen levels by response surface methodology. *Cereal Chem.* 71, 594–601.

Vemulappali, V. and Hoseney, R. C. (1998). Glucose oxidase effects on gluten and water solubles. *Cereal Chem.* 75, 859–862.

Veraverbeke, W. S. and Delcour, J. A. (2002). Wheat protein composition and properties of wheat glutenin in relation to breadmaking functionality. *Crit. Rev. Food Sci. Nutr.* 42, 179–208.

Wagner, M. J., Lucas, T., Le Ray, D., and Trystram, G. (2007). Water transport in bread during baking. *J. Food Eng.* 78, 1167–1173.

Wang, L., Miller, R. A., and Hoseney, R. C. (1998). Effects of $(1\rightarrow3)$ $(1\rightarrow4)\rightarrow\beta$-D-glucan of what flours on breadmaking. *Cereal Chem.* 75, 629–633.

Ward, F. M. and Andon, S. A. (2002). Hydrocolloids as film formers, adhesives and gelling agents for bakery and cereal products. *Cereal Food World* 47, 52–55.

Williams, P. A. and Phillips, G. O. (2000). Introduction to hydrocolloids. In: Phillips, G. O. and Williams, P. A. eds. *Handbook of Hydrocolloids*. Cambridge: Woodhead Publishing, pp. 1–19.

Xu, J., Bietz, J. A., and Carriere, C. J. (2007). Viscoelastic properties of wheat gliadin and glutenin suspensions. *Food Chem.* 101, 1025–1030.

Ylimaki, G., Hawrysh, Z. J., Hardin, R. T., and Thomson, A. B. R. (1991). Response surface methodology in the development of rice flour yeast breads: Sensory evaluation. *J. Food Sci.* 5, 751–759.

Zobel, H. E. and Kulp, K. (1996). The staling mechanisms. In: *Baked Goods Freshness: Technology Evaluation, and Inhibition of Staling*. New York: Marcel Dekker, pp. 1–64.

Formulation and nutritional aspects of gluten-free cereal products and infant foods

14

Eimear Gallagher

Introduction

As awareness of gluten allergy/intolerance increases, and with better diagnostic tools to detect celiac disease, the requirement for gluten-free products is increasing world-wide, especially in developed countries. However, the formulation of gluten-free products is usually more challenging than that of gluten-containing products, as the major structure-forming component, namely gluten, is absent. In addition, the nutritional profile of gluten-free foods may also be a challenge; for example due to their low dietary fiber content. To date, gluten-free biscuits, cakes, pasta, and pizza, which may be included in the diets of patients with celiac disease, are commercially available. However, they are often based on pure starches, resulting in a dry, sandy mouthfeel and poor overall eating quality. This chapter will review different approaches taken in the development of such goods.

Biscuit, confectionery, and pasta products

Biscuits and cookies are popular products all over the world, and their many combinations of texture and taste give them a universal appeal. Innovation and development of new biscuit products is limited only by the imagination of the food technologist. Originally believed to be craft-based, the manufacture of biscuits has developed into a science, with different combinations of textures and flavors being produced to suit any local palate and marketplace. The differences in formulation, processing, and finished product attributes are all a function of the biscuit dough consistency, or dough rheology. Many different types of biscuit exist; however, regardless of the category they belong to, there are certain rheological requirements for all biscuits, i.e. the dough must be adequately cohesive for molding/forming, without excessive stickiness, and the dough must have a short, cuttable texture (Hazelton *et al.*, 2004). The degree of gluten development of the dough is also an extremely important determinant for some of the biscuit types. Soft wheat flour, sugar, and fat are the basic ingredients used for biscuit manufacturing, and in this chapter the approaches that have been taken to replace wheat flour in the formulation of gluten-free biscuits and cookies will be overviewed.

Cakes are chemically leavened batter-based products. The variety and diversity of cake products is large, with formulations varying substantially across the globe (Oritz, 2004). The definition of cake varies, but essentially the term refers to products that are characterized by formulations based on wheat flour, sugar, whole eggs, and other liquids, to which fat or oil may be added. The level of added liquids is such that a low-viscosity batter is formed rather than a dough. There is no significant gluten formation in cake batters, and cake-making technology exploits steps to prevent gluten network formation, whereby the batter is formed by a complex emulsion. The key structure-forming component of cakes is starch, which is present in the wheat flour, and the modification of its gelatinization characteristics through the addition of sugars and liquids (Cauvain, 2003a).

Pasta products have been known to Mediterranean civilizations for many centuries. The current range of products referred to as pasta is vast (e.g. macaroni, spaghetti, lasagne, vermicelli, and noodles) and products vary widely in terms of shape, color, composition, storage requirements, and use (Cubadda and Carcea, 2003). The word "pasta" is Italian for "dough," and is generally used to describe products fitting the Italian style of extruded foods such as those mentioned above (Sissons, 2004). The raw material of choice for pasta production is semolina flour from durum wheat (Cubadda and Carcea, 2003). Durum is a hard-wheat, and the milled semolina has a coarse particle size which is ideal for making pasta. As consumers of pasta are becoming more discerning in their quality requirements, and less accepting of product variability, pasta producers must use the correct raw materials that will have the desired characteristics for processing into pasta. The protein present in the grain significantly affects the processing properties of pasta. The continuity and strength of the protein matrix formed during dough mixing and pasta extrusion is important in determining the textural characteristics of the pasta. Compared with weak gluten of the same protein level, strong gluten wheats exhibit less sticky dough with better extrusion properties and superior cooked textural characteristics. Instant pastas have thinner walls and need more strength during processing, whereas fresh pasta requires a more extensible dough and weaker gluten to improve the sheeting properties. The gluten matrix is thus a vital parameter which dictates the quality of the pasta product.

Pizza is widely believed to have originated in Naples and may be defined as a flat leavened bread that can have a wide variety of toppings. The quality of the base and sauce plays an important role in the overall quality of pizza. The pizza industry has continued to grow with unprecedented momentum in recent decades (Sun and Brosnan, 2003), and the increase in demand is such that food companies show growing interest in the industrial production of pizza dough (Formato and Pepe, 2005). Pizza crust or base is formulated with hard wheat flour, and constitutes a significant part of the overall product. Its appearance, taste and texture are important factors for consumer identification and acceptance (Larsen *et al.*, 1993). However, in comparison to other baked products, pizza crust quality, and in particular gluten-free pizza crust quality remains a less researched area.

Biscuit and cookies

Classification of biscuits and cookies

The term "biscuit" is derived from the Latin term *bis coctus*, which means twice baked. The original process consisted of baking the biscuits in a hot oven and then drying them in a cool oven. These days, however, this technique is extremely rare. "Cookie" originates from a Dutch word *koekje*, which means "little cake;" the sound of a cracker being eaten most likely led to the use of that name (Zydenbos *et al.*, 2004). Biscuits and cookies may be classified by their formulation and method of manufacture. Biscuit dough and baking properties are greatly influenced by the degree of gluten development of the dough. The major classifications of biscuits are discussed below, and a simple diagram of the biscuit-making process is shown in Figure 14.1.

Short dough biscuits

The majority of biscuits and cookies consumed worldwide are made from short doughs. Short dough formulations have high proportions of fat and sugar, which can range up to 100% and 200% of flour weight, respectively. Such high levels of fat result in doughs that are cohesive and plastic, but lack extensibility and elasticity, due to a limited gluten network development. The flour in short doughs is given very little mixing; this also minimizes the development of the protein network. Short dough biscuits are usually formed either by rotary molding, by extruding and cutting or by sheeting and cutting. The dough pieces formed tend to retain their shape until baking but then they spread or flow, becoming thinner. This type of biscuit breaks easily, and examples include digestives, shortbreads, and custard creams.

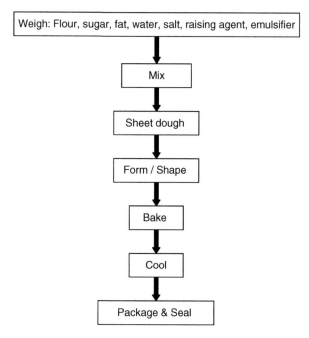

Figure 14.1 Biscuit production.

Hard sweet and semi-sweet biscuits

These biscuits are characterized by a dough which has a stiff consistency, and a more developed gluten network. The levels of fat and sugar in the formulation are low relative to the flour, the doughs are both elastic and extensible, and require extensive mixing. Hard doughs can be similar to bread doughs, except that the sugar and fat contents modify their viscoelastic properties. These doughs are usually laminated and sheeted before cutting or stamping. The formed pieces will generally shrink due to the elastic quality of the gluten. Examples of this type of biscuit include rich tea and petit beurre.

Crackers

Cracker is a generic term used to describe products with low sugar and fat contents (Zydenbos *et al.*, 2004). They may be fermented, as in the case of soda crackers and cream crackers, or chemically leavened, like snack crackers. Cracker doughs have a developed gluten network, and protein quality is important for dough processing (Kent and Evers, 1994). During fermentation/leavening, the protein network is modified. After fermentation/leavening, the dough is laminated, followed by cutting and sheeting. The combination of protein modification and lamination gives rise to the characteristic flaky and blistered appearance of crackers (Manley, 1983).

The major ingredients in biscuit and cookie production

The most prominent ingredients used in the manufacture of biscuits are flour, fat/oil, and sugar.

Flour

Biscuit flour is milled from soft winter wheats, with a low damaged starch content (Hoseney, 1994). The flour strength is a function of the protein content. For biscuits and cookies, flours with a relatively low protein content are used (Millar and Hoseney, 1997) (Table 14.1).

Soft wheat differs from hard wheat in kernel hardness, a basic genetic, directly inherited characteristic. When ground or milled, soft wheat generally fractures into significantly smaller particles than does hard wheat. In an effort to achieve product uniformity and consistency, strict specifications are adopted by millers and bakers, which vary slightly between companies (Gaines, 1990). As their name suggests, hard wheats are harder in nature. Thus, more work is needed to reduce the wheat to a fine particle size. One result of this work is that a large percentage of the starch is damaged during milling. Higher damaged starch values are often viewed as a negative factor, particularly for cookie flour (Hoseney, 1994). Cookies made from hard wheat flour would be undesirably hard in texture. Hence, soft wheat flours with a low damaged starch content are more suitable for the manufacture of biscuits and cookies.

Flour with low water absorption is also desirable. In their study of cookie spreading during baking, Millar and Hoseney (1997) found that flours with a low water retention capacity were superior for cookie baking. Malick and Sheikh (1976) pointed out that intense competition for water amongst flour components contributes significantly to the baking properties of cookies; therefore water content not exceeding 13% is most desirable. The most suitable wheats used in biscuit flour manufacture originate from Britain and Northern Europe, i.e. temperate regions, with a growing season from late autumn to early autumn of the following year.

Fats and oils

Fats are extremely important ingredients in biscuit manufacture. They are obtained from a huge variety of plant (e.g. palm, rapeseed, sunflower, coconut, vegetable and soyabean oils) and animal sources (Manley, 1983). Fats for biscuit manufacture are usually semi-solid at room temperature so that they blend smoothly with other ingredients. The primary function of fat is to create more tender products and shorter doughs. Fat lubricates the structure by being dispersed in the dough during mixing, helping to prevent the starch and protein from forming a continuous network

Table 14.1 Protein requirements of wheat products

End-product	Wheat protein content (%) (14% mb)	Type of wheat
Macaroni products	13.0 and above	Durum
Hearth bread and hard rolls	13.5 and above	Spring
Pan bread	11.5–13.0	Winter
Crackers	10.0–11.0	Soft/hard
Biscuits	9.0–11.0	Soft/hard
Cakes, pies, cookies	8.0–10.0	Soft

From Halverson and Zeleny (1988).

(Glickman, 1991). Recently, Anon. (1997) discussed the role of fats in cookie dough, pointing out that fat and the aqueous phase compete for the surface of the flour particles during mixing. The formation of a gluten network is inhibited if the fat coats the flour before it can be hydrated. After baking, the desired eating properties of these products are: less hard, shorter and more inclined to melt in the mouth. Therefore, the desired consistency of the dough can be achieved by increasing the fat content while decreasing the amount of water.

Sugar

The sugar present in biscuits affects sweet flavor, dimensions, color, hardness, and surface finish. Sugar can inhibit gluten development during dough mixing by competing with the flour for the water. Sucrose is the main sugar utilized in the biscuit industry. It can act as a hardening agent by crystallizing as the cookie cools, thus making the product crisp (Olewnik and Kulp, 1984). However, Venkateswar and Indrani (1989) found that a moderate amount of sucrose could act as a softening agent in cookies, due to the ability of sucrose to retain water. Generally, as the size of the sugar crystals increases, the size and symmetry of the biscuit decreases, while the thickness increases.

Biscuit dough

Biscuit doughs are cohesive, but lack the extensible and elastic characteristics of bread doughs (Maache-Rezzoung et al., 1998). Because of the minimal gluten network that is formed, the texture of the baked biscuit is attributable to starch gelatinization and supercooled sugar rather than a protein/starch network—the development of gluten is minimal and only serves to provide cohesion for handling and subsequent shaping (Olewnik and Kulp, 1984). Biscuit doughs are sufficiently extensible to be easily sheeted, without being so elastic that they prevent the products from retracting after cutting; this aids in the potential of biscuits for packaging. Contamine et al. (1995) studied the relationship between the energy input during mixing and the subsequent dough rheology and biscuit properties, concluding that biscuit dough should be poorly elastic but sufficiently supple and extensible to allow an easy and stable shaping of the products. In addition, they concluded that the gluten network should be slightly developed for the dough to be cohesive without being too elastic.

Of importance to dough consistency and biscuit quality are the proteins of the flour, namely the gliadin and glutenin fractions. Gaines (1990) points out that it is a mistake to view the proteins of soft wheat flour as functionally inert in cookie dough. As a cookie bakes, the dough viscosity decreases, causing spreading and expansion in all directions. It is at this stage of baking that the critical function of soft wheat flour proteins comes into action (i.e. by reducing the spread of the dough). Doescher et al. (1987) put forward the idea that the flour proteins swell when their glass transition temperature is reached—a continuous phase or network is formed which decreases water mobility and increases biscuit dough viscosity, thus stopping expansion of the dough. Weegels and Hamer (1989) backed up this idea that soft wheat proteins affect important quality parameters, including dough consistency.

Formulation of gluten-free biscuits

In the manufacture of gluten-free biscuits, the wheat flour needs to be replaced by other ingredients. These ingredients need to replace not only the starch, which is normally delivered by the wheat flour, but also the protein fractions. The following section reviews approaches that have been taken in replacing wheat flour in gluten-free biscuit and cookie formulations.

Schober *et al.* (2003) produced gluten-free short dough-type biscuits from a range of gluten-free flours. Starches from corn, soya, millet, buckwheat, rice, or potato were combined with different types of fat (palm oil, cream powder, microencapsulated high-fat powder and low-fat dairy powder). Dough characteristics and biscuit texture, color, moisture, dimensions, and sensory attributes were evaluated. It was found that the combination of rice, corn, potato, and soya with a high-fat powder produced biscuit doughs which were sheetable, and the baked biscuits were of comparable quality to wheat biscuits (Figure 14.2).

Recently, many researchers have studied the use of pseudocereals as wheat flour replacers in biscuit, confectionery, and pasta production. Marcilio *et al.* (2005) used a factorial design to study the effects of refined amaranth flour and fat contents on gluten-free biscuits. Overall appearance of the biscuits was affected by the amount of refined flour present, whereas the fat levels used had a positive influence on the flavor of the products. Results from a 39-member sensory panel concluded that amaranth flour showed potential for the manufacture of gluten-free biscuits. Taking into consideration the important nutritional properties of amaranth, i.e. methionine, cysteine, lysine, vitamins, and minerals (Samiyi and Ashraf, 1993; Akingbala *et al.*, 1994; Hozová *et al.*, 1997), the replacement of wheat flour by amaranth flour is desirable for the production of gluten-free baked products with high protein and energy value. Hozová *et al.* (1997) undertook a complete nutritional, sensory and microbiological evaluation of amaranth-containing crackers and biscuits; with a favorable outcome. However, they found that over the four-month trial period, the total bacterial count of the biscuits rose above the acceptable limit of 10^3 CFU/g. Factors such as improper

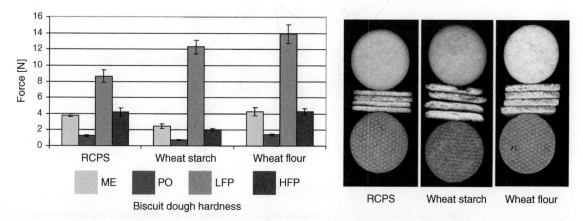

Figure 14.2 Gluten-free biscuits produced with rice, corn, soya and potato (RCPS), microencapsulated fat powder (ME), palm oil (PO), low-fat powder (LFP), and high-fat powders (HFP). From Schober *et al.* (2003).

packaging, or the presence of spores in the flour were pointed out as the possible causes of the microbial spoilage. Nonetheless, the authors concluded that shelf-stable amaranth products should be recommended in a gluten-free diet.

Tosi *et al.* (1996) used wholemeal amaranth flour to develop gluten-free biscuits and found that the addition of 0.1% butylated hydroxytoluene to the fat extended the shelf-life without affecting the product flavor. The protein content of these biscuits (5.7%) was higher than the average content for non-amaranth-containing gluten-free biscuits.

Schoenlechner *et al.* (2006) used amaranth, buckwheat, and quinoa at levels of 25, 50, 75, and 100% in the production of gluten-free biscuits. Biscuit crispiness was in the order buckwheat > quinoa > amaranth, and biscuits containing buckwheat and amaranth were preferred in a sensory panel (Figure 14.3).

Sorghum flour is becoming increasingly common in gluten-free baked goods, mostly because it has similar nutritional properties to wheat, is light in color and bland in flavor (Lovis, 2003). Dahlberg *et al.* (2004) discussed how sorghum, due to its unique phenolic compounds and starch characteristics, is suitable for the development of healthy and nutritional gluten-free foods such as breads and biscuits. Moreover, Taylor *et al.* (2006) stressed that sorghum contains substantial levels of phenolic compounds and antioxidant activity. Cookies from 100% sorghum or pearl millet were produced by Badi and Hoseney (1976), but these were described as "tough, hard, gritty and mealy." Improvements were sought via various additions (wheat flour lipids, unrefined soybean lecithin, hydratation of the flour, increasing dough pH) and the authors concluded that the lack of polar lipids in sorghum is partly responsible

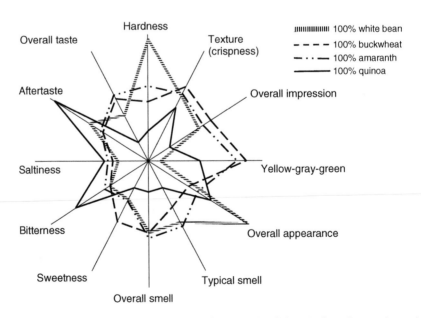

Figure 14.3 Results of sensory evaluation of gluten-free short dough biscuits from the pseudocereals amaranth, quinoa and buckwheat with common beans. From Schoenlechner *et al.* (2006).

for the lower quality of the cookies when compared to those from wheat. However, Morad *et al.* (1984) hypothesized that the above problems could be overcome by altering the extraction rate and particle size of the sorghum flours, and/or by using alternative formulations.

Conclusion

A vast range of biscuit and cookie products are available today in the marketplace. Biscuits are classified by their formulation, with the main types generally derived from short dough formulations (containing high proportions of fat and sugar) or hard sweet and semi-sweet formulation (containing lower levels of fat and sugar than the short dough formulations). Biscuit flour is milled from soft winter wheats with low damaged starch content and low water absorption properties. When producing gluten-free biscuits, the flour and its constituents (starch, protein, etc.) must be replaced by other ingredients. To date, a wide variety of ingredients, including starches, pseudocereals, sorghum, and millet have been studied, with varying degrees of success. Research has also focused on the utilization of ingredients such as amaranth to boost the nutritional properties of gluten-free biscuits.

Cake

Introduction

In general, modern cake batters can be considered as fat, or oil, in water emulsion systems. The aqueous phase contains the dissolved sugar and suspended flour particles. In many batter systems, air bubbles are held in the solid fat rather than in the aqueous phase. As the batter warms during baking, the air bubbles transfer from the fat to the aqueous/foam phase and expand. Later in the baking process, the foam sets to yield the cake structure (Cauvain, 2003b). As already mentioned, soft wheat flour is the main structure-forming component, and there is little gluten network formation in the cake batter.

Major raw materials in cake formulations

Cake flour

For the production of cakes, flours with low protein contents and a low α-amylase activity are usually desirable (Kent and Evers, 1994) (Table 14.2).

Flour is the most important ingredient in cake manufacture, functioning primarily to establish crumb structure (DesRochers *et al.*, 2003). Cake flour is milled from soft wheats which have low protein and ash levels as well as a fine particle size. Although the protein quantity is low, its quality must be high enough to ensure the formation of films for trapping gas in small air bubbles, but not to give the toughness/chewiness commonly obtained with bread flours (Oritz, 2004). Another important constituent of the wheat flour is the starch, the gelatinization of which forms the major structural component of cakes (Cauvain, 2003b). When hydrated, the flour proteins form

Table 14.2 Wheat classes and their general characteristics and principal uses

Class	General characteristics	General uses
Hard red winter (HRW)	High protein, strong gluten, high water absorption	Bread and related products
Soft red winter (SRW)	Low protein, weak gluten, low water absorption	Cakes, cookies, pastries, pie crusts, crackers, biscuits
Hard red spring (HRS)	Very high protein, strong gluten, high water absorption	Bread, bagels, pretzels and related products
Hard white	High protein, strong gluten, high water absorption, bran lacks pigments	Bread and related products
Soft white	Low protein, weak gluten, low water absorption, bran	Noodles, crackers, wafers and other products in which specks are undesirable
Durum	High protein, strong gluten, high water absorption	Pasta

From Atwell (2004).

a significantly weaker gluten structure in cake batters than in bread doughs because of the disruptive effects of the sugars and fats in the cake formulation. In addition, the high level of water in a cake formulation produces a low-viscosity batter which reduces the ability of the mixing action to impart the necessary energy for gluten formation.

Shortening, sugar, and leavening agents

Shortening performs three basic roles in cakes: (1) It aids in aeration or leavening of the batter and baked cake by entrapping air during the creaming process. These minute air cells provide the nucleus for bubble expansion via steam and carbon dioxide during baking. (2) It coats the protein and starch particles, preventing hydration and formation of a continuous gluten-starch network. (3) It is also involved in emulsifying liquids in the batter, which increases the moisture of the crumb and hence affects the subsequent crumb texture (DesRochers et al., 2003).

Sugar provides sweetness in cakes, but also plays a significant role in batter aeration and structure formation. Sugar affects the physical structure of the baked products by regulating the gelatinization of the starch. Delay in starch gelatinization during baking allows air bubbles to expand properly due to vapor pressure build-up by carbon dioxide and water vapor before the cake sets (Kim and Walker, 1992). Sugar delays the gelatinization of starch from 57°C to 92°C, which allows the formation of the desired cake structure. Granulation size of the sugar also contributes to the viscosity of the batter. The most common sugar used in cake manufacture is sucrose.

Leavening is a critical factor in aiding the formation of the aerated structure that is expected in cakes and cake batters. In lieu of yeast, chemically leavened cakes utilize sodium bicarbonate (baking soda) plus an acidic agent to generate carbon dioxide in the presence of water (DesRochers et al., 2003). A significant portion of the final baked cake is, in fact, air. The production and release of carbon dioxide gas as the batter initially enters the oven is an important part of the cake expansion mechanism, and it affects the volume and eating quality of the finished product.

Properties of cake batters

The interaction of the ingredients and their effect on the structure of the batter occur both during mixing and baking. Generally, an uncooked cake batter can be regarded as an oil-in-water emulsion with a continuous aqueous phase containing dissolved sugars and suspended flour particles. Initially, after the mixing stage of cake preparation, the role of fat is important to the aeration of the batter. The occlusion of air cells in the system during mixing gives rise to foam. To give maximum cake volume, the distribution of air in the system should be in a large number of small cells, rather than small number of larger cells. During the baking process, the fluid-like, aerated emulsion of the cake batter is converted to a semi-solid, porous, and soft structure. This is mainly due to starch gelatinization, protein coagulation, and gas bubbles produced from chemicals dissolved in the batter, the occluded air, and the interaction among ingredients (Sahi, 1994).

Formulation of gluten-free cake

As mentioned previously, the presence of wheat flour in cake formulations serves many roles, from altering the viscosity of the cake batter to establishing crumb structure and good eating quality in the final baked product. To date, relatively little work has been published on the use of alternative flour sources for the production of gluten-free cake products.

The rheological properties of batters and cakes formulated with rice flour, gums (xanthan, guar, carrageenan, locust bean) and an emulsifier blend was studied by Turabi *et al.* (2007). Overall, gums in combination with the emulsifier resulted in batters with the ability to entrap more air during the mixing step, and additionally the stability of the emulsion was increased. The authors recommended using of 1% xanthan gum (increased the viscosity of the cake batter and prevented collapse of the cakes in the oven) and 3% emulsifier (increased the volume and porosity of the cakes and increased softness) when formulating rice cakes of acceptable quality.

Cassava starch is a food ingredient which originates in South America. Its chemical properties have been discussed extensively (Cárdenas and de Buckle, 1980; Camargo *et al.*, 1988) but it is only recently that its suitability in gluten-free cereal systems has been investigated. It is sensitive to oxidation (Mat Hashim *et al.*, 1992; Paterson *et al.*, 1994). Demiante *et al.* (2000) discussed the production of chemically oxidized samples of cassava starch and then studied the relationship between the baking properties of this starch (and a range of other cassava starches) and its chemical structure. From their study, the authors concluded that chemically treated cassava starch contains carboxylate groups which are important for dough expansion and baking properties.

Bean *et al.* (1983) produced layer cakes made from 100% rice flour and found that hydration of the flour, coupled with an intense mixing regime, improved cake properties. They hypothesized that the intense mixing of the rice flour and water freed some starch granules from the endosperm and increased their functionality. Moreover, the high speed mixing could possibly bring about the formation of a "gel" protein which would enhance the crumb grain of the cake. Cake-type products containing soy meal, corn flour and rice flour were also formulated by Borowski and Pomianowski (1994), while Jud (1993) discussed the properties of tara gum, with particular reference to its galactomannan composition. This neutral polysaccharide can be applied as a thickening and gelling agent in gluten-free pastry-type products, and is also regarded as dietary fiber.

Von Atzingen and Machado Pinta e Silva (2005) studied the instrumental texture and color of gluten-free cakes made from 10 gluten-free starches and flours (including cassava, rice, and corn products). Cakes with rice flour had the highest compression force. Of the cassava derived starches, an acid starch preparation gave the brightest color whereas color intensity was greatest for cakes made with rice or corn flour.

The microbial aspect of gluten-free muffin mixes based on cassava, soya, sorghum, sage, and potato flours were evaluated by Chauhan et al. (2001). Microbial counting of the predominant bacteria and molds from the raw materials and from the muffins revealed a minimum mold-free shelf-life of muffin mixes of about 3 days. The results are in good agreement with those from wheat-containing muffins.

Conclusion

In cake-making, the wheat flour is milled from soft wheats with low protein contents, low alpha-amylase levels and a fine particle size. The function of the flour is mainly to establish the structure of the cake crumb, giving it a crumbly texture without the "spongy" properties normally associated with bread texture. Cake formulations also contain high levels of fat and sugar. When water is added to the formulation, the flour proteins form a weak gluten network due to interference by the sugar and fat. To date, few studies have been reported regarding the replacement of wheat flour in cake-making. Approaches taken have involved using flours derived from rice, cassava and corn, and gums such as xanthan, guar, carrageenan, tara, and locust bean. High-quality cakes were produced using rice flour, xanthan gum, and an emulsifier, and the application of an intense mixing regime using rice flour was also recommended.

Pasta and extruded products

Production of pasta

All pasta types share the same basic technology that involves the preparation of a dough made by mixing a flour with a liquid (mainly water) which is then processed (by extrusion) to obtain the required shape and dimension of the product itself (Figure 14.4).

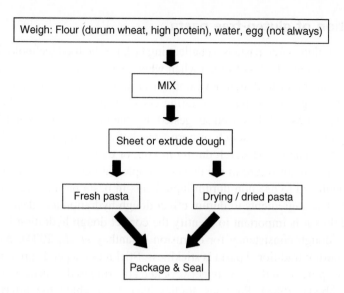

Figure 14.4 Pasta production.

The major raw material in pasta manufacture—durum wheat

Pasta products are made almost exclusively from semolina flour, which is milled from durum wheat. In fact, durum wheat semolina is the only raw material permitted for pasta production by national laws in Italy, France, and Greece. Special pasta products are also produced by adding a variety of other ingredients (e.g. fresh, frozen, or powdered eggs, soy protein, wheat gluten, milk protein, etc.) (Cubadda and Carcea, 2003).

Durum wheat is a hard wheat, and the semolina flour is a granular product composed of evenly sized endosperm particles. The proteins in semolina flour are linked together by disulfide, hydrogen, and hydrophobic bonds to form a matrix which gives cooked pasta its viscoelastic properties. The protein content in durum wheat can range from 9 to 18%. Both protein content and gluten composition are generally considered to be the main factors affecting dough properties and cooking quality of durum wheat pasta (Table 14.2). The development of a protein matrix during pasta cooking is important, as a cohesive pasta dough and product is desirable, one which does not split or break apart while being boiled in water (Feillet and Dexter, 1996). Doughs obtained with common wheat are very extensible, and of medium/low tenacity, while those obtained with durum wheat are generally characterized by high tenacity and minor extensibility. A good gluten tenacity permits to keep starch granules inside the pasta structure, reducing stickiness. It also allows the modulation of water absorption during cooking, preventing excessive expansion of the pasta, and giving it the right firmness. Although the gluten quality and quantity of the durum semolina are the most important factors affecting cooking quality, starch and minor constituents such as soluble and insoluble pentosans, lipoproteins, various enzymes and products of enzyme interactions are also involved (Cubadda and Carcea, 2003).

Formulation of gluten-free pasta

Preparation of gluten-free pasta is a challenging task for the food technologist, because of the lack of gluten which is formed when wheat is used as starting material. Gluten is the main contributor to dough development during mixing and extrusion, and thus prevents disaggregation of the pasta during cooking in boiling water (Feillet, 1984; Abecassis *et al.*, 1989). It has been suggested that the lack of gluten can be overcome by blending pre-gelatinized starch or corn flour before adding water and mixing, or by gelatinizing some of the starch during mixing or extruding (Molina *et al.*, 1975).

Durum pasta cannot figure in the diets of people with celiac disease, and research has been conducted to use non-durum ingredients in pasta production. However, the use of alternative ingredients to durum effect dough hydration and development properties, and thus it is important to identify the correct dough hydration levels needed for a proper dough consistency for extrusion (Manthey *et al.*, 2004). An important property of durum traditional pasta is stickiness. Stickiness depends predominantly on the starch escaping from the protein network and adhering to the surface of the cooked product (Cubbeda, 1988). Previous studies have highlighted the inferior cooking quality of pasta made from non-semolina ingredients/flours. However, this may be overcome by drying at high or ultra-high temperatures, which denatures the protein in the gluten matrix, and subsequently protects the starch granules from rupturing during cooking. Large amounts of damaged starch strongly increase water absorption by semolina and constitute a detrimental condition for the development of the gluten matrix (Manthey and Schorno, 2002). Moreover, starch damage is associated with increased stickiness of the cooked pasta (Grant *et al.*, 1993). These properties must therefore be taken into consideration when formulating gluten-free pasta. Manthey *et al.* (2004) discussed the alteration of pasta properties when non-durum ingredients are used. As the chemical and water-binding properties of non-durum ingredients are varied, dough hydration, dough development, and dough consistency will change. Therefore, the production process has to be adapted accordingly, since the dough strength affects the amount of mechanical energy required to extrude as well as the rate of extrusion (Levine, 2001).

Gluten-free research using pseudocereal flours is becoming increasingly popular due to their important functional properties. Pseudocereals were used by Schoenlechner *et al.* (2004) in their investigations on gluten-free pasta. Using buckwheat, amaranth, and quinoa at 10% levels resulted in products with a high cooking loss and a low stability during cooking. Blending the three ingredients in different ratios by means of an experimental design (along with the addition of albumen, emulsifier, and enzymes) improved the properties of the gluten-free pasta (Table 14.3).

Caperuto *et al.* (2001) used mixtures of quinoa and corn to produce a gluten-free spaghetti-type product. Cooking quality, texture (adhesiveness and elasticity), and viscosity were determined. Acceptability was evaluated by a sensory panel. Milling of the quinoa grain was necessary to make the mixture of the quinoa and corn flours compatible. The products had a mild corn taste and were generally moderately acceptable. In an attempt to produce gluten-free pasta, Marconi and Carcea (2001) used dairy proteins and caroubin (a protein isolated from carob gum) (Table 14.4). The authors discussed how caroubin has similar rheological properties to gluten and

Table 14.3 Comparison of noodle properties produced from the pseudocereals amaranth, buckwheat and quinoa

	Texture firmness (N)	Cooking time (min)	Cooking weight (%/100)	Cooking loss (%)
Amaranth	0.43	4	2.48	10.5
Buckwheat	1.19	10	3.21	11.2
Quinoa	0.92	7	3.22	14.4
Wheat	1.55	10	3.03	6.0

From Schoenlechner *et al.* (2004).

Table 14.4 Formulations and nutritional characteristics of some non-traditional pastas containing wheat flour (WF), chickpea flour (CF), and milk powder (MP)

	% Ingredient in blend				Protein quality	
Blend	WF	CF	MP	Protein content (%)	Chemical score[a]	LAA[b]
WF	100			13.0	0.43	Lysine
CF		100		22.0	1.00	SAA[c]
MP			100	34.0	>1.00	None
Blend 1	70	25	5	16.3	0.79	Lysine
Blend 2	60	30	10	17.8	0.90	Lysine
Blend 3	50	35	15	19.3	0.99	Lysine

[a] Based on FAO/WHO/UNU reference pattern (1991, 1998).
[b] Limiting amino acids.
[c] Sulfur amino acids.
From Marconi and Carcea (2001).

could, therefore, serve as a texturizing agent in a gluten-free pasta formulation, as it would have the ability to form a protein network capable of providing firmness to the pasta and with the ability to hold the starch during pasta cooking.

Feillet and Roulland (1998) studied the rheological similarities between carob protein isolate and wheat gluten. Despite having some biochemical differences, the two proteins show very similar behavior which was attributed to the high content of high molecular weight subunits in the carob protein. Therefore, the authors hypothesized how caroubin could have potential use in gluten-free foods. In a novel approach in gluten-free research, Gobbetti *et al.* (2007) used sourdough lactobacilli to preferment durum wheat semolina under semi-solid conditions. Following fermentation, the dough was freeze-dried, mixed with buckwheat flour and made into "fusilli" type pasta. Two-dimensional electrophoresis and mass spectrometry showed that the durum wheat gliadins were almost totally hydrolyzed during fermentation by lactic acid bacteria. Although the levels of gluten present would still deem the pasta to be unsafe for people with celiac disease, they hypothesize that a mixture which would include 20% fermented durum wheat semolina in a pasta formulation could lead to a novel product with a safe threshold for celiac disease.

Response surface methodology (RSM) was used by Huang *et al.* (2001) in the formulation of non-gluten pasta. They based their optimization procedure on sensory properties and pasta stickiness, and found that gluten-free pasta with characteristics most similar to a wheat-based pasta was obtained when higher levels of modified starch, xanthan gum, and locust bean gum were used. This gave samples with a good "hardness of first bite" and a high level of cohesiveness.

The gluten-free pea flour is higher in protein and lysine than both wheat flour and semolina. The cooking quality of pasta products made by twin screw extrusion of 100% pea flour was evaluated by Wang *et al.* (1999). It was found that pea flour ingredient, coupled with a novel process, exhibited improved texture and flavor after cooking, and less change after overcooking, when compared with the same product prepared using a conventional pasta extruder. Limroongreungrat and Huang (2007) developed novel pasta products from alkaline-treated potato flour which was fortified with soy protein, while Chen *et al.* (2002) discussed the production of gluten-free starch noodles, where the dough is made from 5% pre-gelatinized starch (acting as the gluten) and 95% native starch. In their studies, they used sweet potato starches to make noodles, and although the products had good cooking characteristics and sensory properties, they stressed that large differences between sweet potato varieties exist, and it is important to select the correct variety when preparing such noodles. Mestres *et al.* (1988) investigated the starch networks of gluten-free noodles made from rice flour. In these products, the native starch structures disappeared, but new crystalline organizations were found. Amylose-based structures were present either in the complexed form (in the rice flour noodles) or in the retrograded form (in both product types). The authors also found that both pasta products exhibited good cooking behavior, which was attributed to amylose networks.

Studies on the properties of gluten-free macaroni were conducted by Kovacs and Varga (1995). Cooking quality and sensory properties were assessed with macaroni mixes based on three types of corn starch and two emulsifiers (to improve the water-binding capacity and consistency of the doughs). High-quality products were obtained when using corn starch with high amylose content and low free glucose or other starch decomposition products contents.

Extrusion is a suitable process for producing snack foods for patients with celiac disease, as starch is the main component providing the desirable expanded structure in the final product (Acs *et al.*, 1996). İbanoğlu *et al.* (2006) used an experimental design to study the expansion characteristics, color, and sensory properties of a gluten-free extruded snack based on rice, chickpea, and maize flours. Changes in the feed rate and screw speed of the extruder did not affect the color, flavor, and overall acceptability of the final product. However, increasing the screw speed increased the expansion and firmness of the product, while increasing the feed rate resulted in less hard, yet more expanded products. Extrusion parameters for a rice-based snack product were also studied by Bhattacharya and Choudhury (1994). They concluded that the length-to-diameter ratio (L/D) of the extruder significantly affects the extrusion system parameters and the product attributes. An increased L/D ratio yielded a harder product, while a barrel temperature >150°C was required for a high-quality expanded product from rice flour.

Conclusion

The vast majority of pasta products are derived from semolina flour, which is milled from durum wheat. The protein content of this wheat ranges from 9 to 18%, and both the quantity and quality of the gluten present are important determinants affecting the dough and cooking properties of the pasta. When making pasta, a gluten matrix is desirable, as this aids the pasta retaining its texture whilst being boiled in water. In the formulation of gluten-free pasta, it has been suggested that the production process should be altered, for example by pre-gelatinizing the gluten-free starch during mixing or extruding, or by drying at high or ultra-high temperatures to denature the protein and protect the starch from rupturing during cooking. A novel approach has described how sourdough lactobacilli were used to pre-ferment durum wheat semolina, resulting in hydrolysis of the gliadins.

The use of pseudocereal flours (buckwheat, amaranth, and quinoa) has been reported in studies in which good-quality gluten-free pasta products were obtained using emulsifiers, enzymes, and applying an experimental design. Other successful ingredients that have been used to produce gluten-free pasta include pea flour, potato flour, rice flour, and corn starch, along with hydrocolloids.

Pizza

Pizza flour, pizza dough, and gluten-free pizza dough

In general, two types of pizza exist: deep pan and thin and crispy pizza. Deep pan pizza needs a fairly high protein flour, and is fermented with yeast to produce a bread-like base. Thin and crispy pizza uses a slightly lower protein flour and can be fermented or gas aerated to produce a biscuit-type base. Pizza dough is prepared using a straightforward process (Figure 14.5). Along with wheat flour, the remaining ingredients are salt, water, and baker's yeast as a leavening agent. Different types of lactic acid bacteria and yeast are involved in the leavening process (Coppola *et al.*, 1998).

The overall quality of a pizza depends mainly on the dough, whose properties are affected by the leavening process, in addition to the flour type and preparation procedure. For a good-quality pizza, the dough has to be sheetable, to rise on proving, hold the gas produced by the yeast, as well as to have good textural and sensory attributes. As for bread, strong wheat flour is the principal ingredient of pizza crust. The quality of gluten present in the flour must be such that once the flour is hydrated, a cohesive, extensible dough is formed, that is able to rise during proofing and retain its shape during the sheeting process. In some industrial productions of pizza bases, refrigerated storage (retardation rather than freezing) of yeasted doughs is used to slow intermediate proofing. This process has been adopted to facilitate shipping of the intermediate product.

Commercially available gluten-free pizza bases exist. These are based on ingredients such as wheat starch, maize starch, potato starch, rice flour, corn flour, gums,

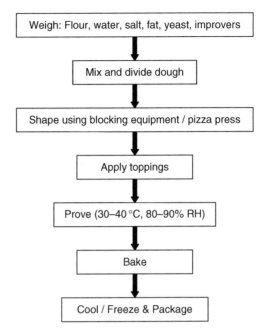

Figure 14.5 Pizza production.

and emulsifiers. However, the topic still remains a little-researched area. Researchers at University College Cork, Ireland have studied the formulation, rheological aspects and baking properties of gluten-free pizza bases. By combining a variety of gluten-free flours and starches, protein sources (egg, soya), or hydrocolloids (guar gum) and a microencapsulated high-fat powder, it was possible to fulfill all the requirements stated above (O'Brien *et al.*, 2002b). Tests such as dough hardness, texture (pizza base hardness), color, and pizza volume confirmed that it is possible to produce a gluten-free pizza product with similar attributes to the wheat-based control (O'Brien *et al.*, 2002a). The influence of the various ingredients on dough rheology of the optimized recipe was tested using fundamental rheology (oscillation tests in the linear viscoelastic region). From these measurements it was very clearly seen that in the corn starch system the biggest increase in elastic modulus was achieved when guar gum was combined with high-fat powder (O'Brien *et al.*, 2002b), and this product had similarities to the wheat-based control pizza base.

Conclusion

Like bread, strong wheat flour is used to produce pizza bases. The pizza dough must also have a fully developed gluten matrix with elastic properties and with the ability to trap carbon dioxide during the fermentation process. Pizza products are consumed throughout the world and by people of all ages. However, few results are available on the development of gluten-free pizza bases. One study reports how corn starch, protein sources, and hydrocolloids, in the correct proportions, can produce sheetable doughs and good-quality pizza base products.

Nutritional aspects of gluten-free products

Due to the fact that gluten-free products generally are not enriched/fortified and are frequently made from refined flour or starch, they may not contain the same levels of nutrients as the gluten-containing counterparts they are intended to replace. Therefore, uncertainty still exists as to whether patients with celiac disease living on a gluten-free diet are ensured a nutritionally balanced diet (Gallagher *et al.*, 2004). Kunachowicz *et al.* (1996) carried out a comprehensive study on the nutritive value of a range of gluten-free ingredients and products. Results showed that the protein concentration in typical gluten-free flours based on wheat starch is low (0.4–0.5 g/100 g), while that in buckwheat and millet flours is high, i.e. 14.6 and 11.6 g/100 g respectively. Remarkably, these values are higher than those found in wheat or rye flours (9.2 and 5.5 g/100 g respectively). In addition, buckwheat flour is rich in thiamine, riboflavin, and niacin, and millet flour also has high riboflavin and niacin concentrations; the values obtained are much higher than those found in wheat and rye flours. Wheat starch gluten-free cakes had low protein and vitamin concentrations. Kunachowicz *et al.* (1996) concluded that although many "healthy" gluten-free flours (e.g. buckwheat, rye) do exist, these are not used regularly, and in general, the nutritive value of the majority of gluten-free flours and products examined was generally lower than that of corresponding conventional products.

Thompson (1999, 2000) conducted two comprehensive surveys on the nutritional aspects of gluten-free foods. In the first study, it was found that many gluten-free cereal products contained inferior amounts of thiamine, riboflavin, and/or niacin compared with the enriched wheat products they are intended to replace. Although intakes of these vitamins among people with celiac disease will also depend on the overall diet of the person, the results of this survey indicate that the use of refined, unenriched, gluten-free cereal products contributes to a diet deficient in thiamine, riboflavin, and/or niacin, and therefore celiac diets should be regularly assessed for deficiencies of B vitamins. Thompson's second study (2000) focused on levels of folate, iron, and dietary fiber contents in gluten-free foods (breads, pastas, and cold cereals were assessed). Again, it was found that gluten-free cereal products generally provide lower amounts of folate and iron than their enriched/fortified gluten-containing counterparts. Therefore, if gluten-free cereal products provide a substantial portion of total energy intake for people with celiac disease, it is probable that a gluten-free diet will contain inadequate levels of these nutrients.

Grehn *et al.* (2001) screened the intake of nutrients and foods of 49 adults diagnosed with celiac disease and following a gluten-free diet; results indicated a lower intake of fiber when compared to a control group of people on a normal diet. As dietary levels of nutrients were found to be low in a range of typical gluten-free foods, Mariani *et al.* (1998) concluded that adherence to a strict gluten-free diet worsens the already nutritionally unbalanced diet of adolescents.

Gluten-free infant foods

Good nutrition is essential during the critical period of infancy to promote optimal growth and development (Räihä and Axelsson, 1995). Cereals, incorporated as flours, are the most common foods recommended by pediatricians because of their high energy load, based on carbohydrate and protein contents (Pérez-Conesa *et al.*, 2002). Recent epidemiological studies indicate that celiac disease is common in children, and its prevalence is between 1:300 and 1:80 (Korponay-Szabo *et al.*, 1999). Conflicting studies exist as to whether infant dietary habits affect the overall occurrence of celiac disease. Ivarsson *et al.* (2002) suggested that prolonged breastfeeding coupled with the introduction of a small amounts of gluten can reduce the risk of celiac disease. However, Ziegler *et al.* (2003) found a trend towards increased risk of anti-transglutaminase positivity (a specific serological marker of celiac disease) in children who received gluten-containing supplements before the age of three months. Therefore, a definitive conclusion has not been reached.

Comparing the nutritional aspects of gluten-free versus gluten-containing ingredients and foods for infants and children

During the pre-weaning period, the diet is paramount to the health of the child, and is based mainly on cereal flours and baby milk formulae. Phytic acid is found in high concentrations in cereals. Phytate cannot be absorbed by humans, and phytic acid has an adverse effect on the bioavailability of important minerals. In addition, the phosphorus of the phytate is not nutritionally available. Therefore, the negative properties of phytates could have an effect on the health of the baby. Febles *et al.* (2001) studied the phytic acid content of 400 cereal products (gluten-containing as well as gluten-free) and found that the gluten-free products had significantly lower levels of phytic acid (average 3.3 mg/g) in comparison with the gluten-containing samples (average >20 mg/g).

Non-protein nitrogen has essential components and functions in infant nutrition. However, industrial processing conditions can impair the nutritional value and safety of foods (Man and Bada, 1987). Pérez-Conesa *et al.* (2002, 2005) examined the effects of industrial processing, i.e. heat followed by enzymatic treatment, on the non-protein nitrogen (NPN) levels of a range of wheat and gluten-free commercial infant cereals. Both heat and enzymatic treatments significantly modified the NPN of all cereals studied. In particular, the gluten-free cereals had lower NPN levels than gluten-containing cereals, and the complete absence of taurine in these cereals (another essential component for newborns) was most likely due to its transformation into ammonia during the heat treatment step.

Conclusion

As gluten-free flours are generally highly refined and not fortified, it is possible that people on a gluten-free diet are nutritionally imbalanced. It has been reported that nutritive gluten-free flours do exist, for example buckwheat or millet. These have high protein and mineral levels. However, such flours are not used extensively, with manufacturers most often opting for refined, unenriched gluten-free flours, which are low in dietary fiber, iron, and folate.

A healthy diet is essential for infants to promote optimal growth and development. As cereals (especially wheat) form the basis of foods for infants and young children, their replacement must be carefully selected so that the nutritive intake is not compromised. As well as ingredient selection, the processing of these ingredients also needs to be monitored, as studies have reported how high heat and enzymatic treatments can decrease some of the essential components of the flours.

Conclusions

Gluten replacement in gluten-free cereal products remains a significant technological challenge. To date, the use of starches and hydrocolloids represent a widespread approach used to mimic gluten in the manufacture of gluten-free bakery products. The soft flour in biscuit-making has successfully been replaced by a range of starches such as corn, soya, sorghum, and buckwheat. Gluten-free cakes have been produced by blending gums and hydrocolloids with flours derived from rice, cassava, and corn, and making gluten-free pasta products has been monitored from the initial selection of ingredients to the production process (where parameters such as starch pre-gelatinization or drying techniques have been introduced). However, the majority of gluten-free flours are highly refined and are of low nutritive value. Steps to enhance the "healthy" aspect of gluten-free products are gradually being addressed, most notably through the introduction of pseudocereals and other nutritive flours in gluten-free cereal products. This is especially important for the production of high-quality, healthy infant and baby foods, where growth and development is of utmost importance.

References

Abecassis, J., Faure, J., and Feillet, P. (1989). Improvement of cooking quality of maize pasta products by heat treatment. *J. Sci. Food Agric.* 47(4), 475–485.

Acs, E., Kovacs, Z., and Matuz, J. (1996). Bread from corn starch for dietetic purposes. I Structure formation. *Cereal Res. Commun.* 24, 441–449.

Akingbala, J. O., Adeyemi, I. A., Sangodoyin, S. O., and Oke, O. L. (1994). Evaluation of amaranth grains for ogi manufacture. *Plant Foods Hum. Nutr.* 46, 19–26.

Anon. (1997) Reduced fat in cookie products. *Food Market. Technol.* 11, 52–53.

Badi, S. M. and Hoseney, R. C. (1976). Use of sorghum and pearl millet flours in cookies. *Cereal Chem.* 53, 733–738.

Bean, M. M., Elliston-Hoops, E. A. and Nishita, K. D. (1983). Rice flour treatment for cake-baking applications. *Cereal Chem.* 60, 445–449.

Bhattacharya, S. and Choudhury, G. S. (1994). Twin extrusion of rice flour: effect of extruder length-to-diameter ratio and barrel temperature on extrusion parameters and product characteristics. *J. Food Process. Preserv.* 18, 389–406.

Borowski, J. and Pomianowski, J. F. (1994). Baked confectionery with addition of soy flour. *Przegl. Pierkarski Cukierniczy* 42, 11–12.

Camargo, C., Colonna, P., Buleon, A., and Richard-Molard, D. (1988). Functional properties of sour cassava (*Maiaaahot utilissima*) starch, polvilho azedo. *J. Sci. Food Agric*. 45, 273–289.

Caperuto, L.C., Amaya-Farfan, J., and Camargo, R.O. (2001). Performance of quinoa (Chenopodium quinoa Willd) flour in the manufacture of gluten-free spaghetti. *J. Sci. Food Agric*. 81, 95–101.

Cárdenas, O. S. and de Buckle, T. S. (1980). Sour cassava starch production: a preliminary study. *J. Food Sci*. 45509–45512.

Cauvain, S. P. (2003a). Cakes: Nature of Cakes. In: *Encyclopedia of Food Sciences and Nutrition*. London: Elsevier Science, pp. 751–756.

Cauvain, S. P. (2003b). Cakes: Methods of manufacture. In: *Encyclopedia of Food Sciences and Nutrition*. London: Elsevier Science, pp. 756–759.

Chauhan, S., Lindsay, D., Rey, M. E. C., and von Holy, A. (2001). Microbial ecology of muffins baked from cassava and other nonwheat flours. *Microbios* 105, 15–27.

Chen, Z., Sagis, L., Legger, A., Linssen, J. P. H., Schols, H. A., and Voragen, A. G. J. (2002). Evaluation of starch noodles made from three typical Chinese sweet-potato starches. *J. Food Sci*. 67, 3342–3347.

Contamine, A. S., Abecassis, J., Morel, M.-H., Vergnes, B., and Verel, A. (1995). Effect of mixing conditions on the quality of dough and biscuits. *Cereal Chem*. 72, 516–522.

Coppola, S., Pepe, O., and Mauriello, G. (1998). Effect of leavening microflora on pizza dough properties. *J. Appl. Microbiol*. 85, 891–897.

Cubadda, R. and Carcea, M. (2003). Pasta and macaroni: Methods of manufacture. In: *Encyclopedia of Food Sciences and Nutrition*. London: Elsevier Science, pp. 4374–4378.

Cubbeda, R. (1988). Evaluation of durum wheat, semolina, and pasta in Europe. In: Fabriani, G. and Lintas, C. eds. *Durum Chemistry and Technology*. St. Paul, MN: American Association of Cereal Chemists, pp. 217–235.

Dahlberg, J. A., Wilson, J. P., and Snyder, T. (2004). Sorghum and pearl millet: health foods and industrial products in developed countries. Alternative uses of sorghum and pearl millet in Asia. Proceedings of an expert meeting ICRISAT Patancheru Andhra Pradesh India CFC Technical Paper, pp. 42–59.

Demiante, I. M., Dupuy, N., Huvenne, J. P., Cereda, M. P., and Wosiacki, G. (2000). Relationship between baking behaviour of modified cassava starches and starch chemical structure determined by FTIR spectroscopy. *Carbohydr. Polym*. 42, 149–158.

DesRochers, J. L., Seitz, K. D., and Walker, C. E. (2003). Cakes: Chemistry of baking. In: *Encyclopedia of Food Sciences and Nutrition*. London: Elsevier Science, pp. 760–765.

Doescher, L. C., Hoseney, R. C., Millikan, G. A., and Rubenthaler, G. L. (1987). Effects of sugars and flours on cookie spread evaluated by time-lapse photography. *Cereal Chem*. 64, 163–167.

Febles, C. I., Arias, A., Hardisson, A., Rodriguez-Alvarez, C. and Sierra, A. (2001). Phytic acid level in infant flours. *Food Chem*. 74, 437–441.

Feillet, P. (1984). The biochemical basis of pasta cooking quality. Its consequences for durum wheat breeders. *Sci. Aliments* 4, 551–566.

Feillet, P. and Dexter, J. E. (1996). Quality requirements of durum wheat for semolina milling and pasta production. In: Kruger, J. E., Matsuo, R. B., and Dick, J. W. eds. *Pasta and Noodle Technology*. St. Paul, MN: American Association of Cereal Chemists, pp. 95–133.

Feillet, P. and Roulland, T. M. (1998). Caroubin: A gluten-like protein isolated from carob bean germ. *Cereal Chem.* 75, 488–492.

Formato, A. and Pepe, O. (2005). Pizza dough differentiation by principal component analysis of alveographic, microbiological and chemical parameters. *Cereal Chem.* 82, 356–360.

Gaines, C. S. (1990). Influence of chemical and physical modification of soft wheat protein on sugar snap cookie dough consistency, cookie size and hardness. *Cereal Chem.* 67, 73–77.

Gallagher, E., Gormley, T. R., and Arendt, E. K. (2004). Recent advances in the formulation of gluten-free cereal-based products. *Trends Food Sci. Technol.* 15, 143–152.

Glickman, M. (1991). Hydrocolloids and the search for the "oily grail." *Food Technol.* 45, 94–101.

Gobbetti, M., Rizzello, C. G., Di Cagno, R., and De Anglis, M. (2007). Sourdough lactobacilli and celiac disease. *Food Microbiol.* 24, 186–196.

Grant, L. A., Dick, J. W. and Shelton, D. R. (1993). Effects of drying temperature, starch damage, sprouting and additives on spaghetti quality characteristics. *Cereal Chem.* 70, 676–684.

Grehn, S., Fridell, K., Lilliecreutz, M., and Hallert, C. (2001). Dietary habits of Swedish adult coeliac patients treated by a gluten-free diet for 10 years. *Scand. J. Nutr.* 45, 178–182.

Hazelton, J. L., DesRochers, J. L., Walker, C. E., and Wrigley, C. (2004). Cookies, biscuits and crackers: Methods of manufacture. In: *Encyclopedia of Grain Science*. London: Elsevier, pp. 307–312.

Hoseney, R. C. (1994). Rheology of dough and batters. In: *Principles of Cereal Science and Technology*. St. Paul, MN: American Association of Cereal Chemists, pp. 213–228.

Hozová, B., Buchtová, V., Dodok, L., and Zemanovič, J. (1997). Microbiological, nutritional and sensory aspects of stored amaranth biscuits and amaranth crackers. *Nährung* 41, 155–158.

Huang, J. C., Knight, S., and Goad, C. (2001). Model prediction for sensory attributes of nongluten pasta. *J. Food Qual.* 24, 495–511.

Ibanoğlu, S., Ainsworth, P., Ozer, E. A., and Plunkett, A. (2006). Physical and sensory evaluation of a nutritionally balanced gluten-free extruded snack. *J. Food Eng.* 75(4), 469–472.

Ivarsson, A., Hornell, O., Stenlund, H., and Peterson, A. (2002). Breast-feeding protects against celiac disease. *Am. J. Clin. Nutr.* 75, 914–921.

Jud, B. (1993). Tara gum, a new galactomannan. *Food Ingred. Eur.* 342–346.

Kent, N. L. and Evers, A. D. (1994). Flour quality. In: Kent, N. L. and Evers, A. D. eds. *Kent's Technology of Cereals*. Oxford: Pergamon, pp. 170–189.

Kim, C. S. and Walker, C. E. (1992). Interactions between starches, sugars and emulsifiers in high ratio cake model systems. *Cereal Chem.* 69, 206–212.

Korponay-Szabo, I. R., Kovacs, J. B., Czinner, A., Goracs, G., Vamos, A., and Szabo, T. (1999). High prevalence of silent celiac disease in preschool children screened with IgA/IgG antiendomysium antibodies. *J. Pediatr. Gastrointest. Nutr.* 28, 26–30.

Kovacs, E. and Varga, J. (1995). Examination of macaroni dough quality on carbohydrate basis. *Tecnica Molitoria* 46, 1206–1211.

Kunachowicz, H., Nadolna, I., Klys, W., and Ruthowska, U. (1996). Evalutation of the nutritive value of some gluten-free products. *Zywienie Czlowieka i Metabolizm* 23, 99–109.

Larsen, D. M., Setser, C. S., and Faubion, J. M. (1993). Effects of flour type and dough retardation time on the sensory characteristics of pizza crust. *Cereal Chem.* 70, 647–650.

Levine, L. (2001). Extruder screw performance. *Cereal Foods World* 46, 169.

Limroongreungrat, K. and Huang, Y.-W. (2007). Pasta products made from sweet potato fortified with soy protein. *Lebensm. Wiss. Technol.* 40, 200–206.

Lovis, L. J. (2003) Alternatives to wheat flour in baked goods. *Cereal Foods World* 48, 61–63.

Maache-Rezzoung, Z., Bouvier, J. M., Allaf, K., and Patras, C. (1998). Study of mixing in connection with the other properties of biscuit dough and dimensional characteristics of biscuits. *J. Food Eng.* 35, 43–56.

Malick, S. K. and Sheikh, A. S. (1976). Composition and technology of biscuits. *Pakistan J. Sci. Res.* 28, 90–94.

Man, E. H. and Bada, J. L. (1987). Dietary d-amino acids. *Annu. Rev. Nutr.* 7, 209–225.

Manley, D. J. R. (1983). Short dough biscuits. In: *Technology of Biscuits*, *Cookies and Crackers*. Chichester, UK: Ellis Horwood, 204–214.

Manthey, F. A. and Schorno, A. L. (2002). Physical cooking quality of spaghetti made from whole wheat durum. *Cereal Chem.* 79, 504–510.

Manthey, F. A. Saujanya, R. Y., Dick, T. J., and Badaruddin, M. (2004). Extrusion properties and cooking quality of spaghetti containing buckwheat bran flour. *Cereal Chem.* 81, 232–236.

Marcilio, R., Amaya-Farfan, J., da Silva, M. A. A. P., and Spehar, C. R. (2005). Evaluation of amaranth flour for the manufacture of gluten-free biscuits. *Braz. J. Food Technol.* 8, 175–181.

Marconi, E. and Carcea, M. (2001). Pasta from non-traditional raw materials. *Cereal Foods World* 46, 522–530.

Mariani, P., Grazia, V. M., Montouri, M. *et al.* (1998). The gluten-free diet: a nutritional risk factor for adolescents with celiac disease. *J. Pediatr. Gastroenterol. Nutr.* 27, 519–523.

Mat Hashim, D. B., Morthy, S. N., Mitchell, J. R., Hill, S. E., Linfoot, K. J., and Blanshard, J. M. V. (1992). The effect of low level of antioxidants on the swelling and solubility of cassava starch. *Starch/Stärke* 44, 471–475.

Mestres, C., Colonna, P., and Buleon, A. (1988). Characteristics of starch networks within rice flour noodles and mungbean starch vermicelli. *J. Food Sci.* 53, 1809–1812.

Millar, R. A. and Hoseney, R. C. (1997). Factors in hard wheat flour responsible for reduced cookie spread. *Cereal Chem.* 74, 330–336.

Molina, M. R., Mayorga, I., Lachance, P. A., and Bressani, R. (1975). Production of high-protein quality pasta products using a semolina-corn-soy flour mixture. I. Influence of thermal processing of corn flour on pasta quality. *Cereal Chem.* 52, 240–247.

Morad, M. M., Doherty, C. A., and Rooney, L. W. (1984). Effect of sorghum variety on baking properties of US conventional bread, Egyptian pita "Balady" bread and cookies. *J. Food Sci.* 49, 1070–1074.

O'Brien, C. M., von Lehmden, S., and Arendt, E. K. (2002a). Development of gluten free pizzas. *Irish J. Agric. Food Res.* 42, 134.

O'Brien, C. M., Schober, T., and Arendt, E. K. (2002b). Evaluation of the effect of different ingredients on the rheological properties of gluten-free pizza doughs. *Proceedings: American Association of Cereal Chemists International Annual Meeting.*

Olewnik, M. C. and Kulp, K. (1984). The effect of mixing time and ingredient variation on farinograms of cookie doughs. *Cereal Chem.* 61, 532–537.

Ortiz, D. E. (2004). Cakes, pastries, muffins and bagels. In: *Encyclopedia of Grain Science,* pp. 134–140.

Paterson, L. A., Mat Hashim, D. B., Hill, S. E., Mitchell, J. R., and Blanshard, J. M. V. (1994). The effects of low level of sulphite on the swelling and solubility of starches. *Starch/Stärke* 46, 288–291.

Pérez-Conesa, D., Ros, G., and Periago, M. J. (2002). Protein nutritional quality of infant cereals during processing. *J. Cereal Sci.* 36, 125–133.

Pérez-Conesa, D., Periago, M. J., Ros, G., and López, G. (2005). Non-protein nitrogen in infant cereals affected by industrial processing. *Food Chem.* 90, 513–521.

Räihä, N. C. R. and Axelsson, I. E. (1995). Protein nutrition during infancy. An update. *Pediatr. Nutr.* 42, 745–764.

Sahi, S. S. (1994). Interfacial properties of the aqueous phases of wheat flour doughs. *J. Cereal Sci.* 20, 119–127.

Samiyi, M. and Ashraf, H. R. L. (1993). Iranian breads supplemented with amaranth flour. *International Journal of Food Science and Technology.* 28(6), 625–628.

Schober, T. J., O'Brien, C. M., McCarthy, D., Darnedde, A., and Arendt, E. K. (2003). Influence of gluten-free flour mixes and fat powders on the quality of gluten-free biscuits. *Eur. Food Res. Technol.* 216, 369–376.

Schoenlechner, R., Jurackova, K., and Berghofer, E. (2004). Pasta production from the pseudocereals amaranth, quinoa and buckwheat. *Proceedings: 12th ICC Cereal and Bread Congress, Harrogate, UK.*

Schoenlechner, R., Linsberger, G., Kaczyk, L. and Berghofer, E. (2006). Production of short dough biscuits from the pseudocereals amaranth, quinoa and buckwheat with common bean. *Ernährung* 30, 101–107.

Sissons, M. (2004). Pasta. In: *Encyclopedia of Grain Science.* London: Elsevier, pp. 409–418.

Sun, D.-W. and Brosnan, T. (2003). Pizza quality evaluation using computer vision—part 1. Pizza base and sauce spread. *J. Food Eng.* 57, 81–89.

Taylor, J. R. N., Schober, T. J., and Bean, S. R. (2006). Novel food and non-food uses for sorghum and millets. *J. Cereal Sci.* 44, 252–271.

Thompson, T. (1999). Thiamin, riboflavin and niacincontents of the gluten-free diet: Is there cause for concern? *J. Am. Diet. Assoc.* 99, 858–862.

Thompson, T. (2000). Folate, iron and dietary fiber contents of the gluten-free diet. *J. Am. Diet. Assoc.* 100, 1389–1396.

Tosi, E. A., Ciappini, M. C., and Masciarelli, R. (1996). Utilisation of whole amaranthus (*Amaranthus cruentus*) flour in the manufacture of biscuits for coeliacs. *Alimentaria* 34, 49–51.

Turabi, E., Sumnu, G., and Sahin, S. (2007). Rheological properties and quality of rice cakes formulated with different gums and an emulsifier blend. *Food Hydrocolloids* (in press).

Venkateswar, R. G. and Indrani, D. (1989). Studies on the use of artificial sweeteners in sweet bread and biscuits. *J. Food Sci. Technol. India* 26, 142–144.

Von Atzingen, M. C. and Machado Pinta e Silva, M. E. (2005). Evaluation of texture and colour of starches and flours in preparations without gluten. *Cienca y Technol. Aliment.* 4, 319–323.

Wang, N., Bhirud, P. R., Sosulski, F. W., and Tyler, R. T. (1999). Pasta-like product from pea flour by twin-screw extrusion. *J. Food Sci.* 64, 671–678.

Weegels, P. L. and Hamer, R. J. (1989). Predicting the baking quality of gluten. *Cereal Foods World* 34, 210–212.

Ziegler, A. G., Schmid, S., and Huber, D. (2003). Early infant feeding and risk of developing type I diabetes-associated autoantibodies. *J. Am. Med. Assoc.* 290, 1721–1728.

Zydenbos, S., Humphrey-Taylor, V., and Wrigley, C. (2004). Cookies, biscuits and crackers: The diversity of products. In: *Encyclopedia of Grain Science*. London: Elsevier, pp. 313–317.

Malting and brewing with gluten-free cereals

15

Blaise P. Nic Phiarais and Elke K. Arendt

Introduction

History of brewing

The art and craft of beer-making can be traced back almost 5000 years, as documented by excavations undertaken in many parts of the world. Depictions of the ancient methods of brewing are to be found among the tomb paintings in Ancient Egypt and mentions of beer are contained in Mesopotamian writings from 2800 BC (Arnold, 1911). Beer, together with bread, was the most important ingredient in the diet of the ancient cultures. In addition to being a foodstuff, beer played a central role in religious belief and ritual practice (Corran, 1975). It was a well-loved drink of the Scythians, the Celts, and the Germanic tribes, where it was brewed as daily household food by the women, as baking and brewing were women's work in all cultures. Changes to the brewing "industry" occurred in the breweries of Christian religious foundations, where beer was not only for their own consumption, but supplied to others for payment (Arnold, 1911; Kunze, 1996a). It subsequently became an occupation for men, as it tends to be today (Bamforth, 2003).

Historically, it is reported that beer was produced from barley and since the introduction of the beer purity law in 1516, barley has been traditionally used as the main ingredient of beer (Arnold, 1911; Rich, 1974). It was once believed that beer could not be produced without barley; however it is well documented that opaque beers made from cereals like sorghum (Owuama, 1997, 1999; Igyor *et al.*, 2001; Goode *et al.*, 2003; Nso *et al.*, 2003), millet (Eneje *et al.*, 2001; Agu, 1991, 1995) and maize (Ilori *et al.*, 1991; Lanares, 1992; Shephard *et al.*, 2005) have the potential to be alternative substrates for conventional beer brewing in the tropics. To facilitate consumer requirements, other cereals (rice and maize) and pseudocereals

Gluten-Free Cereal Products and Beverages
ISBN: 9780123737397

(buckwheat, quinoa, and amaranth) have been investigated as brewing ingredients because of the absence of gluten and the presence of compounds that are claimed to have positive effects on health (Zarnkow *et al.*, 2005; Kreisz *et al.*, 2005).

Overview of the malting and brewing process

The malting process

The purpose of malting is to produce enzymes in the grain kernel and to cause defined changes in its chemical constituents (Kunze, 1996b). The malting process involves the cleaning and grading of stocks of barley, steeping the grain in water, germinating the grain and finally drying and curing it on the kiln (Figure 15.1).

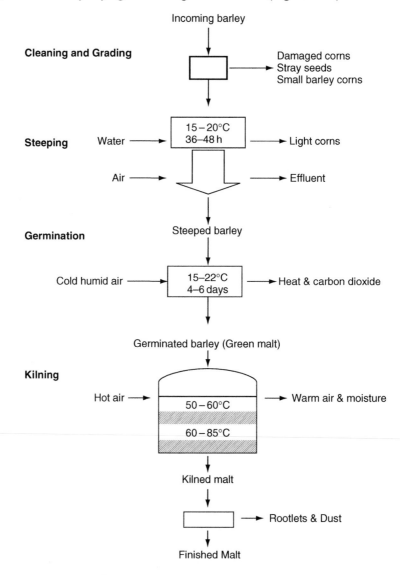

Figure 15.1 Flow diagram of the malting process.

During steeping, the grain absorbs water and increases in volume. After 4–6 hours, the first steep water is drained off and the grain is intensively aerated to allow for the removal of CO_2. Steeping is ceased when the moisture content of the grain reaches 43–46% (Briggs *et al.*, 1981c).

During germination, the kernel develops rootlets and an acrospire in the presence of sufficient water. During this growth phase, starch-degrading, cytolytic, and proteolytic enzymes are formed and activated. These enzymes are essential for the break down of large molecules such as starch, proteins, and β-glucans occurring during mashing. As a result of the intensive respiration, the steeped grain must be provided with sufficient aeration to allow for cooling and the removal of CO_2. Germination is generally terminated when the acrospire length is equal to approximately two-thirds to three-quarters of the corn length and the grain is known as green malt (Kunze, 1996b).

During kilning water is removed from the green malt. Malt is kilned to produce a friable, stable-on-storage product, from which roots can be easily removed. Kilning consists of passing a flow of warm dry air through a bed of malt at various rates and at increasing temperatures to dry the malted grain. The survival of enzymes in malt is greatly influenced by the temperature and time of the kilning regime (Briggs *et al.*, 1981d).

The brewing process

The two most important processes in beer production are the degradation of starch to sugar during mashing followed by the fermentation of these sugars to form alcohol and CO_2 (Kunze, 1996d). Brewing in its simplest form involves seven steps (Figure 15.2):

1. Crushing malted barley to form very coarse flour (i.e. grist) in the grist mill.
2. In the mash tun, malt grist is mixed (mashed) with warm water to form a porridge-like, viscous mash. Malt enzymes, which were produced during malting, are encouraged to solubilize the degraded endosperm of the ground malt at their optimum temperatures to give as much soluble extract as possible.
3. In the lautertun, the soluble extract in the wort is separated from the insoluble spent solids (grain husk). Furthermore water is sprayed from the top of the tank onto the mash to increase extract.
4. The wort is then boiled in the wort kettle with hops. This halts enzyme action, sterilizes the wort, coagulates some proteins and imparts distinctive flavors and aromas to the wort from the hops.
5. The hot wort is separated from the precipitated particles (i.e. trub) in a whirlpool.
6. Cooling and aeration of the wort follows, so that an ideal medium for yeast fermentation is produced (Briggs *et al.*, 1981b; Hough, 1985b; Kunze, 1996d).
7. Yeasts are then added to ferment the wort and the carbohydrates present are converted into alcohol and CO_2. Other yeast metabolites contribute to flavor and aroma.

Maturation and clarification of the beer follows, during which the flavor, aroma, and keeping qualities of the beer are modified. Finally the beer is packaged, usually after it has been sterile-filtrated or pasteurized (Hough, 1985a; Kunze, 1996c).

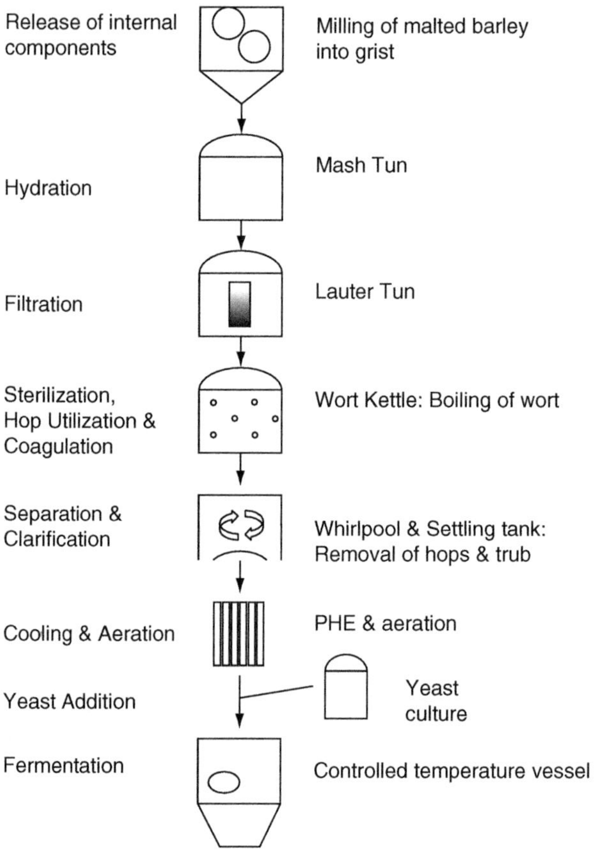

Figure 15.2 Flow diagram of the brewing process.

Gluten-free cereals

Celiac disease prevalence has been estimated to be 1 in about 100 people worldwide (Hamer, 2005; Sollid and Khosla, 2005). Such a rate establishes celiac disease as one of the most common food intolerances known. This disease is caused by an immune-mediated response in the small intestine triggered by the ingestion of gluten in genetically-susceptible individuals (Fasano and Catassi, 2001). The only effective treatment is a strict adherence to a diet that avoids ingestion of cereals (wheat, spelt, triticale, rye, and barley) that contain gluten and their products throughout the patient's lifetime (Ellis et al., 1990).

Gluten is the general term used to describe the protein fraction in wheat. Gluten proteins can be divided into two main fractions according to their solubility in aqueous alcohols: the soluble gliadins and the insoluble glutenins, collectively known as the prolamins (Lewis, 2005). Wheat, rye, and barley are all members of the grass family (Poaceae) and are taxonomically closely related. All these cereals and their prolamins [gliadins (wheat), hordein (barley), secalin (rye) and possibly avenin (oats)] are toxic to people with celiac disease (Kasarda, 2001).

Cereals not containing gluten include: rice (*Oryza sativa*), maize (*Zea mais*), sorghum (*Sorghum bicolor*), and millets (e.g. *Panicum miliaceum*, *Setaria italica*, *Pennisetum typhoideum* and *Eleusine coracana*). Other carbohydrate-rich pseudocereals without gluten are buckwheat (*Fagopyrum esculentum*), quinoa (*Chenopodium quinoa*), and amaranth (*Amaranthus*) (Zarnkow *et al.*, 2005).

Although the consumption of oat products in the gluten-free diet is discouraged in the United States and in many other countries, recent studies have shown that patients with celiac disease can consume moderate amounts of uncontaminated oats without harmful effects on the intestinal mucosa, even after long-term use (Janatuinen *et al.*, 1995; Srinivasan *et al.*, 1996; Hoffenberg *et al.*, 2000; Thompson, 2001; Peraaho *et al.*, 2004). The Finnish Coeliac Association (FCA) includes oats in the gluten-free diet (Kanerva *et al.*, 2003). In 1981 and 2000, revised drafts of the Codex standard for gluten-free foods by the World Health Organization (WHO) and the Food and Agricultural Organization (FAO) stated that so-called gluten-free foods are described as: (i) consisting of, or made from ingredients which do not contain any prolamins from oats with a gluten level not exceeding 20 ppm and (ii) consisting of oats, which has been rendered gluten-free, with a gluten level not exceeding 200 ppm (Gallagher *et al.*, 2004). Nonetheless, many practitioners and organizations such as the Celiac Sprue Association (CSA), Celiac Disease Foundation (CDF), and the American Dietetic Association (ADA) are reluctant to change their views on oats in the absence of additional evidence on their safety (Thompson, 2000).

One of the most investigated gluten-free cereals is sorghum, which was originally used not to produce beer for people with celiac disease, but to overcome the 1988 ban on importation of barley malt into Nigeria. Despite some inherent problems associated with sorghum (low amylolytic enzyme levels, low extract yields, and slow mash filtration) (Aisen, 1988; Dale, 1990), several studies have investigated the possibility of substituting barley malt with sorghum malt (Okafor and Aniche, 1980; Goode, 2001; Odiboa *et al.*, 2002; Goode *et al.*, 2003). Recently, the amount of research carried out in the area of gluten-free products has increased significantly (Gallagher *et al.*, 2004).

On the question of whether buckwheat, quinoa, and amaranth can be included in a gluten-free diet, both the Gluten Intolerance Group (GIG) (Seattle, Washington) and the CDF (Studio City, California) consider these plant foods acceptable, while the CSA (Omaha, Nebraska) lists them as unacceptable (Shewry, 2002). However, the CSA does not claim that these grains actually do contain gluten (Thompson, 2001). Hence, current evidence more strongly supports the conclusion that these grains can be used for the production of gluten-free product.

Recent studies have focused on the production of malt and beer from gluten-free cereals such as rice, maize, millet, and pseudocereals such as buckwheat, quinoa, and amaranth (Bauer *et al.*, 2005; Nic Phiarais *et al.*, 2005, 2006b; Wijngaard and Arendt, 2006; Wijngaard *et al.*, 2006). These studies independently came to the conclusion that the raw material with the most prevalent potential for brewing appeared to be buckwheat. For this reason, buckwheat malting and brewing is covered in more detail.

Research on malt and beer based on gluten-free raw materials has focused on sorghum and the objective of this review is to focus on the use of gluten-free cereals such as rice, maize, and millet, as well as pseudocereals such as buckwheat, quinoa, and amaranth as alternatives to sorghum.

Malting of gluten-free cereals

Cereals

Sorghum

Since barley cultivation is not feasible in tropical areas, beer production in Africa requires the costly import of barley malt from temperate regions (Dufour *et al.*, 1992). As a result of the 1988 ban on importation of barley malt into Nigeria, sorghum and maize malts have been produced to brew traditional African beverages (Igyor *et al.*, 2001; Okungbowa *et al.*, 2002; Nso *et al.*, 2003; Ogbonna *et al.*, 2004). Sorghum has shown a great potential as a substitute for barley malt (Aisen, 1988; Ilori *et al.*, 1991). However, breweries who have had experience with the use of sorghum as an adjunct in commercial beer production encountered the following problems: slow and incomplete saccharification of mash, poor wort separation, and poor bright beer filtration (Aisen, 1988). Nonetheless, while not claiming the same effect on the palate as barley, 100% sorghum malt beer is consistent in quality and acceptable to a large proportion of beer drinkers in Nigeria and other parts of the world (Aisen and Muts, 1987).

The malting process increases the relative nutritional value of sorghum, and malted sorghum has been used to make low-viscosity grists for feeding children at weaning (Briggs, 1998a). However sorghum malt had been primarily produced for the brewing of Kaffir beer and similar products (Aisen and Muts, 1987). One serious problem that usually arises in experimental studies of brewing with sorghum malt is the insufficient enzyme levels. Early studies reporting little or no β-amylase for saccharification in sorghum malt led to the incorrect conclusion that sorghum malt was unsuitable for brewing lager-type beers (Kneen, 1945, 1944). However, several workers argued that the low detection of amylolytic enzymes was due to the use of the enzyme assay for barley which is unsuitable for research studies on sorghum (Novellie, 1962; Okon and Uwaifo, 1984). In support of this view, Taylor and Robbins (1993) reported that when malted, sorghum had β-amylase activity of less than 25% of the level in barley malt when different methods were used for the determination of diastatic power. The authors concluded that β-amylase in sorghum is in the active soluble form, unlike in barley where almost all of the β-amylase is in a bound inactive form. Therefore it can be assumed that the level of active β-amylase is similar in both sorghum and barley.

Detailed studies have been carried out to optimize the malting procedure for sorghum (Agu and Palmer, 1997b; Obeta *et al.*, 2000; Ogbonna *et al.*, 2004). To date, an optimized malting procedure for sorghum involves steeping for 8 hours at 20–25°C, followed by a 2 hour air-rest and a further 14 hour wet-steep, with an out of steep moisture of 34–36% and germination at the same temperatures for 120 hours.

This is then followed by kilning for 24 hours at 50°C (Agu and Palmer, 1999; Igyor *et al.*, 2001). Sorghum, raw and malted, is now widely used in European-type lager beer brewing in many developing countries of the tropics due to economic difficulty. In Nigeria, there is already a total replacement of imported barley malts with sorghum which is locally produced (Okolo, 1996).

Rice

Rice is widely used for the production of alcoholic beverages. Alcoholic beverages such as sake in Japan, *shaoshinshu* in China, and miscellaneous alcoholic drinks in south-eastern Asia are produced with rice as the main ingredient, often as the sole cereal source (Yoshizawa and Kishi, 1985). In addition, rice is used as an adjunct in the production of alcoholic beverages such as beer (Coors, 1976). Beer and sake are the most popular rice-based alcoholic beverages and are produced in huge amounts (approximately 9×10^{10} L of beer and 15×10^8 L of sake, annually). As a brewing adjunct, rice has a very neutral flavor and aroma, and when properly converted in the brewhouse yields a light, clean tasting beer (Canales, 1979). The more recent interest in malting rice has been triggered by its possible use in foodstuffs, due to its increased free sugar and amylase content and decreased viscosity after germination (Malleshi and Desikachar, 1986b), and by the need in some parts of Africa to find alternatives to imported barley. In addition, germination has been shown to increase the level of nutrients present in rice (Capanzana and Buckle, 1997).

Early studies suggest that beer can be brewed from malted rice (Malleshi and Desikachar, 1986b; Okafor and Iwouno, 1990; Aniche and Palmer, 1992). In these trials, the Californian kernel paddy rice was used. The grain was steeped for 48–60 hours at 15°C, germinated for 72 h at 17.8–18.9°C and finally kilned for 48 hours, with temperatures rising from 32.2 to 65.5°C. These low temperatures were used to prevent vitrification of the grain. However, the product was poor in enzymes, expensive to produce, and the endosperm material was poorly modified. Nonetheless, in contrast to unmalted rice, liquefaction in the cooker was rapid and occurred at a comparatively low temperature. In addition, the high extract yield is an advantage and the flavors and aromas of products made using this material are excellent (Briggs, 1998a).

Okafor and Iwouno (1990) report that malted rice is not satisfactory as a brewing material as malting losses are often quite high. When the rice malts are finely ground and mashed in the conventional way, saccharification is incomplete and wort run-off is slow. In addition, rice malt was found to be very bitter (Malleshi and Desikachar, 1986b). The incongruence of the studies on rice indicates that there is a strong need to optimize the mashing regimes in order to increase the extract recoveries and improve flavor.

Maize

Maize has been used to make the traditional Tesuino maize beer of the Tarahumara Indians in Mexico (Lanares, 1992). In addition, the production and consumption of home-brewed Xhosa maize beer is a widespread traditional practice in South Africa

(Shephard *et al.*, 2005). As a brewing adjunct, maize must be de-germed to limit beer foam damaging effects.

Maize is considered to malt less well than sorghum and most of the reported trials have given unsatisfactory results from the brewers' viewpoint (Briggs, 1998a). To our knowledge little is known about the malting characteristics of this cereal. Wang and Fields (1978) noted that germinating maize can, in some respects, enhance its nutritive value. However, since the gelatinization temperatures of maize starches are reported to be high it is likely that, as with sorghum malts, improved extract recovery would be obtained if modified mashing regimens were used (Hough, 1985b). Singh and Bains (1984) recommends a malting regime whereby the grains are pre-dried for 12 hours at 36°C before steeping to 40% moisture at 25°C, germinated for 168 hours at 25°C and finally kilned for 24 hours at 45°C. Therefore, as with sorghum, maize needs to be malted "wet and warm," resulting in the extreme likelihood of mold infestation development. Green malts should be dried at low temperatures to favor enzyme survival. Increased extract values, enzyme levels, Kolbach indices, and malting losses can be obtained. Moreover, the addition of gibberellic acid elevates α-amylase, protease, extract, and Kolbach index values (Singh and Bains, 1984; Malleshi and Desikachar, 1986a). Currently, there is no apparent effort being made to select maize varieties for superior malting qualities.

Millet

The millets are a heterogeneous group of small-grained "tropical" cereals. They have been malted for use in foodstuffs, in opaque beers and, at least experimentally, clear beers. There are considerable differences in malting qualities between the different species of millets and between individual varieties of one species (Briggs, 1998b). Earlier studies have shown that malting and brewing, and hence the production of a lager beer from millet, is possible, however extensive work is still needed to improve the flavor and color of the beer (Eneje *et al.*, 2001; Pelembe *et al.*, 2002).

In southern Africa, pearl millet is traditionally processed by malting and fermentation. Malted pearl millet is used to make weaning foods for infants with reduced viscosity (Pelembe *et al.*, 2002). Unlike barley and sorghum, little is known about the technology of millet malting. An optimal malting procedure for pearl millet, which involves steeping at 25°C, with a cycle of 2 hours wet and 2 hours air-rest for a total of 8 hours, germination at 25–30°C for 72–96 hours and finally a kilning regime at 50°C for 24 hours has been suggested (Pelembe *et al.*, 2002). These conditions resulted in high diastatic power, α- and β-amylase activities, good free amino nitrogen and moderate malting loss. In some instances, additions of gibberellic acid enhanced extract yields and causes enzyme activities to peak sooner, resulting in higher α-amylase activities (Agu and Okeke, 1991).

The limited work on millet malting has been mainly carried out on finger millet (Chandrasekhara and Swaminathan, 1953; Nout and Davies, 1982; Malleshi and Desikachar, 1986a, 1986b; Nirmala and Muralikrishna, 2003). Finger millet makes very good quality malt, which is used in brewing traditional African opaque beer and making digestible liquid foods. Finger millet malt is reported to have a highly agreeable flavor and an acceptable taste, although during short periods of storage, these malts begin to develop bitter flavors (Malleshi and Desikachar, 1986a). Unmalted finger millet grain possesses very little amylase, protease, and phosphatase activities, but the activity of these enzymes increases considerably when the grain is germinated (Chandrasekhara and Swaminathan, 1953). In addition, finger millet is a rich source of calcium and dietary fiber. Malleshi and Desikachar (1986b) recommend a malting regime whereby the grains are steeped for 24 hours at 25°C, germinated for 96 hours at 25°C, and finally kilned for 24 hours at 45°C. Studies performed so far indicate that pearl and finger millet malt has the potential for replacing barley malt in brewing lager-type beer.

Oats

Food uses of oat have traditionally been restricted mainly to oatmeal, oat flakes, and breakfast cereals (Wilhelmson *et al.*, 2001). Native oat is known as a health-promoting cereal with a high content of soluble dietary fiber, beneficial fatty acids, and a large selection of vitamins, minerals, sterols, and antioxidants (Peterson, 2001). Moreover oat has gained increased interest among consumers due to Food and Drug Administration (FDA)-approved health claims indicating that soluble fiber from oat-meal, as part of a low saturated fat, low-cholesterol diet, may reduce the risk of heart disease (Anderson and Chen, 1986). Oat is usually associated with a pleasant nutty and grainy aroma, and germination can be used to improve the flavor and sensory characteristics (Heydanek and McGorrin, 1986), along with mineral bioavailability (Larsson and Sandberg, 1995). Currently, oat malts are rarely used and consequently there are few data available (Larsson and Sandberg, 1995; Peterson, 1998, 2001; Wilhelmson *et al.*, 2001). Oat malt is a minor product when compared with barley malt. Malted oats were used sporadically by European brewers for many centuries, but their use now is infrequent. Sometimes oat malt is used for ale and stout brewing in the UK and for special food ingredients (Briggs *et al.*, 1981a; Little, 1994).

Avena sativa and *Avena byzantina* are the two most widely grown oat species (Schrickel, 1986) but only a small proportion is used in malting and brewing (Little, 1994). The main oat grain species used for malting is *A. sativa* (Wilhelmson *et al.*, 2001) and, for brewing *A. gramineae* (Little, 1994). Research has clearly established that starch is slowly hydrolyzed during germination, and the starch content of germi-nated oats is the same as or slightly lower than that of ungerminated oats (Peterson, 1998). In addition, germination of oats leads to an increase in essential amino acids (in particular, lysine and tryptophan) content, and a slight decrease in the prolamine content (Dalby and Tsai, 1976).

Oats are noted among the cereals for their high mineral content (Anderson and Chen, 1986). However recent studies in humans using a radioisotope technique, demonstrated a low degree of iron and zinc uptake from breakfast meals containing oat porridge and oat bran bread (Sandstrom *et al.*, 1987; Rossander-Hulten *et al.*, 1990). The high phytate content of oats, together with a low phytase activity, is considered to be the explanation for the low iron absorption (Zhou and Erdman, 1995). Hence, to obtain a better mineral availability, researchers have investigated appropriate con-ditions for phytate degradation during processing and found that germination can be used to lower the phytate content of oat (Larsson and Sandberg, 1995). The optimal malting conditions for oats are as follows: steeping for 16 hours at 16°C, germination for 144 hours at 16°C, followed by kilning for 22 hours at 49–85°C (Peterson, 1998). With the basic knowledge of hydrolysis of macromolecules, controlled utilization of amylolytic, glucanolytic, and proteolytic activities of germinating grains, a lager-type beer from oat malt could be achieved.

Pseudocereals

Buckwheat

Buckwheat is an alternative crop belonging, unlike major cereals, to the Polygonaceae family. Buckwheat is not related to wheat and its name is probably based on its triangular seeds, which resemble the much larger seed of the beechnut, and the fact that it is used like wheat (Biacs et al., 2002). The buckwheat grain is highly nutritious, being a rich source of protein. The nutritive value of buckwheat is superior to millet or even cereals such as rice and wheat (Marshall and Pomeranz, 1982). Since its constituents are favorable from a nutrition–biological point of view, it can be fitted into a health-preserving diet. A constant consumption of buckwheat can prevent some "nutrition-born civilization diseases" (indigestion, obesity, constipation, cholesterol, obesity, diabetes, hypertension, etc.) (Qian and Kuhn, 1999b; Li and Zhang, 2001; Prestamo et al., 2003; Sun and Ho, 2005).

Buckwheat can be obtained either hulled or unhulled (Biacs et al., 2002), but recent studies have demonstrated that the use of unhulled buckwheat is advantageous over hulled material, since the water uptake is slower and the resulting malt is improved (Wijngaard et al., 2005b). In addition to lowering malting loss, another advantage of using unhulled material is improved filterability. Investigation of the impact of steeping time and temperature on the quality of buckwheat malt has revealed that the optimal moisture content at the end of steeping is 35–40% and the recommended steeping time is 7 to 13 hours at a temperature of 10°C (Wijngaard et al., 2005a, 2005c). At these moisture levels the malting loss falls within an acceptable range and malt quality is optimized. Optimal enzymatic activity in buckwheat malt can be obtained when buckwheat is germinated for 96 hours at 15°C (Wijngaard et al., 2005b, 2006). At this time, the grains are sufficiently modified and nutrients have not yet been exhausted. Moreover, rutin, a polyphenol with functional properties, is increased significantly during malting.

Several optimal conditions have been recently proposed for buckwheat malting. Response surface methodology (RSM) has been used to optimize the malting conditions of a wide range of gluten-free cereals including buckwheat using unhulled buckwheat (Zarnkow et al., 2005). The authors recommended a steeping time of 96 hours, degree of steeping 47%, and a germination time of 120 hours at 19°C. These malting conditions varied significantly to those reported by Wijngaard et al. (2005b, 2006) and Nic Phiarais et al. (2006b), where a steeping time of only 10 hours, a degree of steeping between 35 and 40% and a germination time of 96 hours at 15°C is recommended. Bauer et al. (2005) performed extensive flavor analysis on different gluten-free malts and found that buckwheat crystal malt has a striking toffee, malty, and nutty aroma and has the potential to be used as a brewing ingredient for ale production. A study by Nic Phiarais *et al.* (2005) investigated the impact of kilning on the enzymatic activity in buckwheat malt. Results indicated that after prolonged kilning at 40°C, inactivation of hydrolytic enzymes occurred. A multistage drying process has therefore been suggested, since the enzymatic activities during the kilning process appears to be dependent on the temperature at which kilning is performed. For the optimization of enzyme levels in buckwheat, buckwheat malt should be kilned for 5 hours at 40°C, 3 hours at 50°C, followed by 3 hours at 60°C (Nic Phiarais *et al.*, 2006b). Under these conditions, the highest levels of amylolytic enzymes and of total soluble nitrogen and free amino nitrogen are observed. Results collected so far strongly suggest that buckwheat, when optimally malted, shows potential as a health-preserving, gluten-free alternative to sorghum malt for brewing purposes.

Quinoa

Quinoa grain, a pseudocereal of the Chenopodiaceae family is appreciated for its significant contribution to good nutrition and as a weaning food (Caperuto *et al.*, 2000). Quinoa is said to have a high content of lysine and methionine (Mahoney *et al.*, 1975). Much research has been carried out worldwide on the agricultural aspects of quinoa (Sigstad and Garcia, 2001), but little has been done on a physiological level or its malting and brewing potential. Quinoa seeds have the advantage of fast germination *in vitro*, although they germinate very poorly in soil (Aufhammer *et al.*, 1996). When malted for 36 hours, the α-amylase activity of quinoa increased 4-fold (Atwell *et al.*, 1988), however the starch granules of the perisperm appear not to be extensively degraded by amylase during germination (Varriano-Marston and De Francischi, 1984). While this is not advantageous for the malting process, it may provide some benefit for mashing and brewing. If the level of α-amylase is such that it is not exhausted during malting, the residual enzyme will be available for the mashing process. Kunze (1996d) advocates that without the presence of α-amylase during mashing, saccharification will be incomplete and low extract wort will result. Using RSM, Zarnkow *et al.* (2005) optimized the malting conditions of quinoa as follows: steeping time of 36 hours, degree of steeping 54%, and a germination temperature of 8°C for a time of 144 hours.

Malting quinoa grain also improves nutrient availability. During germination phytate is reduced by 35–39%, whilst iron solubility under physiological conditions (and *in vitro* estimation of iron availability) increases 2- to 4-fold (Valencia *et al.*, 1999). Since customers are looking for innovative products and additional health benefits while consuming food or drinks, the nutritive properties of quinoa make attempts to investigate this raw material for malting and brewing purposes worthwhile.

Amaranth

Amaranth is a species of the Amaranthaceae family, found mostly in subtropical and tropical regions (Berghofer and Schoenlechner, 2002). The plant, which has the potential as a source of dietary nutrients, is eaten as a vegetable, and its grains are used as cereals (Irving and Becker, 1985). There is little information currently available in the literature on amaranth as a brewing material.

The germination or malting of seeds is commonly used to hydrolyze partially (pre-digest) substances, synthesize desired substances, and break down undesired substances by utilizing the endogenous enzymes in the raw material (Kunze, 1996b). Paredes-Lopez and Mora-Escobedo (1989) first described in detail the use of this technology for amaranth. After a 10-minute soaking time, the amaranth seeds were left to germinate at 35°C for 72 hours. Crude protein and true protein were found to increase and fat content to decrease. After 48 hours of germination lysine did not change and after 72 hours a slight decrease in lysine was observed along with increased protein digestibility. In contrast, Balasubramanian and Sadasivam (1989) soaked amaranth seeds for 12 hours and germinated them for up to 192 hours. Under these conditions, a decrease in protein between 48 and 182 hours germination and an increase in lysine of 31% after 24 hours of germination was observed (Paredes-Lopez and Mora-Escobedo, 1989). These malting conditions varied significantly from those reported by Zarnkow et al. (2005), who used RSM to optimize the malting conditions of amaranth for brewing purposes, and recommended a steeping time of 36 hours, degree of steeping 54% and a germination temperature of 8°C for a time of 168 hours. Therefore for malting and brewing purposes, conditions reported by Zarnkow et al. (2005) should be used for malting amaranth.

On account of its excellent nutritional composition before and after germination, amaranth presents an interesting alternative to sorghum as a gluten-free raw material (Colmenares De Ruiz and Bressani, 1990). However more research is still required to evaluate the use of amaranth as a malting and brewing ingredient.

Brewing of gluten-free cereals

Cereals

Sorghum

The potential of sorghum as an alternative substrate for lager beer brewing has been recognized for over five decades and several comprehensive reviews on brewing lager beer from sorghum and in particular from malted sorghum have appeared recently (Owuama, 1997; Agu and Palmer, 1998c; Owuama, 1999). Optimization of conditions for mashing and fermentation are necessary for the production of acceptable sorghum lager beer. Beer has been brewed successfully at both laboratory- and pilot-scale levels from sorghum malt without the need to supplement endogenous enzymes with external heat-stable enzymes (Agu and Palmer, 1998b; Nso et al., 2003). However, there is as yet no commercial production plant based on such a process of lager beer brewing. An adjusted mashing procedure is required to extract sorghum malt because of the high gelatinization temperature (>70°C) of sorghum starch (Agu, 2005), thus complicating the attempt to project sorghum malt as a brewing material.

Several problems are associated with the brewing process from malted sorghum, including the development of insufficient diastatic power, limited protein modification, high malting losses and costs, together with the need to supplement mashes with exogenous enzymes. These factors, combined with the lack of sorghum malting capacity in native sorghum countries, have led some brewers to favor the use of sorghum as an unmalted brewing material in combination with the necessary exogenous enzymes (Bajomo and Young, 1992; Agu and Palmer, 1998a; Goode and Arendt, 2003). Malted sorghum can develop sufficient hydrolytic enzymes required for the production of commercially acceptable levels of sugars/proteins for brewing the continental-type beers. However, an optimized mashing procedure is required to extract the sorghum malt (Agu and Palmer, 1998b). When mashing with 100% malted sorghum without the addition of commercial enzymes, a decantation process, where separated enzyme-active worts are used to convert the gelatinized starch of the mash, is recommended (Agu and Palmer, 1997a; Nso *et al.*, 2003).

Various mashing programs have been proposed when mashing with unmalted sorghum. In all cases, exogenous enzymes are required to ensure saccharification of the unmalted sorghum mash. This is achieved by the addition of a heat-stable α-amylase (Hallgren, 1995). When mashing with 100% unmalted sorghum, a single infusion mash (with temperature stands at 50°C, 80–90°C, and 60°C) is suggested (MacFadden and Clayton, 1989; Little, 1994; Hallgren, 1995). Various combinations of enzymes have been investigated, however, a heat-stable α-amylase, a protease, and a fungal α-amylase are all that is required to successfully produce a beer from sorghum (Little, 1994).

The 1988 ban on the importation of sorghum into Nigeria has since been lifted. In countries such as South Africa where sorghum (malted and unmalted) is being used in the production of traditional opaque beer and malted barley is being used for the production of lager beers, there is current interest in the production of lager beers from locally grown sorghum crops (Agu, 2005).

Rice

Rice in sake production
Sake, a traditional alcoholic beverage in Japan, is produced from rice and water (Yoshizawa and Kishi, 1985). The process of sake-making entails *koji*-making and alcohol fermentation. During the *koji*-making stage, spores of *Aspergillus oryzae* are inoculated into steamed rice and incubated for 48 hours. At the alcohol fermentation stage, *Saccharomyces cerevisiae* is inoculated onto the *maromi* mush, which contains *koji* and steamed rice and incubated for approximately one month (Iemura *et al.*, 1999). In *koji*, molds serve as one of the main sources of hydrolytic enzymes (e.g. amylases, proteases, and lipases). The steamed rice in the *maromi* mush is digested by these extracellular enzymes during fermentation (Fujita *et al.*, 2003). In beer brewing, fermentation takes place after filtration of the mash, whereas in the sake mash, sugars released from rice grains are fermented successively by yeast, and the content of fermentable sugars formed, mainly glucose, regulates the fermentation by the yeast (Yoshizawa and Kishi, 1985).

Before preparing the steamed rice used to make sake, brown rice is polished. As a by-product of this process, rice polish is produced at an annual rate of approximately 126 000 t in the Japanese sake industry. To help reduce the amount of rice polish that is produced as a by-product of the sake-brewing process, Iwata *et al.* (1998) developed a new type of alcoholic beverage, called "*nuka-sake,*" which is only made from uncooked rice polish, lactic acid, water, and sake yeast with supplementation of any enzyme sources such as *koji* or crude enzymes sold commercially. During *nuka-sake* brewing rice starch is saccharified by the amylolytic enzymes contained in the rice itself, mainly α-glucosidase. *Nuka-sake* brewing is an energy-saving process, given that the rice steaming process can be eliminated, and *nuka-sake* made from the rice polish of the inner part of the rice kernel has a distinctive taste compared with ordinary sake or beer and is of acceptable quality (Iwata *et al.*, 2002). Therefore, from an economical and environmental point of view, *nuka-sake* brewing is a sound way to produce alcohol.

Rice in beer production

Rice is the second most used adjunct in the United States and in Japan after maize (Juliano, 1994). As an adjunct, rice is preferred by some brewers because of its lower protein and lipids contents as compared with those of corn grits. Rice used for brewing is usually in the form of rice grits or broken rice, which is obtained as a by-product of the edible rice milling industry (to produce whole grain rice for culinary use) and rice harvesting (Yoshizawa and Kishi, 1985). Rice possesses a neutral flavor and aroma, and when converted efficiently to fermentable sugars, yields a clean tasting, light beer (Coors, 1976).

With regard to 100% rice beer, Moonjai (2005) found that a 72-hour germinated, fully malted rice with rice adjunct was able to produce beer wort. However, results showed that the alcohol content of the 100% rice beer was rather low (2.0 ± 0.2 wt %) but increased alcohol levels (4.9 ± 0.1 wt %) were observed when sorghum was added. The beer made from rice malt with rice adjunct was rated low with regard to color, foam, aroma, carbonation, and alcoholic taste. The low alcoholic taste detected correlated to high scores for sweet taste in the 100% rice beer. When sorghum was used as an adjunct, the resulting beer had high scores for bitterness and a sour taste. In conclusion, organoleptic assessment confirmed that the beer made from rice malt with sorghum adjunct was generally acceptable. Thus the use of rice malt in brewing can only be successful when an adjunct such as sorghum is used (Moonjai, 2005). Without further research to find an optimized mashing program for 100% malted rice and an investigation into the use of commercial enzymes to increase extract and alcohol content, the current status in the brewing industry of the use of rice as an adjunct will remain the same.

Maize

The high gelatinization temperature of maize starch indicates that it cannot be converted between 63 and 67°C, like the starches of wheat and barley, but must be heated to temperatures which approach or exceed 100°C in order to ensure that endosperm disruption and starch gelatinization occur (Ilori *et al.*, 1991). In addition, maize must be invariably processed to remove the oil-rich germs and bran before being used to manufacture adjuncts (Briggs *et al.*, 1981b). Even so, along with rice, maize appears to be the most popular adjunct (Briggs *et al.*, 1981a).

The suitability of two maize varieties for brewing purposes have been investigated (Aderinola, 1992). Studies with 100% maize malt, 90% maize malt with 10% barley malt, and 80% maize malt with 20% barley malt showed that saccharification could only be achieved with 80% maize malt and 20% barley malt. Malted barley was used as a source of enzyme complement rather than industrial enzymes because the latter is very expensive. Poor foam formation/head retention was noted with all three beers. This is most likely due to the high fat content of the maize varieties. These preliminary trials indicate that malted maize could be a future brewing material though further work is needed to investigate the beer color intensity, poor foam formation/head retention, and the addition of a proper combination of commercial enzymes.

Millets

Research studies have suggested that millet could be used in brewing European-type lager beer (Nout and Davies, 1982; Agu, 1991, 1995). Pearl millet is used in Mozambique for brewing traditional beer called *uphutsu* (Pelembe *et al.*, 2002) and other low-alcohol beverages also made from millet malt include *braga*, *darassum*, and *cochate* (Chavan and Kadam, 1989). In general, products made from millet do not keep for long periods because millet has a high fat content, making products made from them rancid after a few days. However, fermentation of finger millet at 30°C has been shown to decrease the starch and long-chain fatty acid content, giving millet products a longer shelf-life (Anthony *et al.*, 1996). Moreover, a combination of germination and fermentation has been found to decrease the phytate and tannin content of finger millet, thus increasing nutrient bioavailability and enhancing digestibility (Nzelibe and Nwasike, 1995; Sripiya *et al.*, 1997). Protein contents in most of the millets are comparable to those of wheat, maize, and rice, but finger millet is nutritionally superior because of its high levels of methionine, making it the best material for malting and brewing (Shewry, 2002).

It has been shown that millet malt produces wort that filters faster than sorghum malts wort and produces beers that have better foam properties than beers brewed from sorghum malt (Agu, 1995). Moir (1989) attributes beer quality to color, clarity, foam appearance, and flavor and comparative studies of barley, sorghum, and millet showed that beer brewed from millet malt met these qualities (Agu, 1995).

Millet has some physical properties that are similar to those of sorghum, in particular the gelatinization temperatures of the starches (Palmer, 1989). The fact that a suitable mashing program has been developed for extracting sorghum malt, whose starch, like that of millet, gelatinizes at a high temperature, suggests that millet malt can be extracted in a similar way (Palmer, 1989). Eneje *et al.* (2001) evaluated whether similar mashing methods developed for extracting sorghum malt would be suitable for extracting millet malt. Highest extract recovery was obtained using the decantation mashing system because using this mashing procedure, the enzymes of the millet malt are protected and the starch adequately gelatinizes. However, the decantation mashing method produced wort with lower values of soluble nitrogen and free amino nitrogen, and wort that filtered more slowly than the infusion mashing. It can be concluded that it is possible to produce a lager beer from millet although extensive work is needed to improve the flavor and color of the beer.

Oats

Oats have large proportions of extract-poor husk, about 30% compared with barley's 10%, and so inevitably oat malts have relatively low extract values, approximately 70–75% of those of barley malts (Kreisz *et al.*, 2005). In addition, oat malts are deficient in both α- and β-amylase, which results in low extract recovery. Since the raw grain is rich in β-glucans, it is necessary to sufficiently modify the grain during germination to avoid producing highly viscous, slow-draining worts (Briggs, 1998a). In addition, results of malt analysis of malted oats reveal a very low level of nitrogen modification, which leads to a low ratio of soluble to total nitrogen (SNR) (Taylor, 2000).

It is fair to say that oats do not play a significant role in beer production today (Taylor, 2000). However, studies have shown the addition of oats as an adjunct may benefit flavor properties (Heydanek and McGorrin, 1986). Taylor (2000) advocates a pronounced toasted, biscuity aroma, and palate, combined with a creamy and relatively intense mouth-feel in beer. These flavors may be apparent at less than 10% replacement with oats, depending on overall strength required. It is for these reasons that oats appear to be unsuitable as an ingredient for brewing and should be limited to an optimized level as an adjunct to enhance flavor.

Pseudocereals

Buckwheat

The first step in the production of buckwheat beer that needs to be optimized is mashing. Wort derived from malted buckwheat showed low fermentability values and high viscosity levels in comparison to wort derived from barley malt (Nic Phiarais *et al.*, 2005; Wijngaard *et al.*, 2005b). These worts were obtained by congress mashing which did not appear to be optimal for buckwheat malt. However, the decantation mashing method also appears to be unsuitable for buckwheat malt (Nic Phiarais *et al.*, 2006b). Recently a wide range of tests have been performed to characterize the action of the various enzymes during mashing. Optimization of mashing procedures were performed combining rheological tests with traditional mashing experiments (Goode *et al.*, 2005a, 2005b; Wijngaard and Arendt, 2006). It was found that the grist should be milled as fine as possible and a grist-to-liquor ratio of 1:4 is recommended. With the help of isothermal mashing experiments, a mashing program is recommended with a temperature/time profile consisting of mashing in at 35°C for 15 minutes, followed by 45°C for 15 minutes, 65°C for 40 minutes, 30 minutes at 72°C and mashing off at 78°C for 10 minutes. Preliminary brewing experiments using a 50 L pilot-scale brewery revealed that it is possible to produce gluten-free beer from buckwheat (Wijngaard and Arendt, 2006). Improved lautering performance of the mash was observed when unhulled buckwheat was used instead of hulled buckwheat.

Maccagnan *et al.* (2004) recently used buckwheat, mainly as an unmalted adjunct in micro brewing, for the production of gluten-free beer. Results of this study revealed that buckwheat has suitable beer-making properties with regard to both appearance and taste. However, all studies performed so far have shown that the enzymatic content of buckwheat and its malt is significantly lower than that of barley malt (Nic Phiarais *et al.*, 2005, 2006b; Wijngaard *et al.*, 2005b; Zarnkow *et al.*, 2005). Moreover, buckwheat contains polysaccharides, which causes a high viscosity of the wort (Wijngaard *et al.*, 2005c). These problems can be overcome by the addition of commercial enzymes (MacFadden and Clayton, 1989; Bajomo and Young, 1992). Investigation of the effectiveness of a wide range of commercial enzymes on buckwheat malt for brewing purposes was recently performed (Nic Phiarais *et al.*, 2006a). It was found that the addition of increasing levels of α-amylase to the buckwheat mash increased color, extract levels, wort filtration, fermentability, and total fermentable extract, along with decreasing viscosity values. Furthermore, the addition of increasing levels of amyloglucosidase to buckwheat mashes resulted in corresponding increases in fermentability and total fermentable extract, along with increases in total soluble nitrogen, free amino nitrogen, and Kolbach index. Therefore, these studies show that with the aid of commercial enzymes, buckwheat malt has the potential for replacing barley malt as a gluten-free material for those people with celiac disease. Still more extensive work is required to optimize fermentation performance and beer characteristics (e.g. flavor, aroma, and foam development).

Quinoa

To date, little research has been carried out on quinoa as a brewing ingredient, and mainly studies on the properties of quinoa starch are available (Atwell *et al.*, 1983; Qian and Kuhn, 1999a). Quinoa starch, being high in amylopectin, gelatinizes at a low temperature, comparable with the temperate cereals wheat and barley and rather lower than the tropical cereals such as maize and sorghum (Hoseney, 1994). Gelatinization temperature ranges of 57–64°C (Atwell *et al.*, 1983) and 60–71°C (Qian and Kuhn, 1999a) have been reported. This suggests that an adjusted mashing procedure would not be required to extract quinoa malt. Quinoa starch exhibits a much higher viscosity than wheat (Atwell *et al.*, 1983) and amaranth (Qian and Kuhn, 1999a). In contrast to maize starch, quinoa starch exhibits a single-stage starch swelling in the temperature range 65–95°C and lower viscosity (Ahamed *et al.*, 1996). These characteristics are presumably due to the very small size of the granules.

With regard to the use of quinoa as a brewing ingredient, Kreisz *et al.* (2005) performed malt analysis on optimally malted quinoa and found a slightly higher extract than barley malt. A subsequent study by Zarnkow *et al.* (2005) showed that beer made from quinoa malt contained a similar alcohol level to barley beer and therefore has the potential to be used as a brewing ingredient.

Amaranth

To the best of our knowledge, only limited data on amaranth brewing is available in literature. Fenzl *et al.* (1997) examined whether products pre-gelatinized through extrusion cooking are suitable as a partial substitute for barley malt in the production of lager beers. It was found that a 20% substitution is technically feasible without problems. Compared with the pure barley malt beer, the beer produced with amaranth was judged better on smell, taste, bitterness quality, and full body taste and was judged worst on two of the evaluated characteristics (bitterness intensity and freshness of flavor).

Due to the rather high starch content, amaranth is not only an interesting raw material for the production of beers, but also for the production of spirits. For the production of spirit, amaranth seeds can be milled and continuously hydrolyzed according to a hot mashing process (Sarhaddar, 1992) by adding thermostable α-amylase. After cooling, 10% barley malt is added for saccharification of the mash followed by fermentation and distillation. Following this procedure an excellent spirit with distinct sensory attributes specific to amaranth has been obtained (Berghofer *et al.*, 1997). However, to be suitable for safe consumption by people with celiac disease, brewing with 100% unmalted or malted amaranth is required.

Amaranth has been investigated as a potential gluten-free brewing material (Bauer *et al.*, 2005; Zarnkow *et al.*, 2005). Small brewing experiments with amaranth malt, which had an extract content of 79.9% resulted in a very low alcohol beer (0.64%) (Zarnkow *et al.*, 2005). Considering the literature available and without further studies into its brewing potential, amaranth could be promoted as a low-alcohol innovative functional beverage.

Conclusion

The search for new gluten-free brewing materials is still in its infancy. Limited studies are opening a new area of brewing and once process conditions are adjusted to accommodate gluten-free raw materials, the production of satisfactory gluten-free beers and products will be more realistic and should lead to a greater variety of products for people with celiac disease.

Currently only sorghum, millet, and buckwheat appear to be successful gluten-free beer ingredients, while others have only shown adjunct possibilities. Initial research on sorghum was not to find gluten-free alternatives but was in response to the 1988 ban on importation of barley malt into Nigeria. While acceptable to a large proportion of beer drinkers in Africa, the taste and flavor of sorghum beer may not be acceptable to countries outside this region. Further extensive research work is necessary to develop products that meet the tastes and consumer habits of the industrialized countries.

A search of the internet reveals that there are a number of micro-breweries producing gluten-free beer. However, a detailed analysis of the ingredient list of some of those so-called gluten-free beers shows that a small percentage of malt was included in the recipes and this contamination would certainly not be suitable for patients with celiac disease. Results collected so far indicate that buckwheat beer shows the most promise as a gluten-free alternative to sorghum beer.

In addition, thorough marketing efforts are needed to increase the knowledge and popularity of these cereals and pseudocereals, as at present only a small percentage of the population are familiar with or consumes these cereals. Successful commercial exploitation of these materials is tightly bound to the aspects reported above.

References

Aderinola, A. V. (1992). The brewing studies of two malted Nigerian maize varieties: Pool-16 (white) and EVTZXR-Y-1 (yellow). *Discovery and Innovation* 4, 103–108.

Agu, R. C. (1991). Studies on beer production from Nigerian millet. *J. Food Sci. Technol.* 28, 81–83.

Agu, R. C. (1995). Comparative study of experimental beers brewed from millet, sorghum and barley malts. *Process Biochem.* 30, 311–315.

Agu, R. C. (2005). Some relationships between enzyme development, extract recovery, and sugar profile in malted sorghum. *Tech. Q. Master Brew. Assoc. Am.* 42, 120–124.

Agu, R. C. and Okeke, B. C. (1991). Studies on the effect of potassium bromate on some malting properties of Nigerian millet (Pennisetum maiwa). *Process Biochem.* 26, 89–92.

Agu, R. C. and Palmer, G. H. (1997a). Effect of mashing procedures on some sorghum varieties germinated at different temperatures. *Process Biochem.* 32, 147–158.

Agu, R. C. and Palmer, G. H. (1997b). The effect of temperature on the modification of sorghum and barley during malting. *Process Biochem.* 32, 501–507.

Agu, R. C. and Palmer, G. H. (1998a). Effect of mashing with commercial enzymes on the properties of sorghum worts. *World J. Microbiol. Biotechnol.* 14, 43–48.

Agu, R. C. and Palmer, G. H. (1998b). Enzymatic modification of endosperm of barley and sorghum of similar total nitrogen. *Brew. Dig.* 73, 30–36.

Agu, R. C. and Palmer, G. H. (1998c). A reassessment of sorghum for lager-beer brewing. *Bioresour. Technol.* 66, 253–261.

Agu, R. C. and Palmer, G. H. (1999). Comparative development of soluble nitrogen in the malts of barley and sorghum. *Process Biochem.* 35, 497–502.

Ahamed, N. T., Singhal, R. S., Kulkarni, P. R., and Pal, M. (1996). Physicochemical and functional properties of *Chenopodium quinoa* starch. *Carbohydr. Polym.* 31, 99–103.

Aisen, A. O. (1988). Soghum: a suitable source for brewing beer. *Brew. Distill. Int.* 3, 20–22.

Aisen, A. O. and Muts, G. C. J. (1987). Micro-scale malting and brewing studies of some sorghum varieties. *J. Inst. Brew.* 93, 328–331.

Anderson, J. W. and Chen, W. J. (1986). Cholesterol-lowering properties of oat products. In: Webster, F. H. ed. *Oats Chemistry and Technology*. St. Paul, MN: American Association of Cereal Chemists, pp. 309–327.

Aniche, G. N. and Palmer, G. H. (1992). Influence of gibberellic acid (GA 3) on the development of amylolytic activities in rice during germination. *Process Biochem.* 27, 291–297.

Anthony, U., Sripiya, G., and Chandra, T. S. (1996). Effect of fermentation on the primary nutrients in finger millet (*Eleusine coracana*). *J. Agric. Food Chem.* 44, 2616–2618.

Arnold, J. P. (1911). Asia and Africa. In: *Origin and History of Beer and Brewing*. Chicago, Illinois: Wahl Henius Institute of Fermentology, pp. 67–200.

Atwell, W. A., Hyldon, R. G., Godfrey, P. D. *et al.* (1988). Germinated quinoa flour to reduce the viscosity of starchy foods. *Cereal Chem.* 65, 508–509.

Atwell, W. A., Patrick, B. M., Johnson, L. A., and Glass, R. W. (1983). Characterization of quinoa starch. *Cereal Chem.* 60, 9–11.

Aufhammer, W., Kaul, H.-P., Kruse, M., and Lee, J. H. (1996). Dry matter and nitrogen accumulation and residues of oil and protein crops. *Eur. J. Agron.* 5, 137–147.

Bajomo, M. F. and Young, T. W. (1992). Development of a mashing profile for the use of microbial enzymes in brewing with raw sorghum (80%) and malted barley or sorghum malt (20%). *J. Inst. Brew.* 98, 515–523.

Balasubramanian, T. and Sadasivam, S. (1989). Changes in carbohydrate and nitrogenous components and amylase activities during germination of grain amaranth. *Plant Food Hum. Nutr.* 39, 327–330.

Bamforth, C. W. (2003). A brief history of beer. In: *Beer: Tap into the Art and Science of Brewing*, 2nd edn. New York: Oxford University Press, pp. 25–29.

Bauer, J., Walker, C., and Booer, C. (2005). Of pseudocereals and roasted rice, the quest for gluten-free brewing materials. *The Brewer and Distiller* 4, 24–26.

Berghofer, E. and Schoenlechner, R. (2002). In: Belton, P. S. and Taylor, J. R. N. eds. *Pseudocereals and Less Common Cereals-Grain Properties and Utilization Potential*. Springer-Verlag, Berlin, Germany, pp. 219–253.

Berghofer, E., Marques, E., Robic, F., and Epalle, G. (1997). Herstellung von Amaranth-brand. *Praktikumsprotokoll*, Unpublished.

Biacs, P., Aubrecht, E., Leder, I., and Lajos, J. (2002). Buckwheat. In: Belton, P. S. and Taylor, J. R. N. eds. *Pseudocereals and Less Common Cereals-Grain Properties and Utilization Potential*. Springer-Verlag, Berlin, Germany, pp. 123–147.

Briggs, D. E. (1998a). Types of malt. In: Briggs, D. E. ed. *Malts and Malting*. London: Blackie Academic & Professional, pp. 699–740.

Briggs, D. E. (1998b). Grains and pulses. In: Briggs, D. E. ed. *Malts and Malting*. London: Blackie Academic & Professional, pp. 35–73.

Briggs, D. E., Hough, J. S., Stevens, R., and Young, T. W. (1981a). Adjuncts, sugars, wort syrups and industrial enzymes. In: Briggs, D. E., Hough, J. S., Stevens, R., and Young, T. W., eds. *Malting and Brewing Science*. London: Chapman & Hall, pp. 222–253.

Briggs, D. E., Hough, J. S., Stevens, R., and Young, T. W. (1981b) The chemistry and biochemistry of mashing. In: Briggs, D. E., Hough, J. S., Stevens, R., and Young, T. W., eds. *Malting and Brewing Science*. London: Chapman & Hall, pp. 254–300.

Briggs, D. E., Hough, J. S., Stevens, R., and Young, T. W. (1981c). The biochemistry of malting grain. In: Briggs, D. E., Hough, J. S., Stevens, R., and Young, T. W., eds. *Malting and Brewing Science*. London: Chapman & Hall, pp. 57–107.

Briggs, D. E., Hough, J. S., Stevens, R., and Young, T. W. (1981d). The technology of malting and kilning. In: Briggs, D. E., Hough, J. S., Stevens, R., and Young, T. W., eds. *Malting and Brewing Science*. London: Chapman & Hall, pp. 145–192.

Canales, A. M. (1979). Unmalted grains in brewing. In: Pollock, J. R. A. ed. *Brewing Science*. London: Academic Press, pp. 233–278.

Capanzana, M. V. and Buckle, K. A. (1997). Optimisation of germination conditions by response surface methodology of a high amylose rice (*Oryza sativa*) cultivar. *Lebensm.-Wiss. U.-Technol.* 30, 155–163.

Caperuto, L. C., Amaya-Farfan, J., and Camargo, C. R. O. (2000). Performance of quinoa (*Chenopodium quinoa* Willd) flour in the manufacture of gluten-free spaghetti. *J. Sci. Food Agric.* 81, 95–101.

Chandrasekhara, M. R. and Swaminathan, M. (1953). Enzymes of ragi (*Eleusine cora-cana*) and ragi malt—Amylases, proteases and phosphatases. *J. Sci. Indust. Res.* 12B, 51–56.

Chavan, J. K. and Kadam, S. S. (1989). Nutritional inprovement of cereals by fermentation. *Crit. Rev. Food Sci. Nutr.* 28, 349–400.

Colmenares De Ruiz, A. S. and Bressani, R. (1990). Effect of germination on the chemical composition and nutritive value of amaranth grain. *Cereal Chem.* 67, 519–522.

Coors, J. (1976). Practical experience with different adjuncts. *Tech. Q. Master Brew. Assoc. Am.* 13, 117–119.

Corran, H. S. (1975). Ancient beer. In: *A History of Brewing*. Newton Abbot: David & Charles, pp. 50–110.

Dalby, A. and Tsai, C. Y. (1976). Lysine and tryptophan increases during germination of cereal grains. *Cereal Chem.* 53, 222–226.

Dale, C. J. (1990). Small scale mashing experiments with grists containing high proportions of raw sorghum. *J. Inst. Brew.* 96, 403–409.

Dufour, J. P. Melotte, L., and Sebrnik, S. (1992). Sorghum malts for the production of a lager beer. *J. Am. Soc. Brew. Chem.* 50, 110–119.

Ellis, H. J., Freedman, A. R., and Ciclitira, P. J. (1990). Detection and estimation of the barley prolamin content of beer and malt to assess their suitability for patients with coeliac disease. *Clin. Chim. Acta* 189, 123–130.

Eneje, L. O., Obiekezie, S. O., Aloh, C. U., and Agu, R. C. (2001). Effect of milling and mashing procedures on millet (*Pennisetum maiwa*) malt wort properties. *Process Biochem.* 36.

Fasano, A. and Catassi, C. (2001). Current approaches to diagnosis and treatment of celiac disease: an evolving spectrum. *Gastroenterology* 121, 636–651.

Fenzl, G., Berghofer, E., Silberhummer, H., and Schwarz, H. (1997). Einsatzmoglichkeiten extrudierter, starkereicher Rohstoffe zur Bierherstellung. *Tagungsband Oesterr Brauforum* 1, 1–6.

Fujita, J., Shigeta, S., Yamane, Y.-I. *et al.* (2003). Production two types of phytase from *Aspergillus oryzae* during industrial koji making. *J. Biosci. Bioeng.* 95, 460–465.

Gallagher, E., Gormley, T. R., and Arendt, E. K. (2004). Recent advances in the formulation of gluten-free cereal based products. *Trends Food Sci. Technol.* 15, 143–152.

Goode, D. L. (2001). Pilot scale production of a lager beer from unmalted sorghum. MSc thesis, University College Cork, Ireland.

Goode, D. L. and Arendt, E. K. (2003). Pilot scale production of a lager beer from a grist containing 50% unmalted sorghum. *J. Inst. Brew.* 109, 208–216.

Goode, D. L., Halbert, C., and Arendt, E. K. (2003). Optimization of mashing conditions when mashing with unmalted sorghum and commercial enzymes. *J. Am. Soc. Brew. Chem.* 61, 69–78.

Goode, D. L., Rapp, L., Schober, T. J., and Ulmer, H. M. (2005a). Development of a new rheological laboratory method for mash systems—Its application in the characterization of grain modification levels. *J. Am. Soc. Brew. Chem.* 63, 76–86.

Goode, D. L., Wiltschko, E. A., Ulmer, H. M., and Arendt, E. K. (2005b). Application of the rapid visco analyser as a theological tool for the characterisation of mash viscosity as affected by the level of barley adjunct. *J. Inst. Brew.* 111, 165–175.

Hallgren, L. (1995). Lager beers from sorghum. In: Dendy, D. A. V. ed. *Sorghum and Millets—Chemistry and Technology*. St. Paul, MN: American Association of Cereal Chemists, pp. 283–298.

Hamer, R. J. (2005). Coeliac disease: Background and biochemiacl aspects. *Biotechnol. Adv.* 23, 401–408.

Heydanek, M. G. and McGorrin, R. J. (1986). Oat flavour chemistry: Principles and prospects. In: Webster, F. H. ed. *Oats Chemistry and Technology*. St. Paul, MN: American Association of Cereal Chemists, pp. 335–369.

Hoffenberg, E. J., Haas, J., Drescher, A. *et al.* (2000). A trial of oats in children with newly diagnosed celiac disease. *J. Pediatr.* 137, 361–366.

Hoseney, R. C. (1994). Starch. In: Pomeranz, Y. ed. *Principles of Cereal Science and Technology*, 2nd edn. St. Paul, MN: American Association of Cereal Chemists, pp. 147–148.

Hough, J. S. (1985a). Fermentation—the fundamental process. In: Baddiley, Sir J., Carey, N. H., Davidson, J. F., Higgins, I. J., and Potter, W. G. eds. *The Biotechnogy of Malting and Brewing*. Cambridge: Cambridge University Press, pp. 114–134.

Hough, J. S. (1985b). Sweet wort production. In: Baddiley, Sir J., Carey, N. H., Davidson, J. F., Higgins, I. J., and Potter, W. G. eds. *The Biotechnogy of Malting and Brewing*. Cambridge: Cambridge University Press, pp. 54–71.

Iemura, Y., Takahashi, T., Yamada, T., Furukawa, K., and Hara, S. (1999). Properties of TCA-insoluble peptides in Kimoto (Traditional seed mash for sake brewing) and conditions for liberation of the peptides from the rice protein. *J. Biosci. Bioeng.* 88, 531–535.

Igyor, M. A., Ogbonna, A. C., and Palmer, G. H. (2001). Effect of malting temperature and mashing methods on sorghum wort composition and beer flavour. *Process Biochem.* 36, 1039–1044.

Ilori, M. O., Ogundiwin, J. O., and Adewusi, S. R. A. (1991). Sorghum malt brewing with sorghum/maize adjuncts. *Brew. Distill. Int.* 3, 10–13.

Irving, D. W. and Becker, R. (1985). Seed structure and composition of potential new crops. *Food Microstructure*, 4, 43–53.

Iwata, H., Nagano, T., Shiokawa, K., and Suzuki, A. (1998). Development of a new alcoholic beverage made from uncooked rice bran. *J. Brew. Soc. Jpn.* 93, 139–147.

Iwata, H., Suzuki, T., Takahashi, K., and Aramaki, I. (2002). Critical importance of a-glucosidase contained in rice kernal for alcohol fermentation of rice polish. *J. Biosci. Bioeng.* 93, 296–302.

Janatuinen, E. K., Pikkarainen, P. H., Kemppainen, T. A. *et al.* (1995). A Comparison of diets with and without oats in adults with celiac disease. *N. Engl. J. Med.* 333, 1033–1037.

Juliano, B. O. (1994). Production and utilization of rice. In: B. O. Juliano, ed. *Rice-Chemistry and Technology*. St. Paul, MN: American Society of Cereal Chemists, pp. 1–14.

Kanerva, P., Sontag-Strohm, T., and Lehtonen, P. (2003). Determination of prolamins in beers by ELISA and SDS-PAGE. *J. Inst. Brew.* 111, 61–64.

Kasarda, D. D. (2001). Grains in relation to celiac disease. *Cereals Foods World* 46, 209–210.

Kneen, E. (1944). A comparative study of the development of amylases in germinating cereals. *Cereal Chem.* 21, 304–314.

Kneen, E. (1945). Sorghum amylase. *Cereal Chem.* 27, 483–500.

Kreisz, S., Zarnkow, M., Kebler, M. *et al.* (2005). Beer and innovative (functional) drinks based on malted cereals and pseudo-cereals. In: *Proceedings of the 30th European Brewery Convention*, Prague, Czech Republic, Contribution 103, pp. 1–8.

Kunze, W. (1996a). Beer—the oldest drink for the common man. In: *Technology Brewing and Malting*, 7th edn. Berlin: Versuchs- und Lehranstalt fur Brauerei, pp. 19–25.

Kunze, W. (1996b). Malt production. In: *Technology Brewing and Malting*, 7th edn. Berlin: Versuchs- und Lehranstalt fur Brauerei, pp. 88–155.

Kunze, W. (1996c). Beer production. In: *Technology Brewing and Malting*, 7th edn. Berlin: Versuchs- und Lehranstalt fur Brauerei, pp. 323–448.

Kunze, W. (1996d). Wort production. In: *Technology Brewing and Malting*, 7th edn. Berlin: Versuchs- und Lehranstalt fur Brauerei, pp. 171–316.

Lanares, J. P. (1992). Das bier der Tarahumara. *Brauwelt* 132, 970–974.

Larsson, M. and Sandberg, A.-S. (1995). Malting of oats in a pilot-plant process. Effects of heat treatment, storage and soaking conditions on phytate reduction. *Cereal Chem.* 21, 87–95.

Lewis, M. J. (2005). Celiac disease, beer, and brewing. *Tech. Q. Master Brew. Assoc. Am.* 42, 45–48.

Li, S. Q. and Zhang, Q. H. (2001). Advances in the development of functional foods from buckwheat. *Crit. Rev. Food Sci. Nutr.* 41, 451–464.

Little, B. T. (1994). Alternative cereals for beer production. *Ferment* 7, 163–168.

Maccagnan, G., Pat, A., Collavo, F., Ragg, G. L., and Bellini, M. P. (2004). Gluten-free beer containing rice malt and buckwheat. European Patent no. 0949328B1.

MacFadden, D. P. and Clayton, M. (1989). Brewing with sorghum-Use of exogenous enzymes *Brew. Beverage Ind. Int.* 1, 71–81.

Mahoney, A. W., Lopez, J. G., and Hendricks, D. G. (1975). An evaluation of the protein quality of quinoa. *J. Agric. Food Chem.* 23, 190–193.

Malleshi, N. G. and Desikachar, H. S. R. (1986a). Influence of malting conditions on the quality of finger millet malt. *J. Inst. Brew.* 92, 81–83.

Malleshi, N. G. and Desikachar, H. S. R. (1986b). Studies on comparative malting characteristics of some tropical cereals and millets. *J. Inst. Brew.* 92, 174–176.

Marshall, H. G. and Pomeranz, Y. (1982). Buckwheat: description, breeding, production and utilization. In: Pomeranz, Y. ed. *Advances in Cereal Science and Technology*. St. Paul, MN: American Association of Cereal Chemists, pp. 157–210.

Moir, M. (1989). Effect of raw material on flavour and aroma. *Brewers' Guardian* 118, 64–71.

Moonjai, N. (2005). Comparative study of experimental beers brewed from rice malt with some starchy adjuncts. In: *Proceedings of the 31th Congress on Science and Technology of Thailand*, Nakhon Ratchasima, Thailand. Contribution 84, pp. 1–3.

Nic Phiarais, B. P., Wijngaard, H. H., and Arendt, E. K. (2005). The impact of kilning on enzymatic activity of buckwheat malt. *J. Inst. Brew.* 111, 290–298.

Nic Phiarais, B. P., Schehl, B. D., Oliveira, J. C., and Arendt, E. K. (2006a). Use of response surface methodology to investigate the effectiveness of commercial enzymes on buckwheat malt for brewing purposes. *J. Inst. Brew.* 112, 324–332.

Nic Phiarais, B. P., Wijngaard, H. H., and Arendt, E. K. (2006b). Kilning conditions for the optimisation of enzyme levels in buckwheat. *J. Am. Soc. Brew. Chem.* 64, 187–194.

Nirmala, M. and Muralikrishna, G. (2003). Three α-amylases from finger millet (Ragi, Eleusine coracana, Indaf-15)—purification and partial characterization. *Phytochemistry* 62, 21–30.

Nout, M. J. R. and Davies, B. J. (1982). Malting characteristics of finger millet, sorghum and barley. *J. Inst. Brew.* 88, 157–163.

Novellie, L. (1962). Kaffircorn malting and brewing; Effect of malting conditions on malting losses and amylase activities of kaffircorn malts. *J. Sci. Food Agric.* 13, 121–126.

Nso, E. J., Ajebesome, P. E., Mbofung, C. M., and Palmer, G. H. (2003). Properties of three sorghum cultivars used for the production of Bili-Bili beverage in Northern Cameroon. *J. Inst. Brew.* 109, 245–250.

Nzelibe, H. C. and Nwasike, C. C. (1995). The brewing potential of acha (*Digitaria exilis*) malt compared with pearl millet (*Pennisetum typhoides*) malts and sorghum (*Sorghum bicolor*) malts. *J. Inst. Brew.* 101, 345–350.

Obeta, J. A. N., Okungbowa, J., and Ezeogu, L. I. (2000). Malting of sorghum: further studies on factors influencing alpha-amylase activity. *J. Inst. Brew.* 106, 295–304.

Odiboa, F. J. C., Nwankwoa, L. N., and Agu, R. C. (2002). Production of malt extract and beer from Nigerian sorghum varieties. *Process Biochem.* 37, 851–855.

Ogbonna, A. C., Obi, S. K. C., and Okolo, B. N. (2004). Optimization of proteolytic activities in malting sorghum. *Process Biochem*, 39, 711–716.

Okafor, B. N. and Aniche, G. N. (1980). Brewing a lager beer from Nigerian sorghum. *Brew. Distill. Int.* 10, 32–35.

Okafor, N. and Iwouno, J. (1990). Malting and brewing qualities of some Nigerian rice (Oryza sativa L.) varieties and some thoughts on the assessment of malts from tropical cereals. *World J. Microb. Biot.* 6, 187–194.

Okolo, B. N. (1996). Enhancement of amylolytic potential of sorghum mlats by alkaline treatment. *J. Inst. Brew.* 102, 78–85.

Okon, E. U. and Uwaifo, A. O. (1984). Partial purification and proportions of beta-amylase isolated from *Sorghum bicolor* (L) Moench. *J. Agric. Food Chem.* 32, 11–15.

Okungbowa, J., Obeta, J. A. N., and Ezeogu, L. I. (2002). Sorghum beta-amylase production: relationship with grain cultivar, steep regime, steep liquor composition and kilning temperature. *J. Inst. Brew.* 108, 362–370.

Owuama, C. I. (1997). Sorghum: a cereal with lager beer brewing potential. *World J. Microbiol. Biotechnol.* 13, 253–260.

Owuama, C. I. (1999). Brewing beer with sorghum. *J. Inst. Brew.* 105, 23–34.

Palmer, G. H. (1989). Cereals in malting and brewing. In: Palmer, G. H. ed. *Cereal Science and Technology*. Aberdeen, Scotland: Aberdeen University Press, pp. 61–242.

Paredes-Lopez, O. and Mora-Escobedo, R. (1989). Germination of amaranth seeds: effects on nutrient composition and colour. *J. Food Sci.* 54, 761–762.

Pelembe, L. A. M., Dewar, J., and Taylor, J. R. N. (2002). Effect of malting conditions on pearl millet malt quality. *J. Inst. Brew.* 108, 7–12.

Peraaho, M., Collin, P., Kaukinen, K., Kekkonen, L., Miettinen, S., and Maki, M. (2004). Oats can diversify a gluten-free diet in celiac disease and dermatitis herpetiformis. *J. Am. Diet Assoc.* 104, 1148–1150.

Peterson, D. M. (1998). Malting Oats: effects on chemical composition of hull-less and hulled genotypes. *Cereal Chem.* 75, 230–234.

Peterson, D. M. (2001). Oat antioxidants. *Cereal Chem.* 33, 115–129.

Prestamo, G., Pedrazuela, A., Penas, E., Lasuncion, M. A., and Arroyo, G. (2003). Role of buckwheat diet on rats as prebiotic and healthy food. *Nutr. Res.* 23, 803–814.

Qian, J. and Kuhn, M. (1999a). Characterization of *Amaranthus cruentus* and *Chenopodium quinoa* starch. *Starch/Staerke* 50, 7–13.

Qian, J. and Kuhn, M. (1999b). Evaluation on gelatinization of buckwheat starch: a comparative study of Brabender viscoamylography, rapid visco-analysis, and differential scanning calorimetry. *Eur. Food Res. Technol.* 209, 277–280.

Rich, H. S. (1974). The beer of ancient and medieval times. In: *One Hundred Years of Brewing*. New York: Arno Press, pp. 11–167.

Rossander-Hulten, L., Gleerup, A., and Halberg, L. (1990). Inhibitory effect of oat products on non-heme iron absorption in man. *Eur. J. Clin. Nutr.* 44, 788–791.

Sandstrom, B., Almgren, A., Kivisto, B., and Cederblad, A. (1987). Zinc absorption in humans from melas based on rye, barley and oatmeal, triticale and wholewheat. *J. Nutr.* 177, 1898–1902.

Sarhaddar, S. (1992). Method for hydrolysing starch to produce saccharified mash. US Patent no. 5114491.

Schrickel, D. J. (1986). Oats production, value and use. In: Webster, F. H. ed. *Oats-Chemistry and Technology*. St. Paul, MN: American Association of Cereal Chemists, pp. 1–11.

Shephard, G. S., van der Westhuizen, L., Gatyeni, P. M., Somdyala, N. I., Burger, H. M., and Marasas, W. F. (2005). Fumonisin mycotoxins in traditional Xhosa maize beer in South Africa. *J. Agric. Food Chem.* 53, 9634–9637.

Shewry, P. R. (2002). The major seed storage proteins of spelt wheat, sorghum, millets and pseudocereals. In: Belton, P. S. and Taylor, J. R. N. eds. *Pseudocereals and Less Common Cereals-Grain Properties and Utilization Potential*). Berlin: Springer-Verlag, pp. 1–20.

Sigstad, E. E. and Garcia, C. I. (2001). A microcalorimetric analysis of quinoa seeds with different initial water content during germination at 25°C. *Thermochim. Acta* 366, 149–155.

Singh, T. and Bains, S. S. (1984). Malting of corn: Effect of variety, germination, gibberellic acid, and alkali pretreatments. *J. Agric. Food Chem.* 32, 346–348.

Sollid, L. M. and Khosla, C. (2005). Future therapeutic options for celiac disease. *Nat. Clin. Pract. Gastroenterol. Hepatol.* 2, 140–147.

Srinivasan, U., Jones, E., Kasarda, D. D., Weir, D. G., O'Farrelly, C. and Feighery, C. (1996). Absence of oats toxicity in adult coeliac disease. *B. M. J.* 313, 1300–1301.

Sripiya, G., Anthony, U., and Chandra, T. S. (1997). Changes in carbohydrate, free amino acids, organic acids, phytate and HCL extractability of minerals during germination and fermentation of finger millet. *Food Chem.* 58, 345–350.

Sun, T. and Ho, C.-T. (2005). Antioxidant activities of buckwheat extracts. *Food Chem.* 90, 743–749.

Taylor, D. G. (2000). Brewing ales with malted cereals other than barley. *Ferment* January, 18–20.

Taylor, J. R. N. and Robbins, D. J. (1993). Factors influencing beta-amylase activity in sorghum malt. *J. Inst. Brew.* 99, 413–416.

Thompson, T. (2000). Questionable foods and the gluten-free diet. *J. Am. Diet Assoc.* 100, 463–465.

Thompson, T. (2001). Case Problem: Questions regarding the acceptability of buckwheat, amaranth, quinoa and oats from patient with celiac disease. *J. Am. Diet Assoc.* 101, 586–587.

Valencia, S., Svanberg, U., Sandberg, A.-S., and Ruales, J. (1999). Processing of Quinoa (*Chenopodium quinoa* Willd): effects on in vitro iron availability and phytate hydrolysis. *Int. J. Food Sci. Nutr.* 50, 203–211.

Varriano-Marston, E. and De Francischi, A. (1984). Ultrastructure of quinoa fruit (*Chenopodium quinoa* Willd). *Food Microstructure* 3, 165–173.

Wang, Y. D. and Fields, M. L. (1978). Germination of corn and sorghum in the home to improve nutritive value. *J. Food Sci.* 43, 1113–1115.

Wijngaard, H. H. and Arendt, E. K. (2006). Optimisation of a mashing program for 100% malted buckwheat. *J. Inst. Brew.* 112, 57–65.

Wijngaard, H. H., Nic Phiarais, B. P., Ulmer, H. M., Goode, D. L., and Arendt, E. K. (2005a). Gluten-free beverages based on buckwheat. In: *Proceedings of the 30th European Brewery Convention Congress*, Prague, Czech Republic. Contribution 78, pp. 1–11.

Wijngaard, H. H., Ulmer, H. M., and Arendt, E. K. (2005b). The effect of germination temperature on malt quality of buckwheat. *J. Am. Soc. Brew. Chem.* 63, 31–36.

Wijngaard, H. H., Ulmer, H. M., and Arendt, E. K. (2006). The effect of germination time on the final malt quality of buckwheat. *J. Am. Soc. Brew. Chem.* 64, 214–221.

Wijngaard, H. H., Ulmer, H. M., Neumann, M., and Arendt, E. K. (2005c) The effect of steeping time on the final malt quality of buckwheat. *J. Inst. Brew.* 111, 275–281.

Wilhelmson, A., Oksman-Caldentey, K.-M., Laitila, A., Suortti, T., Kaukovirta-Norja, A., and Poutanen, K. (2001). Development of a germination process for producing high beta-glucan, whole grain food ingredient from oat. *Cereal Chem.* 78, 715–720.

Yoshizawa, K. and Kishi, S. (1985). Rice in brewing. In: Juliano, B. O. ed. *Rice-Chemistry and Technology*. St. Paul, MN: American Society of Cereal Chemists, pp. 619–636.

Zarnkow, M., Kebler, M., Burgerg, F., Kreisz, S. and Back, W. (2005). Gluten free beer from malted cereals and pseudocereals. In: *Proceedings of the 30th European Brewery Convention*, Prague, Czech Republic. Contribution 104, pp. 1–8.

Zhou, J. R. and Erdman, J. W. (1995). Phytic acid in health and disease. *Crit. Rev. Food Sci. Nutr.* 35, 495–508.

Cereal-based gluten-free functional drinks

16

Stefan Kreisz, Elke K. Arendt, Florian Hübner,
and Martin Zarnkov

Introduction

In the big industrial countries (USA, Europe, Japan) the interest of consumers in the relations between nutrition and health has grown substantially in recent years. The realization that a healthy lifestyle, including nutrition, reduces the risk of disease and increases health and well-being has received a huge amount of publicity. This is associated with an increased interest in food such as fruits, vegetables, and wholegrain products or in industrially manufactured products that offer additional health benefits. According to a study by Frost and Sullivan (management consultancy, www.frost.com), the market for functional drinks in Europe has grown from US$2.35 billion in 1999 to US$5.73 billion in 2006. Functional drinks, like any other beverages, in general have the advantage of being easily available and quick to consume.

The concept of "functional foods" comes from Japan, after the introduction during the 1980s of special food to aid health and to decrease the risk of diseases. In general, a functional food is regarded as a food that is suitable to be consumed as a part of the common nutrition and that contains biologically active components with the ability to increase health and to lower the risk of diseases. This definition reflects the advantages for the consumer and the possibility of innovation for the food industry.

However, the consumer must be protected against wrong or misleading claims concerning the constitutional effects. Japan has had a leading role in the production of functional foods. In 1991 the Foods for Specified Health Use (FOSHU) concept was introduced. A food can only be accredited as FOSHU after extensive scientific investigation by the Minister for Health and Welfare. In Europe uniform legislation is lacking, but in November 1995 the European Commission introduced concerted action on Functional Food Science in Europe (FUFOSE), which aimed to establish a science-based approach for concepts in functional foods science. Under FUFOSE the currently known facts will be determined and necessary additional investigations decided. Ten study groups, involving 54 researchers from ten EU countries, are cooperating, concerning themselves with six fields considered most important for human physiology: gastrointestinal system, defense from reactive oxygen radicals, cardiovascular system, complete metabolism and metabolism illnesses, development, growth and differentiation as well as psychological functions and behaviors. As an outcome of this work a new definition of functional food has been produced:

> A food can be regarded as functional, when it has proved satisfactory that it influences positively one or several physical functions beyond a nutritive value, in a way that it has relevance for the well-being or the reduction of disease risks.

FUFOSE conclusions and principles have been taken to the next logical stage (i.e. application of the principles). The project Process for the Assessment of Scientific Support for Claims on Foods (PASSCLAIM) started with, and built upon, the principles defined within the publications arising out of the FUFOSE project. The objectives of PASSCLAIM are:

- To produce a generic tool with principles for assessing the scientific support for health-related claims for foods and food components which are eatable or drinkable;
- To evaluate critically the existing schemes which assess the scientific substantiation of claims;
- To select common criteria for how markers should be identified, validated and used in well-designed studies to explore the links between diet and health.

PASSCLAIM is divided into two phases. All papers and further information about the structure and the development of the project are published from the International Life Sciences Institute (http://europe.ilsi.org).

According to an older definition, functional drinks also belong in the category "functional food" and can be divided into four types:

- Sport drinks
- Energy drinks
- Wellness drinks
- Nutrient enriched drinks.

The specified categories with the definitions stated above do not match up entirely. Even if the authors specify their definitions for functional drinks in different ways, a functional drink should contain biological substances in order to bring about real additional benefits. Constitutional effects are ascribed to biological substances. According to Elmafada (1998), biological substances are to be distinguished from essential nutrients, which the human being needs for the preservation of life activities and cannot be synthesized or are synthesized only in low amounts. Currently, six groups of biological food substances are known:

- Secondary plant substances
- Prebiotic carbohydrates
- Omega-3 fatty acids
- Conjugated linoleic acid
- Peptides from milk protein
- Maillard products.

Nevertheless, additional groups probably exist.

Cereals, usually Poaceae (true grasses) such as barley, oat, maize, rice, rye, sorghum, triticale, emmer, einkorn, wheat, kamut, and millet as well as pseudocereals such as buckwheat, amaranth, and quinoa, are vital constituents of the human nutrition. According to the Food and Agriculture Organization of the United Nations, around 620 million tons of wheat were produced in 2005 worldwide. Apart from alcoholic beverages like beer or kvass, grain is mainly used for the production of solid food, such as bread, animal feed, or recently for the production of energy in unprocessed form. According to the above definition, gluten-free beverages based on cereals or pseudocereals could already be considered to be functional drinks in the broader sense. In principle the designation "gluten-free" only narrows down the possible cereals that are considered as a starting basis for functional beverages. Thus, the standard procedure is not changed, although it has not yet been decided unanimously what kinds of cereals, beside the pseudocereals, which are basically gluten-free, can be used as a raw material for beverages. Table 16.1 shows the approved cereals (regional differences are taken into account).

Using cereals or pseudocereals for drinks makes it necessary to deal with one major characteristic of most grains. Most grain components have a high molecular weight and they are not, or only in a minor part, water soluble. Most of the cereals or pseudocereals have a starch content of above 60% and mixing a certain amount of milled grains with water results in more or less highly viscous doughs. The starch will gelatinize (according to the temperature) but not hydrolyze. High molecular weight proteins and cell wall polysaccharides will remain mainly unsolubilized, therefore the yield for a drink is insufficient and the soluble part is not suitable for further processing. To design a drink based on cereals or pseudocereals, the following steps, mainly already invented for barley and therefore for beer productions, have to be adapted:

- Evaluation (specification) of cereals or pseudocereals for the beverage production
- Malting (optional)
- Milling
- Substrate production with or without exogenous enzymes
- Alcoholic or non-alcoholic fermentation (optional)
- Blending and stabilization.

Table 16.1 Categorization of cereals according to their approval for gluten-free products

	AFDIAG	Celiac Sprue Association	The Celiac Group	St. Johns University	Association Canadienne	DZG
	France	USA	USA	USA	Canada	Germany
Amaranth		Not accepted	Accepted under reserve	Accepted under reserve		Accepted
Buckwheat	Accepted	Not accepted	Accepted under reserve	Accepted under reserve		Accepted
Quinoa	Accepted	Not accepted	Accepted under reserve	Accepted under reserve		Accepted
Spelt	Not accepted	Not accepted	Not accepted	Not accepted	Not accepted	Not accepted
Barley	Not accepted	Not accepted	Not accepted	Not accepted	Not accepted	Not accepted
Oat	Not accepted	Not accepted	Not accepted	Not accepted	Not accepted	Not accepted
Kamut	Not accepted	Not accepted	Not accepted	Not accepted	Not accepted	Not accepted
Maize	Accepted	Accepted	Accepted	Accepted	Accepted	Accepted
Rice	Accepted	Accepted	Accepted	Accepted	Accepted	Accepted
Rispenhirse	Accepted	Not accepted	Accepted under reserve	Accepted under reserve		Accepted
Rye	Not accepted	Not accepted	Not accepted	Not accepted	Not accepted	Not accepted
Sorghum		Accepted	Accepted	Accepted		Accepted
Triticale	Not accepted	Not accepted	Not accepted	Not accepted	Not accepted	Not accepted
Wheat	Not accepted	Not accepted	Not accepted	Not accepted	Not accepted	Not accepted

Evaluation of cereals or pseudocereals as base for (functional) drinks

Almost all standards, analysis, and technological know-how in malting, substrate production as well as fermentation are based on research and experience with barley and barley malt. Barley was selected a long time ago as the brewing cereal, and in the last 100 years intensive breeding efforts have led to a highly specialized product. No other cereals or pseudocereals have been optimized for malting, substrate production, or fermentation, and their breeding programs are sometimes counterproductive as they are focused on high protein levels and low enzyme activities (e.g. wheat). In addition, the harvesting and trading condition standards are rarely as high as those for barley so it is not always possible to get authentic information about the traded varieties and the homogeneity of the batches. Nevertheless, the general proceedings in evaluating the raw material for the beverage production may be adapted from barley to any cereal or pseudocereal.

The analytical parameters should be as follows.

Protein content

Very high protein content is a quality attribute for some cereals (e.g. rye and wheat). On the other hand there are cereals with very low protein contents (< 8%), such as rice or maize. As some proteins are positive for human nutrition, high protein content is desirable, but in terms of production the high content of protein may cause problems. In drinks, proteins are responsible for foaming, and in combination with polyphenols (which are available in most cereals or pseudocereals) they form haze in the bottled beverage.

Fat content

The fat content of maize, oat, or rice germs is higher than that of wheat or barley. Off-flavors and unstable beverages may result from the use of high-fat content ingredients.

Germination capacity/energy

The germination capacity should be at least 95% if the grains are to be malted. It is advisable to use seed quality grain for beverage malt production.

Husk

Most cereals are traded without the husk (e.g. wheat, spelt, einkorn) or have a totally different seed coat (rice, maize, sorghum, and millet). The typology of the seed coat influences the water uptake during malting and determines the substrate production strategy (lautering or separation).

Basic enzyme supply

A major improvement in barley breeding has been the enrichment of enzymes needed for malting and mashing. As other cereals or pseudocereals are not selected for brewing purposes, their enzymatic background is unknown and may not necessarily be suitable for substrate production.

Size of the kernels and homogeneity of the batch

Inhomogeneous size distribution causes difficulties during milling. Small kernels tend to have a higher protein level and water uptake, while malting is faster. Therefore, processability and the final quality of the beverage are less predictable.

Contamination

The grain should be inspected for infesting insects, mixture of varieties, the presence of toxin-producing fungi, or any sign of heavy fungal attack or heavy microbial contamination, as well as for fungicides or pesticides according to local regulations.

Sampling and analysis of grains are partly standardized and published by organizations like the European Brewery Convention (van Erde, 1998), the

Mitteleuropäische Brautechnische Analysenkommision (MEBAK) (Anger, 2006), or the International Association for Cereal Science and Technology (ICC, 2004).

Germination and drying (malting, optional)

Defined germination and drying could be advantageous for the composition of a functional drink based on cereals and/or pseudocereals. The main advantages are:

- The process includes cleaning and classifying which makes the further processes easier and more predictable.
- Grain enzymes necessary for the hydrolysis of starch, proteins and cell wall polysaccharides are activated/released.
- Positive aroma and flavor components are formed, mainly by Maillard reactions occurring while drying.
- Negative volatile aroma and flavor components are reduced.
- Functional components may be enhanced (Maillard products as well as phenolic components).
- The water content is reduced for better storage of the grains and the number of microorganisms on the grain surface may be reduced by higher drying temperatures (e.g. kilning).

Overall, the same process as that used for standard barley malting can be used, but it has to be adapted for each cereal or pseudocereal. The parameters to be taken into account are vegetation time and temperature as well as moisture content. To adapt the malting conditions for each cereal or pseudocereal, it is advisable to check their malting performances by micro-malting trials. Different procedures have already been published and evaluated. Moreover, it is advisable to use software programs to calculate a model that shows the reaction of each cereal or pseudocereal to the variation of the parameters, thus allowing prediction of the resulting quality (Kreisz *et al.*, 2007).

Substrate production

The quality of the beverage and the number of functional components available in the drink are mainly influenced by substrate production from either raw grains or malt. The main goal is to dissolve as much functional components as possible, and hydrolyze polysaccharides and proteins to the desired level for fermentation (if fermentation is part of the concept). Starch is the major constituent of most grains. The starch is mostly contained in cells surrounded by cell walls containing other polysaccharides, such as β-glucans or xylans, as well as proteins. To receive any economical reasonable yield, the cell walls must be degraded. This can be accomplished by exogenous or endogenous enzymes. The starch also has to be gelatinized. The gelatinization temperature of starch from cereals or pseudocereals varies, but the minimum temperature is most likely above 60°C. The basic process is to mix ground grains or malt at the gelatinization temperature and add exogenous enzymes or, alternatively, to hydrolyze to the desired degree all high molecular components by using endogenous enzymes. Different temperature and time profiles can be used, and basic equipment is required to heat and stir the mash.

The resulting mixture of solid residues and solubilized components must then be separated by using different systems according to the particle size distribution of the solid phase. A single separation using a separator or centrifuge has the disadvantage that the residual solid part cannot be eluted again. Existing systems in breweries such as lauter tuns and mash filters are designed to rinse the residual solid part two or three times with water, which results in a better yield of substrate. After separation, the substrate should be heated to high temperatures or boiled to inactivate the enzymes, fix the composition of the substrate, and sterilize it. The heat treatment has other positive effects like building aroma components and reducing undesired volatile aroma components. However, a longer heating or cooking time may have an impact on the quality/benefit of the final product, since some functional components are not heat stable (Kreisz *et al.*, 2005).

Fermentation (optional)

According to the production technology used, the substrate may be rich in fermentable sugars and amino acids. Here, a controlled fermentation is advisable because of the high calorie content and the microbiological risk of direct use. A health claim as part of the concept excludes alcoholic fermentations since most local regulations do not permit the advertising of a health claim if the beverage contains a certain amount of alcohol (most likely above 0.5% vol) and alcoholic beverages may be banned for religious reasons. In the last 20 years a great deal of work has been carried out to screen microorganisms for lactic acid fermentation (sometimes together with acidic acid fermentation) (Caplice and Fitzgerald, 1999; Blandino *et al.*, 2003; Idler *et al.*, 2005). The fermentation significantly reduces the calorie content and increases microbiological stability by processing sugar and amino acids, and therefore lowering the supply of substrate for other microorganisms, as well as lowering the pH to <4. The low pH may have a major influence on the protein content of the beverage. As the low pH matches the isoelectric point of some water-soluble proteins of cereals or pseudo-cereals, these proteins may precipitate, forming haze in the drink. If the fermenting microorganisms are not part of the concept, they have to be removed by filtration.

Blending and stabilization

The result of the fermentation is most likely a sour plain base with an individual aroma and a certain number of functional components. This base may be used for a wide range of different blending strategies:

- Carbonization
- Dilution with water or juice

- Addition of aroma and flavors
- Stabilization of specific proteins.

As the field of research of functional drinks based on cereal or pseudocereal is still in its infancy, only a few studies on the stability of these drinks have been carried out.

Potentially functional compounds in cereals and pseudocereals

In the following sections a short overview of compounds which have the potential to act as functional ingredients in cereal-based beverages is given. Possibilities for further enhancing functional properties such as the addition of probiotics are shortly discussed.

Dietary fiber

Dietary fiber is a term used for the edible part of plants or analogous carbohydrates that resist hydrolysis by alimentary tract enzymes. Wholegrain cereal products are one of the main sources of dietary fiber in the human nutrition. Other important sources are legumes, while fruit and vegetables provide limited amounts. For applications in functional beverages, water-insoluble sources of fiber (cellulose, hemicellulose, resistant starch, insoluble arabinoxylans, and lignin) are less important, but they are very important in breakfast cereals and bakery goods.

Water-soluble sources of fiber such as β-glucans and arabinoxylans are suitable compounds for the incorporation into functional beverages. Both β-glucans and arabinoxylans form viscous solutions, slowing the transit of food through the intestine and thus delaying gastric emptying. This leads to a reduced absorption of glucose and sterols as well as serum cholesterol. As a result, post-prandial blood glucose and insulin content in the body is decreased. β-Glucan consists of glucanopyranosil units with either $1 \rightarrow 3$ or $1 \rightarrow 4$ linkages. Many studies have looked at the health-promoting effects of mixed-link β-glucans from oats. β-Glucan is degraded during the germination process and is found in beverages made from malted cereals only in small amounts. However, attempts have been made to produce high content β-glucan products either by malting (Wilhelmson et al., 2001) or by the use of β-glucan concentrates (Temelli et al., 2004).

The basic structure of arabinoxylans is a non-branched chain of $1 \rightarrow 4$-β-D-xylopyranosyl units to which α-L-arabinofuranosyl substituents are attached. Arabinoxylans form more viscous solutions than β-glucans, due to the ferulic acid residues that are attached to the arabinose side-chains. They are found in much higher amounts than β-glucan in malt-based beverages. Arabinoxylan-degrading enzymes are released late in the germination process, which makes under-modified malts a rich source of arabinoxylans (Li et al., 2005). However, the health effects of arabinoxylans have not been studied as intensely as those of β-glucans.

All types of dietary fiber can be substrates for the microbiota in the intestinal tract. When the fibers are selectively fermented by health-promoting bacteria, i.e. lactobacilli and bifidobacteria, they are called prebiotics (Roberfroid, 2007). β-Glucan and its breakdown products have been shown to stimulate the growth of bifidobacteria and/or lactobacilli (Kontula *et al.*, 1998; Snart *et al.*, 2006). Products from the fermentation of dietary fiber in the lower intestine include short-chain fatty acids, which lower the pH in the lower intestine and thereby limit the growth of many undesirable organisms. In order to utilize the full health-promoting potential, the content of dietary fiber in the beverages should be high. The US Food and Drug Administration (FDA) regards the intake of 3 g of β-glucan as part of a total dietary fiber intake of 30 g per day as beneficial for gut health.

However, in traditionally produced malt and malt-based beverages the dietary fiber content is relatively low. Therefore, the malting process should be optimized in order to limit the degradation of substances such as β-glucan and/or arabinoxylans. However, the incomplete breakdown of cell wall material would lead to incomplete breakdown of other compounds, i.e. starch and proteins, which can thus cause problems during the fermentation of such raw materials.

The content of β-glucan and related compounds in beverages could be increased by addition of ingredients rich in fiber. Such ingredients can be produced using special milling techniques, since β-glucans are concentrated in certain sections of the grain, i.e. the outer layers of the grain, especially the aleuron cells and the sub-aleuron endosperm walls.

Alternative sources for dietary fiber which could be added to functional beverages are β-glucans isolated from seaweed, bacteria, and/or yeast. These types of β-glucans have different properties from those of β-glucans isolated from cereal grains such as barley or oats (i.e. different chain length and/or linkages in the β-glucan molecule). Unfortunately, the purification procedure is very expensive, making this alternative non-realistic from a commercial point of view. Fiber can also be found at high levels in other plants, for example in psyllium husk derived from the seeds of *Plantago psyllium* and *Plantago ovata*. Health claims for psyllium products were approved by the FDA in 1998.

In conclusion, the health-promoting effects of dietary fiber are well known and a higher uptake of these compounds is recommended by nutritionists. Functional beverages can be an additional source for soluble dietary fiber. However, there are a number of technological problems associated with high concentrations of fibers in beverages. Most soluble fibers are known to form gels and thereby increase the viscosity of the products. While this might be tolerated or even be desired, up to a certain degree, to create certain organoleptic properties, very high viscosities can be problematic for downstream processing such as filtering and might be rejected by the consumer.

Antioxidants

Antioxidation is one of the most important mechanisms for preventing or delaying the onset of major degenerative diseases of aging, including cancer, heart disease,

Table 16.2 Antioxidants (μmol Trolox equivalent/g sample, dry weight) in sorghum measured by three different methods

Sorghum type	ORAC[a]	ABTS[b]	DPPH[c]
White grain	22	6	6
Red grain	140	53	28
Black grain	220	57	41
High tannin grain	450	108	118
Sumac grain, brown	870	226	206

[a] ORAC, oxygen radical absorbance capacity, fluorescein used as a probe.
[b] ABTS, 2,2′-azinobis (3-ethyl-benzothiazoline-6-sulfonic acid); activity was measured in pH 7.4 phosphate buffer saline.
[c] DPPH, 2,2-diphenyl-1-picrylhydrazyl; activity measured in methanol.
Adapted after Rooney and Awika (2005).

cataracts, and cognitive dysfunction. Antioxidants are believed to exert their effect by blocking oxidative processes and free radicals that contribute to the causation of these chronic diseases. The most potent oxidative substances in food are oxygen, especially in the form of singlet oxygen, and other free radicals. Therefore, scavenging free radicals and quenching of singlet oxygen are desired actions for antioxidative substances because they limit the amount of oxidizing molecules. Heavy metal ions act as catalysts for oxidizing reactions. This effect can be minimized by metal complexing compounds which act as synergists with the other antioxidants. Finally, reducing agents can reduce oxidized material to a certain extent, and thereby reverse the damaging effect of free radicals and oxygen. Preventing oxidation of cell membrane material in the human body is thought to prevent cancer and cardiovascular diseases.

Several methods to assess total antioxidant activity and radical scavenging capacity have been developed. However, comparison between the different studies is complicated by the use of different analytical methods. For example, Rooney and Awika (2005) used three different methods to evaluate the antioxidant properties of sorghum grain (Table 16.2). Results of this study clearly showed that the antioxidant capacities found in specific grains is dependent on the method used for analysis.

The most important classes of substances found in plant material with antioxidative properties are phenolic compounds, some vitamins, phytate, and sterols. In addition, compounds with antioxidant activity can be formed during food processing by the Maillard reaction. The different classes are discussed below.

Phenolic compounds

Phenolic compounds are synthesized by plants as products of secondary metabolism. Many different classes of phenolic substances have been described with a wide range of antioxidant capacities. Some grain species (e.g. sorghum) contain a large amount of phenolic acids, while others contain comparatively small amounts (e.g. rye). Studies have shown that the content of phenolic compounds increases during malting (Dicko et al., 2005). However, this might be due to biosynthesis or better extractability of those compounds after breakdown of cell wall components. Other rich sources of phenolic antioxidants are fruits and green tea (Dimitrios, 2006).

Polyphenols like tannins have long been regarded as detrimental for nutrition because they interact with proteins and decrease their biological availability. However, the positive effects of polyphenols in a well-balanced diet seem to prevail.

Vitamins with antioxidant properties in grains
Cereals are the main source of tocopherols (vitamin E) in human diet. In addition, maize contains considerable amounts of carotinoids (pre-vitamin A). Ascorbic acid is not found in grains but can be detected in pseudocereals. However, vitamins only have a minor impact on the total antioxidant capacity of cereals.

Sterols
Sterols are structural components of the plant cell membrane and can be found as free sterol alcohols, steryl esters, or steryl glycosides. Germination can increase levels of sterols, as recently observed in oats (Oksman-Caldentey *et al.*, 2001). In addition to their antioxidative function, plant sterols have been proven to lower LDL cholesterol levels and are therefore used to enrich products such as margarine. The addition of plant sterols to food products has become common practice in recent years, and a range of beverages such as fruit juices as well as dairy-based functional drinks have been marketed. However, concerns about the benefit of the cholesterol-lowering effect for part of the population (i.e. children and pregnant women) have been raised recently by the Health Council of the Netherlands (The Health Council of the Netherlands, 2001).

Maillard products
Maillard products are derived from heat-induced reactions of reducing sugars with amino acids or simple peptides. The Maillard reaction involves three basic phases. Initially, an amino group undergoes a condensation reaction with the carbonyl function of a sugar. The imine formed is unstable and undergoes a reaction (Amadori rearrangement), in which stable amino-ketoses are formed. In the second step, three different desoxyosones with strong reductive activity can be formed by dehydration reactions. From the desoxyosones a wider range of aroma and flavor compounds can be formed by subsequent dehydration, fission, and/or polymerization reactions. Diacetyl, acetol, pyruvaldehyde, and similar compounds can be formed, which can undergo the Strecker degradation with amino acids to aldehydes, or condensate to aldols. In the final steps, heterocyclic compounds such as furans, furanones, and pyrones like isomaltol and maltol are formed. The specific amino acid and sugar composition, together with pH and temperature, are key factors influencing the nature of the end-products. In cereal-based beverages, Maillard products are mainly formed during the kilning step in malting, or, to a lesser extent, during heating of the beverage (e.g. during mashing or heat treatment for preservation). Unfortunately, the Maillard reaction can also lead to the formation of mutagenic compounds and loss of essential amino acids, which makes it less than optimal as a method to enrich antioxidant compounds in beverages.

Conclusions

The main difficulty in assessing the advantages of a diet rich in antioxidants is the difficulty in connecting the assumed or proven effects to a single compound. A number of methods assessing the antioxidative capacity and radical scavenging properties have been published, but considerable differences in the results delivered have been observed. Even if the properties of a single substance are clear *in vitro*, the reactions *in vivo* are not fully understood. Although epidemiological studies have shown positive effects of high antioxidant intakes, this has not been confirmed by interventional studies (Dimitrios, 2006). Some of the observed health-promoting effects may not be caused by the antioxidants but rather by reactions or interactions with other food compounds.

Minerals

Positive effects on human health have been reported for an increased uptake of some minerals. Calcium is claimed to protect against osteoporosis and help to maintain bone density. Zinc is related to the possibility of preventing or curing the common cold, even if the data are not conclusive. Nonetheless, these minerals are commonly enriched in functional food. The zinc and calcium contents of some cereals are shown in Table 16.3.

An excessive intake of some minerals might have negative effects on the human health. Excessive intake of calcium has been linked to increased risk of some types of cancer (Giovannuci *et al.*, 1998), and high intake of sodium is reported to have a negative effect on blood pressure. Like all plant materials, cereals are typically low in sodium, and the use of water with low sodium contents, either naturally or due to treatment, can ensure a product without negative health effects.

In many plants, a range of compounds can be found which interact with minerals, decreasing their bioavailability (e.g. tannins). The negatively charged side-groups of tannins can interact with positive charged metal ions, and thereby hinder their uptake in the intestine. Tannins can be degraded by some lactic acid bacteria, releasing the formerly complexed ions into the medium. Phytate is the major compound responsible for the low uptake of minerals from grains. Phytate is myo-inositol, with each hydroxyl group esterified with a phosphate group, and it can be found in the outer

Table 16.3 Content of zink and calcium in cereals and pseudocereals

	Calcium (mg/100 g)	Zink (mg/100 g)
Rice[a]	7.3	1.1
Maize[a]	7.2	1.5
Finger millet[a]	325	1.7
Sorghum[a]	14	2.2
Wheat[a]	37	1.6
Buckwheat[b]	18–22	2.3

[a] According to Hemalatha *et al.* (2007)
[b] According to Mazza (1988).

Table 16.4 Phytate content of grains and activity of phytase

Grain	Phytate content (mg/g)	Phytase activity (U/g)
Wheat	12.4	180
Triticale	12.9	650
Rye	11.8	2800
Barley	11.9	350
Oat	11.3	48
Maize	9.2	9

According to Belitz *et al.* (2001).

layer of the grains. Phytate contains about 70% of the grain's total phosphorus content and limits its availability. Typical contents of some grains as well as the activities of the phytate-hydrolyzing enzymes are shown in Table 16.4. Phytate forms very stable complexes with zinc, iron, calcium, and/or magnesium ions, thereby hindering their uptake. However, dephosphorylation of phytate during germination can significantly improve the availability of minerals (Larson and Sandberg, 1995).

Because of its interaction with minerals, phytate has long been considered anti-nutritional. However, some positive health effects have been observed, such as its ability to lower blood glucose content during the consumption of starchy food, reduction of plasma levels of cholesterol and triacylglycerols, as well as reduction of cancer risk (Rickard and Thompson, 1997).

Vitamins

Vitamins are substances that are essential for normal cell functions, growth and development of an organism. They cannot be synthesized in sufficient amounts by human metabolism, and therefore need to be taken up from food. The recommended daily allowances for vitamins and their content in some cereals are shown in Table 16.5. In general, cereal products are a major source of B vitamins in the typical Western diet, and these should be preserved during the production of a beverage to fully utilize the nutritional potential of the grains. The contents of riboflavin, thiamine, biotin, pantothenic acid, tocopherols, and folates increase during germination (Merx *et al.*, 1994; Plaza *et al.*, 2003). In addition, fermentation could increase the content of some vitamins, as reported for riboflavin, thiamine, and niacin in oat-based products (Sanni *et al.*, 1999).

Excessive intakes of some vitamins can, however, cause health problems. Classically known is hypovitaminosis for vitamin A which may occur following intakes of more than 7.5 mg/day retinol. However, beta-carotene, which is present in cereals and can be metabolized to retinol, does not cause hypervitaminosis. Intakes of 20 mg/day of beta-carotine increased the risk of lung cancer in smokers (Alpha-Tocopherol Beta Carotene Cancer Prevention Study Group, 1994). However, most of the vitamins known to cause negative effects when consumed in large quantities are not found in high amounts in grains. Therefore, negative health effects will not be seen in cereal-based products unless vitamins from external sources are added at very high levels.

Table 16.5 Vitamin contents in grains and recommended daily intake

Vitamin	Recommended daily intake	Content in grains (mg or μg per 10 g edible portion)				
		Wheat	Rye	Corn	Oat	Rice unpolished
A	0.8–1.0 mg Retinol equivalent	*	*	*	*	*
Carotinoids (mg)		0.02	*	1.3	*	*
D (μg)	5	*	*	*	*	*
E (mg)	14	1.4	2.0	2.0	1.5	0.74
K (μg)	60–70	40	*	*	63	*
B1 (μg)	1.0–1.3	0.48	0.35	0.36	0.59	0.41
B2 (mg)	1.2–1.4	0.09	0.17	0.20	0.15	0.09
Niacin (mg)	13–16	5.1	1.8	1.5	1.0	5.2
Pantothenic acid (mg)	6	1.2	1.5	0.7	1.1	1.7
B6 (mg)	1.2–1.5	0.27	0.23	0.4	0.16	0.28
Biotin (μg)	30–60	6.0	5.0	6.0	20	12
Folic acid (μg)	400	87	143	26	87	16
B12 (μg)	30	*	*	*	*	*
C (mg)	100	*	*	*	*	*

* Not detected or trace amounts.
According to Belitz *et al.* (2001).

Probiotics

Probiotic beverages have been traditionally based on milk, but recently cereal-based probiotic drinks have appeared in the market. Probiotics are defined as live microorganisms which when consumed in adequate numbers confer a health benefit to the host (Isolauri *et al.*, 2004). Many such benefits have been described, including suppression of potentially harmful organism in the intestine, stimulation of immune response or prevention of cancer. Some of those effects might be caused by bioactive metabolites, so-called biogenic substances (e.g. certain vitamins, bioactive peptides, or short-chain fatty acids). Key properties required for the use of organisms as probiotics are:

- Safety with regard to human use
- Ability to survive passage through the digestive tract
- Colonization potential in the human intestinal tract or the respective target organ
- Stability during storage.

Most known probiotic organisms are part of the normal mammalian biota or have been isolated from fermented food. Many are lactobacilli or bifidobacteria, but other organisms, e.g. *Saccharomyces boulardii*, have also been used. Probiotics have been applied in recent years both in medical applications as well as in functional food. For medical purposes viable cells have been dried and applied as powders. Different strategies for preserving the cells' viability, including microencapsulation in cereal-based substances, have been developed. To date, probiotics as ingredients in functional food have been mostly applied in fermented dairy-based products. For beverages, a dose of about 10^7 cells per mL at the moment of consumption is considered functional (Gomes and Malacta, 1999) and is demanded as minimum by the Fermented Milk and Lactic Acid Bacteria Beverages Association in Japan (Ishibashi and Shimamura, 1993). According to Kurman and Rasic (1991), the minimum therapeutic dose per day is 10^8–10^9 cells.

The usual way to obtain the requested concentrations is to grow the probiotic directly in the beverage, most commonly milk. However, milk represents a limited substrate for the growth of fastidious organisms such as lactobacilli and bifidobacteria. In fact, the nutritional requirement of such bacteria is very complex; it includes fermentable sugars, peptides, free available amino acids, fatty esters, salts, nucleic acid derivatives, and/or vitamins, which are often not present simultaneously in the beverage. Typically, cereal flour slurries (e.g. doughs, mashes) contain low concentrations of fermentable sugars, but processing of the grain, especially by malting or addition of exogenous enzymes, can greatly improve the amount of available sugars and amino acids. Cereal-based beverages contain a wide range of sugars such as glucose, maltose, maltotriose, and various pentoses, making grain substrates a good alternative to milk. The type and amount of the sugars depends on the grain species and processing conditions. Phosphorus and vitamins, which are essential for the growth of lactobacilli, are found in higher amounts in grains than in milk. In addition, substances such as arginine, fructose, citric, and malic acid, which might be present in cereals, are sources of slow metabolizing energy and enhance the viability during the storage period.

The medium in which the bacteria are consumed has an impact on the microorganisms' ability to survive the passage through the intestine. Large amounts of sugars as found in malt extracts have been shown to increase significantly the chance of survival during the passage through the stomach (Charalampopoulos *et al.*, 2002). Free amino acids may also contribute to this effect. In addition, high buffering capacity and pH values of the cereal-based beverage can increase the pH of the stomach and protect the probiotics.

Probiotics are Generally Recognized As Safe (GRAS) organisms. However, some strains might be responsible for one or more of the following negative effects, as pointed out recently by Hoesl and Altwein (2005):

- Systemic infections
- Deleterious metabolic activities
- Excessive immune stimulation in susceptible individuals
- Gene transfer.

To date, adverse effects have mostly been observed in immunocompromised individuals, and since this indicates a possible danger, the intake of probiotics in immunocompromised patients should be regulated by medical experts. In conclusion, cereal-based beverages represent possible alternatives as carrier of probiotics. Their nutrient compositions might be more favorable for the growth as well as the survival of the probiotics during storage.

Possible additives for functional beverages

Besides utilizing the functional components in the cereals or pseudocereals, and enriching them by means of germination and/or fermentation, functional components can be directly added into the beverage. A list of herbs and herbal extracts, their health claims and the authors' opinion of efficacy has been published by Katan and de Roos (2004) and is shown in Table 16.6. Note that the described effects have been shown for the use of the substances as supplements, whereas the amounts added to the food are usually much lower. Furthermore, the intake of herbal drugs might interfere with other drugs and reduce their efficiency; or the herbal drugs might interact with food components and lower either the nutritive value or the functional effect.

Table 16.6 Herbal products with potential to be added to functional foods and opinion of efficacy

Ingredient	Product examples	Health effect or claim	Evidence in humans
Guarana (*Paulina cupana*)	Drinks	Extra energy, improved cognitive performance	++ Efficacy due to the high amounts of caffeine
Ginkgo (from *Ginkgo biloba*)	Drinks, cereals	Enhances memory and alertness	+ for halting cognitive decline in senile dementia (Beaubrun and Gray, 2000; Ernst, 2002; Le Bars and Kastelan, 2000) − for enhancing memory or alertness in healthy individuals (Ernst, 2002; Le Bars and Kastelan, 2000; Salomon *et al.*, 2002)
Kava (from *Piper methysticum*)	Drinks, cereals	Relaxation, mental balance, reduces stress	+/0 for anxiolytic effect (Beaubrun and Gray, 2000; Ernst, 2002), though not at levels found in drinks and cereals
St John's Wort (*Hypericum perforatum*)	Drinks, cereals	Mental balance, lifts the spirits, reduces anxiety	+/0 for mild to moderate depression (Gaster and Holroyd, 2000), but effect smaller in recent trials (Shelton *et al.*, 2001)
Echinacea	Drinks	Supports immune system, antibiotic	+/0 for common cold but trial data weak and inconclusive (Ernst, 2002)
Ginseng	Drinks, teas, cereals	Extra energy, reduces body weight, mind-supporting	0 Inconclusive (Kitts and Hu, 2000; Bucci, 2002); no evidence of efficiency for any condition (Ernst, 2002)
Red yeast rice	Not used in food	Lowers cholesterol	Effect largely due to lovastine produced by the yeast (Havel, 1999)

Conclusions

Cereal-based beverages have a huge potential as functional food. They can serve as carriers for a range of functional compounds, for example antioxidants, dietary fiber, minerals, probiotics, and vitamins. However, more research is needed to fully understand the impact of some of the functional components (e.g. antioxidants) present in cereal-based beverages.

The health effects of dietary fibers are well known, but technological issues such as increased viscosities must be considered. The addition of fibers from external sources is possible, but in many cases too expensive. Cereals and pseudocereals are a considerable source of vitamins. In order to fully utilize this potential, measures should be taken to protect these vitamins during the processing or to increase their content by means of germination or fermentation, where possible. In order to fully utilize the minerals present in cereals, it is necessary to break down metal-complexing agents such as phytate. Special attention should be given to the application of probiotics in cereal-based beverages, as cereal malt extracts represent excellent media for the growth of microorganisms.

In conclusion, it will not be possible to produce a beverage containing all potentially beneficial compounds found in cereals and having acceptable organoleptic properties. Consequently, when the production of a functional beverage is planned, raw materials and processing steps need to be carefully assessed in order to fulfill the demands of the consumer with regard to taste, aroma, and appearance as well as ensuring that the desired functional properties are available and active.

References

Alpha-Tocopherol Beta Carotene Cancer Prevention Study Group (1994). The effect of vitamin E and beta-carotene on the incidence of lung cancer and other cancers in male smokers. *N. Engl. J. Med.* 330, 1029–1035.

Anger, H. (2006). *Band Rohstoffe*. 1. Auflage ed. Brautechnische Analysenmethoden. (2006). Freising, Selbstverlag der MEBAK.

Asplund, K. (2002). Antioxidant vitamins in the prevention of cardiovascular disease, a systematic review. *J. Intern. Med.* 251, 372–392.

Beaubrun, G. and Gray, G. E. (2000). A review of herbal medicines for psychiatric disorders, *Psychiatr. Serv.* 51, 1130–1134.

Belitz, H.-D., Grosch, W., and Schieberle, P. (2001). *Lehrbuch der Lebensmittelchemie*, 5th edn. Berlin: Springer.

Blandino, A., Al-Aseeri, M. E., Pandiella, S. S., Catero, D., and Webb, C. (2003). Cereal-based fermented foods and beverages. *Food Res. Int.* 36, 527–543.

Bucci, L. R. (2000). Selected herbals and human exercise performance, *Am. J. Clin. Nutr.* 72, 624S–636S.

Caplice, E. and Fitzgerald, G. F. (1999). Food fermentations, role of microorganisms in food production and preservation. *Int. J. Food Microbiol.* 50, 131–149.

Centers of Disease Control and Prevention (1993). Recommendations for use of folic acid to reduce number of spina bifida cases and other neuronal tube defects. *JAMA* 269, 1233–1238.

Charalampopoulos, D., Wang, R., Pandiella, S. S., and Webb, C. (2002). Application of cereals and cereal components in functional foods, a review, *Int. J. Food Microbiol.* 79, 131–141.

Dicko, M. H., Gruppen, H., Traore, A. S., van Berkel, W. J. H., and Voragen, G. J. (2005). Evaluation on the effect of germination on phenolic compounds and antioxidant activities in sorghum varieties. *J. Agric. Food Chem.* 53, 2581–2588.

Dimitrios, B. (2006). Sources of natural phenolic antioxidants, *Trends Food Sci. Nutr.* 17, 505–512.

Elmafada, I. (1998). *Ernährung des Menschen*, 3rd edn. Stuttgart, p. 409.

Ernst, E. (2002). The risk-benefit profile of commonly used herbal therapies: Ginko, St. John's Wort, Ginseng, Echinacea, Saw Palmetto and Kava, *Ann. Intern. Med.* 160, 152–156.

Gaster, B. and Holroyd, J. (2000). St. John's wort for depression: A systematic review, *Arch. Intern. Med.* 160, 152–156.

Giovannuci, E., Rimm, E. B., Wolk, A., Ascherio, A., Stampfer, M. J., and Colditz, G. A. (1998). Calcium and fructose intake in relation to risk of prostate cancer. *Cancer Res.* 58, 442–447.

Gomes, A. M. P. and Malacta, X. F. (1999). Bifidobacterium ssp and Lactobacillus acidophilus, biological, technological and therapeutical properties relevant for use as probiotics. *Trends Food Sci. Technol.* 10, 139–157.

Hathcock, J. N. (1997). Vitamins and minerals: Efficacy and safety. *Am. J. Clin. Nutr.* 66, 427–437.

Havel, R. J. (1999). Dietary drug supplement or drug? The case of cholestin. *Am. J. Clin. Nutr.* 66, 175–176.

Hemalatha, S., Platel, K., and Srinivasan, K. (2007). Zinc and iron contents and their bioaccessability in cereals and pulses consumed in India. *Food Chem.* 102, 1328–1336.

Hoesl, C. E. and Altwein, J. E. (2005). The probiotic approach, an alternative treatment option in urology. *Eur. Urol.* 47, 288–296.

ICC (International Association for Cereal Science and Technology) (2004). *ICC-Standards*. Wien: Internationale Gesellschaft für Getreidewissenschaft.

Idler, F., Seidel, B., Kreisz, S., Kurz, T., and Fleischer, L.-G. (2005). Development of functional drinks based on cereals by lactic fermentation. In: *30th EBC Congress*. Prague.

Ishibashi, N. and Shimamura, S. (1993). Bifidobacteria, research and development in Japan. *Food Technol.* 47, 126–135.

Isolauri, E., Salminen, S., and Ouwehand, A. C. (2004). Microbial-gut interactions in health and disease. Probiotics. *Best Pract. Res. Clin. Gastroenterol.* 18, 299–313.

Katan, M. B. and de Roos, N. M. (2004). Promises and problems of functional foods. *Crit. Rev. Food Sci. Nutr.* 44, 369–377.

Kitts, D. and Hu, C. (2000). Efficacy and safety of ginseng, *Public Health Nutr.* 3, 473–485.

Kontula, P., Jaskari, J., Nollet, L. *et al.* (1998). The colonization of a simulator of the human microbial ecosystem by a probiotic strain fed on fermented oat bran product, effects on the gastrointestinal microbiota. *Appl. Microbiol. Biotechnol.* 50, 246–252.

Kreisz, S., Zarnkow, M., Back, W., and Kurz, T. (2005). Beer and innovative (functional) drinks based on malted cereals and pseudocereals. In: *30th EBC Congress*. Prague.

Kreisz, S., Zarnkow, M., and Back, W. (2007). A new statistical method to evaluate the malting performance of new barley varieties. In: *31st EBC Congress*. Venice.

Kurman, J. A. and Rasic, J. L. (1991). The health potential of products containing bifidobacteria. In: Robinson, R. ed. *Therapeutic Properties of Fermented Milks*. Oxford: Elsevier Science Publishers, pp. 117–158.

Larson, M. and Sandberg, A.-S. (1995). Malting of oats in a pilot plant process. Effects of heat treatment, storage and soaking conditions on phytate reduction, *J. Cereal Sci*. 21, 87–95.

Le Bars, P. L. and Katelan, J. (2000). Efficiacy and safety of Ginko biloba extract, *Public Health Nutr*. 3, 495–499.

Li, Y., Lu, J., Gu, G., Shi, Z., and Mao, Z. (2005). Studies on water-extractable arabi-noxylans during malting and brewing. *Food Chem*. 93, 33–38.

Mazza, G. (1988). Lipid content and fatty acid composition of buckwheat seed. *Cereal Chem*. 65, 122–126.

Merx, H., Seibel, W., Rabe, E., and Menden, E. (1994). Influence of germination para-meters on the vitamin content and the microbiological quality of sprout cereals (rye and wheat). Part 1 State of research. *Getreide Mehl Brot* 48, 17–20.

Oksman Caldentey, K.-M., Kaukovirta-Norja, A., Heiniö, R.-L., Kleemola, T., Mikola, M., Sontag-Strom, T., Lehtinen, P., Pihlava, J.-M. and Poutanen, P. (2001). Kauran biotekninen prosessointi uusiksi elintarvikkeiksi (Biotechnological process-ing of oat for novel food ingredients). In: Salovaara, H. and Sontag-Strom, T. eds. Kaurasta elinvoimaa. EKT-sarja 1221. Helsingin yliopisto, Elintarviketeknologian laitos. pp. 85–108 (in Finnish).

Plaza, L., de Ancos, B., and Cano, M. P. (2003). Nutritional and health-related compounds in sprouts and seeds of soybean (*Glycine max*), wheat (*Triticum aestivum*. L) and alfalfa (*Medicago sativa*) treated by a new drying method. *Eur. Food Res. Technol*. 216, 138–144.

Rickard, S. E. and Thompson, L. U. (1997). Interactions and psychological effects of phytic acid. In: Shahidi, F. ed. *Antinutritients and Phytochemicals in Foods*. Symposium Series 662. Washington DC: American Chemical Society, pp. 294–312.

Roberfroid, M. (2007). Prebiotics, the concept revisited. *J. Nutr*. 137, 830S–7S.

Rooney, L. W. and Awika, J. M. (2005). Speciality sorghums for healthful foods. In: Abdel-Aal, E. and Wood, P. eds. *Specialty Grains for Food and Feed*. St. Paul, MN: American Association of Cereal Chemists, pp. 283–312.

Sanni, A. I., Onilude, A. A., and Ibidabpo, O. T. (1999). Biochemical composition of infant weaning food fabricated from fermented blends of cereals and soybean. *Food Chem*. 65, 35–39.

Shelton, R. C., Keller, M. B., Gelenberg, A., Dunner, D. L., Hirschfeld, R. and Thase, M. E. (2001). Effectiveness of St. John's Wort in major depression: A ran-domized controlled trial, *JAMA* 285, 1978–1986.

Snart, J., Bibiloni, R., Grayson, T. *et al*. (2006). Supplementation of the diet with high-viscosity beta-glucan results in enrichment for lactobacilli in the rat cecum. *Appl. Environ. Microbiol*. 72, 1925–1931.

Solomon, P. R., Adams, F., Silver, A., Zimmer, J. and De Veaux, R. (2002). Ginko for memory enhancement: A randomized controlled trial, *JAMA*, 288, 835–840.

Temelli, F., Bansema, C., and Stobbe, K. (2004). Development of an orange-flavored barley β-glucan beverage. *Cereal Chem*. 81, 499–503.

The Health Council of the Netherlands (2001). Committee on the Safety Assessment of Novel Foods. Phytosterols (2). Publication no. 2001/04VNV.2001. The Hague: Health Council of the Netherlands.

van Erde, P. (1998). *Analytica-EBC*. Nürnberg: Hans Carl.

Wilhelmson, A., Oksman-Caldentey, K.-M., Laitila, A., Suortti, T., Kaukovirta-Norja, A., and Poutanen, K. (2001). Development of a germination process for producing high β-glucan, whole grain food ingredients from oat. *Cereal Chem.* 78, 715–720.

The marketing of gluten-free cereal products

17

Joe Bogue and Douglas Sorenson

Introduction

Developing and marketing new products in today's fiercely competitive business environment is an essential, and integral, part of a firm's business strategy and is widely acknowledged as being loaded with risk. Consumer demand for gluten-free cereal products is rising steadily with the increase in celiac disease and other allergic reactions to gluten from wheat-, rye- or barley-based foods. Gluten-free cereal products represent a growing market opportunity, within the global health and wellness market, for food manufacturers that develop consumer-led new products with high added-value levels, which ultimately gain consumer acceptance. This chapter examines the marketing of gluten-free cereal products and highlights some of the strategic issues that will be relevant to the successful marketing of such products. Following the introduction, an overview of the gluten-free market is outlined and emerging trends are discussed. Key strategic issues of importance to the health and wellness market are outlined and then applied to gluten-free cereal products, which presents a framework to guide manufacturers developing and marketing consumer-led gluten-free cereal products.

Gluten-Free Cereal Products and Beverages
ISBN: 9780123737397

Overview of the gluten-free market

The market for gluten-free cereal products is expected to rise significantly as consumer demand increases in reaction to increased levels of diagnosis of celiac disease and also as specific consumers make the conscious choice to remove gluten from their diets. Furthermore, the level of diagnosis is set to rise and it is thought to be the most under-diagnosed disease in the United States (Palmer, 2004). The market for gluten-free food products is growing to satisfy the needs of people with celiac disease and those who wish to exclude gluten from their diets, and is also, importantly, a market that is here to stay.

The United States market for gluten-free cereal products stood at US$700 million in 2006, and is estimated to continue to grow at 25% per year to reach US$1.7 billion by 2010 (Gourmet Retailer, 2006). The UK gluten-free foods market in 2006 was estimated at €225 million (Food Production Daily, 2006). According to the Gourmet Retailer (2006), most gluten-free cereal products are alternatives to traditional grain-based goods, including bakery products, pasta and cereals made with alternative grains and flours, such as rice and corn.

The difficulty associated with the development of the gluten-free market has been attributed to the strict processing requirements of the sector and also the perceived size of the market (Mark, 2006). However, according to Food Production Daily (2006) there was a 37.1% growth of gluten-free cereal products on the market in the UK between 2000 and 2002. It is estimated that 1 in 300 people in both Europe and the United States are intolerant to gluten, with higher figures reported in Germany (1 in 200) and in the UK (1 in 100) (Food Navigator USA, 2006a). Where markets are saturated, there are market opportunities for food firms who target the food intolerance segment of the market by focusing on this segment and thus gain a competitive advantage over rival firms.

What are the market requirements? When consuming a gluten-free diet is a necessity for consumers, they are looking for gluten-free cereal products with the same appearance and texture as conventional products. However, the majority of the gluten-free bakery products on the market are of very poor quality, particularly when compared with their wheat counterparts (Arendt et al., 2002). The increasing number of people with celiac disease being diagnosed each year and their desire for more better-tasting and better-textured products offers great market opportunities for food manufacturers (Shinsato, 2006). As Coeliac UK (2007) noted, the market for gluten-free cereal products is growing in scale and also, significantly, in sophistication. However, as noted by Food Navigator USA (2006b) some major food corporations have not as yet entered the market, as they are reluctant to invest in research and development until fixed regulations for gluten-free cereal products are in place. The Food and Drug Administration (FDA) was required to propose a regulation in 2006, and a final regulation will be issued in 2008 to define the term "gluten-free" for voluntary use in food labeling.

Gluten-free market trends

The market for gluten-free cereal products is expected to grow enormously over the next few years and this will provide many product development opportunities for firms to market new gluten-free cereal products that are tasty and affordable (Reeves, 2006). Across food markets, according to Milton (2003), the key food areas for future new product development (NPD) include: convenience foods, foods with perceived health benefits, low fat and organic products, range extensions, extending brands, product improvements, new categories, and premium quality foods. These areas offer opportunities for manufacturers to develop gluten-free cereal products that gain consumer acceptance. In addition, this market opportunity can be illustrated further as neither drugs nor surgery treat the celiac condition, it is only treatable through the consumption of gluten-free foods (Reeves, 2006).

A strong food trend for gluten-free cereal products was evident in 2006 and this is likely to continue towards 2010 with other ingredient/nutrient drivers including: omega-3 oils, specialized proteins, probiotics, and prebiotics (Stagnito Communications, 2006a). These specific food ingredient platform drivers are expected to drive NPD activities across all food, beverage, and supplement markets. It is estimated that the market for gluten-free foods and beverages will grow by a factor of 10 up to 2010, which will provide opportunities for the development and marketing of new gluten-free food and beverages (Food Navigator USA, 2006b). For example, in the beverage sector Anheuser-Busch has developed a sorghum beer called Redbridge targeted at consumers following a wheat-free or gluten-free diet (Nutra Ingredients USA, 2007). Opportunities in the gluten-free market can be viewed in light of the fact that 97% of people with celiac disease remain undiagnosed and go untreated (Gourmet Retailer, 2006). As Reeves (2006) noted, in countries where there is more testing undertaken (e.g. Italy), the figure for those with celiac disease is notably higher.

The product range of gluten-free products includes: dressings, drinks, pizza, frozen entrées, baking mixes and flours, gluten-free sweetener syrups, beer, and confectionery products. According to Palmer (2004), popular gluten-free cereal products requested by consumers include: bread products, pizza crusts, pastas, snack foods, gluten-free flours and baking mixes, cakes, cookies, and bars. This product range is likely to broaden as busy people with celiac disease seek food products that fit in with their lifestyles, including: on-the-go foods, fast foods, snack foods, ready meals, and functional beverages. These lifestyle trends offer significant new product opportunities for "forward thinking product developers" to produce products that fit in with changing consumer lifestyles (Reeves, 2006). For example, Wennström and Mellentin (2003) report that consumers are increasingly accommodating meals when they can, rather than planning their activities around meals. This has resulted in more flexible, rather than fixed meals, for consumers with busy lifestyles. Furthermore, the convenience trend has meant that consumers seek everyday meals that are simple and easy to prepare.

One major trend noted in the gluten-free market is the move from the prescription gluten-free product market to the speciality and mainstream gluten-free food markets and this is growing steadily every year (Nutra Ingredients USA, 2004; Food Production Daily, 2006). In addition, the production of gluten-free cereal products offers opportunities for craft industries such as bakeries to increase sales and levels of added-value within their product portfolios. This market opportunity also provides a means of product differentiation from mass-produced goods from industrial bakeries (Food Navigator USA, 2006a). This product differentiation can be seen in

Anheuser-Busch's sorghum beer Redbridge that was developed as a hand-crafted specialty beer made without wheat or barley (Nutra Ingredients USA, 2007). This niche market product is clearly targeted at those consumers who wish to exclude gluten from their diets. A niche market like this can be attractive to firms as it typically attracts fewer competitors (Kotler, 2000). Sources of new product ideas for gluten-free cereal products can be generated from consumers, competitors, distribution channels, employee suggestions, and management.

Marketing issues and novel foods

Wennström and Mellentin (2003) identified five strategies for entering the health and wellness market: leveraging hidden nutritional assets, new category creation, new segment creation, category substitution, and a food product make-over. These market entry strategies can be applied to firms when entering the gluten-free market as firms aim to exploit market opportunities across food and beverage categories. This applies to the way in which innovative ingredients can be used in the development of gluten-free cereal products whereby consumers may need to be convinced of the benefits of these ingredients and that they are safe to use, have improved sensory qualities, and a long shelf-life. In that context, there have been many barriers to consumer acceptance of novel foods in the health and wellness market. Consumers are often unfamiliar with novel ingredients, and are also unaware of the benefits from consumption of these ingredients. For example, in terms of nutrient-enriched foods there is a certain risk associated with the purchase of these foods. One such risk is that of consumers exceeding the recommended daily intake of essential nutrients, which may render them toxic (Frewer et al., 2005). Similarly, for people with celiac disease there is a risk involved in the purchase of gluten-free cereal products, and as Reeves (2006) suggested, celiac disease patients had become "hyper-vigilant about food, reading labels religiously during long, slow grocery shopping trips." For most consumers food safety is a priority and there is an expectation that food supplied for human consumption is safe and nutritious to eat (Frewer et al., 2005). However, for people with celiac disease there is a risk perception associated with the foods they consume and whether they contain traces of gluten that may adversely impact on their health.

While there have been some very notable failures in the health and wellness market, strategic marketing lessons have been learnt from products that have been successful in the marketplace. Wennström and Mellentin (2003) studied the strategic success of specific health and wellness foods and beverages and concluded that successful product design and positioning in this market was related to four key success factors: the consumer needing the product; the consumer accepting the ingredient; the consumer understanding the benefits of the ingredients; and the consumer trusting the brand. In relation to gluten-free cereal products the four success factors can also be applied to the target market for gluten-free cereal products. This target market includes both patients with celiac disease and those who do not have celiac disease but also wish to consume gluten-free cereal products. If Wennström and Mellentin's (2003) four success factors are applied to the gluten-free market, the consumer initially must want the food or beverage that is being developed and it must fit in with their lifestyle. In addition, they must also accept, and understand, the ingredients that substitute for gluten in terms of what people with celiac disease can tolerate and those ingredients they cannot tolerate. Finally, they must also trust the brand and feel that the brand can deliver in terms of product quality and safety. Wennström and Mellentin's (2003) four success factors can play an important role in guiding the marketing strategy for the development of successful gluten-free foods.

The gluten-free target market

Heasman and Mellentin (2001) recounted the difficulties faced by food and beverage manufacturers in the identification of consumer groups to target with new and innovative health and wellness products. Overall, the key factors for new product success in the health and wellness market have been summarized as: overcoming consumer acceptance issues; proof of efficacy; legislative issues concerning the promotion of health and wellness products making health claims; product promotion and consumer education; and importantly, the identification and selection of key target markets (Hilliam and Young, 2000; Heasman and Mellentin, 2001; Bistrom and Nordstrom, 2002).

A firm entering a new market must identify a segment, target the segment, and then position its products within the market. Market segmentation is the management tool that enables a total market to be divided into consumer groups that can be served by specific market programs (Hisrich and Peters, 1991). This means that firms serve groups of consumers with specific but similar wants, needs, and behaviors with respect to a product (Meulenberg and Viaene, 2005). A market segment can be defined as: "a consumer group whose expected reactions will be similar when faced with a given marketing mix. A segment seeks a unique set of benefits from the product or service purchased" (Bradley, 1995). Therefore, a key strategic decision associated with gluten-free foods is initially the identification of a suitable market segment in terms of whom these products are targeted at in the marketplace. The identification of a suitable segment in the gluten-free market can be carried out through a behavioral segmentation process whereby consumers are grouped according to the likely benefit they would seek or derive from a product (Kotler, 2000). In the case of the gluten-free market, consumers seek a very specific benefit to be had from the product that they purchase, i.e. that the product is gluten-free and delivers quality and safety to the purchaser.

There are various groups of consumers that comprise the gluten-free market segment, i.e. those consumers with food intolerances and also those who wish to exclude gluten from their diets. According to Food Navigator USA (2006b) gluten-free foods are not only consumed by those who have celiac disease but also other family members who may wish to avoid buying different versions of the same products. In addition, as celiac disease is known to be hereditary, family members may consume gluten-free cereal products for preventative reasons, as celiac disease is often under-diagnosed and it may often not be obvious to the patient. Furthermore, according to Mark (2006) gluten-free cereal products can also be beneficial for other medical conditions such as autism or attention deficit disorder, and for those consumers who are allergic to wheat, eggs, soy, and milk. Therefore, if other members of a family are consuming gluten-free cereal products there is a need for food manufacturers to ensure the intrinsic aspects of the products, in terms of taste, texture, and appearance

Furthermore, another distinct target market may also be the consumer who just wishes to exclude gluten from their diet. According to the Gourmet Retailer (2006) these consumers may be migrating from the organic and natural foods markets and profiles these as middle- to upper-class consumers. They may try to avoid consumption of foods with allergens that they may feel may exacerbate other health conditions, such as migraine and menstruation. In addition, there are also consumers who do not have celiac disease but wish to avoid wheat for other reasons. However, targeting such attitudinally differentiated market segments presents challenges for firms pursuing opportunities in the healthy foods market, particularly where such foods meet with poor consumer acceptance (Verbeke, 2004; Saher *et al.*, 2004). This suggests that consumer acceptance issues have been either ignored or poorly understood by firms. As Wennström and Mellentin (2003) argue: "often technology is used to create value for the producer and this can sometimes be a very different matter from creating customer value." Therefore, as the consumer base widens for gluten-free cereal products there will be an increased emphasis on improving the texture and taste profiles of gluten-free cereal products (Food Production Daily, 2006).

The broadening market base can be illustrated by Sunstart Bakery in the UK who provide gluten-free cookies to multiple retail outlets. They noted that their market was split between people with celiac disease and people with mild wheat intolerance or simply consumers who wished to avoid wheat (Food Production Daily, 2006). The development of new gluten-free cereal products of higher quality than those currently on the market, such as bread, with increased nutritional value, longer shelf-life, and a similar texture to regular bread may cater for the broadening market for gluten-free cereal products (Medical News Today, 2006).

Product positioning in the gluten-free market

According to Hisrich and Peters (1991) a lack of understanding of a brand's position in the marketplace is the major cause of product failure for about 80% of all new products introduced. Kotler (2000) defined positioning as: "the act of designing the company's offering and image to occupy a distinctive place in the target market's mind." All elements of the marketing strategy therefore help define a product's position in the marketplace, and firms need to identify a good position for a product in the market and then promote this position to consumers. By having a clear position in the marketplace, brands can develop a strong source of competitive advantage. In terms of product positioning consumers live in an environment where they are exposed to a continuous onslaught of marketing messages and as a defense against marketing information overload, consumers screen or reject much of the marketing information offered to them (Trout and Ries, 1995). Therefore, how can a food firm ensure that its message is heard and received by the consumer? Firms carry out market research to determine positioning options and how consumers might react to various marketing strategies. One of the most important strategic marketing issues is that the positioning strategy differentiates the product from its competitors in the minds of the target market.

For firms marketing gluten-free cereal products a specific marketing question then becomes obvious: how should these products be positioned as an essential part of a successful marketing strategy? Market positioning initially concerns segmenting a market, and then targeting a group of consumers with a product. The last step of the positioning process is to position the specific product in the mind of those consumers in relation to the competitors' products. In the gluten-free market firms can develop positioning strategies based on their product's benefits (gluten-free) or based on their product's attributes (superior sensory qualities), or both. For example, Stagnito Communications (2006b) reported that new health label claims were appearing on several brands or categories of established products, as companies tried to reposition them with narrowly targeted wellness appeals. For example, Unilever, through Slim-Fast its line of dietetic shakes and energy bars, had launched easy-to-digest versions of its products that contained no dairy or gluten-derived ingredients. These products were intended for consumers with gluten allergies or lactose intolerance.

For people with celiac disease one of the major product requirements of gluten-free products is that they do not contain any traces of gluten. The positioning of these products in terms of their assured gluten-free reliability will therefore be critical to people with celiac disease as the market is very purity conscious (Barr, 2004). This means that manufacturers must produce products that meet the needs of the target market in terms of the recipes and processes used to produce gluten-free cereal products. In this regard the development of strong, trustworthy brands is likely to be very significant to these consumers, in terms of these brands being synonymous with quality but also reducing the risk involved in the purchasing decision. To respond to the purity conscious consumer with celiac disease, certain firms have opened food facilities to exclusively manufacture gluten-free cereal products. This prevents cross-contamination between gluten-free grains and the ordinary varieties, thus offering reassurance and reducing the risk when purchasing new products.

In addition, there is a need to develop products that are tasty and exciting, just like the conventional products they will replace. In terms of food choice there are many influences on the products consumers choose, but product taste, texture, and appearance are still central to product acceptance, even in the gluten-free market and particularly as this market becomes more sophisticated.

The marketing mix and gluten-free cereal products

The development and successful marketing of new products have emerged as two key critical strategic concerns of firms. The development of successful marketing strategies for gluten-free cereal products is essential to these products gaining overall consumer acceptance and for people with celiac disease to integrate these products into their lifestyles as part of a strategy to keep their condition under control. Patients with celiac disease can live a normal life as long as they avoid gluten and so the extrinsic attributes (marketing) of these products in terms of brand, labeling, perceived quality, and packaging will be central to initial consumer purchase and subsequent repurchase. For a company developing a new or modified gluten-free product a key ingredient is formulating a marketing strategy that contains the correct mix of product features, branding and packaging, price, distribution channel, and promotion for the particular product/market situation (Hisrich and Peters, 1991). This clearly necessitates the integration of consumer information into the process of developing and marketing new gluten-free foods and beverages.

Most activities in the product and market development process are conducted in a probabilistic setting, and that uncertainty is characteristic of the early stages of this process in terms of identifying new product concepts, segments, and market strategies that would be most promising, and ultimately gain consumer acceptance (Kim and Wilemon, 2002). Slater and Narver (1996) and Moorman (1995) argue that a market-oriented culture, i.e. an organization's culture and associated activities that is consumer-focused, reduces many of the risks associated with the process of developing and marketing new and innovative products. As Calantone *et al.* (1996) note: "it is important to collect and assess market and competitive information in order to understand consumers' needs, wants and specifications for a product in order to understand consumers' purchase decisions, and to learn about competitors' strategies." Therefore, market-oriented organizations continuously monitor their external environments for both market opportunities and threats from competitors. By focusing on consumers' needs, market-oriented firms are well positioned to recognize emerging needs and rapidly assess consumers' responses to new products. Indeed, through their market-scanning efforts, market-oriented firms are able to discover under-developed market niches and segments, and are also capable of identifying opportunities created by competitors' miscues (Slater and Narver, 1996). A firm's intelligence-generation systems and processes will therefore heavily influence the outcomes of the product and market development process.

Many firms fail to implement and manage formal intelligence-generation processes and neglect critical stages of the product and market development process (Harmsen, 1994; Bogue, 2001). Although market intelligence can be generated throughout the product and market development process, researchers such as Bogue (2001), Urban and Hauser (1993), and Cooper (1988) argue for the integration of "voice of the consumer" information, particularly at the early stages of the process, where consumers' unmet needs and wants can be identified. The early stages of the product and market development process is the period when opportunities are first considered and move through the stage-gate process for further development. Indeed Cooper (1993) stresses the importance of proficiency in these early stages and argues against avoiding front-end activities, as oversights in relation to front-end activities would increase the risk of product failure. Companies therefore need to gain a greater understanding of the "voice of the consumer" in order to develop and market successful new products.

So how does a market-oriented approach to business benefit firms that wish to develop and market gluten-free cereal products? Factors such as rapidly changing technologies, new market entrants, changing consumer trends, and shortening product life cycles are amongst some of the drivers that have increased both the amount of new product activity, and also the inherent risk involved. Acknowledging this risk, pertinent questions are as follows: are there ways in which the marketer can improve the chances of new product success in the gluten-free market? What product development and marketing strategies are likely to be successful in the gluten-free market?

In increasingly competitive markets, marketers are under pressure not only to reduce product development times, but also to improve product quality as perceived by the consumer. One widely acknowledged means of achieving both of these requirements is to pay more attention to understanding consumers' needs at the product design stage, and to translate those needs into products that satisfy the consumer. Theoretically, incorporating the consumer into the new food product development process at an early stage should aid in the understanding of consumer needs, satisfying those needs, and ultimately developing better products and designing more effective marketing programmes. For example, in many high-technology industries, and business-to-business markets, the consumer has been utilized at an early stage of the product development process with high levels of success. However, in more complex markets, such as consumer food markets, utilizing the consumer for product development is more difficult, where there is increasing separation between the manufacturer and the consumer along the food supply chain.

The development of gluten-free foods and beverages, that gain consumer acceptance, poses many technical and marketing challenges for NPD personnel in terms of optimizing the marketing (extrinsic) and technical/sensory (intrinsic) attributes of these foods. These extrinsic and intrinsic attributes have a strong influence on consumer acceptance of such foods. In a market-oriented NPD process consumers are viewed as co-designers of products since they can make an effective contribution to new food product design, and the integration of the consumer with the NPD process can best be achieved at the pre-development stages of concept ideation, concept screening, and optimization (Cooper, 1993; Bogue, 2001; Sorenson and Bogue, 2005). In addition, consumers can play a role in the design of effective and efficient marketing strategies by firms. For example, Anheuser-Busch when they developed their Redbridge sorghum beer worked closely with the National Foundation for Celiac Awareness (NFCA) to get a better understanding of the needs of consumers who were leading gluten-free or wheat-free lifestyles (Nutra Ingredients USA, 2007). Thus, the idea is to stay close to the consumer when developing and marketing new gluten-free cereal products to include lifestyle factors in a successful marketing strategy. This is a point noted by the managing director of the leading gluten-free marketing firm Schär from Italy: "I think that only by remaining close to consumers can we be sure that we are going in the right direction" (Schär, 2006). Grunert (2005) points out that we may not be able to ask consumers what new products they want, but by understanding consumer behavior and purchase motivations we can help reduce the failure rates for new products.

So how can marketers and technical R&D personnel integrate the consumer with the NPD process in the design of optimal gluten-free cereal products? Market-oriented consumer research techniques, which focus on the early stages of the NPD process, lead to a more systematic and multidisciplinary approach to product development. There are a family of consumer research techniques that can utilize both technical R&D and marketing information. These techniques promote closer integration between the marketing and technical functions.

Gluten-free cereal products: the application of consumer research techniques

One of the ways of integrating the consumer at the early, or pre-development stages of the NPD process, is through the use of various market research techniques such as: ethnography, in-depth interviews, focus groups, conjoint analysis and sensory analysis. The purpose of these research techniques is to use information generated from the consumer, through consumer-oriented research methodologies, to identify product ideas and concepts as well as related consumer segments, and finally to aid in the design and marketing of gluten-free products that gain consumer acceptance. The market information generated can then be used to model consumer preferences for new product concepts and predict consumer acceptance, and identify potential consumer segments and market opportunities for these concepts. This consumer-led approach is vital if new gluten-free products are to prove commercially successful and gain consumer acceptance. In addition, through consumer integration it is likely to enable firms increase speed to market of new or modified gluten-free cereal products.

Qualitative consumer research techniques such as ethnography and in-depth interviews have a critical role to play at the concept ideation and generation stages of the NPD process for gluten-free foods. For example, ethnography and in-depth interviews generate rich data on life experiences and reveal a wealth of information on informants (McDaniels and Gates, 1991). As Hill (1993) comments: "living through the highs and lows of informants allows the researcher to know the phenomenon under investigation in a way that few other methodologies permit." Elliott and Jankel-Elliott (2002) argue that it can be difficult to truly participate with consumers in a situation without disturbing the authenticity of the behaviors. In that sense, ethnography can be used to study consumers' behavior as it occurs in their natural setting, and specifically, the problems and challenges that people with celiac disease or food intolerances experience in their everyday lives, in terms of food preparation and decisions. On the other hand, emerging behavior patterns can be recognized better and earlier through in-depth interviews, which provide greater insights into multi-faceted behaviors, attitudes, and motivations of respondents, and can lead to the identification of new product and market opportunities (Kiener, 1995; Krueger and Casey, 2000).

In terms of understanding consumers' needs for product design, focus groups are an obvious and most suitable initial NPD starting point (van Trijp and Steenkamp, 1998; Sultan and Barczak, 1999). One of the main benefits of focus groups for NPD is that they allow consumers to express themselves in their own language, and allow for social interaction in a group setting. Product developers must then interpret this "voice of the consumer" information. The focus group technique therefore has a unique strength in that it is well suited to: determining innovation possibilities; exposing consumers to new technologies and products; providing perceptions on different products; gathering consumers' impressions of new concepts, services, and products; and stimulating new ideas (McDonagh-Philip and Bruseberg, 2000; Sorenson and Bogue, 2005). Focus groups can be used to gain unique insights into consumers' perceptions of gluten-free product ideas and concepts, including an assessment of: packaging alternatives, health claims, acceptance of novel substitution ingredients, and alternative positioning and pricing strategies. Overall, the data generated through qualitative consumer research techniques can therefore assist product developers in the design of gluten-free cereal products that meet consumer expectations and the development of successful marketing strategies.

Differentiated markets are now recognized as more commonplace in recognition of the heterogeneous nature of consumers' values, needs, beliefs, and preferences (Moriarty and Reibstein, 1986). Conjoint analysis is a multivariate technique that models purchase decision-making processes through an analysis of consumer trade-offs among hypothetical multi-attribute products (Green and Srinivasan, 1978). Conjoint analysis has a number of commercial applications of relevance to both marketers and R&D personnel involved in the marketing and development of gluten-free cereal products. From a marketing perspective, conjoint analysis has been used extensively: to estimate the value that consumers associate with particular value-added product features; to segment markets based upon the differing benefits sought out by consumers; and to design effective product pricing and positioning strategies based on consumers' trade-off decisions among alternative design features (Green and Krieger, 1991; Herrmann et al., 2000; Sorenson and Bogue, 2005). From an R&D perspective, conjoint analysis is becoming increasingly important in terms of: defining consumer-led new product concepts with the optimal combination of features; predicting consumers' preferences for new concept features; and identifying viable market opportunities for new products not presently on the market (Hair et al., 1998; Kamakura, 1998; Sorenson and Bogue, 2006). Conjoint analysis is therefore an extremely significant consumer research technique, which can bring R&D personnel closer to understanding the "voice of the consumer" and facilitate more effective integration of marketing and R&D personnel to aid in the new product design process for gluten-free foods and beverages.

The development of the health and wellness market depends upon sensory acceptance of such products by consumers in terms of taste parity with conventional products (Gray et al., 2003). However, Wennström (2000) reported that many new health and wellness foods and beverages met with poor consumer acceptance. In that sense, sensory analysis is important for the development of gluten-free cereal products due to the role sensory perception plays in food choice, and in increasingly competitive markets, provides firms with "actionable" information on their products and those of their competitors. More specifically, the sensory properties of foods are extremely important to food product developers and marketers because they relate directly to product quality and consumer acceptance. Bogue et al. (1999) argue that integrated market and sensory analysis recognizes the demands of consumers and their own sense of quality. In that context, sensory analysis provides a natural link between marketing and technical R&D, which provides a more complete picture of consumers' preferences (Biedekarken, 1993). Sensory analysis therefore provides a link between the consumer and the product. In addition, it provides direction for the development of products with superior performance and for effective positioning strategies (Goldman, 1994). Sensory analysis has naturally developed from its traditional role in quality control, to being used in product development, by contributing to the understanding of consumers' sensory preferences and the evaluation of competing products from a sensory perspective (Bogue *et al.*, 1999). In recent times, sensory science has developed to provide detailed information on consumer acceptance of foods and has become a significant market research technique (Bogue *et al.*, 1999). Sensory analysis can therefore provide marketers with: a greater understanding of the sensory quality of gluten-free cereal products; direction for consumer relevant product quality; sensory evaluation of new gluten-free product concepts by consumers; and sensory profiles of competitors' gluten-free cereal products from a consumer perspective.

Branding and product promotion strategies for the gluten-free market

On average, consumers take only 12 seconds to make a brand selection (Moorman, 1995). Branding helps consumers identify a product while assisting the firm in image building and planning other elements of the marketing mix. When consumers hear brand names it helps them to visualize images that are different from competitors, as brand names inspire acceptance, preference, and loyalty among consumers. In that sense, consumers do not regard these products to be the same as competitive products because they value the reputation of the company or brand name (Hisrich and Peters, 1991). Jones (2001) visualized today's marketplace as a war of ideas and noted that if firms did not stand for something then they would not stand out. In particular, he felt that firms needed something deeper than a brand: an emotional big idea that would give firms space and keep rivals at bay. Therefore, the importance of a brand to the purchase decision-making process cannot be understated when people with celiac disease seek reassurances in terms of what they are purchasing and try to reduce the risk associated with purchase decisions. For people with celiac disease the permanent consumption of gluten-free cereal products for their whole life illustrates the significance of the purchase decision in relation to food products. Very low levels of gluten, as little as 0.1 g per day, can adversely affect them and so the strategies used to market gluten-free cereal products are critical to consumer acceptance (Palmer, 2004). In addition, the risk associated with an incorrect purchase decision also illustrates the importance of brand awareness to people with celiac disease as well as how these products are positioned and promoted in the marketplace.

For those marketing gluten-free cereal products there is an opportunity for firms to distinguish themselves in terms of what their products offer for people with celiac disease, and for those that wish to consume gluten-free cereal products.

Jones (2001) felt that firms could develop a competitive advantage and differentiate their offering in the marketplace by having a big idea behind their brand.

Consumers' attitudes and responsiveness towards branding and product promotion strategies can vary across cognitive and socio-demographic groups, with different branding and promotional strategies appealing to different consumers. For example, while on-pack promotions and information regarding health benefits generally appeal to adults aged 35–59 years in the health and wellness market, younger adults aged 20–34 years tend to be more influenced by honest and informative positioning and communication messages (Boyle and Emerton, 2002). It is important, however, to communicate the benefits of gluten-free cereal products to consumers without marginalizing other potential segments within the market. This is particularly true for the gluten-free market where firms might wish to develop gluten-free products with greater mainstream appeal, and not exclusively target people with celiac disease. To achieve this, a firm would need to position gluten-free products at different points in the product life cycle by meeting the immediate needs of people with celiac disease first, and then, reposition these products to appeal to mainstream consumers through communicating their benefits in a credible manner. As Wennström and Mellentin (2003) posited:

> the key to a winning strategy [for health and wellness foods] is to identify a single bridgehead of pragmatic consumers in a mainstream market and to accelerate the formation of 100 per cent of their whole product. The goal is to win a niche foothold in the mainstream as quickly as possible.

This necessitates servicing the needs of different consumers, segment by segment, through tailoring the marketing mix in order to develop the market for gluten-free products with greater mainstream appeal.

Distribution channels and gluten-free cereal products

A good distribution channel is essential for product success, and channel systems must be at all times accessible to existing customers, and developed for potential new target markets (Hisrich and Peters, 1991). There are five key distribution channels for health and wellness foods and beverages, which are often specific for particular product categories. These key distribution channels are: online channels; mail order/direct sales; pharmacy; health food and specialist stores; and supermarkets (Moosa, 2002). Traditionally, specialty shops, health shops, and organic stores pioneered gluten-free cereal products. In the United States in 2006 sales of gluten-free cereal products were still dominated by specialist marketing channels: health food and natural food stores (40%); speciality food websites or catalog purchases (20%); and mainstream supermarkets (14%). For example, the United States firm 'Cause You're Special! offers a range of gourmet gluten-free foods on-line and also at health food, specialty, and grocery stores across the United States, Asia, and Europe (Chase, 2006).

One of the major issues with gluten-free distribution channels is moving the products from niche to mainstream retail outlets. However, the increasing number of supermarkets that stock gluten-free selections highlights the potential for mainstream distribution of gluten-free cereal products. In that context, Moosa (2002) stated that the development of the health and wellness market, which includes gluten-free products, from specialist small-scale to mainstream distribution channels would depend on: increased demand by mainstream consumers; interest by large-scale food and beverage firms and ingredients companies to stimulate growth within niche categories; retailers seeking to differentiate themselves from their competitors; and importantly, the identification of sustainable pricing strategies.

Pricing gluten-free foods

Challener (2000) and Hasler (1996) characterize health and wellness products, which include functional, organic, and gluten-free food and beverages, as "breakthrough" products that on one hand can provide value to consumers, while on the other hand potentially deliver new product success in the marketplace. The health and wellness market has indeed the potential to realize strategic competitive advantages for both manufacturers and retailers in terms of value creation for long-term growth and profitability (Mark-Herbert, 2004). Weststrate *et al.* (2002) and Shah (2001) affirm that the health and wellness market has proven attractive to firms with an average growth rate ranging from 15 to 20% per annum, in comparison to growth rates of 2–4% per annum for the general foods market. The real attraction of the health and wellness market therefore lies in adding value to otherwise conventional foods and beverages in reaction to the downward pressure on price, where consumers are increasingly seeking greater value for money in their food and beverage choices (Longman, 2001; Moosa, 2002).

In fact, one of the main reasons for the lack of development of the gluten-free market, and firms not entering it, is the cost of producing gluten-free cereal products. The pricing strategy of gluten-free foods is linked to the technology, and often, the long ingredient list involved in the development of such foods. Therefore, gluten-free cereal products tend to be priced at a premium level and this can often prevent people with celiac disease strictly adhering to a gluten-free diet (Food Navigator USA, 2006b). More so, as the pricing of foods that are free from gluten remains high, the target market for gluten-free foods and beverages remains limited, particularly when consumers do not perceive value or immediate benefits from the elimination of gluten from their diet. For example, Heasman and Mellentin (2001) and Hilliam and Young (2000) attributed the comparatively poor performance or withdrawal of many high-profile health and wellness brands, e.g. Novartis' Aviva, Raisio's Benecol, and General Mills' Maval, to over-pricing, and specifically, the pursuance of a mass-marketed product through a premium pricing strategy.

Hilliam and Young (2000) therefore questioned the sustainability of pricing strategies that sought exceedingly high premiums above standard conventional products. On that basis Heasman and Mellentin (2001) and Von Alvensleben (2001) argued that, in future, consumer tolerance of premium prices for health and wellness products would depend upon: the intended target market; the strength of the health proposition; the positioning strategy; and issues related to the product format such as naturalness, convenience, or sensory pleasure. Wennström and Mellentin (2003) concluded that, in future, firms would need to identify the optimal pricing strategy or premium that consumers would be willing to pay for specific health and wellness products, in order to remain competitive in the market. In terms of gluten-free foods, if the products are to go more mainstream it is likely that less premium pricing strategies will be more successful.

Conclusions

The increasingly competitive nature of the health and wellness market, and the inherent risks associated with the new food product development process, highlight the importance of involving the consumer in the development and strategic marketing of gluten-free foods and beverages. The gluten-free market should be viewed more as a long-term strategy for future growth in the health and wellness market and less as a short-term strategy for high profitability, owing to the niche market nature of the gluten-free market presently. However, it is likely that the target market of gluten-free cereal products will extend in the future to not just include people with celiac disease but also those who desire products without allergens or other ingredients that may negatively influence their health. In this regard the identification of key extrinsic (marketing) and intrinsic (sensory) product attributes that influence consumer acceptance will need to be identified and incorporated into the development of new gluten-free foods. The development and marketing of gluten-free cereal products will therefore need to change in the future to reflect changing consumer lifestyles, and ensure that gluten-free cereal products are designed with the consumer in mind.

Involving the consumer in the process of developing and marketing gluten-free foods through market research provides for a more systematic means of managing consumer knowledge in new food product development. Firms that adopt this market-oriented approach to business will benefit from a deeper understanding of consumers' choice motives and value systems. Being more market-oriented will assist R&D personnel become more consumer-led in terms of product design, assist marketers to identify new and emerging market segments, and more accurately inform strategic marketing decision-making when bringing gluten-free foods and beverages to the market. This in turn can improve the competitiveness of both food and beverage manufacturers and retailers, and increase the chances of new product success in the gluten-free market.

Sources of further information and advice

Brody, A. L. and Lord, J. B. (2000). *New Food Products for a Changing Marketplace*. Pennsylvania: Technomic Publishing Company.

Earle, M. and Earle, R. (2000). *Building the Future on New Products*. Management Series. Surrey: Leatherhead Publishing.

Jongen, W. M. F. and Meulenberg, M. T. G. (2005). *Innovation in Agri-food systems: Product Quality and Consumer Acceptance*. Wageningen, The Netherlands: Wageningen Academic Publishers.

Journal of Product Innovation Management: http://www.pdma.org/journal/*Just-food*: http://www.just-food.com/

MAPP—Aarhus School of Business: http://www.asb.dk/research/centresteams/centres/mapp.aspx

Moskowitz, H. R., Sebastiano, P., and Silcher, M. (2005). *Concept Research in Food Product Design and Development.* Malden, MA: Blackwell Publishers.

Padberg, D. I., Ritson, C., and Albisu, L. M. (1997). *Agro-food Marketing.* Oxford: CAB International.

Product Development and Management Association—*Visions Magazine*: http://www.pdma.org/visions

Stagnito Communications: http://www.stagnito.com/

UCC—Department of Food Business and Development: http://www.ucc.ie/academic/foodecon/

References

Arendt, E., O'Brien, C., Schober, T., Gormley, T., and Gallagher, E. (2002). Development of gluten-free cereal products. *Farm and Food* 12, 21–27.

Barr, W. (2004). Value-added corner: amazing grains. *Rural Co-operative Magazine* July–August.

Biedekarken, O. (1993). Positionierung neuer bzw. Modifizierter nahrungs- und genumittel durch integrierte Markt- und Sensorikforschung (Positioning of new or modified products through integrated marketing and sensory analysis). In: Benz, K. ed. *Successful Food Marketing to the European Consumer.* European Sensory Network Symposium, 23–24 September, Rome, 1999.

Bistrom, M. and Nordstrom, K. (2002). Identification of key success factors of functional dairy food product development. *Trends Food Sci. Technol.* 13, 372–379.

Bogue, J. (2001). New product development and the Irish Food Sector: a qualitative study of activities and processes. *Irish J. Manage. incorporating IBAR* 22, 171–191.

Bogue, J. C., Delahunty, C. M., Henry, M. K., and Murray, J. M. (1999). Market-oriented methodologies to optimise consumer acceptability of Cheddar-type cheeses. *Br. Food J.* 101, 301–316.

Boyle, C. and Emerton, V. (2002). *Food and Drinks through the Lifecycle.* Surrey: Leatherhead International.

Bradley, F. (1995). *Marketing Management.* Hertfordshire: Prentice Hall.

Calantone, R., Schmidt, J., and Song, M. (1996). Controllable factors of new product success: a cross-national comparison. *Marketing Sci.* 15, 341–358.

Challener, C. (2000). Functional foods market offers promise and risk. *Chem. Market Report.* 257, 16.

Chase, K. (2006). Cause You're Special. www.glutenfreegourmet.com

Coeliac UK (2007). About gluten-free on the go. www.gluten-free-onthego.com

Cooper, R. G. (1988). Pre-development activities determine new product success, *Indust. Market. Manage.* 17, 237–247.

Cooper, R. G. (1993). *Winning at New Products*, 2nd edn. Reading: Addison-Wesley.

Elliott, R. and Jankel-Elliott, N. (2002). Using ethnography in strategic consumer research, Discussion Paper No. 02/02. Exeter, UK: School of Business and Economics.

Food Navigator USA (2006a). Gluten-free the key to boost craft bakery sales? www. foodnavigator-usa.com

Food Navigator USA (2006b). Gluten-free market set to boom, says report. www. foodnavigator-usa.com

Food Production Daily (2006). New opportunities for gluten-free market. www. foodproductiondaily.com

Frewer, L., Fischer, A., Scholderer, J., and Verbeke, W. (2005). In: Jongen, W. M. F. and Meulenberg, M. T. G. eds. *Innovation in Agri-food systems: Product Quality and Consumer Acceptance.* Wageningen, The Netherlands: Wageningen Academic Publishers.

Goldman, A. (1994). Gaining competitive edge: sensory science in the marketplace. *Cereal Foods World* 39, 822–825.

Gourmet Retailer (2006). Gluten-free market set to explode. *Gourmet Retailer* September 2006, 27. 13–3.

Gray, J., Armstrong, G., and Farley, H. (2003). Opportunities and constraints in the functional food market. *Nutr. Food Sci.* 33, 213–218.

Green, P. E. and Krieger, A. M. (1991). Product design strategies for target-market positioning. *J. Product Innovation Manage.* 55, 20–31.

Green, P. E. and Srinivasan, V. (1978). Conjoint analysis in consumer research: issues and outlook. *J. Consumer Res.* 5, 103–123.

Grunert, K. (2005). Consumer behaviour with regard to food innovations: Quality perception and decision-making. In: Jongen, W. M. F. and Meulenberg, M. T. G. *Innovation in Agri-food systems: Product Quality and Consumer Acceptance.* Wageningen, The Netherlands: Wageningen Academic Publishers.

Hair, J. F., Anderson, R. E., Tatham, R. L., and Black, W. C. (1998). *Multivariate Data Analysis*, 5th edn. New Jersey: Prentice-Hall.

Harmsen, H. (1994). Improving market-oriented product development in Danish food companies. In: *Managing the R&D Process.* University of Twente, School of Management Studies: Enschede; TQC, Twente Quality Centre; Milano: Politecnico di Milano Dipartimento di Economia e Produzione.

Hasler, C. M. (1996). Functional foods: the western perspective. *Nutr. Rev.* 54, 6–10.

Heasman, M. and Mellentin, J. (2001). *The Functional Foods Revolution. Healthy People, Healthy Profits?* Surrey: Leatherhead International.

Herrmann, A., Huber, F., and Braunstein, C. (2000). Market-driven product and service design: bridging the gap between customer needs, quality management and customer satisfaction. *Int. J. Product. Econ.* 66, 7–96.

Hill, R. (1993). *Ethnography and Marketing Research: A Post-modern Perspective.* Chicago: American Marketing Association.

Hilliam, M. A. and Young, J. (2000). *Functional Food Markets, Innovation and Prospects: A Global Analysis.* Surrey: Leatherhead International.

Hisrich, R. D. and Peters, M. P. (1991). *Marketing Decisions for New and Mature Products*, 2nd edn. Columbus, Ohio: Merrill Publishing Company.

Jones, R. (2001). *The Big Idea.* London: HarperCollinsBusiness.

Kamakura, W. (1998). A least squares procedure for benefit segmentation with conjoint experiments. *J. Market. Res.* 25, 157–167.

Kiener, S. (1995). The future of mail order. *Direct Marketing* 57, 17–22.

Kim, J. and Wilemon, D. (2002). Strategic issues in managing innovation's fuzzy front-end, *Eur. J. Innovation Manage.* 5, 27–39.

Kotler, P. (2000). *Marketing Management.* Millenium Edition. Saddle River, NJ: Prentice Hall.

Krueger, R. A. and Casey, M. A. (2000). *Focus Groups: A Practical Guide for Applied Research.* California: Sage Publications.

Longman, B. (2001). *Future Innovations in Food 2001: Forward-focused NPD and Maximizing Brand Value.* London: Reuters Business Insight.

Mark, K. (2006). A healthy habit, *Food in Canada,* June.

Mark-Herbert, C. (2004). Innovation of a new product category—functional foods, *Technovation* 24, 713–719.

McDaniels, C. and Gates, R. (1991). *Contemporary Marketing Research.* Minnesota: West.

McDonagh-Philip, D. and Bruseberg, A. (2000), *Using Focus Groups to Support New Product Development.* London: Institution of Engineering Designers.

Medical News Today (2006). New gluten-free bread developed by CeRPTA. www.medicalnewstoday.com

Meulenberg, M. T. G. and Viaene, J. (2005). Changing agri-food systems in Western countries: a marketing approach. In: Jongen, W. M. F. and Meulenberg, M. T. G. eds. *Innovation in Agri-food systems: Product Quality and Consumer Acceptance.* Wageningen, The Netherlands: Wageningen Academic Publishers.

Milton, J. (2003). *New Product Development Strategies in Food to 2007.* London: Reuters Business Insight.

Moorman, C. (1995). Organisational information processes: cultural antecedents and new product outcomes. *J. Market. Res.* 32, 318–335.

Moosa, S. (2002). *Wellbeing: A Cross Category Approach to Nutrition, Health and Beauty.* London: Reuters Business Insight.

Moriarty, R. T. and Reibstein, D. J. (1986). Benefit segmentation in industry markets. *J. Business Res.* 14, 463–486.

Nutra Ingredients USA (2004). Whole Foods Market goes gluten-free. www.nutrain- gredients-usa.com/news

Nutra Ingredients USA (2007). Anheuser-Busch launches sorghum beer. www.nutrain- gredients-usa.com/news

Palmer, S. (2004). A growing need for gluten-free foods. July, www. foodproduct design.com

Reeves, C. (2006). Getting the gluten out, *Food In Canada* October.

Saher, M., Arvola, A., Lindeman, M., and Lahteenmaki, L. (2004). Impressions of functional food consumers. *Appetite* 42, 79–89.

Schär (2006). www.schar.com

Shah, N. P. (2001). Functional foods for probiotics and prebiotics. *Food Technol.* 55, 43–46.

Shinsato, E. (2006). Tapping into the gluten-free market. www.naturalproductsinsider.com

Slater, S. F. and Narver, J. C. (1996). Competitive strategy in the market-focused business. *J. Market Focused Manage.* 1, 159–174.

Sorenson, D. and Bogue, J. (2005). Market-oriented new product design of functional orange juice beverages: a qualitative approach. *J. Food Products Market.* 11, 57–73.

Sorenson, D. and Bogue, J. (2006). Modelling soft drink purchasers' preferences for stimulant beverages. *Int. J. Food Sci. Technol.* 41, 704–711.

Stagnito Communications (2006a). Top 10 Health And Wellness Trends of 2006. www.stagnito.com

Stagnito Communications (2006b). Food and Drug Packaging. www.stagnito.com

Sultan, F. and Barczak, G. (1999). Turning marketing research high-tech. *Marketing Management* Winter.

Trout, J. and Ries, A. (1995). *The Future of Positioning.* Chicago: NTC Publishing Group, pp. 47–52.

Urban, G. L. and Hauser, J. R. (1993). *Design and Marketing of New Products.* Englewood Cliffs, New Jersey: Prentice Hall.

Van Trijp, J. C. M. and Steenkamp, J. E. B. M. (1998). Consumer-oriented new product development: principles and practise. In: Jongen, W. M. F. and Meulenberg, M. T. G. eds. *Innovation of Food Production Systems.* Wageningen: Wageningen Press.

Verbeke, W. (2004). Consumer acceptance of functional foods: socio-demographic, cognitive and attitudinal determinants. *Food Quality and Preference*, 16, 1:45–57.

von Alvensleben, R. (2001). Beliefs associated with food production. In: Frewer, L., Risvik, E., and Schifferstein, H. eds. *Food, People, and Society—A European Perspective of Consumer's Food Choices.* Berlin: Springer-Verlag.

Wennström, P. (2000). Functional foods and the consumer's perception of health claims. *Scand. J. Nutr.* 44, 30–33.

Wennström, P. and Mellentin, J. (2003). *The Food and Health Marketing Handbook.* London: New Nutrition Business.

Weststrate, J. A., van Poppel, G., and Verschuren, P. M. (2002). Functional foods, trends and future. *Br. J. Nutr.* 88, 233–235.

New product development: the case of gluten-free food products

18

Alan L. Kelly, Michelle M. Moore, and Elke K. Arendt

Introduction to new product development

New product development (NPD) is a key activity for the global food industry, with all markets undergoing significant change in short time-scales, e.g. new products appearing, products which have reached the end of their life cycles being withdrawn, and many existing products undergoing modifications in their processing, formulation, or packaging, even if not apparent to consumers. Development of new or existing food products and processes, considered here as NPD activities, has several key drivers. Most commonly, food companies are driven by changing consumer requirements and market trends. A secondary driver is changing technological developments in food processes, ingredient functionality, and scientific understanding of food formulation and processing, which can offer new product possibilities. Changes in legislation or food regulations may also drive companies to modify or develop products, for example if it becomes recommended that certain ingredients should be reduced or avoided in food formulations (e.g. Sudan red food dye, dietary salt intake).

Gluten-Free Cereal Products and Beverages
ISBN: 9780123737397

Table 18.1 Typical reasons why new food products succeed and fail	
Reasons why products succeed	**Reasons why products fail**
Integration of business and technology functions and personnel	Lack of innovation
Good management and discipline of NPD project	Lack of clear benefit to consumers
Clear marketing strategy and good market research	Poor market research (e.g. overestimation of market size)
Clear benefit to consumer	Insufficient marketing
Superior and differentiated product and/or package	
Rapid and responsive process of NPD	

Consumer demand for food products is complex, responsive to many external factors, and changes over short and sometimes unpredictable time-frames. Such shifts in consumer demand require food companies to be responsive, through their marketing specialists, to opportunities for new food products, which must then be delivered by the scientific and technological capabilities of the company. Responses must be rapid, to ensure competitiveness, and effectively navigate the stages and objectives of NPD which will be discussed in this chapter.

NPD is an extremely risky enterprise, and huge numbers of new food products fail on launch. Some of the factors classically associated with the success or failure of new food products are listed in Table 18.1. Failure of a new product costs a company significant time and money wasted, and may undermine consumer confidence in the company or its other brands. Overall, to succeed, a new product must achieve two results:

- Consumers must change their buying behavior to spend money on the new product, having been made aware of it and/or convinced that it is in their best interest to try the product. This step may be shortened or removed by provision of free samples of product or in-store tasting;
- Having tried it once, consumers must make a decision to continue purchasing the product, ensuring its long-term sustainability.

In this chapter, the key principles and stages of NPD will be discussed, with specific reference to the development of gluten-free food products.

NPD in the gluten-free sector

Development of gluten-free products is clearly a specialized category of NPD, which differs in some key respects from broader NPD activities in other food product categories. The case of development of gluten-free food products thus arguably represents a unique set of factors relative to NPD for various reasons, including:

- The market is defined and "captive," i.e., consumers with celiac disease who are actively seeking gluten-free products.

- The market is small but growing worldwide, as more and more individuals are diagnosed with celiac disease. The increased rate of diagnosis is due to improved diagnostic procedures (anti-gliadin antibody serological tests) as well as increased awareness of celiac disease.
- The product range is limited in many countries, with many opportunities for innovation, or at least "relative innovation," as discussed below.
- Many products on the market are viewed by consumers as being of inferior quality (especially sensory) compared with their traditional (i.e. non gluten-free) counterparts.
- Economic factors play a significant role, as the limited choice is accompanied by product pricing far above traditional counterparts, leading to a perception in many cases of poor value for money.

Innovation is a key concept in NPD, and may refer to real or perceived "newness" in a new or existing product. A genuinely new product for a company may be defined as one which has never been manufactured before by that company and which is currently not available in the market in which it is intended to be sold. Interestingly, the term "innovation" has a particular significance in the gluten-free sector, as products may be quite conventional in terms of availability in gluten-containing forms but simply not be available for people with celiac disease. A familiar product may hence possess "relative innovation" by virtue of its introduction in a gluten-free form; an example in many markets would be the development of gluten-free breaded meat or fish convenience food products.

Gluten-free products may also be differentiated into (i) products that are marketed to both celiacs and non-celiacs, and which are effectively traditional products, the formulation of which happens to be certifiably free of gluten, and (ii) products that are marketed solely at celiac consumers, and which it is highly unlikely that non-celiac consumers will purchase, for reasons or quality of price. Many products will fall into the former category, and a key consideration is the labeling of such products, such that consumers have unequivocal confidence in the gluten-free status of the product. In this chapter, the focus will be largely on the latter category, which is arguably more challenging in terms of marketing and technological development.

Patients with celiac disease are unable to consume some of the most common products on the market today, namely breads, baked goods, and other food products made with wheat flour (Lovis, 2003). Hidden ingredients (e.g. by-products or processed foods that contain wheat and gluten-derivates as thickeners and fillers) must also be avoided. These include hamburgers, salad dressings, cream sauces, dried soup mixes or canned soups, and processed cheese (see Chapter 1). Some medications that include wheat proteins as binders must also be excluded. Other cereals such as rye, barley, malt, kamut, einkorn, dinkel, and spelt are prohibited. Since celiac disease can in fact result in lactose intolerance due to the lack of lactase production (Murray, 1999), many patients with celiac disease must also avoid cow's milk.

Food products may also exist on a marketplace, but new to the company in question, be an existing product of the company being introduced to a new market, or be an existing product being re-launched with some degree of modification or repositioning.

Existing products may thus undergo NPD-like research process, including changes in the technology of production, the raw materials used, the ingredients added or the package in which the product is ultimately presented. Changes in any of these characteristics of a product require effort and cost by a company, obviously to varying extents, and may be invisible to consumers (e.g. changes in process technology).

Overall, there are several categories of new food products, as summarized in Table 18.2. New products clearly differ in their degree of innovation, and the related effort expended in their development. The degree of complexity of development of a product will also be heavily influenced by whether it has clearly defined antecedents, or is being developed *de novo*. For an example of the former, the development of a gluten-free breaded chicken product will draw heavily on models of such products in conventional form to provide information on basic product formulation, cooking

Table 18.2   Categories of new food products

Category	Description	Example
Line extensions	Minor changes in flavor, color, etc. Little effort required in development and relatively minor marketing challenges required	New flavor of product, where most formulation and process variables remain unchanged
Clones	Essentially copies of existing successful products produced by competitors, based on intent to gain share of identified market. Development may involve reverse engineering and variable levels of scientific challenge	Emergence of successful innovative new product followed by number of imitations offering little advantage, such as energy drinks
Reformulations	Changes in formulation of existing products made by the same or another company (e.g. new low-fat, high-fiber, gluten-free versions of traditional antecedents)	Many gluten-free products clearly fall into this category
New forms of existing products	May involve making products miniature in size, instant, frozen, part-cooked etc. Technological challenges may vary widely, as may marketing required	Development of part-baked or microwaveable gluten-free bread rolls
Repackaging of existing products	Redesign of package appearance, size or nature of package. Technological challenges may vary widely once again	Extension of product shelf-life through use of modified-atmosphere packaging or division of larger packages into individually-wrapped single-serve portions
Creative/innovative products	Genuinely novel product or use of novel raw materials, where there are no clearly identified antecedents. These may be the most technologically challenging of products for development, and require good marketing to ensure that the new idea will meet an existing or emerging consumer need	Use of novel cereals for production of new gluten-free bakery products

conditions, etc., thereby presenting the opportunity for significant savings in terms of effort (and money) required to determine these *ab initio*. An example of the latter would be the case of development of an innovative product where there is no conventional antecedent, and probably represents a very small proportion of gluten-free products.

There are a number of other factors by which NPD projects can be classified. For example, as stated earlier, NPD may be driven by:

- Scientific or technological factors (e.g. identification of new ingredients for modification of texture of gluten-free products, ideas to apply scientific breakthroughs in this area to development of product, or availability of new equipment or processes which offers the possibility of making a product which is new, whether to the company or the market).
- Market research (e.g. identification of market opportunities that have a clearly defined group of consumers who would purchase the product, if available).

With these two possible drivers for NPD, it is clear that integration of, and effective communication between, marketing, science, and technology functions within an organization is critical. There has to be confidence that scientific personnel can make reliable judgments as to whether marketing ideas can be translated into successful products, and vice versa. A commonly recognized contributing factor in failure of new products is lack of integration between these business functions of an organization.

The principal stages of NPD will be discussed in the particular context of gluten-free product development. There are many different published schemes for stages of new product development, but several factors are common to each scheme, such as:

- Screening, i.e. the rigorous and analytical evaluation of ideas and prototypes to ensure their worthiness for further progress through the later stages of the NPD process. The NPD process is frequently schematically depicted as a horizontally or vertically oriented inverted triangle (e.g. the Innovation Funnel), as shown schematically in Figure 18.1. The wide end of the figure represents a large number of initial ideas which are progressively screened and narrowed down to a small number of final products actually launched. In theory, the screening process applied should have maximized the likelihood that the "surviving" products are those most likely to meet consumer expectations and succeed in the marketplace;
- Stages/gates: The funnel-like structure of NPD is frequently divided into discrete stages at which specific aspects of the NPD process are undertaken. Each stage is connected to the next by a "gate" where rigorous testing (i.e. screening) is applied to determine whether the product has progressed sufficiently well (as measured against carefully-selected "kill"/"go" criteria, e.g. consumer reaction, product performance, shelf-life achieved) to be allowed to progress to the next stage. These gates represent points at which decisions may be made as to whether the effort, time and cost required in the next stage of the process are justified based on the likelihood that the final product will be a success. Products failing to pass gates may either be abandoned, or rerouted back up the process to earlier stages for re-evaluation or

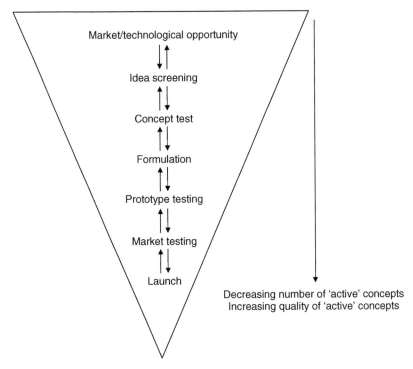

Market/technological opportunity

Idea screening

Concept test

Formulation

Prototype testing

Market testing

Launch

Decreasing number of 'active' concepts
Increasing quality of 'active' concepts

Figure 18.1 Schematic representation of the inverted pyramid concept of new product development, also known as the innovation funnel.

modification, e.g. in a feedback loop-like process. In addition, there has to be strong organizational management to determine whether ideas are realistic and consistent with organizational strategy, finance, and facilities.

The common stages through which an NPD process must flow are discussed individually below.

Stage 1: The idea stage

The starting point for any new food product must be an idea, which can come from several sources, including consumers (as evaluated by the marketing arm of a company), other markets (e.g. overseas), and the scientific and technological resources of the company. In the case of gluten-free products, a key source of new product ideas is inevitably the range of non-gluten-free products, representing a pool of products which potentially could be developed in a form suitable for people with celiac disease. Food companies are also becoming increasingly aware of the key role of consumer research in driving NPD strategy, and increasingly sophisticated tools (from focus groups to conjoint analysis) are being used to garner from consumers their needs (conscious or subconscious) in terms of new food product availability and likelihood of purchase. For gluten-free products, another obvious source of information will be consumers with celiac disease themselves; in many countries, there are also consumer associations representing these consumers who are eager to liaise with food companies.

At the idea stage, one of the key concepts of NPD is first applied: screening, as discussed above. Screening criteria which may be applied include the marketability of the product, its technical feasibility, the manufacturing capabilities of the company and the financial soundness of the development and manufacturing costs of the product. As at all stages of NPD, a company must be prepared to cease developing ideas which fail to meet strictly-applied criteria such as these. Product ideas or concepts which meet the organization's pre-determined screening criteria should then be refined and expanded to allow product definition early in the process, providing goals against which progress may be measured. Attributes of the product such as its target consumers, its expected price and its key quality, appearance, nutritional and convenience attributes may all be roughly defined at this point, informing the requirements of the NPD process to follow.

Stage 2: The formulation stage

As mentioned earlier, the challenges involved in developing any new food product depend on the extent to which pre-existing technological know-how regarding the product, its formulation and processing requirements is available. There are very different challenges in cases where there are clear antecedents for a product relative to approaching design and formulation of a product where no baseline information exists.

In the context of classifications of new products listed in Table 18.2, many gluten-free products are obviously reformulations, where standard products are reproduced using a formulation excluding gluten-containing ingredients. In other areas of food NPD, having a baseline comparator product to benchmark against, and perhaps reverse engineer, is a major advantage, as discussed above. In the case of gluten-free products, however, reformulation can present very significant technological and scientific challenges, particularly if the ultimate aim is to produce a product that matches key sensory properties (e.g. flavor, texture) of the comparator. To take a simple example, the key defects of many commercial gluten-free bread products arise from the simple fact that omitting gluten from the formulation of the bread removes a key functional protein which contributes significantly to textural attributes which are most sought in bread by consumers (e.g. perceived springiness).

Ingredients for a new food product may be selected based on a number of key criteria, such as:

- Function in the product and, critically, consequences of omission
- Cost and availability
- Interaction with other ingredients

- Susceptibility to change due to processes applied to the product (e.g. denaturation, change in color etc.)
- Nutritional or other benefit conferred on product which may make the product more attractive to consumers (e.g. vitamin-enriched)
- Labeling concerns (e.g. allergies, ethical concerns such as vegetarianism).

In the case of gluten-free products, the last factor listed above is key, as all ingredients must be screened first and foremost for their lack of likely allergic reaction by consumers with celiac disease. Gluten is very common, and in many countries, unlabeled ingredients in the human diet present a major challenge for patients with celiac disease (Gobbetti *et al.*, 2007).

In all aspects of NPD, empirical methods for testing and optimizing product parameters such as levels of ingredients and processing parameters are less favored today than statistically based tools which allow identification of optimal parameters with maximum confidence and minimum numbers of trials and tests required. For example, experimental designs such as response surface methodology (RSM) may be used to identify combinations of levels of ingredients or cooking parameters (e.g. time, temperature) to be tested for subsequent measurement of appropriate responses (e.g. color, volume, acceptability) and building of models using these data to identify local maxima and minima, i.e. select optimal conditions to be used in the process. These tools allow multiple parameters to be varied simultaneously in a manner which nonetheless yields reliable data to the manufacturer with a minimum number of trials (and hence minimized cost and time input).

Successful application of RSM in the production of different types of wheat bread has been reported (Lee and Hoseney, 1982; Malcolmson *et al.*, 1993; Clarke *et al.*, 2002, 2004; Gallagher *et al.*, 2003, 2004). Recently, Schober *et al.* (2005) used RSM to study the differences among a range of sorghum hybrids used in the production of gluten-free breads. RSM was applied using two sorghum hybrids with different characteristics to investigate the effect of added ingredients on bread quality. Addition of xanthan gum, skim milk powder and varying water levels were tested using a central composite design. The authors concluded that increasing the water level increased loaf-specific volume, while increasing xanthan gum levels decreased the volume. As skim milk powder levels increased, loaf height decreased. Quality differences between the hybrids were maintained throughout the RSM. This study showed that certain sorghum hybrids have higher intrinsic bread quality than others.

Other statistical tools such as regression analysis are used to investigate the significance of relationships between continuous data sets for product attributes (e.g. between level of lactose in a formulation and instrumentally measured brown color after baking) while analysis of variance (ANOVA) may be used to determine the significance of measured differences in responses (i.e. measured parameters) for different discrete formulations or product types (e.g. comparison of newly developed product to competitors' products for sensory or instrumental quality attribute data). For instance, Moore *et al.* (2004) successfully used ANOVA to compare two newly developed gluten-free breads to a wheat and a commercial gluten-free bread, and the authors found significant differences between the breads.

Stage 3: Process development

The development of a new product involves the optimization of a process for conversion of the selected raw materials and ingredients into a final product. Any product is the sum of the ingredients used in its manufacture plus the process applied to these. As discussed above, different scenarios may be involved, depending on whether the product being developed has clear precedents and prior technological specifications for required processing equipment. A process developed for a product may have a number of objectives:

- Rendering the product safe for consumption by inactivation or prevention of microbiological, chemical (e.g. pesticides, chemical residues) or physical (e.g. glass, metal) hazards
- Change in properties of the product to yield a form more acceptable to consumers (e.g. baking, freezing, etc.)
- Extension of the shelf-life of the product, enhancing distribution possibilities for the company and increasing convenience for the consumer, and perhaps preserving the functional or nutritional properties of a product.

The first-listed objective is a *sine qua non*. The second two differ in their priority and requirements for different food products, and there can often be a compromise between retaining acceptable characteristics for consumers and maximizing shelf-life. A prime example of this compromise involves ultra-high-temperature (UHT)-treated milk, which has a long shelf-life at room temperature, but is less preferred than pasteurized milk by consumers in many countries due to its cooked flavor, despite the latter having an almost 10-fold shorter shelf-life, and requiring refrigeration.

The challenge in developing gluten-free products will also depend on whether the organization involved has existing know-how and equipment which can transfer to the new product line. For companies starting production *de novo*, even setting up quite traditional bakery processes still requires significant investment in equipment and processing facilities, and an inevitable period of optimization of operation and training of personnel. Many gluten-free products will utilize conventional bakery equipment; however, operating parameters may not be automatically transferable from traditional product manufacture. For example, baking or proofing conditions may need to be optimized for gluten-free doughs due to their ingredients reacting and interacting in a different manner to those in gluten-containing products.

Par or pre-baked bread represents half of the volume of frozen breads exported by French industrialists to Northern Europe (Millet and Dougin, 1994). Par-baking bread production has a great market potential in gluten-free cereal processing, as the process provides an opportunity to supply fresh bread with a simple bake-off stage at any desired time. Par-baked products are convenient foods meant to be finished-baked before consumption and have sufficient moisture for the development of desirable quality characteristics. Thus par-baked bread is fully baked and has edible properties when re-baked. The second baking phase is necessary for the production of a consumer-ready product (Leuschner *et al.*, 1997). At present, a large variety of par-baked gas-packaged products are on the market, and the number of par-baked products is still increasing steadily. The first products (the French baguette, petit pan, and other morning goods) still have a large market share. Most of the gluten-free bread and rolls on the market are par-baked; however, there are no publications in this area. Par-baking technology has been mainly applied in wheat bread production; however it may be assumed that the same conditions can be applied for the production of gluten-free par-baked products. Par-baked bread can be stored under a wide range of conditions and is re-baked to give it its final characteristics just before it is sold to the consumer or just before consumption (Sluimer, 2005).

The effect of processing on shelf-life will be discussed later in this chapter, as will the role of packaging as a key step in stabilizing modern food products.

Stage 4: Initial testing and viability assessment

The stages of development of formulation and process for a new product may proceed sequentially or to some extent simultaneously, but should result in a set of prototype products which can again be assessed by screening criteria to determine their fitness for further development, with it being again vital to cease work on products that do not demonstrate likelihood of eventual success, or to return these products to earlier stages of the NPD process for further development. Prototypes passing the screening criteria at this stage then go forward for further development and more detailed analysis and testing. The complexity and cost of testing applied at progressive stages of the NPD process increases significantly, which emphasizes the fact that only products with high probability of success justify such investment.

As the quality of prototype products thus, at least theoretically, becomes higher, a key assessment criterion for viability will be sensory evaluation of the prototypes. Sensory analysis is another area where rapid and significant scientific development has happened in a relatively short space of time, and a number of sophisticated tools exist for generation of data relating to different attributes of new or existing food products. It is beyond the scope of this chapter to discuss in detail modern sensory analysis techniques as applied in the food area, but Table 18.3 summarizes some typical sensory analysis designs that may be used in this regard.

An interesting question in the specific case of gluten-free foods is the use of gluten-containing (conventional) products for comparison; obviously, while this would yield useful comparison, it could only be undertaken with non-celiac assessors, who may have different reference frames and expectations to celiac tasters. In recent years, much research effort has focused on the use of instrumental methods to compliment or replace sensory analysis using human panellists. Examples of such methods include measurement of color using systems that give data on the Hunter LAB scales, and measurement of a wide range of textural parameters (e.g. viscosity, springiness, hardness, fracturability) using instrumental systems such as rheometers or Texture Profile Analysis systems. In the context of bakery products, microscopy tools (light, laser, and/or electron) have been increasingly used to analyze loaf structure, and quantitative data obtained by use of image analysis software. Confocal laser scanning

Table 18.3 Typical sensory analysis tools which may be used in NPD

Method	Principle	Example of use
Triangle test	Three products tested; assessors asked to identify odd one out	Determining whether replacement of an ingredient in a product gives a perceptible change
Ranking test	Assessors are asked to place a set of samples in order of intensity of a particular attribute, or overall preference	Making rapid analyses of a batch of prototypes, or placing new products relative to competitor products
Profiling	Assessors, with or without prior training, provide a complete evaluation of the sensory attributes (possibly including appearance, aroma, flavor and texture) for a product, using an agreed vocabulary of descriptive terms	Gaining detailed overview of sensory characteristics of one or a small number of products
Preference tests	Assessors are asked to indicate which of a pair or larger set of samples they prefer and why	Determining whether a new product would succeed in affecting consumer choice relative to an existing competitor product

microscopy is a useful tool to characterize the structure of gluten-free bakery products, e.g. after transglutaminase treatment (Figure 18.2) of gluten-free flours (Moore *et al.*, 2005; Renzetti *et al.*, 2007), fermentation by lactic acid bacteria (Moore *et al.*, 2007), or addition of dairy ingredients (Gallagher *et al.*, 2006). In addition, scanning electron microscopy has been successfully applied to characterize gluten-free breads and beers (Wijngaard *et al.*, 2005).

Digital image analysis (DIA) is widely used for characterization of the crumb structure and pore characteristics of gluten-free breads (Gallagher *et al.*, 2003, 2006; Moore *et al.*, 2005). The most common characteristics studied are mean cell area, total number of cell and number of cells per square centimeter. Application of DIA to measure the cell size and degree of cross-linking with the application of transglutaminase revealed significant differences between three protein sources and four enzyme addition levels in gluten-free breads containing transglutaminase (Moore *et al.*, 2005). In another study, Gallagher *et al.* (2003) used DIA to examine the effect of the addition of dairy and rice powder on loaf and crumb characteristics. The authors found that the number of cells decreased with the addition of dairy and rice powders, which led to an improvement in crumb structure. These results were in agreement with those of Crowley *et al.* (2002) for wheat bread. Differences in crumb grain characteristics were also detected among a range of sorghum hybrids used in the production of gluten-free breads (Schober *et al.*, 2005). Therefore, it can be concluded that methods such as DIA are excellent tools to optimize product formulations, since they allow quantification of the visual appearance of breads.

Techniques such as descriptive sensory analysis typically yield a huge amount of data for each product and assessor, and for tests of multiple products (e.g. prototypes being screened or new products compared to competitors' products) analysis of

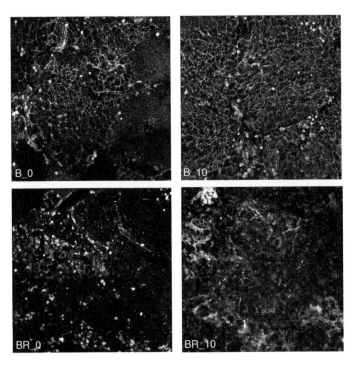

Figure 18.2 Confocal scanning-laser microscopy of gluten-free breads treated with transglutaminase. Buckwheat bread untreated (B_0) and treated with 10 U/g flour transglutaminase (B_10); brown rice bread untreated (BR_0) and treated with 10 U/g flour transglutaminase (BR_10) (Renzetti and Arendt, unpublished data).

such large and complex data sets can present significant challenges. A particularly useful statistical technique in this regard is principal component analysis (PCA), in which complex data sets are processed mathematically to yield, for example, two-dimensional maps in which spatial closeness of products indicates similarity on two or more broad major indices differentiating products in the set of samples being tested (i.e. principal components). Product maps may be overlaid with representation of locations of key sensory or other attributes on the same plane to yield additional information, and combination of PCA with tools such as hierarchical cluster analysis can yield very useful analysis of broad similarities or differences between groups of products. Hierarchical cluster analysis has not been widely used for gluten-free products; however, it is used when identifying celiac disease (Diosdado *et al.*, 2004).

Sensory analysis is a very important attribute in relation to gluten-free products. Sanchez *et al.* (2002, 2004) applied a scoring system to gluten-free breads and they found that the addition of protein, cornstarch, rice, and cassava starch to gluten-free breads affected the sensory attributes. Korus *et al.* (2006) studied the influence of prebiotic additives on gluten-free breads and they also assessed the effectiveness of gluten-free bread supplementation with the selected prebiotics. Using a sensory scoring system it was found that the best effects on sensory features of gluten-free bread were observed when medium doses of prebiotics were applied. Gallagher *et al.* (2003) also carried out sensory analysis on gluten-free breads with rice flour and dairy protein addition. They found that both the addition of rice flour and dairy protein improved the sensory attributes of gluten-free bread. Overall, it can be concluded that sensory analysis is vital in assessing the quality of gluten-free products.

Stage 5: Shelf-life testing

One of the key steps in NPD is the determination of the shelf-life of the product, and the identification of information which must be displayed on the product package in guiding the consumer regarding the acceptability or safety of the product at future points in time. Product quality and safety must both be considered during storage, as presence or increase in level of safety hazards (e.g. levels of pathogenic bacteria, germination of bacterial spores) must be considered with a higher degree of priority than changes in non-hazardous quality parameters. A product may be apparently acceptable to consumers but still not suitable for consumption due to determined threat from associated hazards at that point in storage.

A fundamental decision to be made about any new food product, likely made as part of concept definition at the start of the NPD process, is the intended temperature of storage; storage at sub-zero (e.g. frozen products), refrigerated, or ambient temperatures will be probably the biggest single influence on the rates of changes and reactions in a food product, and hence its shelf-life. The next issue to consider when evaluating the shelf-life of a food product is the characterization of the parameters that change in a negative manner during storage. Implicit in the concept of shelf-life is the assumption that food products lose quality during storage and eventually become either unacceptable to consumers or unsafe. Exceptions to this generalization include alcoholic beverages such as wine and whisky, where flavors develop over long time-scales and the high level of alcohol renders the product extremely microbiologically stable, or ripened cheeses, where biochemical and microbiological activities yield an optimal product flavor and texture, through processes such as lipolysis and proteolysis, at some point weeks to months, or even years, after manufacture, generally followed by deterioration in sensory quality. For the majority of products, however, the assumption of progressive loss of quality does hold, and for a new product the key deteriorative reactions must be determined, means of quantifying the extent of the reaction identified, and the extent which renders the product either unsafe or unacceptable to consumers specified.

Key factors that affect the shelf-life of products include:

- Intrinsic factors, e.g. pH, a_w, presence of natural or added antimicrobial or antioxidant substances
- Extrinsic factors, e.g. temperature, humidity, and gaseous atmosphere during product storage
- Implicit factors, e.g. microbes present in raw materials and their metabolism
- Processes applied to raw materials.

Changes that negatively affect the quality of food products may be enzymatic (e.g. proteolysis or lipolysis leading to changes in flavor or texture), chemical (e.g. oxidation of lipids, retrogradation or crystallization of starch, migration of water) or biological (e.g. growth of bacteria or fungi which impair the quality of food). For example, for a gluten-free bread roll product, loss of quality on storage at room temperature may result from growth of mold and/or development of staleness due to starch recrystallization (although both may be slowed significantly by use of modified-atmosphere packaging, as discussed below). If, under room temperature storage, growth of mold, as measured by microbiological methods, reached unacceptable levels after 52 days and staleness by instrumental texture profile analysis indicated a product hardness which was unacceptable to consumers after 28 days, the shelf-life of the product, based on these parameters, may be estimated to be 28 days at room temperature. Indeed, for many bakery products, physical changes in texture due to chemical processes occurring during storage are more likely to limit shelf-life than microbiological concerns.

Despite its importance, shelf-life testing may represent a major challenge for product development, as it implicitly requires a significant investment of time for study and verification; resultant significant delays in advanced stages of NPD may present difficulties for organizations, particularly in markets where competing organizations may gain advantage by reaching market first. For this reason, much attention has focused on development of techniques for rapid establishment of shelf-life parameters. These include the following:

Accelerated shelf-life testing

The time required to obtain reliable shelf-life data may be shortened significantly by storage of a food product under conditions which result in deterioration of quality at a rate which, while shorter than that observed under normal conditions, is mathematically correlated with, and can be used to calculate, that rate, by application of classical kinetic techniques (e.g. use of Q_{10} values, which quantify the change in rate of a reaction with a 10°C change in temperature). Typically, this involves holding samples of the product at a number of temperatures higher than that expected to be used in normal handling. For the example of gluten-free bread rolls mentioned above, samples of the rolls could be held at 30, 40, and 50°C and rates of growth of mold measured by microbiological methods, and staleness by instrumental texture profile analysis. Plotting the measured parameters versus time allows calculation of rate constants, and extrapolation to expected reaction rate at 20°C, and hence shelf-life estimation.

Challenge tests

Much process development for products at high potential risk for microbiological hazards will concern establishment of steps (e.g. critical control points in a hazard analysis and critical control points (HACCP) scheme) which will prevent or eliminate these hazards. However, during development, if good-quality raw materials are used, these hazards will likely not be encountered, and the efficacy of the step as a guarantee of product safety cannot reliably be established. Hence, trials may be performed where batches of raw materials and/or ingredients are deliberately contaminated ("spiked") with agents of concern, and the survival or persistence thereof monitored following processing, with the process being modified if necessary to ensure adequate consumer protection based on the results of such studies. These trials must, of course, be undertaken in controlled laboratory conditions, due to concern about deliberately introducing potential pathogens into a food environment. Much progress has also been made in terms of use of predictive microbiology to avoid such tests by prediction of survival of pathogens following processing, based on knowledge of the heat resistance (e.g. D- and z-values) and other factors such as acid tolerance of target microorganisms.

Shelf-life extension may be a key influence on selection of packaging conditions for gluten-free products, as discussed below.

Stage 6: Scale-up and consumer testing

One of the key challenges in NPD is the transfer of product prototypes to increasing scales of production as the NPD process advances. In general, the early stages of NPD, where many options for formulation and process may be tested, are performed on a small scale, in kitchen or laboratory facilities, with only successful prototypes being passed for pilot-scale production and eventually commercial-scale production. The later stages of NPD must involve consumers in tasting and evaluating the prototype products, and the scale of such testing will influence the scale at which prototypes may be produced.

Stage 7: Packaging and labeling

Development of any new food products eventually involves the consideration of a suitable package design and material for the distribution and sale of the product. The functions of any food package are as follows:

- Contain the product
- Protect the product (e.g. from physical damage on handling)
- Protect the consumer (e.g. by prevention of recontamination)
- Preserve the product (e.g. by providing a barrier to certain gases or moisture)
- Communicate information
- Market the product
- Disperse and dispense the product.

All food packages are made from a relatively small number of basic materials, e.g. polymer films, aluminum foil or cans, glass, or paperboard. However, many food products are packaged using combinations of these materials, e.g. milk cartons where plastic layers add barrier properties and paperboard provides physical strength, while foil may be used to withstand temperature fluctuations. Food packages may also consist of multiple different discrete elements, e.g. products held in a modified atmosphere in a plastic pouch, contained in an external paperboard box. Many of the considerations for selecting packaging strategies for gluten-free products are identical to those for conventional food products:

- What portion size will consumers expect or purchase (including subdivided packages, with individually wrapped product portions in a multi-pack assembly)?
- Are there specific requirements for the food in terms of physico-chemical stabilization during distribution and storage (e.g. control of moisture transport, gas atmosphere, stability to light)?
- Is the package required to be stable to physical treatments subsequent to filling (e.g. freezing, heating in microwave or conventional ovens)?
- What information will be required by the consumer (e.g. cooking instructions, allergy information) and what is required to be consistent with appropriate legislation?
- How can the product be made attractive to consumers, particularly those making a first-time purchase (e.g. appearance, convenience)?
- How available is the desired package and is there significant capital investment required for installation of the packaging system?
- How much will the package add to the retail cost of the product (both in terms of initial equipment cost and subsequent per-unit cost)?

Modified-atmosphere packaging (MAP) is a commonly used technique for extension of shelf-life of gluten-free bread. Carbon dioxide suppresses the development of yeasts and bacteria and has a fungicidal action against molds. Therefore carbon dioxide is the main gas used to package bakery products, sometimes with nitrogen as a support gas to decrease gas diffusion from the package. Additional protection is achieved by decreasing the oxygen content in the package to less than 1%, since yeasts and many bacteria grow more slowly under anaerobic conditions while moulds are fully inhibited (Sluimer, 2005). Gas mixtures of 40% carbon dioxide and 60% nitrogen (Moore *et al.*, 2004, 2005) and 80% carbon dioxide and 20% nitrogen (Gallagher *et al.*, 2003) are widely used for gluten-free breads. The shelf-life of gluten-free breads in general is very poor. However, Gallagher *et al.* (2003) found that with the addition of dairy powder and rice flour and packaging using 80% carbon dioxide and 20% nitrogen, the shelf-life of gluten-free bread improved considerably. The shelf-life of the resulting breads was found to be up to 23 days. Similar results were found for Rasmussen and Hansen (2001), where the maximum crumb firmness of MAP wheat bread was attained at 35 days.

Labeling of gluten-free food products is becoming a key consideration in many countries. It is critical for consumers with celiac disease to be informed as to whether a product is suitable for consumption or not, and explicit labeling which does not depend on interpretation of ingredients listings (which may be ambiguous, for example in the case of modified starches, which may be gluten-free or may not) is the ideal. In Europe, the new allergen labeling Directive 2003/89/EC became mandatory on November 25, 2005. The new Directive will make it much easier for people with celiac disease and food allergies to identify allergens in foods. All allergens and ingredients derived from allergens will have to be specified, for example, "vegetable oil" will need to be specified as "peanut oil" or "wheat germ oil" (Food Labelling Regulations UK, 2004).

Recently in the US, the Food Allergen Labeling and Consumer Protection Act proposed a Bill that called for the Food and Drug Administration to issue final regulations defining "gluten-free" and permitting the voluntary labeling of products as "gluten-free" no later than 2008. Irish and EU legislation in relation to food labeling relies mainly on the Codex Standard. The Codex Standard for gluten-free foods was adopted by the Codex Alimentarius Commission of the World Health Organization (WHO) and by the Food and Agriculture Organization (FAO) in 1976. In 1981 and 2000, draft revised standards stated that so-called gluten-free foods are described as:

> (a) consisting of, or made only from, ingredients which do not contain any prolamins from wheat or all *Triticum* species such as spelt, kamut or durum wheat, rye, barley, oats or their crossbred varieties with a gluten level not exceeding 20 ppm; or (b) consisting of ingredients from wheat, rye, barley, oats, spelt or their crossbred varieties, which have been rendered gluten-free; with a gluten level not exceeding 200 ppm; or (c) any mixture of two ingredients as in (a) and (b) mentioned, with a level not exceeding 200 ppm.

In this context, the WHO/FAO standard gluten was defined as a protein fraction from wheat, rye, barley, oats, or their crossbred varieties (e.g. *Triticale*) and derivatives thereof, to which some people are intolerant and that is insoluble in water and 0.5 mol/L NaCl. However, there is much debate on whether or not oats may be used in the production of gluten-free products (see Chapters 1, 3, and 8). The prolamin content of gluten is generally taken as 50%. In the United States and Canada, the gluten-free diet is devoid of any gluten, and is based on naturally gluten-free ingredients such as rice. However, in the UK and most European countries, products labeled as being gluten-free may still contain an amount of wheat starch.

Required label information also includes an indication of the time-frame for consumption of the product, either in terms of best-before, use-by or expiry dates (see Chapter 2).

Conclusions

New product development (NPD) is a complex activity for the food industry. The specific challenges vary for each product and market, but general stages and principles can be outlined. The development of specifically gluten-free food products forms a very interesting case for NPD, as innovation and novelty can be high for the defined and growing market of celiac customers, despite the availability of traditional counterparts, and there are highly complex technological barriers to successful development of products of high organoleptic quality. In conclusion, transformation of even familiar food products to gluten-free formulations, or *ab initio* development of gluten-free products, offer particular challenges for food processors.

References

Clarke, C. I., Schober, T. J., and Arendt, E. K. (2002). Effect of single strain and traditional mixed strain starter cultures on rheological properties of wheat dough and on bread quality. *Cereal Chem.* 79, 640–647.

Clarke, C. I., Schober, T. J., Dockery, P., O'Sullivan, K., and Arendt, E. K. (2004). Wheat sourdough fermentation: Effects of time and acidification on fundamental rheological properties. *Cereal Chem.* 81, 409–417.

Codex Alimentarius Commission (1983). Codex standard for "gluten-free foods," Codex Stan 118-1981 (amended 1983).

Codex Alimentarius Commission (2000). Draft revised standard for gluten-free foods. CX/NFSDU 00/4.

Crowley, P., Schober, T. J., Clarke, C. I., and Arendt, E. K. (2002). The effect of storage time on textural and crumb grain characteristics of sourdough wheat bread. *Eur. Food Res. Technol.* 214, 489–496.

Diosdado, B., Wapenaar, C., Franke, L. *et al.* (2004). A microarray screen for novel candidate genes in coeliac disease pathogenesis. *Gut* 53, 944–951.

Food Labelling (2004). (Amendment) (England) (No. 2) Regulations.

Gallagher, E., Kunkel, A., Gormley, T. R., and Arendt, E. K. (2003). The effect of dairy and rice powder addition on loaf and crumb characteristics, and on shelf life (intermediate and long-term) of gluten-free breads stored in a modified atmosphere. *Eur. Food Res. Technol.* 218, 44–48.

Gallagher, E., Gormley, T. R., and Arendt, E. K. (2004). Crust and crumb characteristics of gluten-free breads. *J. Food Eng.* 56, 153–161.

Gobbetti, M., Rizzello Giuseppe, C., Di Cagno, R., and De Angelis, M. (2007). Sourdough lactobacilli and celiac disease. *Food Microbiol.* 24, 187–196.

Korus, J., Grzelak, K., Achremowicz, K., and Sabat, R. (2006). Influence of prebiotic additions on the quality of gluten-free bread and on the content of inulin and fructooligosaccharides. *Int. J. Food Sci. Technol.* 12, 489–495.

Lee, C. C. and Hoseney, R. C. (1982). Optimization of the fat-emulsifier system and the gum-egg white-water system for a laboratory scale single stage cake mix. *Cereal Chem.* 59, 392–395.

Leuschner, R. G. K., O'Callaghan, M. J. A., and Erendt, E. K. (1997). Optimisation of baking parameters of part-baked and rebaked Irish brown soda bread by evaluation of some quality characteristics. *Int. J. Food Sci. Technol.* 32, 487–493.

Lovis, L. J. (2003). Alternatives to wheat flour in baked goods. *Cereal Foods World* 48, 61–63.

Malcolmson, L. J., Matsuo, R. R., and Balshaw, R. (1993). Textural optimisation of spaghetti using response surface methodology: effects of drying temperature and durum protein level. *Cereal Chem.* 70, 417–423.

Millet, P. and Dougin, Y. (1994). Boulangerie industrielle francaise. L'age adulte. *Industries Céréales* 87, 45–50.

Moore, M. M., Schober, T. J., Dockery, P., and Arendt, E. K. (2004). Textural comparison of gluten-free and wheat based doughs, batters and breads. *Cereal Chem.* 81, 567–575.

Moore, M. M., Heinbockel, M., Dockery, P., Ulmer, H. M., and Arendt, E. K. (2005). Network for in gluten-free bread with the application of transglutaminase. *Cereal Chem.* 83, 28–36.

Moore, M. M., Schober, T. J., Juga, B. *et al.* (2007). Effect of lactic acid bacteria on the properties of gluten-free sourdoughs, batters and the quality and ultrastructure of gluten-free bread. *Cereal Chem.* 84, 357–364.

Murray, J.A. (1999). The widening spectrum of celiac disease. *Am. J. Clin. Nutr.* 69, 354–365.

Rasmussen, P. H. and Hansen, A. (2001). Staling of wheat bread stored in modified atmosphere. *Lebensm. Wiss. Technol.* 34, 487–491.

Renzetti, S., Dal Bello, F., and Arendt, E. K. (2007). Microstructure, fundamental rheology and baking characteristics of batters and breads from different gluten-free flours treated with a microbial transglutaminase. *J. Cereal Sci.* (in press).

Sanchez, H. D., Osella, C. A., and de la Torre, M. A. G. (2002). Optimisation of gluten-free bread prepared from cornstarch, rice flour and cassaca starch. *J. Food Sci.* 67, 416–419.

Sanchez, H. D., Osella, C. A., and de la Torre, M. A. G. (2004). Use of response surface methodology to optimize gluten-free bread fortified with soy flour and dry milk. *Int. J. Food Sci. Technol.* 10, 5–9.

Schober, J. T., Messerschmidt, M., Bean, S. R., Park, S. H., and Arendt, E. K. (2005). Gluten-free bread from sorghum: quality differences among hybrids. *Cereal Chem.* 82, 394–404.

Sluimer, P. (2005). Prebaked bread. In: *Principles of Breadmaking Functionality of Raw Materials and Process Steps*, American Association of Cereal Chemists, St. Paul, Minn. Ch 7, pp. 165–182.

Wijngaard, H. H., Ulmer, H. M., and Arendt, E. K. (2005). The effect of germination temperature on malt quality of buckwheat. *J. Am. Soc. Brewing Chem.* 63, 31–36.

Index

Food Science and Technology
International Series

Maynard A. Amerine, Rose Marie Pangborn, and Edward B. Roessler, *Principles of Sensory Evaluation of Food.* 1965.

Martin Glicksman, *Gum Technology in the Food Industry.* 1970.

Maynard A. Joslyn, *Methods in Food Analysis,* second edition. 1970.

C. R. Stumbo, *Thermobacteriology in Food Processing,* second edition. 1973.

Aaron M. Altschul (ed.), *New Protein Foods:* Volume 1, *Technology, Part A*—1974. Volume 2, *Technology, Part B*—1976. Volume 3, *Animal Protein Supplies, Part A*—1978. Volume 4, *Animal Protein Supplies, Part B*—1981. Volume 5, *Seed Storage Proteins*—1985.

S. A. Goldblith, L. Rey, and W. W. Rothmayr, *Freeze Drying and Advanced Food Technology.* 1975.

R. B. Duckworth (ed.), *Water Relations of Food.* 1975.

John A. Troller and J. H. B. Christian, *Water Activity and Food.* 1978.

A. E. Bender, *Food Processing and Nutrition.* 1978.

D. R. Osborne and P. Voogt, *The Analysis of Nutrients in Foods.* 1978.

Marcel Loncin and R. L. Merson, *Food Engineering: Principles and Selected Applications.* 1979.

J. G. Vaughan (ed.), *Food Microscopy.* 1979.

J. R. A. Pollock (ed.), *Brewing Science,* Volume 1—1979. Volume 2—1980. Volume 3—1987.

J. Christopher Bauernfeind (ed.), *Carotenoids as Colorants and Vitamin A Precursors: Technological and Nutritional Applications.* 1981.

Pericles Markakis (ed.), *Anthocyanins as Food Colors.* 1982.

George F. Stewart and Maynard A. Amerine (eds.), *Introduction to Food Science and Technology,* second edition. 1982.

Malcolm C. Bourne, *Food Texture and Viscosity: Concept and Measurement.* 1982.

Hector A. Iglesias and Jorge Chirife, *Handbook of Food Isotherms: Water Sorption Parameters for Food and Food Components.* 1982.

Colin Dennis (ed.), *Post-Harvest Pathology of Fruits and Vegetables.* 1983.

P. J. Barnes (ed.), *Lipids in Cereal Technology.* 1983.

David Pimentel and Carl W. Hall (eds.), *Food and Energy Resources.* 1984.

Joe M. Regenstein and Carrie E. Regenstein, *Food Protein Chemistry: An Introduction for Food Scientists.* 1984.

Maximo C. Gacula, Jr. and Jagbir Singh, *Statistical Methods in Food and Consumer Research.* 1984.

Fergus M. Clydesdale and Kathryn L. Wiemer (eds.), *Iron Fortification of Foods.* 1985.

Robert V. Decareau, *Microwaves in the Food Processing Industry.* 1985.

S. M. Herschdoerfer (ed.), *Quality Control in the Food Industry,* second edition. Volume 1—1985. Volume 2—1985. Volume 3—1986. Volume 4—1987.

F. E. Cunningham and N. A. Cox (eds.), *Microbiology of Poultry Meat Products.* 1987.

Walter M. Urbain, *Food Irradiation.* 1986.

Peter J. Bechtel, *Muscle as Food.* 1986. H. W.-S. Chan, *Autoxidation of Unsaturated Lipids.* 1986.

Chester O. McCorkle, Jr., *Economics of Food Processing in the United States.* 1987.

Jethro Japtiani, Harvey T. Chan, Jr., and William S. Sakai, *Tropical Fruit Processing.* 1987.

J. Solms, D. A. Booth, R. M. Dangborn, and O. Raunhardt, *Food Acceptance and Nutrition.* 1987.

R. Macrae, *HPLC in Food Analysis,* second edition. 1988.

A. M. Pearson and R. B. Young, *Muscle and Meat Biochemistry.* 1989.

Marjorie P. Penfield and Ada Marie Campbell, *Experimental Food Science,* third edition. 1990.

Leroy C. Blankenship, *Colonization Control of Human Bacterial Enteropathogens in Poultry.* 1991.

Yeshajahu Pomeranz, *Functional Properties of Food Components,* second edition. 1991.

Reginald H. Walter, The Chemistry and Technology of Pectin. 1991.

Herbert Stone and Joel L. Sidel, Sensory Evaluation Practices, second edition. 1993.

Robert L. Shewfelt and Stanley E. Prussia, *Postharvest Handling: A Systems Approach.* 1993.

R. Paul Singh and Dennis R. Heldman, *Introduction to Food Engineering,* second edition. 1993.

Tilak Nagodawithana and Gerald Reed, *Enzymes in Food Processing,* third edition. 1993.

Dallas G. Hoover and Larry R. Steenson, *Bacteriocins.* 1993.

Takayaki Shibamoto and Leonard Bjeldanes, *Introduction to Food Toxicology.* 1993.

John A. Troller, *Sanitation in Food Processing,* second edition. 1993.

Ronald S. Jackson, *Wine Science: Principles and Applications.* 1994.

Harold D. Hafs and Robert G. Zimbelman, *Low-fat Meats.* 1994.

Lance G. Phillips, Dana M. Whitehead, and John Kinsella, *Structure-Function Properties of Food Proteins.* 1994.

Robert G. Jensen, *Handbook of Milk Composition.* 1995.

Yrjö H. Roos, *Phase Transitions in Foods.* 1995.

Reginald H. Walter, *Polysaccharide Dispersions.* 1997.

Gustavo V. Barbosa-Cánovas, M. Marcela Góngora-Nieto, Usha R. Pothakamury, and Barry G. Swanson, *Preservation of Foods with Pulsed Electric Fields.* 1999.

Ronald S. Jackson, *Wine Science: Principles, Practice, Perception,* second edition. 2000.

R. Paul Singh and Dennis R. Heldman, *Introduction to Food Engineering,* third edition. 2001.

Ronald S. Jackson, *Wine Tasting: A Professional Handbook.* 2002.

Malcolm C. Bourne, *Food Texture and Viscosity: Concept and Measurement,* second edition. 2002.

Benjamin Caballero and Barry M. Popkin (eds.), *The Nutrition Transition: Diet and Disease in the Developing World.* 2002.

Dean O. Cliver and Hans P. Riemann (eds.), *Foodborne Diseases,* second edition. 2002. Martin

Kohlmeier, *Nutrient Metabolism.* 2003.

Herbert Stone and Joel L. Sidel, *Sensory Evaluation Practices,* third edition. 2004.

Jung H. Han, *Innovations in Food Packaging.* 2005.

Da-Wen Sun, *Emerging Technologies for Food Processing.* 2005.

Hans Riemann and Dean Cliver (eds) *Foodborne Infections and Intoxications,* third edition. 2006.

Ioannis S. Arvanitoyannis, *Waste Management for the Food Industries.* 2008.

Ronald S. Jackson, *Wine Science: Principles and Applications,* third edition. 2008.

Da-Wen Sun, *Computer Vision Technology for Food Quality Evaluation.* 2008.

Kenneth David, *What Can Nanotechnology Learn From Biotechnology?* 2008.

Elke K. Arendt and Fabio Dal Bello, *Gluten-Free Cereal Products and Beverages.* 2008.

Da-Wen Sun, *Modern Techniques for Food Authentication.* 2008.

Debasis Bagchi, *Nutraceutical and Functional Food Regulations in the United States and Around the World.* 2008.

R. Paul Singh and Dennis R. Heldman, *Introduction to Food Engineering,* fourth edition. 2008.

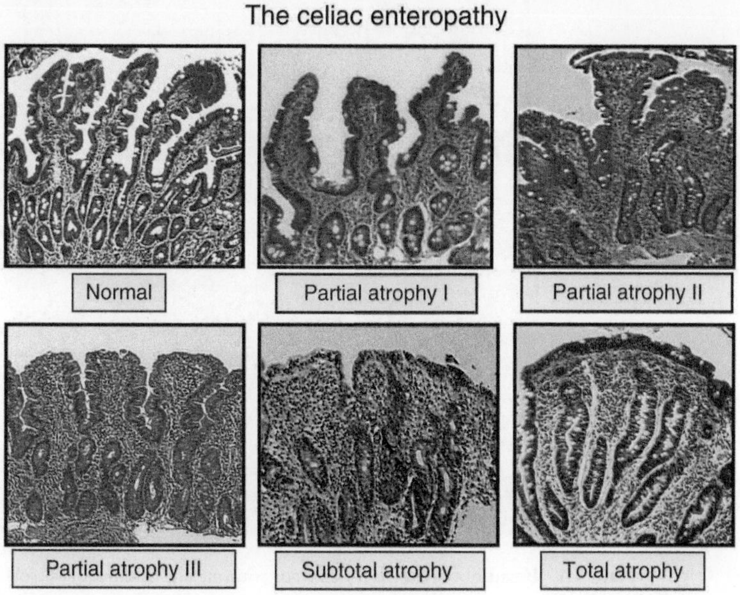

The celiac enteropathy

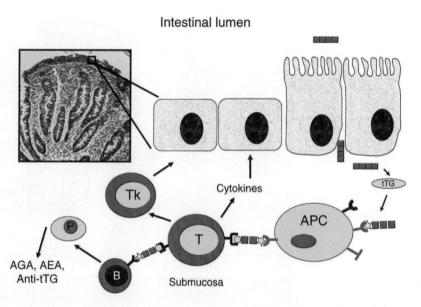

| Normal | Partial atrophy I | Partial atrophy II |

| Partial atrophy III | Subtotal atrophy | Total atrophy |

Plate 1.1 The variable picture of the celiac enteropathy.

Intestinal lumen

Cytokines

tTG

Tk

P

T

APC

B

Submucosa

AGA, AEA,
Anti-tTG

Plate 1.2 The adaptive T cell-mediated response to gluten peptides in the intestinal mucosa, leading to the celiac enteropathy. tTG, transglutaminase; APC, antigen-presenting cell; T, T cell; B, B cell; P, plasma cell; AGA, anti-gliadin antibody; AEA, anti-endomysial antibody; anti-tTG, anti-tissue transglutaminase.

Plate 4.1   Bread obtained from rice flour without any additive (control) (photo by Cristina Marco).

Plate 4.2   Bread obtained from rice flour in the presence of 4% HPMC (photo by Cristina Marco).

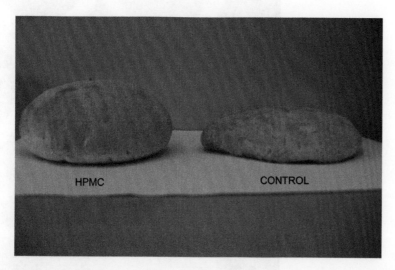

Plate 4.3 Bread obtained from rice flour without additive (control) and in the presence of 4% (flour basis) hydroxypropylmethylcellulose (HPMC) (photo by Cristina Marco).

Plate 6.1 Various millets. Top left to right: Pearl millet (Zimbabwe), finger millet (Ethiopia). Midd[le] to right: Fonio (Senegal), teff (Ethiopia), teff (South Africa). Bottom left to right: Decorticated foni[o] (Senegal), decorticated proso millet (Australia).

Plate 6.2 Various millet-based foods. Top left to right: Finger millet flour (Kenya), finger millet flour (Tanzania), finger millet-based acidified uji mix (Kenya), finger millet and soya mix (Tanzania). Middle left to right: Pearl millet flour (Senegal), pearl millet thiacri steamed couscous (Senegal), puffed proso millet (USA). Bottom left to right: Pearl millet arraw non-steamed couscous, tamarind acidified instant pearl millet kunun (Nigeria).

Plate 7.1 Amaranthe (*Amaranthus cruentus*), quinoa (*Chenopodium quinoa*) (photos by Georg Dobos), an buckwheat (*Fagopyrum esculentum*) (photo by Heinrich Grausgruber).

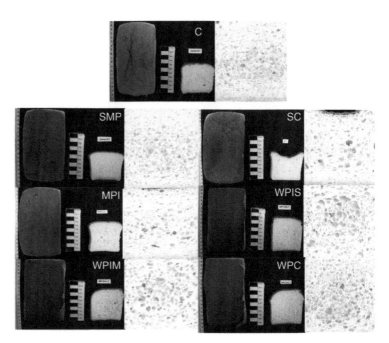

Plate 13.1 Impact of low lactose dairy powders on the quality of gluten-free bread: control bread (C skim milk powder (SMP); sodium caseinate (SC); milk protein isolate (MPI); whey protein isolate, spr dried (WPIS); whey protein isolate, membrane technology (WPIM); whey protein concentrate (WPC).

Printed and bound by CPI Group (UK) Ltd, Croydon, CR0 4YY

08/05/2025

01864821-0002